D1739279

Soldiers and Settlers

University of New Mexico Press
Albuquerque

SOLDIERS and SETTLERS

Military Supply in the Southwest, 1861-1885

Darlis A. Miller

Library of Congress Cataloging-in-Publication Data

Miller, Darlis A., 1939–
Soldiers and settlers : military supply in the Southwest, 1861–1885 / by Darlis A. Miller.
p. cm.
Bibliography: p.
Includes index.
ISBN 0-8263-1159-8
1. United States. Army—Procurement—Economic aspects—Southwest, New—History—19th century. 2. Southwest, New—Economic conditions. I. Title.
UC263.M468 1989
355.6'212'0340979—dc20 89-14637
CIP

First edition.

Dedicated to my parents,
Helen Krutsinger Miller *and* Joseph Edwin Miller

Contents

Tables

Maps

Illustrations

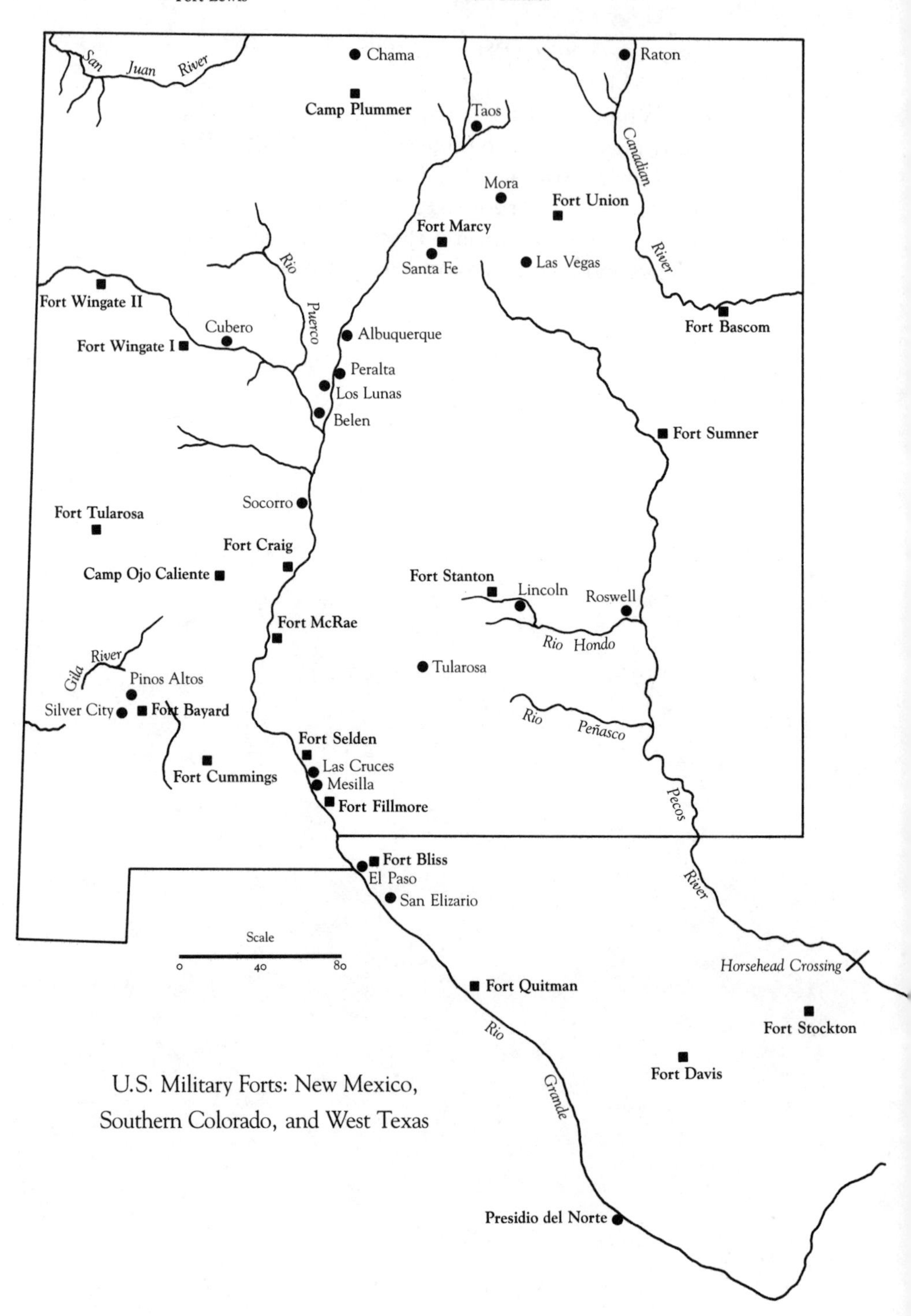

U.S. Military Forts: New Mexico, Southern Colorado, and West Texas

Preface

DURING the past thirty years American frontier military history has enjoyed a renaissance. The scholarly studies that Francis Paul Prucha, Robert M. Utley, and William H. Goetzmann produced in the 1950s and 1960s inspired a large number of other historians to reexamine the frontier military experience so that the army's role in western expansion could more clearly be understood.[1] Scholars have delved into nearly all facets of the Indian-fighting army, producing detailed studies of individual Indian campaigns and of the officers who commanded them. Others have concentrated on enlisted men, black soldiers, Indian scouts, and military scientists and writers. Still others have focused on military dependents and life at frontier garrisons.

Indeed, scholars in the field have long voiced the need to examine more closely the noncampaign aspects of the western army. Prucha provided a model in *Broadax and Bayonet, The Role of the United States Army in the Development of the Northwest, 1815–1860.*[2] Here he suggested the importance of soldiers as farmers, explorers, and road builders in western development. Several other writers have recognized the economic importance of the army to frontier settlements and have called for more investigation into the topic. Robert W. Frazer's pathbreaking *Forts and Supplies, The Role of the Army in the Economy of the Southwest, 1846–1861* published in 1983 is the first book-length study of the army's economic impact on one region of the United States.[3] Frazer focuses on the army's experience in the Southwest between the Mexican and Civil Wars. In *Soldiers and Settlers* I have attempted to carry on the story, though from a slightly different perspective, which will be apparent to those who are familiar with Frazer's work. I examine the economic facet of the army's presence in the Southwest between 1861—the start of the Civil War—and 1885, when the Indian wars were winding down. By the closing date of this study, railroads crisscrossed the nation, forever changing American society and the manner in which the army supplied its western garrisons.

Shortly after General Stephen Watts Kearny occupied Santa Fe in 1846, the army administratively created a military Southwest by establishing Military Department No. 9 and then its successor, the Military Department of New Mexico. This department included all of New Mexico, much of Arizona, and parts of Texas and Colorado. Military geography would change in later years, but the focus of my study remains on this geographical area—a region that shared problems of aridity, long distances, Apache wars, and—to a limited extent—a common market. Military contractors in New Mexico supplied garrisons in all four areas of this study: New Mexico, Arizona, west Texas, and southern Colorado.

The primary aim of the frontier army was to subdue hostile Indians so that white settlers could find new homes and develop western lands. This book, however, is not strictly military history. It provides no detailed account of the Indian wars or of military commanders and the forts they built and commanded. Rather it focuses upon the economic interaction between the military and civilians. Its intent is to document the intricacies of military contracting and to assess the army's economic impact on civilian society. The focus therefore is as much on civilians as it is on military personnel. Since the army was the single largest purchasing and employment agency in the Southwest, it benefited a wide assortment of civilians, including poorer residents who on occasion sold burro loads of hay and wood to the army. By carefully sifting through military documents, it has been possible to recover important history of a segment of society usually neglected in regional studies. In addition, the study underscores the prominence of the frontier army in western society.

Each military garrison in the Southwest deserves a written history of its own, documenting its interaction with nearby settlers and its contribution to economic development. Records for such a detailed analysis for each post, however, are not available. Nonetheless, since army procurement policies were generally uniform throughout the West, a study of the army's economic importance in the Southwest is possible without a detailed history of each post.

During the twenty years following the Civil War the United States experienced tremendous economic growth and by the turn of the century had become the world's greatest industrial power. This growth was possible, in part, because of the economic exploitation of western resources. During these same years the Southwest witnessed a series of devastating Indian wars, development of its mining and ranching industries, and modest agricultural expansion. The coming of railroads in the 1880s, accompanied by an inflow of investment capital, sparked major economic growth.

Still, during the quarter-century covered by this study, the military was a major force in the region's economy. Government subsidies in the form of supply contracts provided funds crucial to social and economic development. Residents came to rely upon the military for investment funds. "Defense" contracts, then as now, were big business, and military spending became a cornerstone of the economy. The army has always been an essential institution in the Southwest right down to the present. Federal expenditures today play an even greater role in the region's economy than they did in the nineteenth century. But the habit of relying on federal money was established soon after Kearny's occupation. Unlike the military-industrial complex that exists today, however, the partnership of federal government and private enterprise forged at that time spread economic benefits widely through the community. And because large numbers of small entrepreneurs won government contracts, no clique of economic royalists emerged to greatly influence either military or civilian policies.

More than a decade ago Francis Paul Prucha challenged historians to support their broad generalizations about the army's contribution to western development with precise details.[4] It is hoped that the detailed information provided in *Soldiers and Settlers* meets this challenge. Chapter 1 sets the stage, describing southwestern society, the crisis of the Confederate invasion, and the army's attempts to secure supplies locally. Thereafter a topical approach has been followed, with separate chapters devoted to agriculture, forage and fuel, flour and other commissary supplies, the cattle industry, construction and civilian employees, and transportation. A final chapter investigates the problem of fraud, examines the character of government contractors, and assesses the army's economic impact in the Southwest.

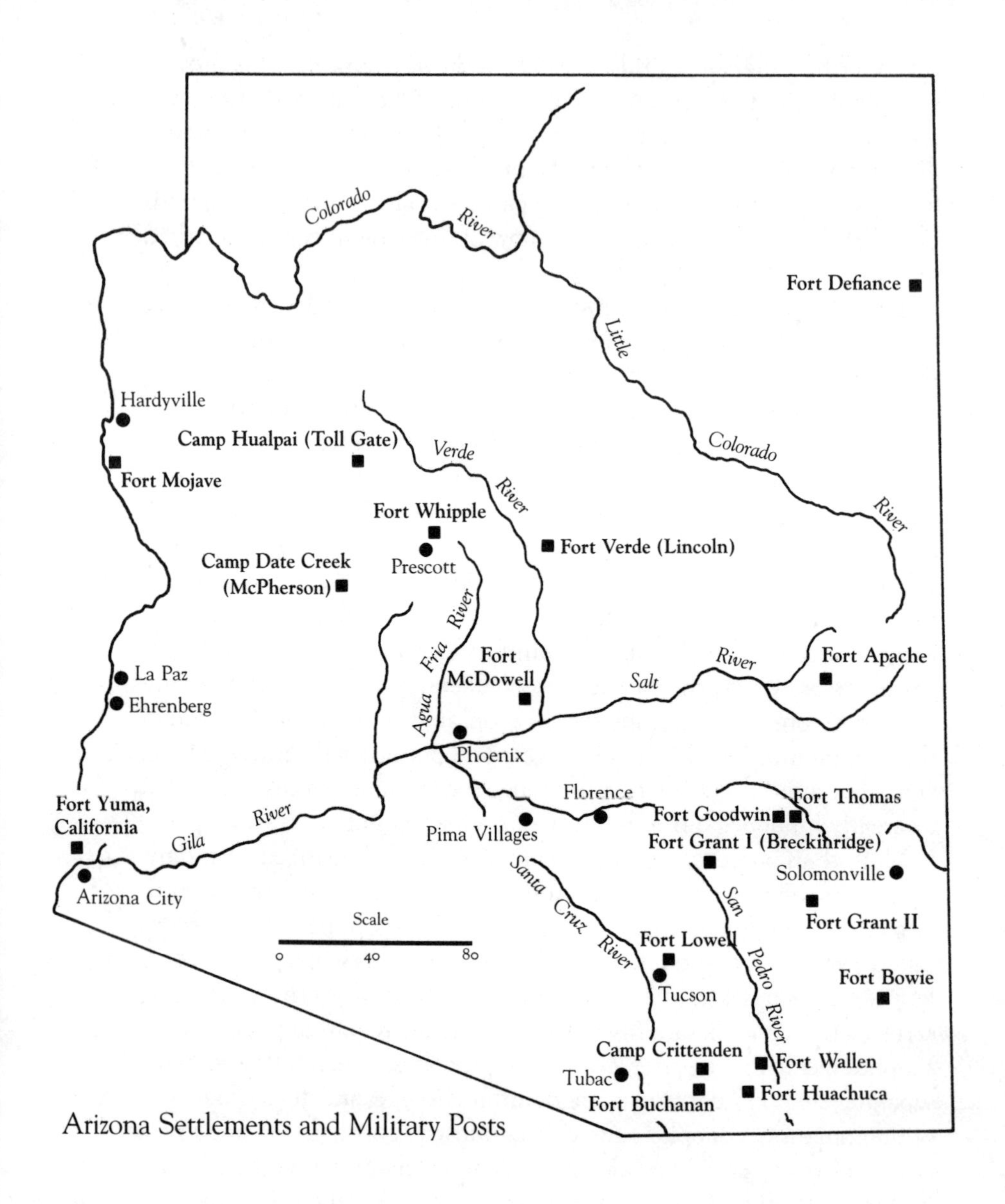

Arizona Settlements and Military Posts

Acknowledgments

THIS book is dedicated to my parents Helen Krutsinger Miller and Joseph Edwin Miller, pioneers in their own generation. During the seven or more years I have spent working on the manuscript for this book, I have received generous help and support from a number of people. I would like to thank Richard N. Ellis for suggesting a need for such a study years ago and for his continued support. I am deeply grateful to Robert W. Frazer for reading the entire manuscript. His extensive research and writing on military topics have always been an inspiration. Others who gave generously of their time to read portions of the manuscript and whom I now thank are Constance W. Altshuler of Tucson, Harwood P. Hinton of the University of Arizona, Myra Ellen Jenkins of Santa Fe, Richard Lowitt of Iowa State University, Mary Williams of Fort Davis National Historic Site, and John P. Wilson of Las Cruces, New Mexico.

In addition to the above, several other individuals have offered valuable suggestions or information during the writing of this book, and I am in their debt: Lori Davisson of the Arizona Historical Society, Charles Kenner of Arkansas State University, Janet Lecompte of Colorado Springs, Fred Nicklason of the University of Maryland, Terry Reynolds of Denver, Marc Simmons of Cerrillos, New Mexico, Patricia Stallard of Knoxville, and Mary Taylor of Mesilla, New Mexico. For their many courtesies extended over several years, I would also like to thank the fine professional staffs in special collections at the universities of Arizona, New Mexico, Texas at El Paso, and the Rio Grande Historical Collections at New Mexico State University. My special thanks also to members of the staffs at Arizona Historical Society, Fort Davis National Historic Site, New Mexico State Records and Archives, Museum of New Mexico, National Archives, Texas State Library, and the Washington National Records Center.

Two special people at the National Archives deserve further recognition since they were present at both the beginning and end of this study, offering helpful suggestions along the way as well as friendship. My warm thanks to Sara D. Jackson and Elaine Everly, who made working in the archives a daily pleasure. And I will always be indebted to my good friend Linda J. Lear of Bethesda, Maryland. By including me in her family circle, she made each research trip to Washington a memorable event and was always present to restore my enthusiasm and lift my spirits.

The research was aided by grants from the National Endowment for the Humanities, the American Philosophical Society, and the Arts and Sciences Research Center at New Mexico State University. I also extend thanks to Juanita Graves for overseeing the typing of the manuscript in several versions and to Toni Murdock and Liz McKethen for their hospitality during research trips to Tucson. My special thanks to David V. Holtby of the University of New Mexico Press for his unflagging interest in this project and for overseeing the editing of this book. And, finally, I want to thank my husband August Miller for his strong support and good humor during the years I have been pursuing military contracts.

Soldiers and Settlers

1

Civil War

In 1861 the United States embarked upon a deadly civil war. Most of the fighting and carnage took place east of the Mississippi, but western states and territories also were sucked into the maelstrom. The southern Confederacy envisioned a coast-to-coast republic, but before it could gain access to ports on the Pacific, Confederate soldiers would have to take by force the Military Department of New Mexico, which then encompassed much of the present Southwest—New Mexico, Arizona, and southern Colorado.

Most civilians residing within this military department had little understanding of the political and social issues that had led to war. The great bulk of the people were Hispanos who had been citizens of the United States for less than fifteen years. In 1860 the population of New Mexico, exclusive of nonfriendly Indians, was 93,516. This included 2,442 non-Indian residents of Arizona County, a region extending west of Apache Pass near the present Arizona–New Mexico border to the Colorado River. It also included more than 2,500 individuals who lived in the San Luis Valley north of the present Colorado–New Mexico border but then part of Taos County, New Mexico. To the south in Texas, 428 people resided in the community of Franklin (now El Paso), about twenty miles from the Texas–New Mexico border. During the 1850s Fort Bliss, which was located near Franklin, had been attached to the Military Department of New Mexico. It was transferred to the Department of Texas in December 1860, but the post and its neighbors continued to have close military and civil ties to New Mexico.[1]

By 1860 New Mexico had a highly stratified society in which a small wealthy Hispanic class shared social, economic, and political power with the wealthier and more talented members within the Anglo community. Most New Mexicans were illiterate and poor, subsisting in small rural villages or working on large ranches. The territory's capital and principal town was Santa Fe, which claimed a population of 4,635. It was also the leading commercial center, and merchants and entrepreneurs who clustered there engaged in a lucrative trade over the Santa Fe Trail. Mesilla, with a population nearing 2,500, was the second largest town. It was the principal commercial and trading center in southern New Mexico, linking Franklin and El Paso, Mexico (now Ciudad Juárez), to Santa Fe and Tucson. Other towns of substantial size included Albuquerque and Las Vegas, each boasting over 1,000 inhabitants.[2]

Arizona's non-Indian population lived primarily in small towns and ranches south of the Gila River. In 1860 Tucson, the principal town in Santa Cruz Valley, had a population of 915. A few years earlier the town had boasted three stores, two butcher shops, and two blacksmith shops, but it contained neither a hotel nor a saloon.[3] John Spring, who visited Tucson in 1866, left a classic description:

> The Tucson of those days had but one regular street, now called Main Street. The buildings which deserved the names of houses were of adobe with flat mud roofs; those of the poorer class of Mexicans were constructed of mesquite poles and the long wands of candlewood (*Fouquieria splendens*), the chinks being filled in with mud plaster. With the exception of the soldiers and teamsters in transit, there were not over a dozen American men in the town, and not one American woman. The doors of many houses consisted of rawhide stretched over rough frames, the windows being simply apertures in the walls, barred with upright sticks stuck therein.[4]

Other Arizona settlements included Tubac, forty-eight miles south of Tucson with more than 350 residents, and Arizona City, located on the site of modern Yuma with about 130 residents.[5]

Despite the importance of merchants and townspeople, the vast majority of families who lived in the Military Department of New Mexico cultivated the soil and cared for livestock. But unrelenting conflict between local Indians and white settlers had hindered agricultural development since the time of early Spanish settlement. Soon after the close of the Mexican War, the United States constructed several military posts in an attempt to pacify the countryside. The number of posts and men who garrisoned them increased in following years as the army waged war against nonfriendly Indians. By June 1860, 3,104 officers and enlisted men were assigned to the Department of New Mexico, although this number included 1,554 soldiers who were en route from duty in Utah. More than 300 soldiers were in the field pursuing Apaches and Kiowas, and the remaining soldiers garrisoned military posts in Arizona (Forts Buchanan, Breckinridge, and Defiance), New Mexico (Cantonment Burgwin, Albuquerque, Los Lunas, and Forts Garland, Union, Marcy, Craig, Stanton, and Fillmore), and Texas (Fort Bliss).[6]

The army's primary goal in the West was to subdue hostile tribes, but it performed other functions as well. The army built and maintained roads, guarded survey parties, and served as a local police force during outbursts of frontier lawlessness. No less important was the economic role it played, for the army encouraged the growth of agriculture and settlements to reduce the cost of supplies. During the 1850s, as Robert Frazer has documented, the army was "the single most significant factor in the economic development of the Southwest."[7] By purchasing locally produced goods, the army stimulated farming and livestock production, and the protection it offered against Indian attacks allowed further expansion. How military spending and supply requirements influenced this region during and after the Civil War is the subject of this volume.

The army's efficiency, in fact, largely depended upon how smoothly the subsistence and quartermaster's departments carried out their responsibilities of procuring supplies and services. The subsistence department, whose responsibility was feeding the troops, tried to purchase items in the soldier's ration close to the area of consumption, thereby saving the government the cost of transporting goods over long distances. The main articles in the ration, as prescribed by regulations, were beef, pork, bacon, flour, beans, coffee, sugar, and salt. Only a portion of these items could be purchased in New Mexico, and commissary officers frequently made substitutions—either from necessity or to provide variety in the soldier's diet—of other local products, including buffalo meat, mutton, pickles, sauerkraut, vinegar, onions, potatoes, and chile peppers. Even though local farmers increased crop production in the 1850s in response to the military market, the department continued to rely heavily upon Fort Leavenworth—the major supply depot for military posts on the Plains—receiving from the east in a single month (July 1861) such items as ham, bacon, rice, desiccated potatoes, molasses, coffee, tea, sugar, and salt.[8]

Except for commissary, medical, and ordnance stores, the quartermaster's department purchased all other military supplies, including forage, fuel, clothing, camp and garrison equipage, horses, wagons, harness, tools, and so forth. In normal times, most equipment was shipped to New Mexico from Fort Leavenworth, but by 1861 the territory provided much of the forage, fuel, and building materials required by the army. Prior to this date, the War Department procured only a limited amount of either forage or foodstuffs in Arizona, and soldiers who were stationed there received most of their supplies from the Rio Grande. Fort Mojave, established in 1859 on the Colorado River and assigned to the Military Department of California (but later attached to the District of Arizona), was supplied primarily from San Francisco, with military stores being shipped by water via the Gulf of California and the Colorado River. A new sea and land route opened to southern Arizona in 1861 when Governor Ignacio Pesqueira of Sonora allowed American goods that were shipped from San Francisco to be transported overland from the port of Guaymas at reduced import rates.[9] Throughout the Department of New Mexico, subsistence and quartermaster officials procured local supplies through the contract system, advertising for bids and awarding contracts to the lowest bidder. During emergencies, such as the invasion of New Mexico by Confederate soldiers, military goods were purchased in open market.

New Mexicans were unprepared for the disruption caused by the outbreak of the Civil War, but in the months that followed they contributed both manpower and supplies to the Union war effort. The military–civilian economic alliance established in preceding years would contribute to the smoothness with which officials rallied local support.

Nonetheless, when Colonel Edward R. S. Canby assumed command of New Mexico in June 1861, the department was in serious difficulty. Many southerners had resigned to join the Confederate army, leaving a shortage of officers and demoralized enlisted men who had not been paid in several months. The department lacked horses and draft animals and a two-year dry spell had produced such a scarcity of water and forage that Canby described conditions as approaching famine.[10] A shortage of ammunition and clothing as well as camp and garrison equipage plagued the volunteer troops. Fortunately, regular troops were well-equipped and clothed, and rations were sufficient to last through the year.[11]

One month after Canby assumed command of federal forces, Texans under the leadership of Lieutenant Colonel John R. Baylor invaded the territory and quickly forced the surrender of Major Isaac Lynde, commander at Fort Fillmore. While Baylor consolidated his hold in the Mesilla Valley, Union officials in Santa Fe carried out plans to raise two regiments of volunteers among the territory's citizenry. Despite initial hesitancy, approximately 2,800 residents—most of them Hispanos—had enlisted in the New Mexico volunteers by February 1862.[12] For many the chief attraction in joining the army was the prospect of soldier's pay and bounties. During most of the war privates were paid thirteen dollars a month, and men enlisting for three years were given a bounty of one hundred dollars—one quarter of which was paid in advance. The local press as well as recruiting agents continually stressed the fact that laborers received better pay in the volunteers than they could get by working for wages.[13]

Military service also was one way to escape peonage. Soon after taking command, Canby issued a circular stating that enlisted men could not be discharged because of past indebtedness and that petitions to reclaim peons had to be filed with U.S. District courts. The Santa Fe *New Mexican* subsequently reported that large numbers of peons had "extricated themselves from their thraldom as servants by going into the United States Volunteer regiments."[14] Then, too, although patriotism undoubtedly motivated many New Mexicans, the masses were motivated by their hatred for Texans, whose "invasion" of New Mexico in 1841 had bequeathed a legacy of bitterness and ill will.[15]

Government purchasing agents anticipated this build-up in troop strength, and in October 1861 they advertised in the Santa Fe *Weekly Gazette* for large quantities of locally produced subsistence stores: 392,600 pounds of cornmeal, 31,250 pounds of hominy, 15,300 gallons of vinegar, 3,900 gallons of pickles, 12,970 gallons of sauerkraut, 153,400 pounds of onions, 800 bushels of beans, 1,200 bushels of salt, 451,000 pounds of flour, and 3,885 head of beef cattle.[16] Since bidders were required to submit proposals for large amounts of one commodity

(50,000 pounds or more of flour, for example), only wealthy businessmen who had sufficient funds to gather items from small producers would introduce bids.

At least twenty-one New Mexicans subsequently received contracts to furnish the above commodities. Lucien B. Maxwell, an experienced government contractor who owned a large estate on the Cimarron River, agreed to deliver cornmeal at Fort Union at 6 cents per pound, and Ceran St. Vrain, one of the army's most reliable flour contractors, agreed to supply Forts Marcy and Craig with flour at 9½ cents and 14 cents per pound, respectively. Oliver P. Hovey, one-time publisher of the Santa Fe *Republican*, received more contracts than did anyone else. He contracted to deliver flour, cornmeal, hominy, vinegar, and beans to Fort Fillmore; sauerkraut and pickles to Fort Marcy, Fort Union, and Albuquerque; and beef cattle to Albuquerque, Fort Craig, and Fort Lyon. Several others also agreed to supply subsistence stores to more than one post.[17]

The quartermaster's department also invited proposals for corn in lots of at least 5,000 *fanegas*. But only one bid was submitted. Captain John C. McFerran, the department's assistant quartermaster, explained that since the subsistence department had advertised for cornmeal and hominy at the same time the quartermaster's department advertised for corn, many failed to bid on the latter, hoping to receive higher prices by turning their corn into meal or hominy.[18] The quartermaster's department subsequently signed two contracts for corn, both for delivery at Fort Garland in Colorado territory. Frederick W. Posthoff, a Costilla merchant, contracted for 600 *fanegas* at $5.98 per *fanega*, and Charles H. Beaubien and his son-in-law Frederick Mueller, both of Taos, contracted for 1,000 *fanegas* at $6.94 per *fanega*. Two contracts also were signed for hay. In July Pedro A. Baca, a leading merchant in Socorro, agreed to deliver 400 tons to Fort Craig at $9.50 per ton, and in September William H. Moore, sutler at Fort Union, contracted to deliver 100 tons to that post for $45 per ton.[19]

But the quartermaster's department needed additional forage. And in the weeks following Baylor's successful entry into New Mexico, Union agents scoured the territory for both corn and hay. In August, Fort Craig was reinforced with twelve new companies, and corn had to be purchased quickly to feed more than 500 animals. Consequently government agents were authorized to purchase corn in open market and to seize it if citizens refused to sell at reasonable rates. Residents of Paraje near Fort Craig were to receive $3 per *fanega* for corn, but citizens living near Fort Union would be paid $5 per *fanega.* In December the town of Alamosa below Fort Craig was placed under military control and its residents required to cut grama grass for sale to the army. In addition, Union officials ordered wheat seized at $3 per *fanega* in Socorro and nearby villages, to be ground into flour for the troops at Fort Craig.[20]

While government agents searched for supplies, Union officials in Santa Fe coordinated plans to move the department's major supply depot from Albuquerque to Fort Union. In April 1861 McFerran, then commanding the depot, raised serious objections to the Albuquerque location. Since the government owned no land or buildings there, McFerran noted, it incurred enormous expense by renting quarters and making repairs on civilian structures. Forage was expensive, and grazing was entirely lacking within ten miles of the depot. Fort Union, on the other hand, was located within one of the best grazing districts in New Mexico and on the junction of the main roads to the States.[21]

Shortly before orders were issued 20 July 1861 establishing the general depot at Fort Union, McFerran started to dismantle the Albuquerque depot—now designated a subdepot—by sending quartermaster's stores and clothing to Forts Craig and Fauntleroy (shortly thereafter renamed Fort Lyon). He recommended that a year's supply of military stores be sent to Forts Craig, Fauntleroy, and Stanton, and left at Albuquerque, to save retransfer of goods to Fort Union. McFerran believed that these were the most important posts for checking Indian raids or countering any advance by the Texans.[22]

In the weeks that followed, McFerran was hard pressed to find wagons and work animals to transport supplies from Albuquerque to Fort Union and the other posts. Efforts to purchase horses and mules met with little success. Only one good horse was offered for sale in the vicinity of Albuquerque, and the Mexican ponies that Oliver P. Hovey offered failed to meet army specifications.[23] McFerran was forced to hire mule trains from private citizens, but even these were unavailable until the merchant caravans arrived from the States. Eventually, merchants Franz and Charles Huning of Albuquerque, Pinckney R. Tully and Estévan Ochoa of Mesilla, and freighters Charles W. Kitchen of Las Vegas and Anastacio Sandoval of Santa Fe hauled government freight at two cents a pound from Albuquerque to the new depot.[24] To ensure rapid transfer of supplies, soldiers repaired roads in the vicinity of Fort Union, while alcaldes within the Fort Craig military district were held accountable for the repair of roads and acequias in that area. To compound McFerran's problems, some government teamsters at Fort Union attempted to resign because of low pay. They believed they could receive higher wages in the volunteers.[25]

During the summer and fall, the freighting firm of Irwin, Jackman and Company, headquartered at Leavenworth City, transported government stores from Fort Leavenworth to the new depot at Fort Union. The firm's first train of twenty-five ox-drawn wagons, hauling 142,136 pounds of supplies, left Fort Leavenworth on 24 May 1861 and arrived at Fort Union on 18 July. In following weeks the firm sent fourteen trains to Fort Union, most averaging twenty-five wagons and carrying more than 140,000 pounds of government stores. The last train of six wagons arrived at Fort Union on 29 December 1861. The government paid the firm between $1.30 and $1.50 per 100 pounds per 100 miles, depending on the season of the year.[26]

Meanwhile, McFerran was relieved from duties at Albuquerque and placed in command of the supply depot at Fort Union, where an awesome confusion prevailed. Packages from Fort Leavenworth had been mixed with those sent from Albuquerque, making it impossible to know what supplies were on hand. McFerran found government packages stacked in the depot storehouse, some in piles adjoining the storehouse, others piled in different rooms at the corrals, and still others inside and outside the new fieldwork. Some clothing and blankets received from Fort Leavenworth had been damaged because of poor packing, but crowded storerooms made it impossible to examine each shipment carefully as decreed by army regulations. It was to be only a matter of time before thieves took advantage of this confusion. By late December McFerran reported that clothing had been stolen from packages piled in the fieldwork. Commissary stores were as badly scattered. Many were left uncovered and suffered damage from exposure. McFerran tried to restore order at the depot by speeding construction of storehouses, but administrative entanglements caused delays.[27]

The biggest problem facing the quartermaster's and subsistence departments, however, was lack of funds. As early as 23 March 1861, Colonel William W. Loring, then commanding the department, reported that people throughout the territory were beginning to refuse credit to the government.[28] Residents were suspicious of government-issued certificates of indebtedness and demanded gold and silver for army purchases. While in charge of the depot at Albuquerque, McFerran used his personal savings to pay government debts. After moving to Fort Union, he reported that again he was without funds and unable to purchase hay or to pay laborers and mechanics required to work on fortifications. Without specie he quickly would have to stop hiring and purchasing altogether.[29]

Several New Mexicans rose to the emergency by providing specie for government vouchers. William H. Moore, government contractor and sutler at Fort Union, became McFerran's chief source of specie. Moore turned over to McFerran every dollar he received in his store. In addition, Moore gave his personal note for specie to individuals who would not take government drafts. McFerran later said that had it not been for Moore "who furnished us with every dollar we used at the time [of the Texas invasion], the Department would not have been able to meet the demands upon it."[30]

To acquire more funds, Canby requested citizens to furnish monthly installments of specie, but few moneylenders came forward. Taos residents, including Ferdinand Maxwell and Ceran St. Vrain, reported in November that they were unable to furnish any specie. The only citizens who turned money over to the quartermaster at this time were Las Vegas residents Romualdo Baca and Miguel Romero, each furnishing $300, and Francisco López and Trinidad Romero, who each furnished $100.[31]

Conditions in New Mexico became even more critical when General Henry Hopkins Sibley arrived in mid-December 1861 at the head of three regiments of Texas–Confederate soldiers. Their advance up the Rio Grande rekindled efforts by Union officials to recruit volunteers, organize the militia, and collect supplies from the surrounding countryside. Militia officers were authorized to purchase or seize provisions to subsist their men. Residents of Belen later complained that militia officers went into many poor peoples' homes and took their last "mouthful of flour" while wealthy residents were bypassed.[32] Nonetheless, scores of New Mexicans furnished food and supplies. Typical of claims later submitted against the government was one sent in by Jesús María Pacheco of Taos, who stated that in January and February of

1862 he provided a militia company with 120 pounds of buffalo meat, 2 sacks of flour, 10 pounds of sugar, 10 pounds of coffee, 2 *fanegas* of corn, some candles, and a rifle, amounting to $152 in value. In addition, many militia captains contracted sizable debts to outfit and care for their men. Juan de Jesús Martines expended nearly $600 on food, lodging, fuel, equipment, and transportation for his company, and a second captain paid out more than $100 for tin cups, plates, and mess pans, $182 for subsistence, plus additional amounts for fuel and forage.[33]

With Sibley moving north, McFerran's chief concern was to concentrate supplies at Albuquerque and Fort Craig. In early January 1862 he purchased nine eight-mule teams and wagons from H. Biernbaum and Brothers of Mora for $7,660 to accomplish this task. Mule trains of private citizens were forbidden to leave the department until the emergency ended. But McFerran continued to lack transportation, and shipments for Albuquerque were left standing in storehouses for lack of wagons and draft animals. On 10 February he complained that "not a mule wagon and team can be purchased or hired within 100 miles of this point." To add to his frustrations, Navajos raided supply wagons hastening to Albuquerque, causing some wagonmasters to request strong military escorts.[34]

Despite these obstacles, McFerran continued purchasing and forwarding supplies. He made arrangements to purchase William Kroenig's corn raised on the Huerfano in Colorado for $9 per *fanega* and accepted William H. Moore's bid to deliver one million pounds of corn from the States. McFerran was still ordering corn south to Albuquerque as late as 26 February, five days after Confederate forces defeated Canby's command at Valverde ford, six miles above Fort Craig.[35]

On 28 February, after receiving news of Canby's defeat, McFerran began recalling supply trains sent to Albuquerque. He said there was no cause for alarm; with energy and promptness all public property could be saved.[36] But he was mistaken. The Texas advance up the Rio Grande was too rapid and Union mule trains too few to remove all government supplies to safety. What remained at Polvadera, Belen, Albuquerque, and Santa Fe was destroyed. Although this action shattered Confederate hopes of capturing needed supplies, it also entailed hardships for Union soldiers.[37]

The Texans continued their advance up the Rio Grande and on 28 March met defeat near Santa Fe at Glorieta Pass. But even after Sibley decided to evacuate New Mexico, the Union army faced grave problems of supply. McFerran informed Montgomery C. Meigs, quartermaster general of the army, in April that the department was in deplorable condition: "most of our supplies have been consumed, captured, or destroyed." Even though he had just borrowed $19,000 in specie from wealthy Hispanos living near Albuquerque, McFerran warned that the army would be overrun by the enemy unless more funds were supplied from the east.[38] Grain was exhausted in the area of troop movements, and little forage could be obtained for animals. Since most military transportation was broken down and unfit for service, the army seized private trains to reestablish posts and depots. Later José Antonio Baca, Vivian Baca, Esquipulo Pino, Pedro A. Baca, and other Hispanos living in the Fort Craig district received $5 a day for use of their wagons and oxen.[39] To meet demands for clothing, McFerran contracted with local agents to manufacture or purchase stockings and drawers. And to meet demands of cavalry mounts, contracts were made with Charles S. Hinckley of Denver and his partner Charles H. Blake to deliver 200 horses to the territory and with Oliver P. Hovey to deliver 800 horses from the States.[40]

Poorer New Mexicans also profited from the army's need for supplies and laborers. The quartermaster's department purchased hay by the burro load in Santa Fe, paying $3 on 1 June to unidentified "Mexicans" who brought in four loads. During the entire month the government made small daily purchases from "Mexicans" amounting to $452.62—this was in addition to hay turned in by such prominent citizens as Fernando Delgado and Juan Archibeque.[41] Other Hispanos, who as laborers and teamsters helped to move government supplies, received $25 per month, considered good wages by one Union officer who predicted that the richer Hispanos would have to raise their price for labor or "lose their *peones*" to the army.[42]

During the spring, beef was in short supply at Fort Craig, and the army issued orders to seize cattle if residents of nearby villages refused to sell at reasonable prices.[43] Colonel Marshall S. Howe, commanding the post, bitterly condemned contractor Oliver P. Hovey for failing to furnish beef according to his contract. Howe claimed that a contractor who caused suffering and starvation among the troops by failing to deliver provisions was a traitor and should be tried as a traitor.[44] But Hovey's case illustrates the difficulties that contractors faced during the crisis of the Texas invasion.

Hovey had signed a contract on 7 November 1861 with Captain Amos F. Garrison, commissary of subsistence, to deliver beef cattle to various military posts in the territory. During the months of December, January, and February 1862, he delivered between 350 and 400 head of cattle to Fort Craig. When the Texans invaded the territory above the fort, military authorities required that all herds and cattle be driven to the extreme northern and eastern limits of the territory. Thereafter, Hovey found it impossible to meet demands on his contract. In April his agent went to the Rio Abajo settlements and purchased all cattle suitable for beef that could be bought with government drafts. The following month, after Hovey was ordered to deliver 500 more head to Fort Craig, the agent returned to the Rio Abajo where he located cattle either unsuitable for beef or owned by individuals who would sell only for specie. Hovey turned for help to his subcontractors, Hinckley and Blake, who had agreed to deliver between 200 and 750 head of cattle from Colorado, but had also found it impossible to secure cattle. According to his backers, Hovey did everything in his power to fulfill his contract, but the army, nonetheless, labeled him a defaulter.[45]

During the spring and early summer of 1862 Sibley's troops retreated south, living off the land as they had on their drive north. Sheep, cattle, hogs, corn, beans, chickens—nothing was safe from Confederate foragers, not even strings of red peppers or kegs of molasses. Toribio Romero of Los Lunas suffered losses totaling more than $4,000 when Texans confiscated wheat, corn, and other edibles in addition to tobacco, wine, brandy, and twelve barrels of whiskey. At Belen, Josefa Baca y Castillo lost $558 worth of provisions, including twenty *fanegas* of wheat and five *fanegas* each of corn and beans. Candelaria Torres, also of Belen, claimed that she lost to the Texans one revolver and one fat hog, each worth $30.[46]

As Confederate soldiers reentered the Mesilla Valley, many residents drove their cattle across the border into Mexico for safety, but Sibley's confiscations were so extensive on the American side of the Rio Grande that some feared there would be famine among the people. Enraged by this marauding, Hispanic residents took up arms, killed members of Confederate foraging parties, and drove off their livestock.[47]

Canby incurred much criticism for allowing the Confederates to withdraw so easily. The Union commander, however, realized that the territory lacked provisions to feed prisoners of war, and he was in no hurry to capture Sibley's entire army. Moreover, flood conditions on the Rio Grande, delinquent flour contractors, and lack of transportation hampered Canby's efforts to concentrate supplies needed to support troop movements south of Fort Craig.[48] By the time Canby was ready to order troops into the Mesilla Valley, the Confederates had left, and the California Column under General James H. Carleton was approaching the Rio Grande from the west.

A career officer in the regular army, Carleton had served under General John E. Wool during the Mexican War and had commanded troops in New Mexico from 1851 to 1856. At the outbreak of the Civil War he was stationed in California, where he received orders in December 1861 to organize an expedition that would march east across the desert to help Canby expel rebels from the territory. A meticulous organizer, Carleton personally supervised logistics for provisioning the 2,000-man column. Supplies were purchased in San Francisco and shipped by sea either to San Pedro, to be utilized by the men marching to Fort Yuma, or to the fort itself by way of the Colorado River, where they supported the column as it moved farther east. California contractors supplied beef cattle to accompany the marching column.[49]

Before the command left California, Carleton made plans to establish a subdepot on the Gila River 200 miles above Fort Yuma at the Pima villages. Although Carleton expected the column to be self-sustaining, living off supplies stockpiled at government depots, he planned to procure additional forage, wheat, flour, and beef at the Pima villages, in Tucson, and in the neighboring Mexican state of Sonora.

Carleton requisitioned 10,000 yards of manta (cotton cloth) to make purchases from the Pimas and their neighbors, the Maricopas. According to a census that Lieutenant Alfred B. Chapman completed in 1858, the Pimas lived in nine villages and numbered 4,117 individuals. The Maricopas, a much smaller tribe, lived in two villages with a population of 518.[50] By this date the Indians had a reputation for befriending overland travelers. During the gold rush to California, they had sold their surplus produce to migrants on the Gila River route, many of whom were dangerously short of provisions by the time they reached the villages. Not only did the Pimas and Maricopas increase cultivation in response to this new demand, but they also adjusted to a money economy, accepting "either textiles *or* cash for foodstuffs."[51] They were well-acquainted with American business practices by the time Ammi White and other Union agents sought their assistance in provisioning the California Column.

Ammi M. White, a native of Maine, owned a gristmill and trading post located near the Indian villages. In 1861 he began purchasing grain and produce from the Pimas and grinding flour for use by the California soldiers. According to J. Ross Browne, the Indians subsequently sold White 300,000 pounds of wheat, 50,000 pounds of corn, 20,000 pounds of beans, and a large amount of dried and fresh pumpkins.[52] William Walker, who visited White's establishment in February 1862, said that White was buying daily from the Indians between 140 and 180 sacks of wheat at about 2 cents per pound. If Walker's observations can be relied upon, this price compared favorably with the price of wheat in Tucson, which Walker said cost about $2 per *fanega.*[53] Later, however, wheat in Tucson would sell for 6 cents per pound.

In February 1862 Captain Sherod Hunter marched west from the Rio Grande with a company of mounted rifles and took possession of Tucson for the Confederacy. He continued to the Pima villages where he took Ammi White a prisoner, confiscated his supplies, and partially destroyed his mill. Hunter later reported that he distributed 1,500 sacks of confiscated wheat among the Indians because he lacked the means to transport the grain to Confederate lines. His men also destroyed much of the hay that had been stockpiled between the Pima villages and Fort Yuma for use by the California Column.[54] Thereafter, the Union army would rely heavily upon Pima grain to feed their animals on the road to Tucson.

Advance units of the column under Colonel Joseph R. West reached the Pima villages in April, and by early May the Californians had established a post at Ammi White's flour mill, which would be named Fort Barrett. West quickly encountered difficulties in negotiating for supplies. The 10,000 yards of manta had not arrived, and the Indians refused to sell more than "trifling" quantities of flour and wheat on credit. In frustration, West wrote to Carleton: "It is difficult to make this people understand the magnitude of our demands, and further, I have nothing but promises to offer them in payment."[55] By 5 May the Indians had agreed upon a scale of prices. A yard of manta was worth each of the following: 4 quarts of flour, 7 quarts of wheat, 4 quarts of pinole (ground corn), 50 pounds of hay, and 150 pounds of green fodder. The horses and mules in West's command consumed daily 3,945 pounds of wheat and 4,830 pounds of hay, equivalent to 400 yards of manta. By this date, the army was indebted to the Pimas for 3,000 yards of manta, and the 10,000 yards expected daily would be exhausted by 20 May. West told Carleton that if the Indians "do not bring in their wheat more freely," he would go into their fields and cut the grain for forage.[56]

This action was averted by the arrival on 12 May of cotton goods, including old and condemned military clothing, which West used to pay outstanding debts and to buy an additional 30,000 pounds of wheat. He believed that after the Pimas harvested their new crop the army could procure an additional 400,000 pounds, and he recommended that blue drills, blue and orange prints, red flannel, paints, beads, and tobacco be forwarded to make the purchase. Much to his disappointment, the Pimas had few beef cattle to sell and refused offers to sell their work cattle.[57]

West and his command entered Tucson on 20 May, only sixteen days after Hunter had abandoned it to rejoin Sibley on the Rio Grande. But even here West was frustrated in his efforts to procure supplies. With his troops badly in need of flour, he was forced to pay "the exorbitant price" of 6 cents per pound for wheat. And because of Indian raids, beef cattle were scarce. West finally signed a contract with Jesús M. Elías, a prominent Tucson resident, for a thirty-day supply of fresh beef at 10 cents per pound. He later made arrangements with Manuel M. Gándara, ex-governor of Sonora, to bring cattle from across the border. He also located near Tucson a small water-powered mill for grinding flour, but it could manufacture only 1,200 pounds daily.[58] During the remainder of the year, government purchasing agents continued buying wheat from the Pimas, sometimes paying for it with scrip, which was redeemed later with merchandise.

Carleton finally joined the advance units of the column in Tucson on 7 June, and on the following day he declared a state of emergency, decreed martial law, and designated himself military governor of Arizona territory. Since he could not rely on getting supplies in New Mexico, he would have to accumulate large quantities of military stores before making his final advance to the Rio Grande. He subsequently made special arrangements with Ammi White and his partner Cyrus Lennan to manufacture flour for the army. White only recently had returned to the Pima villages following his release by Confederates on the Rio Grande. On Carleton's orders, the army helped him repair the mill and loaned him mules to run it. Even though the mill could grind 200 pounds of flour an hour, Major David Fergusson, acting commissary of subsistence at Tucson, doubted that White and Lennan could meet the army's demand.[59]

Recognizing the need for additional supplies, Carleton sent Fergusson to Magdalena in Sonora to inquire about resources. Fergusson's reconnaissance resulted in a detailed list of ranches and agricultural supplies located in northern Sonora and in the purchase of 3,526 pounds of flour from Francisco G. Toraño of Magdalena, 20,000 pounds from Alejandro Daguerre of San Ignacio, and additional flour from businessmen in Hermosillo. Sonoran residents also supplied the army with unspecified amounts of forage and probably other commissary items as well.[60]

As late as 19 June, Union officials in New Mexico were unaware of Carleton's advance into Arizona. Within thirty days, however, Canby established communications with Carleton and informed him that New Mexico now had abundant military stores. Three government ox trains and one belonging to Irwin, Jackman and Company had arrived from Fort Leavenworth in mid-June, each containing twenty-five wagons. The four trains together carried 520,615 pounds of subsistence, clothing, and camp equipment. In following months Irwin, Jackman and Company dispatched eighteen additional 25-wagon trains to Fort Union, and the government sent two.[61] These supplies would be available to the California volunteers.

Moreover, Captain Amos F. Garrison, commissary of subsistence, had called for proposals in June for 1,275,500 pounds of flour, 4,000 beef cattle, and fresh slaughtered beef "as may be required" for troops at Fort Union, Fort Craig, and Peralta. Because local supplies were nearly depleted, Garrison stipulated that the flour and beef cattle

Table 1
Flour and beef contracts awarded in July 1862

Contractor	*Item*	*Quantity*	*Price*	*Place of delivery*
Ceran St. Vrain	Flour	200,000 lbs.	$10.40 per 100 lbs.	Ft. Union
Ceran St. Vrain	Flour	250,000 "	12.50 "	Peralta
Ceran St. Vrain	Flour	150,000 "	14.50 "	Ft. Craig
Moore & Kitchen	Flour	250,000 "	12.40 "	Peralta
Parker & Co.	Flour	350,000 "	.13⅞ per lb.	Ft. Craig
J. & R. Kitchen	Cattle	500 head	.06¼ "	Ft. Union
J. & R. Kitchen	Cattle	1,000 "	.06¾ "	Peralta
Probst & Co.	Cattle	1,500 "	.07½ "	Ft. Craig
Parker & Co.	Cattle	1,000 "	.08¼ "	Ft. Craig
Hinckley & Blake	Fresh Beef	——	.09½ "	Ft. Union
Hinckley & Blake	Fresh Beef	——	.11½ "	Peralta
Hinckley & Blake	Fresh Beef	——	.12 "	Ft. Craig

come "from beyond the Department."[62] The following month he awarded contracts to six individuals or partnerships (see Table 1).

Charles G. Parker, in addition to government contracting, managed a freighting firm and for a brief time was proprietor of the Exchange Hotel in Santa Fe. Charles Probst and his partner August Kirchner, both young Germans, were butchers in Santa Fe. The Kitchen brothers, Charles, Richard, James, and John, were among early Anglo settlers in Las Vegas. James and Richard held the cattle contract and Charles, in partnership with William H. Moore, the flour contract. St. Vrain was a longtime resident of Taos, and some of the flour that he subsequently delivered on his contracts was manufactured at his Taos mill with an appropriate reduction in price to the government. Other army beef contractors in 1862 included Charles Behler, a Santa Fe butcher, who agreed to keep soldiers at Fort Marcy supplied with fresh beef at 9 cents per pound, and Harvey E. Esterday, a prominent businessman in the San Luis Valley, who would do the same at Fort Garland for 8½ cents per pound. As a result of these and other contracts for pickles, beans, sauerkraut, and cornmeal, military posts in New Mexico would be well-supplied with subsistence for the coming year.[63]

Carleton, nonetheless, was forced to remain in Tucson with the major portion of his troops until summer rains refilled water holes in the desert. On 21 June he sent forward an advance guard of 140 men, who reached the Rio Grande at Fort Thorn on 4 July. By this date most of Sibley's soldiers had started their long, grueling retreat to San Antonio. Carleton arrived at the Rio Grande on 7 August, and in mid-September he relieved Canby of command in Santa Fe. In following months his primary objectives were to guard against possible Confederate invasions and to subdue hostile Indian tribes. His success, in part, would depend upon the army's ability to employ local sources of forage and subsistence. And enterprising businessmen like those mentioned above scurried to fill the government's shopping list.

Although the subsistence department relied heavily upon outside sources for food items in 1862, John C. McFerran, who became chief quartermaster of the department in September, worked diligently to secure forage from local suppliers. During the Texas invasion, he had contracted with William H. Moore to transport 1,000,000 pounds of corn from the States at 8¾ cents per pound. He later signed contracts with John Dold of Las Vegas, who agreed to deliver 500,000 pounds of States corn at 8½ cents per pound, and with Arnold T. Winsor, contractor at Fort Wise, Colorado, to deliver 3,000 bushels at $5.60 per bushel (or 10 cents per pound). But he also made arrangements to secure smaller amounts from Santa Fe and Taos residents at $6.50 per *fanega* (or 4⁶⁄₁₀ cents per pound). The grain from the States began arriving at Fort Union in June, and McFerran believed it would last the department until November or December.[64]

The quartermaster's department was hard pressed, nevertheless, to keep posts south of the Jornada del Muerto supplied with forage. In September, or possibly earlier, Carleton instructed residents near Mesilla and Franklin, Texas, to sell their surplus grain to the army at $3 per *fanega.* Colonel Joseph R. West, commanding troops in this district, empowered William B. Rohmann, a resident of Franklin, to purchase for the government 1,000 *fanegas* of wheat and corn at that price.[65] But grain came in so slowly that the army resorted to seizures. In October soldiers confiscated at Ysleta, Texas, 40 *fanegas* of wheat belonging to Dr. Mariano Samaniego, a citizen of El Paso, Mexico; the army paid him $3 per *fanega.* The following month the army seized about 7,840 pounds of wheat from Joshua S. Sledd of Mesilla, who was also paid about $3 per *fanega* (or 2½ cents per pound). Sledd later complained that a few months earlier the army had purchased his wheat at 4¼ cents per pound, and he saw no reason for being paid less now.[66]

During the fall of 1862 McFerran canvassed much of the territory for additional forage. Ceran St. Vrain, who feared that the army's demand for grain in Taos Valley would interfere with his efforts to fill his flour contracts, offered corn and wheat to the army at double the price he was paying local producers. McFerran rejected this proposal, but he approved St. Vrain's subsequent offer to purchase all the surplus grain in Taos Valley for the government, charging only a small fee for his labors.[67] The quartermaster's department received some grain from Andres Dold of Las Vegas, who delivered 180,000 pounds of wheat at Fort Stanton for 6⅔ cents per pound, and from Frederick W. Posthoff, who agreed to furnish Fort Garland with 1,000 *fanegas* of wheat and corn. Nonetheless, there was insufficient grain in New Mexico to feed government animals. In December McFerran signed at least three contracts to import corn from the East, paying $8.30 per hundred pounds: one with William G. Barkley of Kansas City for 100,000 pounds, another with George W. Brient for 180,000 pounds, and the third with Mesilla Valley freighter Epifanio Aguirre for 250,000 pounds.[68]

Meanwhile, several territorial residents received contracts to supply the quartermaster's department with locally grown hay. St. Vrain agreed to deliver 250 tons of hay at Fort Garland for $19 per ton, and John Lemon, a wealthy resident of Mesilla, agreed to deliver 100 tons at Fort Fillmore and 100 tons at Mesilla for $30 per ton. Four contractors (William Kroenig, C. H. Williams, Lucien B. Maxwell, and B. Archuleta) agreed to deliver a total of 315 tons at Fort Union at $45 per ton. Undoubtedly other residents turned in hay at army posts without signing contracts, as was true of suppliers near San Elizario, a small community below Franklin, who delivered in December one ton of hoe-cut hay for $20.[69]

Efforts to accumulate forage quickened in November when it was rumored that Baylor was in San Antonio raising a force of 6,000 men to reinvade the territory. Carleton thereafter ordered West to buy or seize all surplus grain in the Mesilla Valley and in El Paso County, Texas, so that invading forces could not replenish their dwindling supplies. On 2 December West ordered the people of San Elizario, Ysleta, Socorro, Franklin, Amoles, La Mesa, Sánchez Ranch, Santo Tomás, Mesilla, Las Cruces, and Doña Ana to turn in their grain and flour. Any person found after 17 December with over a two months' supply for his family and animals would be treated as an enemy of the United States.[70]

Many Hispanic residents now abandoned their lands and migrated to Mexico, apparently taking their surplus grain with them. Alcaldes in three Mesilla Valley communities petitioned West to revoke his order, claiming that the people fleeing to Mexico were poor and honest laborers who feared that the army would deprive them of their means of livelihood.[71] The order remained intact. On 7 December, however, West increased the price paid for grain from $3 per *fanega* to $4 per *fanega* for corn and $4.50 for wheat. He would also pay $9 per hundred pounds for flour, although residents later testified that the market price was $12.50.[72]

As their fears subsided, residents returned to their homes. Major William McMullen, commanding officer at San Elizario, described local conditions in a letter to West dated 21 December:

> The citizens are gaining confidence and a number of the runaways have returned with their families and a majority of the residents are actively engaged in repairing the acequias and preparing their lands for another crop. Grain is extremely scarce. I seized 8 *fanegas* of wheat yesterday belonging to Dr. [David] Diffenderfer and when that is consumed I have no idea where I can procure more, unless it may be a few *fanegas* from Martin Lujan Alcalde of Socorro. The people here have no more grain than they absolutely require for their own consumption, and many will be compelled in a short time to draw a supply from elsewhere. This they can procure in Mexico in exchange for salt.[73]

West also turned to Mexico for grain. On 4 February 1863 he sent an agent "secretly" to inquire about prices in the towns of Guadalupe and San Ignacio. Within two weeks the army had eighteen wagons returning from Chihuahua with forage.[74]

Not only did the Union army seize grain and flour during the threat of invasion, but it also required citizens to work in fatigue parties. In December Carleton made a special plea asking citizens to work twenty days without pay to strengthen fortifications at Forts Craig and Union. Under the leadership of Ceran St. Vrain, 100 Taos residents quickly went to work at Fort Union. But elsewhere the appeal for labor largely went unheeded. On 1 January 1863, Carleton established a special Fort Craig military district that embraced eighteen villages extending from Sabinal in the north to Cañada Alamosa in the south. Alcaldes for each village were required to furnish reasonable numbers of laborers to work on the defenses at Fort Craig, and by early February more than 200 civilians were hard at work. The territorial legislature also passed a resolution on 29 January urging every county to donate money or send volunteers to help strengthen the two forts. Few citizens responded. Many living in the Fort Craig district had already given their labor, and most were poor farmers who were busy preparing their fields for planting. To complete the work at Fort Craig, therefore, Carleton authorized employing laborers at 75 cents a day, a ration, and—as additional inducement—one gill of whiskey every two days.[75]

The invasion by the Confederates failed to materialize, but Carleton thereafter closely watched the activities of southern sympathizers. Anxious that none but loyal unionists be employed by the government or receive government contracts, Carleton stopped army purchases from Santa Fe merchants Gustave Elsberg and Jacob Amberg until their loyalty was verified, and he dismissed Charles B. Magruder, a paymaster's clerk, for not being "so warm a friend to the Government" as he should be. On 3 February 1863 Carleton issued General Order No. 4: "All persons who desire to furnish supplies of any kind to the army in the Department must first submit unequivocal evidence of their loyalty to the Government."[76]

Charles T. Hayden, Samuel J. Jones, and Ernest Angerstein were among the residents whose loyalty came under suspicion. Each offered supplies and services that the California volunteers desperately needed, but reportedly each man had given support to the Confederate army during the invasion. Union officers thus faced a dilemma. Events that unfolded in subsequent months illustrate the tenacity with which Carleton upheld his edict and the willingness of some of his subordinates to overlook it in order to secure supplies.

Charles Trumbull Hayden, father of long-term Congressman Carl Hayden, arrived in Arizona in 1856. Already a man of some wealth, this former schoolteacher from Connecticut established a store about ten miles south of Tubac and not far from Fort Buchanan. He soon became even more prosperous as a merchant and freighter. Census records in 1860 show that he was living in Tucson where he had amassed property worth $20,000.[77] In July 1862 he witnessed Sibley's evacuation of New Mexico, and, according to his own testimony, he purchased various kinds of army property from retreating Confederate officers, agreeing to transport the supplies to Mexico to keep them from falling into the hands of Union soldiers. General West, who later investigated Hayden's activities, found no evidence of disloyalty, but he believed that Hayden had "favored his pocket at the expense of his patriotism."[78]

General Carleton continued to doubt Hayden's loyalty. In January 1863 he told Hayden, who was then residing in El Paso, Mexico, to select one side of the Rio Grande or the other on which to live. Secessionist spies and sympathizers had established their headquarters in El Paso, and Carleton would not allow Hayden to move freely between Mexico and the United States.[79] Hayden chose the American side, for he owned a store in Pinos Altos where he intended to make his home. At some time between January and April, however, Hayden, with a pass from West, took his freight train into Chihuahua to buy grain for John Lemon, contractor at Fort West, a new military post under construction near Pinos Altos. According to General West, this was the only means he had of supplying the post with grain.[80]

Carleton did not agree. Several weeks later, he wrote a scathing letter to West, accusing him of disobeying orders by allowing Hayden to cross into Mexico.

> Having no corn at Fort West—or difficulty of procuring corn for that post—was no justification. Other people could have gone for corn as well as Mr. Hayden; but even if they couldn't, the animals could have lived on grass until the matter was reported. If they had all died, the responsibility would not have been upon you.[81]

It is unclear whether Carleton ever changed his mind about Hayden's loyalty. Certainly some of Carleton's ire can be attributed to his displeasure with West rather than to a fear that Hayden would betray the Union. In later months, Hayden took up residence in Tucson where he continued furnishing supplies to the army.

Unlike Hayden, Samuel J. Jones, a one-time proslavery sheriff in Lecompton, Kansas, and former sutler at Fort Fillmore, openly expressed his support for the Confederacy. For this, he was thrown into the guardhouse soon after Union soldiers regained control of the Mesilla Valley. But like Hayden and many other westerners, Jones wanted to capitalize on the army's need for supplies. He had sold pickles and other vegetables to the California Column upon its arrival on the Rio Grande. Thereafter, however, General West proscribed all purchases from Jones. When the army later purchased a few barrels of pickles from Frank DeRyther, a Mesilla resident whose loyalty was also questioned, Jones protested. He wrote to Carleton in September 1863 reminding him that furnishing supplies was a matter of open and fair competition in which contracts were awarded to the lowest bidders. "I have this year put up pickles and cultivated quite a large garden," he continued, "but under the present order of things I cannot sell to government."[82]

Carleton and West both agreed that Jones was a "vile secessionist" and therefore not deserving of government patronage, at least in the matter of pickles. The army subsequently annulled a contract for pickles with John L. May of Las Cruces when it was learned that Jones would supply some of the pickles. Housing was a different matter. West planned to move his headquarters from Franklin to Mesilla, and through an agent he sought to rent Jones's spacious residence in Mesilla. Jones did not overlook this anomaly: "If I was too disloyal to sell pickles I should certainly be too disloyal to rent my house for Headquarters."[83] Sometime later, Jones took the loyalty oath prescribed by President Abraham Lincoln, thus reinstating himself as a U.S. citizen, and Carleton thereafter allowed him to compete for government contracts.

The case of Ernest Angerstein illustrates many of the problems that government purchasing agents encountered in their efforts to secure supplies during these tumultuous days. On 18 November 1863 Captain Samuel Archer, with General West's approval, signed a contract with Ernest Angerstein of El Paso, Mexico, for the delivery of 5,000 *fanegas* of corn at El Paso and 1,000 *fanegas* at the Rio Mimbres for $6.75 per *fanega.* A few days later Carleton disapproved the contract because Angerstein allegedly was a southern sympathizer and had refused to take the oath of allegiance to the United States. In early December Archer signed a new contract, again with West's approval, with Ynocente Ochoa of El Paso who agreed to furnish the same amount of corn at the same price stipulated in the Angerstein contract. In fact, Angerstein and Ochoa were partners, and the corn Ochoa subsequently delivered to the Union army was corn Angerstein had purchased to fill the original contract. When this became known, the army refused Ochoa's corn and launched an investigation.[84]

To recover his good name, Angerstein provided Union authorities with letters and depositions attesting his loyalty. Like many immigrants, Angerstein, a native of Hanover, Germany, had joined the U.S. army. Discharged at Fort Fillmore in 1857, he thereafter opened a mercantile establishment in Mesilla. When the war broke out Angerstein decided to move to Mexico to protect his property. He had removed two thirds of his goods before the Texans arrived in the Mesilla Valley. Baylor appointed him—against his will—treasurer of the Confederate Territory of Arizona, but in September 1861 he closed his Mesilla store and moved to Mexico whence he wrote Baylor declining this appointment. On 6 April 1863, William F. M. Arny, acting governor of New Mexico, administered an "oath of neutrality" to Angerstein at the latter's store in El Paso. Angerstein testified that he could not take a full oath of allegiance to the United States as he would lose property inherited in his native country.[85]

The contract signed on 18 November generated dissatisfaction among residents on both sides of the Rio Grande. Many considered Angerstein a rebel and felt it unfair that the government purchased corn from a secessionist and at a higher price than was paid to loyal citizens. Several testified that prior to the signing of the contract corn could be purchased along the river at between $4 and $4.50 per *fanega.*[86]

But the Angerstein affair was more complicated than local residents realized. During the fall of 1863, advertisements for corn had been published at Las Cruces, Mesilla, and Franklin, resulting in only one bid—Ynocente Ochoa's at $7.20 per *fanega*—which was rejected as too high. Two Americans subsequently offered corn at $7 and $8 per *fanega,* and these offers were likewise rejected. General West finally consented to the Angerstein contract believing that every effort had been made to secure corn at a lower price.[87] Juan N. Zubirán, collector of customs at El Paso, believed that Angerstein could buy corn in Mexican towns for $2 per *fanega* and not pay more than $1 per *fanega* for hauling it to El Paso. Supplying the Union army was potentially a lucrative business for citizens who had the necessary funds to accumulate the large quantities of stores required by the military. But the system hurt small producers. Army Inspector Nelson H. Davis, who investigated the Angerstein contract, observed that most of the local people were small farmers without resources to bid on government contracts and whose surplus crops frequently passed into the hands of speculators after harvest. Davis agreed with other area residents who testified that corn could have been secured at a considerably lower price than Angerstein was asking had it been bought in open market and in quantities the people could furnish.[88]

A board of officers, convened in mid-1864 at Angerstein's request, decided that he had not been disloyal. Rather the board judged him to be "one of those mercenary men" who secured protection from the government but who failed to assist it in its hour of need. Thereafter Carleton refused Angerstein permission to enter the United States until "it is made more clear that he is a friend" to the Union.[89]

Despite the difficulties that Hayden, Jones, and Angerstein encountered, numerous entrepreneurs easily satisfied the loyalty requirement and continued selling forage and subsistence to the army. The amount of money they received is only partly determinable because not all purchasing records have been preserved. Chief Quartermaster John C. McFerran requested funds for his department for the year ending 31 December 1864 amounting to $1,998,062. A significant portion of this would be spent locally for building materials, forage, and hiring civilian transportation and laborers. Payments for forage alone would amount to nearly one half the amount. During the same period, the chief commissary of subsistence estimated his expenses to amount to $1,084,256.[90] More than half this amount would go to contractors who supplied fresh beef and cattle on the hoof. Three million dollars was a fabulous sum for an impoverished society like New Mexico. Moreover, federal dollars would be widely dispersed, further conditioning residents to the government's patronage and strengthening their economic ties with the military.

During 1863 and 1864 the army negotiated few contracts with Arizona residents for furnishing either forage or subsistence. The only contracts registered were for grain, hay, and fresh beef. In the spring of 1863, Berry Hill DeArmitt of Tucson contracted to deliver 30 tons of hay at San Pedro Crossing for $40 per ton, 50 tons at Fort Bowie for the same price, and 120 tons at Tucson for $23 per ton. Contract prices in 1864 were higher. Irishman Jeremiah Riordan, a merchant in Tucson, agreed to furnish Fort Bowie with 50 tons of hay at $47.50 per ton; Hiram S. Stevens, a Tucson merchant, agreed to furnish troops stationed there with 450 tons at $28 per ton; Francisco Gándara, brother of former governor of Sonora Manuel Gándara, would deliver 100 tons at Tubac for $29.50 per ton; and Loren Jenks, a pioneer farmer in what would become Yavapai County, agreed to deliver 250 tons for the new post at Fort Whipple at $50 per ton. Fresh beef contracts were let to Manuel Gándara, who would furnish a three months' supply at Tucson for 12½ cents per pound starting 15 December 1862; Pinckney R. Tully and Estévan Ochoa, prominent freighters and merchants, who would furnish a two months' supply "at points in Arizona" for 13 cents per pound; and Hiram S. Stevens, who agreed to furnish Tucson with a four months' supply starting 15 December 1863 at a little more than 5 cents per pound. The only grain contract awarded was to pioneer merchant John B. Allen on 16 June 1863 who agreed to deliver grain "as required" at 3 cents per pound.[91]

It is not surprising that the army signed so few contracts, for Tucson and vicinity had only limited agricultural resources. General John S. Mason, who later commanded the District of Arizona, reported with only slight exaggeration that as late as mid-1865 every ranch south of the Gila had been abandoned because of the ferocity of Apache raids.[92] Moreover, the opening of mines in central Arizona offered a competing market where top prices were obtained for cattle and grain. Then, too, the army lacked funds to make purchases. By mid-1863 the quartermaster's department at Tucson was indebted $100,000, and many citizens refused to extend further credit. It was also increasingly difficult to purchase supplies in Sonora, for businessmen there demanded gold or silver coins, neither of which were available in Arizona.[93]

By the summer of 1863, troops in southern Arizona were running low on both subsistence stores and forage.[94] The grain shortage was the most critical. No one had responded to government advertisements earlier that year for delivery of 500,000 pounds of grain. Consequently the quartermaster's department entered into a controversial arrangement in June with John B. Allen, who agreed to furnish grain at 3 cents per pound in return for exclusive trading privileges for one year with the Pimas and Maricopas. Ammi White, however, would be allowed to purchase as much wheat from the Indians as his mill could convert monthly into flour.[95] In this manner the army hoped to prevent speculators from buying Pima wheat for resale to the government at exorbitant prices, while at the same time permitting White to carry on his milling.

J. Ross Browne recorded that Pimas furnished the government 600,000 pounds of wheat in 1863, presumably under Allen's contract. Browne also stated that because of "the breakage of their main acequia at a critical period of the season," their crop was smaller than usual.[96] They were nearly out of wheat in January 1864 when Arizona's newly appointed superintendent of Indian affairs, Charles D. Poston, arrived in Tucson to assume his duties. Almost immediately the superintendent locked horns with the army over Pima grain. Poston ordered Allen and his agent off the Pima reservation and authorized his own agent, Ammi White, to purchase a large amount of Pima wheat so that seed could be sent to Yuma Indians for planting. But the army's supply of forage was critically low, and government animals faced starvation. Citing military necessity, Army Inspector Nelson H. Davis ordered military officers in March to secure all surplus wheat at the Pima villages. In carrying out this order soldiers seized some of the wheat that White recently had purchased from the Pimas. Because of Poston's interference, the quartermaster's department was compelled to buy additional grain in open market at 8 and 10 cents per pound. Some grain was secured in Sonora, but the amount and cost to the army is not recorded.[97]

Supply officers in Arizona found it equally difficult to secure beef cattle. Joseph R. Beard of Los Angeles had signed a contract on 12 April 1862 for furnishing beef at 9 cents per pound to all the troops composing the Column from California, driving the cattle from the Pacific Coast to the Rio Grande. His agent subsequently negotiated a subcontract with Manuel Gándara, who agreed to furnish beef in Tucson at 7½ cents per pound. Beard was unable to fulfill his contract, however, and he renounced it in January 1863.[98] Beard's failure forced the army to pay local suppliers much higher prices for its beef: 12½ cents per pound to Gándara for a three months' supply and 13 cents per pound to Tully and Ochoa for a two months' supply. When Hiram Stevens contracted in September to furnish beef at Tucson for 18 cents per pound, the army called a halt to rising prices and ordered an investigation.

Theodore Coult, commanding officer at Tucson, subsequently testified that in September beef had been scarce and of poor quality, the best cattle having been driven to the mines. Many residents who owned cattle refused to sell on credit. Consequently when the army advertised for beef, Stevens had submitted the only bid. Coult concluded that beef could not have been purchased locally for less than 18 cents per pound. By December, however, many adventurers had deserted the mines, and cattle again were plentiful.[99] Thereafter the army renegotiated its contract with Stevens, who agreed to supply beef for a little more than 5 cents per pound, an exceptionally advantageous price for the government.

In contrast to Arizona, New Mexico had more resources for meeting the army's subsistence and forage requirements. Its citizens also had nearly fifteen years' experience in selling supplies and services to the U.S. government. But the territory could not meet all of the army's supply needs. Under Carleton's directions, officers began collecting statistics to show the military strength of the department, including the availability of local supplies. For each military district, census takers were to list the names of its inhabitants; the number of acres planted with corn, wheat, beans, onions, potatoes, and vegetables; an estimate of probable yield; and the number of firearms, gristmills, sawmills, wagons, horses, mules, cattle, sheep, and goats owned by residents.

The census compiled by Captain Hugh L. Hinds for the Fort Craig district in April 1863 is one of the few that has been preserved. If all were as detailed as this one, supply officers would have had an accurate picture of territorial resources. Hinds recorded data for sixteen villages. His description of Lemitar is typical:

> This place contains a population of 103 men, 154 women, and 575 children. There are 5 mills about the place, all of which can grind 2000 pounds in 12 hours. The Navajo Indians have stolen 324 head of cattle, 600 head of sheep and 20 horses during the past year from the people of this place and killed 2 of the citizens. There [are] 60 mules and horses, 90 head of cattle, 1000 sheep and 1000 goats. There will be raised 400 bushels of grain for mar-

ket this season. There is [*sic*] some good gardens around the town. The people are not so indolent as at many other places on the river.[100]

In addition to gathering statistics, the army constantly worked to increase local production of crops and livestock. For example, in February 1863, General Carleton issued improved seed corn to farmers in southern New Mexico in hopes of increasing their yields. Two months later military officers urged residents of Pinos Altos to move near Fort West to raise grain for the army. In August the Santa Fe *Weekly Gazette* published military orders establishing Fort Bascom, giving notice that the garrison would protect citizens who wished to open farms along the Canadian River. And several months later Carleton instructed the commissary officer at Fort Stanton to encourage farmers along the Rios Bonito, Ruidoso, and Tularosa "to put in all the crops they can this Spring."[101] Efforts such as these stimulated local production and inspired New Mexicans to compete for the government's patronage.

During 1863 and 1864 contracts were made with thirty-eight individuals or partnerships for the delivery of a wide variety of commissary stores, many produced locally, to posts in New Mexico and to Forts Bliss and Garland. The greatest quantity of stores was delivered to Fort Union in 1863 and to Fort Sumner in 1864. Commodities supplied at Fort Wingate commanded the highest rates, as transportation costs were included in prices paid to contractors. Of the six flour contractors in 1863, five—Joseph Hersch, Louis Gold, John M. Shaw, Henry J. Cuniffe, and William H. Moore—also provided flour in 1864. The largest flour contracts in each year were let to Moore. In 1863 he and his partners contracted to deliver 200,000 pounds of flour at Albuquerque for 10 cents per pound and 50,000 pounds at 9½ cents per pound; and in 1864 they agreed to deliver a total of 1,250,000 pounds to Forts Union, Sumner, Bascom, and Stanton at prices ranging from 8 to 11 cents per pound. Moore also held contracts in 1864 for delivery at Fort Sumner of 200,000 pounds of shelled corn and 150,000 pounds of wheat meal. Several contractors, in addition to Moore, provided more than one commodity to the subsistence department. Hersch, who operated a steam-powered gristmill in Santa Fe, held contracts in 1863 for flour, potatoes, and fine salt, and in 1864 for flour and fresh beef. Shaw, a Socorro resident and a former Baptist missionary, provided flour, cornmeal, and onions in 1863 and flour and cornmeal in 1864. Gold, a Santa Fe merchant, in addition to flour, contracted for potatoes in 1863 and cornmeal in 1864. Twenty-two contractors, however, furnished only one commodity (see Tables 2 and 3).[102]

The flour contract that Henry J. Cuniffe and Juan N. Zubirán signed on 10 April 1863 did not specify the amount of flour to be delivered. The two contractors resided in El Paso, Mexico, where the former served as U.S. consul and the latter as collector of customs. In partnership with Nathan Webb, they paid the U.S. government one hundred dollars a month to lease Hart's Mill, a water-driven gristmill located on the northeast bank of the Rio Grande near Franklin. The mill's owner, Simeon Hart, had supported the Confederacy and had left the river with Sibley's retreating troops. Thereafter the property fell into Union hands. The mill, which could grind 4,000 pounds of wheat during a twenty-four-hour period, would supply flour for the garrisons at Franklin, Las Cruces, and Forts McRae, West, and Craig. In 1863 the army paid the partners 11 cents per pound for flour delivered at Hart's Mill, and in 1864 paid 8 cents per pound at the mill and 10 cents per pound for flour delivered in Las Cruces. During the summer of 1863 the firm was forced to stop grinding because of lack of water. Too many users were competing for the limited amount of water in the Rio Grande, a problem that would plague the region for many years. In a letter to Carleton, Nathan Webb described this competition: "The people of El Paso show every disposition to favor us by granting all the water, two or three days in each week, but so long as the dams are maintained at Doña Ana and Las Cruces the water will be insufficient." Under the circumstances, the army extended the firm's contract for six months.[103]

Some of the cattle delivered on commissary contracts came from outside the territory. Charles S. Hinckley purchased his cattle in Kansas, which helps explain their high cost to the army. In 1864 he contracted to supply nine garrisons with fresh beef "as required" for 22 cents per pound, and seven garrisons with 4,000 head of cattle at 18 cents per pound—these prices making his beef some of the most expensive ever consumed by soldiers in New Mexico.[104] Cattle also came from Mexico. During the summer of 1864, supply officers in

Table 2
Subsistence contracts awarded in 1863 for Fort Bliss, Fort Garland, and posts in New Mexico

Item	*Quantity*	*Price range*
Flour	590,000 pounds	$ .07$\frac{95}{100}$–.12 per pound
Vinegar	14,500 gallons	1.15–1.85 per gallon
Pickles	3,800 "	1.24–2.50 "
Cornmeal	40,000 pounds	.05¾–.11 per pound
Potatoes	1,800 bushels	3.20–5.60 per bushel
Sauerkraut	4,000 gallons	.73–2.50 per gallon
Onions	15,300 pounds	.04½–.11 per pound
Beans	1,300 bushels	3.90–4.75 per bushel
Fine salt	2,100 "	2.25–5.00 "
Coarse salt	1,000 "	2.00–4.00 "
Fresh beef	unspecified	.09–.16½ per pound
Cattle on hoof	1,800 head	.09¼ "

Table 3
Subsistence contracts awarded in 1864 for Fort Bliss, Fort Garland, and posts in New Mexico

Item	*Quantity*	*Price range*
Flour	2,025,000 pounds	$ .07$\frac{74}{100}$–.13 per pound
Cornmeal	16,000 "	.07–.11 "
Wheat meal	307,000 "	.08½–.14$\frac{60}{100}$ "
Salt	2,000 "	4.00 per bushel
Shelled corn	1,010,000 "	.11 per pound
Fresh beef	unspecified	.12¼–.22 "
Cattle on hoof	5,600 head	.12$\frac{40}{100}$–.18 "

Franklin purchased nearly 300 head of Chihuahuan cattle in private transactions for 12½ cents per pound, the same price paid to Juan Zubirán for 1,000 head of Mexican sheep. Since cattle were scarce in the territory, the army sometimes substituted mutton in the soldier's ration. One of the largest suppliers of sheep was Juan Perea of Bernalillo, who sold more than 3,000 head to the subsistence department in 1863 and 1864.[105]

Most of the contracts that the quartermaster's department let in

New Mexico during 1863 and 1864 were for delivery of hay. Five contractors in 1863 agreed to supply more than 1,100 tons for prices ranging from $11.87½ to $50 per ton. John Lemon, who cut his hay near San Agustín Springs east of Mesilla, received the highest price, as he had to transport hay more than sixty miles to Fort Cummings, a new post being built near Cooke's Peak in southwestern New Mexico. Rafael Chávez, who farmed lands within the military reserve at Fort Wingate, received the lowest price for delivery of 500 tons, which he agreed to cut within ten miles of the post. He failed to complete his contract, however, and the following year the army rejected his much higher price of $47.75 per ton, not because it was too high but because he had failed on his earlier contract. Quartermaster John C. McFerran also purchased hay in Santa Fe from small suppliers, paying them 2 cents per pound. Seven hay contracts were signed in 1864 for 1,445 tons at prices ranging from $15 to $39 per ton.[106]

The quartermaster's department signed only two contracts for provisioning the New Mexico posts with corn during this period—one in 1863 and one in 1864—but it secured large amounts of grain through open market purchases. In both years New Mexico's corn crop was less than anticipated. Quartermaster McFerran's greatest worry in 1863 was that government animals would suffer for lack of forage. His worries escalated in 1864 with the opening of the Indian reservation at Bosque Redondo near Fort Sumner. He now feared that Navajos held captive there would starve. James Hunter of Santa Fe received the only corn contract let in 1863. On 21 December he contracted to deliver at Fort Union 1,000 *fanegas* at $6.50 per *fanega.*[107] By this date McFerran had issued orders to subordinate officers to secure "all the corn and wheat you can in your district, without prejudice to the necessary supply of the inhabitants." Citizens were offered $6 per *fanega* for their corn at Fort Wingate, $7 per *fanega* at Albuquerque, and $8.50 per *fanega* at Fort Sumner. Daily grain rations for horses and mules were reduced from twelve and nine pounds, respectively, to ten and eight pounds.[108]

In a letter to Quartermaster General Montgomery C. Meigs dated 10 December 1863, McFerran outlined the grain situation. He estimated that the corn crop in New Mexico was about two thirds of its normal yield, since the crop had been almost an entire failure in southern New Mexico and substandard from Fort Craig to Santa Fe.

Between 15,000 and 16,000 *fanegas* of corn could be purchased within sixty miles of Fort Union, but this amount would not overcome shortages elsewhere. He engaged Ceran St. Vrain to purchase and store grain in Taos Valley, the major grain center of New Mexico, which produced annually nearly 1,000,000 pounds each of wheat and corn. The problem was that grain could not be moved in the winter because of bad roads. McFerran consequently requested funds from Meigs to improve the roads connecting Santa Fe and Fort Union with Taos Valley. If they were passable during all seasons, McFerran reasoned, grain raised in the valley would supply three fourths of the posts in the department. Improving the roads would also encourage residents to plant more extensively, thereby reducing the price of grain.[109] In subsequent months, the quartermaster's department purchased more than 72,000 pounds of corn from William C. Mitchell of Tecolote, 120,000 pounds of wheat from a Mr. Maxwell of Las Vegas, and perhaps 500 *fanegas* of corn from Dittenhoefer and Cohen, also of Las Vegas. McFerran also accepted corn in small amounts in Santa Fe, paying 4 cents per pound for loose shelled corn and $6.50 per *fanega* for corn in sacks. In late February 1864 officials began assembling wagons at Fort Union to haul St. Vrain's corn from Taos Valley.[110]

The grain would arrive none too soon. General Carleton's campaign against the Navajos, which he initiated during the summer of 1863, put severe strains on local resources. Under the command of Colonel Kit Carson, soldiers had destroyed Navajo crops and had scattered their livestock, making it impossible for the Navajos to store food for the winter. Starting in January 1864, cold and destitute Indians surrendered in such large numbers that the army lacked facilities to handle them. By March more than 5,000 had surrendered and were on their way to Bosque Redondo. This sudden influx of so many captives caught Carleton embarrassingly low on supplies. To ward off mass starvation, he placed soldiers on half rations and allotted the prisoners the minimum amount of food necessary for their survival. On 9 March he admitted, "It will require the greatest effort and most careful husbandry to keep the Indians alive until the new crop matures."[111]

The plight of the Navajos proved a windfall for government contractors. Between 1 March and 23 June the subsistence department contracted for and purchased foodstuffs for the captive Indians that cost nearly $415,000. Contractors furnished (or agreed to furnish)

2,765 head of cattle, more than 11,000 head of sheep, 2,500 pounds of flour, more than 1,500,000 pounds of corn, 544,850 pounds of wheat, 2,688 pounds of cornmeal, 711,534 pounds of wheat meal, more than 46,000 pounds of beans, and 60,000 pounds of salt. Suppliers received top prices: $2 to $3 per head for sheep, $7.25 per *fanega* for wheat, 8 to 9 cents per pound for cornmeal, 8½ to 9 cents per pound for wheat meal, $6.50 to $15.40 per *fanega* for corn (or 4 6/10 to 11 cents per pound), and 8¾ to 16 cents per pound for beef cattle. Some of the higher-priced commodities, such as corn and cattle, probably came from the East.[112]

During the emergency Carleton turned to his chief of staff, Major John C. McFerran, to coordinate the purchasing of subsistence stores for the Indians. McFerran continued looking after the needs of the quartermaster's department as well. On 19 March he contracted with Charles W. Kitchen to deliver to the quartermaster's depot at Fort Union 180,000 pounds of corn from the States at 9 cents per pound. Fortunately, Carleton's faith in McFerran was well deserved. By 3 April the major had located sufficient sustenance for the Indians and grain for livestock to allow Carleton to announce that the food crisis had ended.[113]

But even though subsistence stores poured into Fort Sumner during summer months, crop failures in the fall again raised the specter of starving Navajos. Frost had badly damaged the corn crop in Taos and Rio Arriba counties, and worms had destroyed most of the crops at Bosque Redondo. McFerran, who continued purchasing corn for both the subsistence and quartermaster's departments, canvassed the territory for surplus grain. He instructed officers to "purchase all you can get at a reasonable price" and to reduce forage for work animals to six pounds of grain daily. In October he wrote to Captain William L. Rynerson, quartermaster at Las Cruces, instructing him to find out the amount of corn that could be purchased in Mexico and the price for getting a million pounds delivered at Fort Sumner. "You must use great caution in making your inquiries," he warned, "to prevent capitalists and others from getting wind of this demand and taking advantage of it." A few days later McFerran wrote to Captain Herbert M. Enos, quartermaster at Fort Union, ordering him to purchase all the corn and oats that William Kroenig and others living on the Rio

Mora had for sale, paying $9 per *fanega* or even more. High prices, he reasoned, would induce everyone holding corn to sell. But prices asked by the corn holders in San Miguel County were too high. McFerran ordered his agents there to make no further attempts to purchase grain until residents abandoned the idea that "we will pay *any* price for their grain."[114]

Because the territory could not meet the grain requirements for both the subsistence and quartermaster's departments, the army advertised in November for 1,000,000 pounds of corn to be purchased in the States. Andres Dold of Las Vegas was the successful bidder. He agreed to deliver 500,000 pounds of corn at Fort Sumner by 31 May 1865 for 23½ cents per pound, 250,000 pounds by 30 June for the same price, and the final 250,000 pounds by 15 July for 22¾ cents per pound.[115] But problems continued to plague the army's efforts to purchase supplies locally. Twin disasters hit the territory in 1865. First, the Rio Grande flooded, destroying towns and crops in the Mesilla Valley and at Franklin, Texas. Then grasshoppers destroyed crops in the vicinity of Taos and devoured what few fields escaped the flood in the lower Rio Grande areas. To compensate for anticipated shortages, the army made a second contract with Dold on 28 July to supply Fort Union with 1,000,000 pounds of States corn at 18½ cents per pound.[116]

Bosque Redondo continued to be a bonanza for contractors until the government abandoned it in 1868 and the Navajos returned to their homeland. By this date the Indian department had assumed control of the Indians, including the letting of contracts for their subsistence. The five-year experiment at Bosque Redondo and the Confederate invasion of the Southwest had placed severe strains on the department's agricultural resources. For more than a decade before the Civil War, the army had provided a dependable market for agricultural produce. Army purchases, which reached a peak in 1861, stimulated local corn and wheat production.[117] But the war disrupted local agriculture, and the army had to import large amounts of forage and commissary stores to compensate for shortages. Nor did the army locate sufficient food within the department to feed the more than 8,000 Navajos who were imprisoned at Bosque Redondo. New Mexico contractors sold eastern corn and cattle to the army to feed the starving Indian families. And when government agents tried to secure provisions locally, commodity prices rose, making it difficult for poorer citizens to obtain foodstuffs.

Nevertheless, the army continued to encourage farmers and ranchers to increase their crop and livestock production. Since a *fanega* of corn purchased at Fort Leavenworth and delivered to Fort Union cost the army more than $25 in 1865, it is small wonder that the quartermaster's department labored to secure forage locally.[118] At the war's end, farmers throughout the Southwest brought new fields under cultivation, and stockmen revived a dormant cattle industry. The government would remain for several years their most important market, and government contracting would become a permanent element in the region's economy.

2

Farmers and Contractors

On 18 March 1866 Juan José Romero and seventy-two other Hispanic farmers who cultivated lands along the Rio Bonito in what would become Lincoln County, New Mexico, sent a petition to General James H. Carleton in Santa Fe asking for protection from roving bands of Indians. They claimed that Indians had killed sixteen area residents during the previous seven months and had robbed others of their stock. Romero and his neighbors were too poorly armed to defend themselves: "Some of us have slings, some have bows without arrows and others who are the best armed have fire arms but no ammunition."[1] Without military protection, they would have to leave to find safer homes elsewhere. Carleton assured Romero that help was near at hand. He had ordered two new companies to garrison Fort Stanton, located nine miles above the principal settlement on the Rio Bonito, and this force, he said, would make the countryside safe. He added, "I regard the Rio Bonito country as the finest in New Mexico, and shall do all I can to protect the settlers who desire to open it up to civilization."[2]

Rio Bonito farmers were not the only settlers requesting aid from the army. Angry tribesmen fought hard to prevent white encroachment on lands they claimed as their own. Civilians everywhere clamored for military protection. And like Carleton, most army officers gave whatever aid they could, since they believed white settlement would not only help subdue the tribes and advance civilization but also reduce the cost of maintaining troops in the West.

But the army never had sufficient troops to bring a speedy end to the bloody Indian–white conflicts that enflamed much of the West. In 1867 about 57,000 men were in the army, but reductions came rapidly thereafter. From 1871 to 1890, the entire U.S. army numbered between 24,000 and 29,000, scattered in tiny contingents among some

200 posts.[3] In addition, a parsimonious Congress annually cut army appropriations, making it even more difficult to maintain adequate forces to end hostilities. Economy-minded military officials constantly explored ways to use limited resources more efficiently. Consequently military geography in the Southwest was fluid. For more than twenty years, the army balanced financial considerations against strategic necessities as it constructed, moved, and abandoned military posts.

At the end of the Civil War, New Mexico and Arizona had a combined military force of 4,116.[4] Not until the Geronimo campaign in the 1880s would so many soldiers again garrison southwestern military posts. During 1865 the army transferred Arizona from the Department of New Mexico to the Department of California, and it reduced New Mexico to a district within the Department of the Missouri. Fort Garland remained attached to the District of New Mexico. The four posts in west Texas that are germane to this study—Forts Bliss, Quitman, Davis, and Stockton—became part of the Department of Texas after that state was readmitted to the Union in 1870 during Reconstruction. Also in 1870, the army elevated Arizona to a department that would encompass twelve military posts in Arizona territory, in addition to Drum Barracks and Fort Yuma in California.[5]

Frequent changes in military administrative structures had little effect on civilians. Changes in troop strength, however, had a direct impact, since reductions meant fewer supply contracts and greater insecurity for area residents. In both New Mexico and Arizona, troop strength steadily declined through the seventies but then increased to reach an all-time high in 1886 when the army campaigned against Geronimo and his small band of supporters. In 1870, 1,885 officers and enlisted men were assigned to the Department of Arizona and 1,598 to the District of New Mexico. Six years later, these figures fell to 1,536 and 1,299 respectively, and reached a decade low in 1877 when Arizona counted 1,232 soldiers and New Mexico 884. During the campaign to capture Geronimo, however, the army attached the District of New Mexico, which now included Fort Bliss, Texas, and Fort Lewis, Colorado, to the Department of Arizona, thus placing 4,803 men under one command.[6]

No matter how many troops occupied the Southwest, the army was always forced to economize to stay within its appropriation. Government officials tried to reduce expenditures by purchasing forage and foodstuffs close to the area of consumption. But officials and civilians disagreed on exactly how these supplies were to be obtained. Regulations demanded that contracts be awarded through competitive bidding, but some army officers and many farmers and ranchers wanted to abolish the contract system in favor of purchasing in open market.

Captain William C. Bell, who was chief commissary of subsistence for the Department of New Mexico at the close of the Civil War, best expressed the opinion of those officers who supported open market purchases. He argued that the army would obtain supplies "at a much more reasonable rate" by purchasing directly from producers rather than through contractors. Bell claimed that contractors took advantage of farmers by forcing them to sell their crops at low prices because they did not have the resources to bid on government contracts themselves. Even before harvests were gathered, contractors often secured all the produce that a farmer expected to raise since he lacked any other means of disposing of it. In this manner, contractors controlled the market and forced the government to pay exorbitant prices.[7]

Arizona's legislature also supported open market purchases, and in a memorial adopted in 1867 it requested that post quartermasters be authorized to purchase grain, flour, and beans directly from producers. The legislators noted that government contractors purchased large quantities of these commodities in Sonora, which brought little benefit to the territory. Open market purchases, they argued, would stimulate agricultural production, induce emigrants to settle in Arizona, and prevent a drainage of coin into Mexico.[8]

But the contract system prevailed. Lieutenant Colonel Thomas C. Devin, who commanded the Subdistrict of Southern Arizona in 1869, was just one of many officers who believed that open market purchases increased opportunity for fraud and favoritism. Army officials, nonetheless, were sympathetic to the plight of the small farmers. As early as 1867, Acting Quartermaster General Daniel H. Rucker ordered purchasing agents to advertise for supplies in medium quantities so that small producers could compete with the larger contractors.[9]

The army advertised for these supplies in leading territorial newspapers and in posters distributed in remote settlements. Advertisements in New Mexico were printed in both Spanish and English until 1877, but thereafter in English alone. Colonel Edward Hatch, who commanded the district, explained that advertising in Spanish was no longer necessary since all contracts were taken either by "English-speaking people or by Mexicans who read and write the same."[10] In 1870 only one journal each in Arizona and New Mexico was authorized by the War Department to print government advertisements—the Tucson *Arizonan* and the Santa Fe *New Mexican.* Officers who advertised elsewhere risked having their expenses disallowed.[11] Nevertheless, most supply officers attempted to gain wide publicity for their advertisements, as in the case of New Mexico's chief commissary, who in 1872 sought permission to advertise in four unauthorized newspapers, including the anti-administration Las Cruces *Borderer.*[12] For a brief period, the chief quartermaster in Arizona required successful bidders to pay expenses for placing advertisements in newspapers. This practice was stopped after Rucker ruled in 1867 that the quartermaster's department could not compel "private parties to pay for public advertisements."[13]

Although some purchasing procedures changed between 1865 and 1885, the War Department as a rule required that thirty days' notice be given to the public between the advertisement for contracts and the date for opening bids. These advertisements requested that sealed proposals in quintuplicate (sometimes in duplicate, triplicate, or quadruplicate) be sent to the office of the district or department quartermaster and commissary by noon of a certain date, when they would be opened. Bidders for the fresh beef contract at Fort Marcy, New Mexico, in 1865 were required to be present at opening bids or to be represented by an authorized agent. But in later years the army merely invited bidders to be present.[14]

During the 1870s local proposals could be submitted to the purchasing agent at each of the posts in addition to the main offices in Prescott and Santa Fe. Although all bids were opened on the same day, the name of the successful bidder was not known until several days later, after local bids had been forwarded to the central office. Later in the decade, bids were transmitted by telegraph. But the advent of modern technology, coupled with the lack of standard time zones, created new opportunities for trickery. Citizens living near Camp Bowie, for example, could obtain news of bids as they were announced at noon and then telegraph more advantageous bids to Prescott where noon had not yet arrived. To avoid such "sharp practices," Arizona's chief commissary in 1878 delayed the opening of bids at Bowie for thirty minutes. New Mexico's quartermaster simply forbade relaying bids by telegraph. Still, the lowest bidder did not always receive a contract; most advertisements carried the statement that the army reserved the right to reject any bid deemed unsatisfactory. Not infrequently purchasing agents rejected all bids as being too high and thereafter resorted to open market purchases. Even when a chief commissary or quartermaster accepted a bid, no contract was considered in force until it received approval from the commanding officer of the appropriate military department.[15]

Typical of military supply advertisements was the one appearing in the 4 December 1869 issue of the Tucson *Weekly Arizonan,* calling for sealed proposals to supply flour and beans to Fort Whipple and to Camps Toll Gate, Verde, and Date Creek. These commodities were to be put up in "good, strong and new sacks," each containing one hundred pounds. The amount required at each post (as listed in the advertisement) was to be delivered between 1 July and 31 December 1870, in one delivery at each post. Army officials would decide the time when the articles were to be delivered—giving the contractor at least sixty days' notice. Frequently, advertisements contained the clause that commodities were to be delivered at designated posts "at such times and in such quantities" as the purchasing agents desired. In most contracts, the army reserved the right to increase or diminish by one fourth, and sometimes by one third, the amounts stipulated.[16]

By 1881 bidders had to accompany their proposals with a bond equal to one half the value of the supplies furnished as a guarantee that the successful bidder would enter into a formal contract within sixty days after the opening of bids. Contractors were required to give bonds equal to one third the amount of the contract.[17] At least two men of recognized means had to cosign proposals and contracts as sureties, holding them equally responsible for completion of the agreement. If neither contractor nor bondsmen would deliver the required commodities, the government purchased supplies in open market and charged the contractor the difference between the contract price and the open market price, which was almost always higher. In many instances, when a contractor failed to complete a contract, the government sustained little monetary loss, as bondsmen frequently honored legal obligations and payments could be withheld on goods already delivered on the contract.[18]

Contractors supplied the army with large amounts of a few agricultural products: mainly corn, beans, barley, hay, and flour. On occasion, the army let contracts for potatoes, but as a rule vegetables were acquired through individual transactions. An 1864 regulation stipulated that fresh vegetables were to be issued to troops only "in lieu of any component part of the ration of equal money value."[19] Since officers serving in New Mexico were reluctant to requisition vegetables under this rule, Captain Charles McClure, the district's chief commissary in 1867, allowed them to purchase vegetables with company funds. McClure also made arrangements to secure a winter's supply of fresh vegetables for each of the posts. In October he authorized the commissary officer at Fort Union to purchase 14,000 pounds of onions, 6,000 pounds of beets, 1,000 cabbages, and 25 barrels of sauerkraut. Fort Garland, located in one of the finest potato-growing regions in the Southwest, was to furnish 61,000 pounds of potatoes. McClure instructed the commissary at Fort Selden to purchase additional amounts of onions, beets, turnips, and cabbages in nearby Mesilla, where "there are several fine gardens" and where vegetables could be procured at moderate prices.[20]

But McClure did not repeat this experiment. His department suffered great financial loss because companies failed either to purchase vegetables from the government or to requisition them in lieu of other parts of the ration. Company commanders in 1868 were instructed to make their own arrangements for purchasing vegetables for their men using company funds.[21]

The army's failure to include fresh vegetables as a component part of the ration led to outbreaks of scurvy, a deadly disease when left untreated. Scurvy, however, was not a serious problem at posts where soldiers grew their own vegetables, as they had done in the Department of New Mexico during the 1850s.[22] But soldiers had little time for gardening once the Civil War began, and more and more troops fell victim to this "curse of the army." In 1862 scurvy made its appearance at Fort Craig, where men had been six months without fresh vegetables. Fort Whipple's post surgeon, Eiliot Coues, wrote in 1865 that symptoms of the disease were "becoming daily more apparent," and he recommended issuing extra rations of molasses, vinegar, and desiccated vegetables and potatoes as antiscorbutics.[23] Ten cases of scurvy were reported at Fort Union in 1867, and additional cases at Forts Bayard, Craig, Marcy, Selden, Sumner, and Wingate in New

Mexico and at Camps Mojave and McDowell in Arizona. Patients who were admitted to the hospital at Fort Union quickly responded to "a liberal allowance of buttermilk, curds, and fresh vegetables."[24]

Surgeon James C. McKee, chief medical officer for the District of New Mexico, undoubtedly spoke for other medical officers when he stated that a single case of scurvy was "discreditable" to the service. Officers had long known that fresh vegetables were both a preventative and a remedy. McKee claimed that the outbreak of scurvy in 1867 had occurred, not for lack of potatoes and other vegetables, which had been freely supplied to the posts, but rather because officers had allowed these vegetables to rot as they would not issue them "owing to the present law defining that a portion of the ration of equal value must be dropped in lieu of those purchased." Only a more generous ration, McKee believed, would stop "the prevalence of this disease in the army."[25]

Scurvy was not the most common illness among frontier soldiers. Malarial fevers, diarrhea, dysentery, and venereal diseases were more prevalent.[26] In Arizona sickness sometimes incapacitated entire commands, as at Calabasas near the Mexican border late in 1865 when a "fearful epidemic" raged through the post, killing two officers, a "considerable number" of enlisted men, several civilians, and "leaving over two hundred in a woebegone, emaciated and delicate state of health."[27] A few months later, after new sickness hit the Calabasas encampment, more than thirty soldiers deserted at nearby Fort Wallen, in part because they feared they would suffer the same fate.[28]

Medical officers rightfully attributed some intestinal ailments to impure drinking and cooking water. To improve health at Camp McPherson, located 50 miles southwest of Prescott and thought to be one of the most unhealthy posts in Arizona, Assistant Surgeon Passmore Middleton instructed the post's quartermaster to construct a water purifying system that utilized large wooden tanks and charcoal filtering troughs.[29] Much of the sickness, however, was seasonal; soldiers became ill during and after the rainy season. Medical knowledge was not sufficiently advanced to identify the mosquito as the source of infection, but medical officers associated malarial fevers with heavy rainfall and stagnant water.

Although fevers and dysentery continued to be common ailments in the 1870s, scurvy became less of a problem as company gardens were cultivated and the army imported vegetables where gardens failed to flourish. The District of New Mexico reported only seven cases of scurvy in 1870 and none in 1874, while the Department of Arizona recorded thirteen and one for these years.[30] During the 1870s company gardens were cultivated at all posts in New Mexico, but they were only partially successful at Forts Craig, McRae, and Cummings. Soldiers at Fort Selden in 1871 raised enough vegetables to supply the entire post, but heavy rains in August and September of 1875 destroyed their crops. One company that had invested nearly $75 in its garden salvaged almost nothing.[31] At Fort Garland, where the growing season was short, crops also were only moderately successful. Gardens flourished at Fort Stanton in 1876, where soldiers raised radishes, onions, corn, beets, squash, cucumbers, cabbages, and lettuce. Companies purchased additional fruits and vegetables from Mesilla Valley producers.[32]

By 1870 gardens were being cultivated at all four west Texas posts. The soil at Fort Quitman was too alkaline for vegetables to grow well, but gardens at Forts Davis and Stockton flourished. Corporal Alias Jones of the 9th Cavalry (a black regiment) had charge of the twenty-acre garden at Fort Stockton, where in 1869 more okra, onions, melons, and cucumbers were raised than the troops could eat. The two-acre garden at Fort Bliss produced only a small amount of vegetables. These posts also secured fruits and vegetables from villages on either side of the Rio Grande. Troops at Fort Quitman, for example, procured vegetables, grapes, peaches, pears, and melons from the border towns of San Ignacio, Guadalupe, San Elizario, and El Paso.[33]

Troops in Arizona also cultivated gardens during the 1870s. Soldiers at Camp Mojave, however, had to import vegetables from California since high temperatures and little rainfall made gardening difficult. In 1870 neither Camp Mojave nor Camp Bowie had gardens. According to the commanding officer at Bowie, there was "no place within a less distance than 30 miles of the post suitable for gardens." Yet in 1875 a garden was planted about a quarter-mile from the post, and it produced a small amount of vegetables. Soldiers there purchased additional vegetables from farmers living on the Gila and San Pedro Rivers.[34] Successful gardens were planted at Camps Lowell, McDowell, and Verde in 1875 and at Fort Whipple, where enlisted men raised potatoes, cabbages, turnips, corn, beets, tomatoes, melons, and cucumbers. Soldiers were assigned daily duty as gardeners, but many were inexperienced. An officer at Camp Verde claimed that only one man in his company—a Private Wesley Biggles (6th Cavalry)—knew "anything of gardening" and that only Private Caleb S. Garrett could do the ploughing.[35]

Many post commanders took great interest in these gardens, and some, like Major Eugene A. Carr, commanding officer at Camp McDowell, personally selected the seeds to be planted. Carr planned to cultivate a large post garden, and the list of seeds he ordered prior to the planting season of 1872 was long. Carr's list suggests the breadth of army gardening. He ordered most of the seeds from the San Diego firm of Steiner and Klauber. On his list were vegetables already mentioned as being cultivated at other posts: turnips, radishes, cabbages, beets, lettuce, and okra. He also ordered carrots, cauliflower, peas, celery, spinach, parsnip, marjoram, cress, salsify, sage, savory, thyme, rhubarb roots, and strawberry and asparagus plants. From local suppliers, Carr requested "El Paso or native" onions, and from headquarters in San Francisco he asked for eggplant, sweet corn, cucumbers, lima beans, snap beans, tomatoes, and four varieties of melons. He concluded his order to Steiner and Klauber by stating: "Should there be anything else suitable to this climate, for [a] Post garden, I would like to receive suggestions."[36]

In addition to the 10-acre post garden, Camp McDowell also boasted a 240-acre farm, one of the few government farms cultivated in the Southwest after the war. During the early 1850s government farms had been established at several western posts, including all permanent posts in New Mexico except Fort Marcy. Utilizing enlisted men as farm laborers, the army planned to raise on these farms all forage required by public animals and some of the wheat that would be ground into flour for army bread. But military farming was a failure. Rather than reducing expenses, government farms had added to the cost of maintaining troops in the West. The antebellum farm program was quietly terminated in 1854.[37]

The prospect of cutting costs by cultivating farms, however, tempted some army officers a decade later to resurrect the military farming experiment. Two government farms were established in Arizona in 1866. The one at Fort Goodwin was short-lived. This post, established in 1864 as a base for campaigning aginst Apaches, was located just south of the Gila River about 120 miles northeast of Tucson. In

February 1866 Lieutenant Colonel Robert Pollock made a "preliminary treaty of peace" with the Coyotero Apaches, who agreed to confine themselves to the military reservation at Fort Goodwin. To provide for their subsistence, a government farm was established.[38] The army anticipated that Apaches would perform much of the farm labor, but in fact troops and civilians cultivated the farm under the supervision of a Mr. Cooper. The farm was not a success. Cooper had only thirty-three acres under cultivation in 1867 and perhaps seventy-six acres the following year.[39] Moreover, the Coyoteros failed to remain on the reservation, and in 1870 Colonel George Stoneman, commanding in Arizona, ordered the post discontinued.

The government farm at Camp McDowell was no more successful under army management than were the post farms established in the 1850s, but under civilian control the farm continued operating well into the 1880s. Major General Irvin McDowell, commanding the Department of California, issued orders on 7 February 1866 to open a government farm at Camp McDowell, where corn and sorghum would be harvested to save the army the cost of transporting forage from California. Lieutenant Colonel Clarence E. Bennett, commanding at McDowell, supervised construction of the farm and became its first superintendent after mustering out of the California volunteers in August. Civilians and enlisted men performed the labor. Bennett was authorized to hire three civilians at $50 each per month and twenty civilians at $40 in coin, plus a daily ration, with preference given to discharged volunteer soldiers. The site chosen for the farm was on the west side of the Verde, about one half mile north of the post.[40] By July an acequia four and one third miles in length had been dug and about 240 acres cleared of underbrush and planted in sorghum and corn. In October the post's quartermaster estimated that the harvest would amount to 500,000 pounds of shelled corn and 600 tons of corn fodder.[41] Wheat was planted sometime later, and in March 1867 Bennett requisitioned a reaper and a thresher from California. Four months later, Apaches stole the belting for the threshing machine, leaving the wheat unthreshed. By August William H. Hancock, a recently discharged soldier, had replaced Bennett as superintendent and was receiving a monthly wage of $150.[42]

Two army inspectors examined the government farm at Camp McDowell during the spring and summer of 1867. Major Roger Jones, an assistant inspector in the inspector general's department, was optimistic. He reported that the farm in 1866 had cleared $34,325, calculating the value of forage raised on the farm at the prices that civilian contractors received: $65 per ton for fodder and 7 cents per pound for corn.[43] After carefully examining the farm's records, Colonel James F. Rusling, a special investigator for the quartermaster's department, condemned the project as an economic failure. According to Rusling, the calculations used by Jones and others to justify the farm were erroneous and misleading: several months' expenditures had not been listed, most of the fodder had been fed to the post's beef cattle when they should have been grazed, and the corn was of such poor quality it would never have been accepted from a contractor. Moreover, like other critics of government farms, Rusling opposed using soldiers to dig ditches and to harvest crops. Their proper place was in the field fighting Apaches. He also claimed that government farms discouraged settlement since farmers would be unable to sell their grain to the army. He recommended that the McDowell farm be sold or opened to settlement.[44]

Rusling's arguments won out. On 1 January 1868 the farm at Camp McDowell was leased for one year to George F. Hooper and Company, a mercantile firm with outlets in San Francisco, Fort Yuma, Maricopa Wells, and Camp McDowell, where the company ran the sulter's store. In return for the lease, Hooper and Company agreed to sell to the government "all the products of the farm" at the following prices: corn, barley, and wheat at 2 49/100 cents per pound and long forage at $24.75 per ton. The following year Charles G. Mason leased the farm, delivering grain at 3 cents a pound and long forage at $23 per ton. Prices were higher in 1870 when John Smith had the farm. He sold his barley and oats to the government at 3 3/4 cents per pound and hay at $41.75 per ton.[45] Smith, a former officer in the California volunteers, had been a member of the Arizona legislature in 1868. He again leased the farm in 1871 and 1872 and then became post trader at McDowell. In later years conditions for leasing the government farm changed. Henry F. Hardt was allowed to cultivate the farm in 1878 in return for keeping the dams and ditches on the farm in good repair and for furnishing the post free of charge with potatoes, beets, turnips, cabbages, green corn, lettuce, onions, and a limited supply of "water and muskmelons."[46]

The only military farm cultivated in New Mexico after the war was the one established at Bosque Redondo in 1863 to provide sustenance for captive Indians. Like the Fort Goodwin experiment, farming at Bosque Redondo was a failure, as the farm accumulated more debts than produce during the five years it was tilled.[47] Elsewhere in New Mexico the government occasionally rented lands on or near military reservations to encourage local agriculture. In 1862 the army confiscated—and subsequently leased—the Beckwith and Beach ranches located about seven and fourteen miles, respectively, below Fort Stanton.[48] Federal troops had evacuated the post the previous year, and Charles W. Beach and Hugh M. Beckwith were among the Rio Bonito settlers who had abandoned their ranches to seek safety elsewhere. General Carleton later accused Beach of having illegally acquired a horse and a mule on a campaign against the Mescalero Apaches. "I consider Mr. Beach an improper person to reside in the Mescalero Country," Carleton concluded, "so he will be forbidden under any circumstances from settling there." Shortly thereafter, soldiers planted company gardens on Beach's property, and then in 1866 the army rented the ranch to a Mr. Stoke.[49]

Hugh M. Beckwith and his wife Refugia had cultivated one hundred acres along the Rio Bonito in 1860, raising primarily corn, wheat, and beans. Beckwith, a southerner, soon ran afoul of Union officials. He reportedly joined Confederate troops when they occupied Fort Stanton in August 1861. About a year later, in October, he escaped into Mexico after being detained in Mesilla by military authorities, and that same month the Second District Court attached his property to satisfy a claim for $1,680. In February 1863 he was indicted for treason, although the case never came to trial.[50] The army rented the Beckwith ranch to Frederick Stippick in 1864, furnishing him with seed in return for one third of the crop. The following year José Antonio Ulibarrí and Dolores Malcado leased the ranch, agreeing to deliver to Fort Stanton one third of the grain they would raise. This may have been the same property that Emil Fritz, of Lincoln County War fame, leased from the army in 1867 for one third of the crop and again in 1868 for $350.[51] No record has been found of anyone leasing this land in later years.

The "government farm" at Fort Wingate was described in an Albuquerque newspaper in 1863 as a joint effort in experimental farming undertaken by army officers and Rafael Chávez "to test the relative properties of the soil as a producing region."[52] The army may have provided the seed, but Chávez alone "broke up the whole farm" on which he planted corn and wheat. After Chávez had farmed three years, probably turning in a percentage of the crop, the army decided in 1866 to rent the land to the highest bidder. Chávez submitted the highest bid—ten dollars an acre to be paid quarterly. He later admitted that he was "ignorant as to what an acre was." Imagine his dismay when the farm finally was surveyed and he learned that it contained 139 acres and that he owed the government $1,390. Chávez subsequently paid two quarters of the rent in cash, and then because of crop failures he petitioned for relief. The army thereafter allowed Chávez to cultivate the farm free of rent in 1867 and to turn in enough wheat to pay the balance on the previous year's rent.[53]

Civilians cultivated two farms on the Fort Craig military reservation. Estanislao Montoya of San Antonio, New Mexico, farmed a large tract of land above the garrison for three years beginning in 1864, raising corn, barley, and oats. Montoya's farm workers lived on a section of the farm called Ancon, but gamblers, whiskey dealers, and prostitutes soon joined the settlement. The army ordered Ancon residents to leave the reserve in 1866 after Army Inspector Nelson H. Davis claimed that Ancon was "a nuisance to the Post and demoralizing to the Troops."[54]

The second farm, located two and one half miles south of the post, was cultivated by William R. Milligan, who first dug the acequias and plowed the land in 1863. The army collected no fees until 1866 when Milligan agreed to pay a yearly rent of $300. Milligan later protested paying any rent at all. "I must say," he wrote in 1870, "that I think it unfair and unjust for the government to charge me rent for a piece of land that was nothing but a wilderness at the time I commenced working it and would be worth nothing without the labor and money expended by me upon it."[55] During the following two years, Milligan divided his time between the Fort Craig property and a ranch he had established in Arizona. He also was engaged in a spectacular case of fraud. Without legal claim to ownership, Milligan had "sold" one hundred acres adjoining the Fort Craig farm to twenty-six Hispanic farmers for $2,600, requiring each to pay twenty dollars per annum over a five-year period. He also stipulated they were to deliver to him one half of all corn raised on said land, which they did for two years before the deception came to light. The Secretary of War ordered Milligan removed from the military reserve in 1873.[56]

The government's policy of leasing land on military reserves probably generated more trouble than real income. And certainly rental fees would reduce by only a small fraction the enormous expense of outfitting the western army. General William T. Sherman emphasized this expense when he wrote in an 1869 report that the cost of a soldier's maintenance in Texas, New Mexico, Arizona, Nevada, Idaho, Montana, and Alaska was two or three times as great as on the Kansas and Nebraska frontier. The total cost of maintaining the military establishment in New Mexico during the fiscal year ending 30 June 1865 was $4,433,884, although two years later that figure had dropped to $2,779,294. General Edward O. C. Ord, commanding the Department of California, estimated in 1869 that the cost of maintaining troops in Arizona would amount to almost $3,000,000 annually.[57]

To eliminate costly transportation expenses, as previously noted, the army encouraged local production of foodstuffs and forage. Commissary General of Subsistence Amos B. Eaton stressed in his annual reports the need to purchase supplies from local producers and dealers, not only to reduce costs but also to aid in opening and populating newer regions in the West. In 1867 he sent this order to New Mexico's chief commissary officer: "Go and do all you can to induce the people of your district to be prepared to supply still further the wants of the Subsistence Department therein assuring them they shall have the preference in furnishing supplies."[58]

Before farmers could greatly expand their plantings, however, they had to be protected from Indian attacks. The hit and run techniques of Apache raiders were particularly effective in sparsely settled Arizona. When General John S. Mason arrived there to take command in May 1865, he found that most white residents south of the Gila had deserted their farms and that many had sought refuge in Tucson. North of the Gila, he reported, "the roads were completely blockaded, the ranches with but one or two exceptions abandoned, and most of the settlements were threatened with either abandonment or annihilation." The mere establishment of military posts near settlements, he claimed, was of no practical importance; nor would the pursuit of small raiding parties end hostilities. The "only true way to obtain peace," he said, was to stage a winter campaign against the Apaches, destroy their winter supply of provisions, and in this way compel them to sue for peace. In September he informed his superiors that he would keep his troops scouting for Indians during the winter, and he predicted that by spring "Arizona will be safe in every section."[59]

Mason's prediction fell wide of the mark. Not until the end of George Crook's successful campaign in 1872–73 would Arizonans experience the safety Mason spoke of. Nonetheless, the population of Arizona slowly increased as more and more emigrants arrived seeking new economic opportunities. Those who settled in promising farm areas often pleaded for military assistance, and the army did whatever it could to aid them, thereby encouraging the growth of agriculture. In addition to mounting campaigns, post commanders sometimes dispatched small detachments to guard farmers and their families during planting and harvesting. In June 1867 ten soldiers were sent to Tres Alamos, a small settlement on the San Pedro, to protect women and children while the men were in the fields cultivating 700 acres of corn.[60] In central Arizona Apaches frequently raided agricultural settlements at harvest time, quietly entering fields at night and stripping them of crops. Farmers on the Rio Verde in 1867 reported that Indians systematically carried away each night four or five acres' worth of corn. John P. Osborn on the Lower Agua Fria lost one half his entire crop to Indians in this manner. And during the harvest of 1869, it was estimated that Indians carried off twelve tons of corn from various ranches in Kirkland Valley. In October of that year, the commanding officer at Camp Verde took steps to end this "pilfering" of corn. He stationed two privates at Peter Eich's ranch, located about five miles from the post, and two at the nearby Melvin ranch, ordering them to spend at least three hours each night in the cornfields.[61]

But the army did not have enough soldiers to station details in every cornfield. Captain David Krause, commanding the District of Prescott in 1866, noted that the fifty-eight men stationed at Fort Whipple were kept constantly on duty in the quartermaster and commissary departments, many of them herding the post's 200 head of beef cattle and 100 horses and mules. Sixteen miles northwest from Fort Whipple, thirty-five men garrisoned a post in Skull Valley, where seven or eight ranches were located. The garrison there was so small, Krause said, that it "can only furnish moral but little physical protection to the inhabitants." He further pointed out that ranchers on the Agua Fria and at old Fort Whipple were farming profitably without military protection, and he implied that other ranchers could do the same. With more than a little sarcasm, he remarked: "If persons wish to be pioneers they should be prepared to undergo the hardships incident to that kind of life."[62] But settlers everywhere continued to ask for help. In 1872 General George Crook, who then commanded the Department of Arizona, reported that he was in "constant receipt of applications for protection, by stationing guards over ranches and herds." It was impossible to honor these requests, he said, even though he was sympathetic to the plight of the ranchers.[63]

Military officials in New Mexico found it equally impossible to provide guards for everyone who asked for them. Chief Quartermaster Herbert M. Enos believed it was unreasonable for citizens to expect military protection for every little settlement. He recommended that citizens find less dangerous areas to cultivate. "It seems to be a mania with the Mexicans," he wrote in 1867, "to locate [near] some small stream where a few acres of cultivable land can be found, and where the settlers are in danger every day of losing their lives by Indians."[64]

Nonetheless, the army established pickets in some of these isolated areas. For example, in February or March of 1865 a detachment from Fort McRae was stationed at Cañada Alamosa, a small agricultural settlement located about thirty miles northwest of the post. Two noncommissioned officers and ten privates remained there, living in a brush shed, through the October harvest. The picket was reestablished the following March when farmers again began planting their crops. The settlement at this time consisted of about 150 men, women, and children, mostly Hispanos, who lived in twenty-five or thirty jacales and raised wheat, corn, and vegetables on nearby fields.[65] General Carleton also authorized the commanding officer at Fort Craig in January 1866 to send a small detachment of soldiers to Alamillo in Socorro County, where residents would soon begin cleaning their acequia prior to the planting season. Anastacio García, a resident of Alamillo, had written that the community of ten families was poorly armed. More important, when the men went to clean the opening of the acequia five miles above the village, the women and children would be left unprotected.[66]

Citizens of Rio Palomas, located twenty miles below Fort McRae, sent an equally strong petition to district headquarters in Santa Fe asking for protection. In August or September of 1867 Apaches had entered their corn fields, destroying "a great deal of corn and other produce, breaking the grist mills and scattering the machinery in all directions." The Indians also had stolen fifty goats. If these depredations continued, the twenty-five Hispanic families who lived there would have to leave to find safer places to farm. The petitioners pointed out that their village benefited the public, for it guarded a major pass through the Caballo Mountains once favored by Apache raiding parties. Even though the district lacked sufficient soldiers to establish a picket there in 1867, the army loaned muskets and ammunition to Palomas residents, sent mounted patrols to the village, and furnished a guard during the 1868 harvest.[67]

Despite the army's willingness to aid local farmers, officers maintained an underlying suspicion that some appeals for protection were motivated by economics rather than by fear for personal safety. For example, in 1874 residents in Las Animas Valley of Colorado (near Trinidad) petitioned for military protection against hostile Indians. Colonel J. Irvin Gregg, then commanding the District of New Mexico, traveled to Fort Garland to investigate, interviewing sheepmen and cattle ranchers in Taos and Costilla who had herds grazing near Las Animas. None had been informed of any Indian troubles. Although Gregg did not interview the petitioners, and indeed did not travel to Las Animas Valley, he concluded that "there was no necessity for the presence of troops in that vicinity, and that it was the pecuniary interests of the petitioners rather than their personal safety that instigated the petition."[68]

Even more critical of profit-seeking civilians was Colonel August V. Kautz, who took command of the Department of Arizona in March 1875 and soon became involved in controversy with Arizona Governor Anson P. K. Safford. Their feud centered on the policy of the Indian Bureau to concentrate Apaches on the San Carlos Indian Reservation. Safford, who supported concentration, publicly criticized Kautz and demanded his removal, strongly suggesting that Kautz was unable to prevent renegade Indians from overrunning the countryside. On his part, Kautz claimed that Safford's request for additional troops was stimulated "by the desire to increase the trade of Tucson and vicinity." He also believed that concentration came from pressure by Tucson contractors—the so-called "Tucson Ring" or "Indian Ring"—for economic gains. They found it more profitable to keep Indians at San Carlos rather than to allow some to remain at Camp Apache in the White Mountains where contractors from New Mexico had the price advantage.[69]

Even though settlers and contractors would deny Gregg's and Kautz's allegations, few questioned the economic value of having troops in their midst. Newspaper editors, territorial officials, and army officers all agreed that after the war the government was the sole market for surplus agricultural products in both Arizona and New Mexico.[70] Moreover, military posts that were established in isolated areas soon attracted settlers wanting to capitalize on the new markets. Rarely, however, did these new arrivals produce all the forage and foodstuffs required at the garrisons. The army secured most of its grain, for example, from a limited number of agricultural regions, whence it was hauled to distant posts. The examples of Forts Bascom, Wingate, Cummings, and Bayard in New Mexico illustrate this point.

Established in August of 1863 on the banks of the Canadian River about a hundred miles southeast of Las Vegas, Fort Bascom was designed to protect travelers on the road to Fort Smith, prevent incursions by Kiowa and Comanche raiders, and encourage settlement in the Canadian Valley.[71] The post proved too expensive to maintain, however, and it was abandoned in 1870. During its brief existence, a small community sprang up about five miles west of the post, which the soldiers appropriately called "Liberty." The settlement lived on as a ranching community until the turn of the century, although in 1870 a medical officer reported that no civilians were then living in the vicinity of the post.[72] Probably most settlers turned to ranching rather than farming to take advantage of the area's fine grama grass. Some settlers near Fort Bascom, however, were harvesting corn in 1865, and in October the post quartermaster was authorized to pay them 4 cents per pound for "American" corn and $7 per *fanega* for "Mexican" corn. But the farmers failed to produce the amount needed at the post, and in later months grain was provided from the vicinity of Fort Union.[73]

Beginning in October 1866, the army would sign six contracts for delivery of corn at Fort Bascom before the post was abandoned. Three contractors were Las Vegas merchants, one was a merchant residing in Golondrinas, and two were Mora County farmers. The contractors may have purchased some of their grain near Fort Bascom, but they probably secured larger quantities from the more densely cultivated areas closer to Las Vegas and Fort Union. For example, Frank Chapman of Las Vegas, who signed a contract in January 1870 to deliver 350,000 pounds of corn, made arrangements with three residents of Chaperito, a small Hispanic settlement about seventy miles west of Bascom, to deliver more than 150,000 pounds on his contract. Two of these men, Lorenzo Valdez and José Felix Ulibarrí, were themselves large producers, and they probably sold their own corn to Chapman and perhaps also that of their neighbors.[74]

One of the few documented purchases of corn directly from a Fort Bascom producer occurred after James T. Johnson of Mora County refused to fill his contract in 1867. Nearly out of grain, the post quartermaster purchased about 25,000 pounds at 4 cents per pound from John Watts, who had established a ranch adjacent to the post during the fall of 1865. As soon as a government wagon train arrived from Fort Union with grain, the army stopped purchasing from Watts, since his grain was considered too expensive. The quartermaster also recommended purchasing corn from Robert Hamilton, who raised a large amount on 250 acres at Hatch's Ranch, located about three miles south of Chaperito on the Gallinas River. In explaining to superiors his difficulty in supplying the post with grain, Fort Bascom's quartermaster wrote: "The road from this post to any part of the country where supplies can be obtained is very rough, and I am of the opinion that producers and dealers would rather supply any other post than this."[75]

The pattern that emerged at Fort Bascom was repeated at Fort Wingate: demand for grain stimulated agricultural production in older communities but failed to generate a sizable new farming population close to the post itself. Fort Wingate was established in October 1862 three miles south of present Grants, but in June 1868 it was moved about sixty-five miles northwest to its permanent site east of modern Gallup.[76] Alice Baldwin, who arrived at the first Fort Wingate in December 1867 with her husband Lieutenant Frank Baldwin, later recalled that the post then was "one of the most remote military posts on the frontier." She remembered that Hispanic families living nearby were friendly and that the women worked as laundresses. Although Baldwin did not identify the men, they probably included laborers employed by the quartermaster's department and several who worked on Rafael Chávez's farm. She spoke of the nearest settlement as being Albuquerque and dismissed Cubero, a Hispanic settlement about twenty miles east of the post, as "a lonely spot, composed of but a few ramshackle adobes."[77]

Records suggest that the establishment of the first Fort Wingate did not lead to new settlements or even increase significantly the amount of produce raised by residents of Cubero and other nearby villages. During 1864 and 1865 the post received its grain from the Rio Grande Valley even though the army had attempted to purchase corn at settlements closer to the post, including Cubero, Cebolleta, El Rito, and Laguna Pueblo villages. Major Ethan W. Eaton, commanding at Fort Wingate, reported in November 1864 that he had secured only 100 *fanegas* at El Rito and that producers elsewhere refused to sell on credit. "The supply of grain for this post," he wrote, "must come from the River as there is but little to be had in this vicinity."[78] One valley resident who supplied grain to the post was Antonio José Otero, a large landowner and merchant residing at Peralta, located just below Albuquerque. He received contracts in October 1866 to deliver 300,000 pounds of corn to Fort Wingate, 400,000 pounds to Fort Sumner, and 300,000 pounds to Albuquerque. Otero probably secured most of this grain along the Rio Grande, but he also enlisted the aid of Rafael Chávez, who completed the Wingate contract in May 1867 apparently with corn raised on the Wingate farm.[79]

Alice Baldwin accompanied the troops on their move to New Fort Wingate. Her neighbors there were Navajos, who had just returned from Bosque Redondo and were awaiting final survey of reservation boundaries. Baldwin continued to feel as though she were on the farthest edge of the frontier; she later recalled how eagerly "the isolated inhabitants of this well-nigh forgotten community" awaited the mail carrier.[80] Baldwin returned east early in 1869, failing to record the arrival a few months earlier of Nathan Bibo, who would manage the post tradership for Willi Spiegelberg. Soon the Prussian-born Bibo established a dry goods outlet, in partnership with his brothers Simon and Samuel, in Cebolleta, a small Hispanic village located about twelve miles northeast of Cubero. Bibo stated in his reminiscences that residents of Cubero and Cebolleta began to plant more extensively after 1868, "having a market for their corn when we [Bibo and his brothers] commenced to ship the first grain to Wingate."[81]

The change Bibo observed may have occurred, although he claimed too much credit for himself. With markets for grain at Fort Wingate and Fort Defiance, where the Indian Bureau established the Navajo agency in 1868, local farmers were encouraged to cultivate more land. And between 1860 and 1870 Valencia County increased its corn production from 53,587 to 77,854 bushels, while its population declined from 11,321 to 9,093. The majority of farmers listed in the 1870 agricultural census for Cubero and Cebolleta were producing surplus grain—certainly beyond what their own families would consume.[82] Extant records document some aspects of Nathan Bibo's involvement in the grain market. In 1869 Willi Spiegelberg held a contract to deliver corn at Fort Defiance, and Bibo probably aided in gathering this grain. Several months later, Bibo himself agreed to deliver 100,000 pounds of grain at Camp Apache in Arizona, and although he encouraged residents of Cubero and Cebolleta to increase their plantings, he had to buy last season's corn along the Rio Grande to begin filling the contract. In September 1871 he also contracted to deliver 100,000 pounds of corn at Fort Wingate.[83]

Bibo, however, was only one of thirteen contractors who delivered corn at Fort Wingate during the ten-year period starting in 1868. Twelve of the contractors were merchants, including J. Lorenzo Hubbell, a well-known trader with the Navajos. Another—Tranquilino Luna—was a wealthy young landowner from Los Lunas who would be elected as New Mexico's delegate to Congress in 1880.[84] It is no longer possible to show where each contractor secured grain, but records suggest that the Rio Grande Valley continued to be the major source of supply long after Nathan Bibo arrived on the scene. Salvador Armijo, for example, signed three contracts during these years to deliver grain at Wingate: one in 1868 for 175,000 pounds, another in 1870 for 100,000 pounds, and the third in 1871 for 300,000 pounds. Armijo farmed extensive land north of Albuquerque, where he grew large amounts of wheat, corn, and beans. In addition, he and his son-in-law had formed a freighting and mercantile partnership in 1864 and subsequently opened branch stores at a number of locations, including Cebolleta, Cubero, and two villages south of Albuquerque—Jarales and Peralta. Armijo's customers undoubtedly paid their accounts in commodities, enabling Armijo to fulfill his contracts with corn raised on his own lands and grain he procured from farmers in the above-mentioned villages.[85]

Three other merchants also relied upon Rio Grande Valley farmers to fill their contracts. After Las Vegas merchant Adolph Letcher failed to make regular deliveries at Fort Wingate during the spring of 1869, Stephen B. Elkins, his bondsman, took over the contract. In May, Elkins informed the chief quartermaster that he was "daily buying corn" to deliver at Wingate and that 110,000 pounds were being shipped from San Juan, located about twenty-five miles north of Santa Fe in Rio Arriba County.[86] Another contractor, Jacob Schwartz, said in 1870 he would secure 200,000 pounds of corn near Tomé, a village of 1,035 residents south of Albuquerque where Schwartz was a merchant.[87] One of the largest merchants in Santa Fe, Abraham Staab, held three contracts during these years to supply Wingate with a total of 875,000 pounds of corn. In January 1878 he asked for a reduction on his last contract, which had called for 140,000 pounds. Since the fall harvest, he had traveled twice down the river as far as Belen and had succeeded in purchasing only 115,000 pounds in small lots along the river and in the vicinity of Fort Wingate. Later that same year Lieutenant Stephen R. Stafford, the post quartermaster, wrote in his annual report: "Owing to the altitude and the scarcity of water for irrigating purposes, no considerable amount of grain is grown nearer than the Rio Grande Valley."[88]

Two posts located in southwestern New Mexico received most of their grain from contractors who lived in the Mesilla Valley. Fort Cummings was established fifty-three miles west of the Rio Grande on 2 October 1863, and was designed to protect travelers on the Mesilla–Tucson road as they passed through dangerous Cooke's Canyon.[89] A large spring provided the post with drinking and gardening water, but without another means for irrigation nearby lands remained uncultivated. In 1865 one officer stated that nothing had been planted within sixty miles of the post and another in 1870 reported that there were no inhabitants in the immediate vicinity.[90] Before Fort Cummings was temporarily abandoned in 1873, the army signed contracts for the post's supply of corn with five merchants: John Lemon and James Edgar Griggs of Mesilla, Henry Lesinsky and Louis Rosenbaum of Las Cruces, and Samuel J. Lyons, the post trader at Cummings.[91] Doña Ana County, which embraced the fertile Mesilla Valley, was the second largest corn-producing county in the territory, and these contractors most likely secured grain from farmers in the farming communities of Doña Ana, Las Cruces, Mesilla, La Mesa, Chamberino, and La Union. Indeed, when no one submitted a proposal to deliver corn at Fort Cummings in May of 1869, the post quartermaster at Fort Selden, situated just north of the village of Doña Ana, was ordered to purchase the requisite corn in open market "as you are near the grain market."[92]

Fort Bayard, established on 21 August 1866 to protect the Pinos Altos mining district, was considered an undesirable post because of its remoteness and crude housing facilities. One soldier stationed there in 1871 described it as "a lonely, isolated post," and Frances Boyd, who arrived in 1873 with her officer-husband and two small children, called it "a sorry place."[93] An estimated 800 to 1,000 miners inhabited Pinos Altos in 1867, but even more prospectors swarmed to the region after promising silver strikes were made in 1870. Silver City, located nine miles west of Fort Bayard, soon blossomed as the region's major supply center. Some new arrivals took up farms in the Mimbres Valley, about twenty miles east of the post, where they planted corn, wheat, barley, and potatoes. The town of Rio Mimbres, situated where the overland road crossed the river, was a flourishing agricultural community in 1870, but it had been abandoned by 1876 reportedly because its water supply increasingly was being diverted to irrigate neighboring farms. At the same time, San Lorenzo on the Upper Mimbres witnessed steady growth, and by 1876 had a population of 500.[94]

During the first decade of Fort Bayard's existence, the army awarded the post's grain contracts to eleven different suppliers: six resided either in Mesilla or Las Cruces; two in El Paso, Texas; two in Rio Mimbres; and one John L. Waters was a Pinos Altos merchant. Based on surviving records, it is clear that most of the grain delivered on the above contracts came from the Mesilla Valley, with lesser amounts from Mexico and the Mimbres Valley. The six Mesilla Valley suppliers contracted for a total of 3,370,000 pounds of corn, most of it purchased locally. Contractor James Edgar Griggs, however, reported in 1875 that most of his corn would come from Mexico, some from San Lorenzo, and only a small portion from Doña Ana County. The following year contractor Thomas J. Bull of Mesilla shipped some of his corn from Mesilla and some from the Mimbres.[95]

The highest-priced corn shipped to Bayard was supplied by Albert H. French of El Paso, who in October 1866 agreed to deliver 300,000 pounds at 5⅞ cents per pound. His corn probably was imported from Mexico. The lowest-priced corn was purchased from Louis Rosenbaum, who in November 1872 agreed to deliver 400,000 pounds at 1¾ cents per pound.[96] Where he secured the corn is not known. But when contractors like French and Rosenbaum freighted corn overland from the Mesilla Valley, they encountered special problems. High waters caused by spring thaws frequently delayed wagons in crossing the Rio Grande. Raids by Apaches added additional risks. One Fort Selden officer reported in June 1868 that there was "a panic amongst the people here about freighting to Fort Bayard . . . attacks from Indians are constantly expected."[97]

Despite the threat of hostile Indians, by 1880 Grant County's population had reached 4,539, an increase of almost 300 percent since 1870.[98] Although the county was now the second largest producer of barley in New Mexico and raised modest amounts of corn and wheat, its economic base rested on mining rather than agriculture. Fort Bayard's most important contribution to the county's economic development was the protection it gave to miners and investors, although many local residents found employment at the post and others would supply it with building materials, hay, charcoal, and limited amounts of cattle, grain, and other commodities.

Unlike Forts Bascom, Wingate, Cummings, and Bayard, the establishment of Fort Stanton at the foot of the Capitan Mountains about 100 miles east of the Rio Grande stimulated extensive cultivation in a district that was remote from major supply centers. Because of its rich soils, mild climate, and flowing streams, army officers considered the Fort Stanton region the finest in New Mexico. After the post was reestablished in 1862, settlers slowly returned and began farming on nearby lands. By June 1866 300 Hispanos and 5 Anglos were living along the Rio Bonito where they cultivated crops of corn, wheat, and beans and herded about 500 head of sheep, goats, and cattle. Forty Anglos and eighty Hispanos were raising similar crops and herding a small number of cattle along the Ruidoso and Hondo Rivers.[99]

Continued opposition by Mescalero Apaches, however, slowed agricultural development. Although the army had forcibly removed the Mescaleros to the Bosque Redondo Reservation early in 1863, they had remained there only a short time before quietly slipping away to the mountains of southeastern New Mexico and northern Mexico. Despite the fact that most settlers during the 1860s lost stock to Mescalero raiders, the fertile countryside continued to attract new residents.[100] By 1870 Lincoln County's non-Indian population was 1,803, which included 368 farm families. For the twelve-month period ending 1 June 1870 these families had produced 134,162 bushels of corn, 13,607 bushels of wheat, 2,843 bushels of barley, and 2,430 bushels of oats, making the county the territory's leading producer of both corn and barley.[101]

Army officers frequently alleged that farmers like those in Lincoln County who cultivated lands adjacent to military posts were totally dependent upon the army, not only for protection but also as a market for produce. General John Pope, who commanded the Department of the Missouri, wrote in 1870 that settlers near Fort Stanton found their sole market at the fort. "So far from being self-sustaining," he said, "the settlers could sell nothing except to the post, and if it goes they must go also, and that entirely irrespective of Indians."[102] Clearly the army pumped large sums of money into the local economy. During eleven months in 1869, the firm of Lawrence G. Murphy and Company sold to the quartermaster at Fort Stanton 29,780 pounds of corn, 254,502 pounds of hay, 110,000 pounds of fodder, 275,000 feet of lumber, 2,000 bushels of charcoal, 98 pounds of rope, and 40 pounds of wagon grease at a total cost of $21,876. During the same months, contractor Lehman Spiegelberg, a Santa Fe merchant, delivered 600,000 pounds of corn, receiving in return $13,740. Local farmers undoubtedly supplied much of this corn and would receive a share of the purchase money. The quartermaster bought from other Lincoln County residents forage and building materials amounting to $9,526. In addition, 136 civilians were employed at the post in January, most of them helping to rebuild its decaying buildings. Their total pay for the month came to $4,993.[103]

Certainly this military spending stimulated the local economy, but by 1870 agricultural production in Lincoln County had outstripped its government market. Representing less than 2 percent of New Mexico's population, the county produced 21 percent of the territory's corn. The amount required at Fort Stanton in 1870 represented only 6 percent of the local crop. During the fiscal year ending 30 June 1872, Lawrence G. Murphy and Company held contracts to supply Forts Stanton and Tularosa with 1,200,000 pounds of corn or about 16 percent of the area's total crop. The farmers who sold corn to Murphy while the two contracts were in force represented less than 16 percent of Lincoln County farmers.[104] Starting in 1871 the Mescalero Indian Agency would provide an additional outlet for agricultural produce as the Indian Bureau attempted to persuade Mescaleros to settle on a reservation in exchange for rations.[105] Still, the government provided only a limited market. When farmers produced more than the army and the Indian Bureau could absorb, as they did during the fall of 1875, prices plummeted. A Lincoln correspondent reported in March 1876 that the county had a surplus of more than 1,000,000 pounds of corn and was without a prospective buyer. More would be on hand after the fall harvest.[106] The government later signed contracts in November with Abraham Staab for delivery of 250,000 pounds of corn at Fort Stanton at $1.62½ per 100 pounds and with James J. Dolan for 300,000 pounds at $1.69. At no other time between 1866 and 1885 were contract prices at Stanton as favorable to the army.[107]

But marketing conditions in Lincoln County were extremely unstable, partly because of crop failures but also because of chronic lawlessness. Even experienced suppliers like Lawrence G. Murphy sometimes had difficulty finding grain to fill their contracts. Until his death in 1878, Murphy was a leading figure in the Fort Stanton region. He and his partner Emil Fritz had been stationed at Fort Stanton in 1866 while serving as officers in the New Mexico and California volunteers respectively. With ample opportunity to survey the surrounding countryside, Murphy predicted that it would become "*the* farming section of New Mexico." After their discharge Murphy and Fritz operated the sutler's store at Fort Stanton and soon entered the competition for government contracts. The controversy surrounding Murphy's alleged control over contracts in Lincoln County will be examined in a later chapter. L. G. Murphy and Company was asked to leave Fort Stanton in 1873 after one of Murphy's clerks fired a shot at an officer. Thereafter the firm resumed business in the town of Lincoln, nine miles to the east.[108]

Though evicted from the post, Murphy and Company continued to win government contracts. The contract Murphy signed in October 1873 for delivery of 250,000 pounds of corn to Fort Stanton proved difficult to fill, and he asked the quartermaster's department to cancel the agreement. According to Murphy, the corn could not be purchased locally as the fall harvest had yielded only one fourth the usual amount and increased numbers of cattlemen were placing heavy demands on the crop. Most important, farmers had failed to bring their corn to market because of the outbreak of the Horrell War in December. By the time the Horrell brothers departed from Lincoln County in February 1874, they had twice raided the town of Lincoln and shot and killed at least five of its residents. During these strife-filled weeks, farmers had abandoned their houses, their fields, and their farm equipment. Among those whose grain subsequently was either stolen or destroyed were Manuel Gutiérrez, Juan José Gutiérrez, Stephen Stanley, and Alexander H. Mills, all producers who had agreed to help fill Murphy's contract. Since the government would not release Murphy from his commitment, though it did grant him an extension, he had to purchase corn on the Rio Grande, thereby losing money on the contract.[109]

Crops in Lincoln County again partially failed during the fall of 1877, and the following year contractors Abraham Staab and Charles Ilfeld had to import corn from the States to fill their Fort Stanton contracts.[110] During the summer and fall of that same year of 1878, the bloody Lincoln County War disrupted local commerce. According to two prominent residents of Roswell, entire families had to "abandon their ranches, crops, and flee from the county for safety." John H. Riley, who sided with the Murphy–Dolan faction against the McSween–Tunstall camp, was one who fled, leaving behind an unfilled contract calling for delivery of 100,000 pounds of corn to Fort Bliss. His attorneys, Thomas B. Catron and William T. Thornton, later persuaded the army to terminate the contract, arguing that the government would lose nothing as a new contract had been let at a lower figure.[111]

If agricultural censuses can be relied upon, grain production in Lincoln County had declined precipitously by the end of the decade. The county produced only 41,597 bushels of corn in 1879, less than one third of the amount recorded in the 1870 census. During the same period, the county's population had increased by 39 percent. A decade of lawlessness and intrusion by large cattle herds contributed to the decline in grain production, but the inability of the government to consume local surpluses should also be mentioned. Even so, Lincoln County residents continued to deliver locally grown corn to Fort Stanton long after cheaper Kansas corn was being imported by rail to supply the other posts.[112]

Railroads indeed would change distribution patterns throughout the territory, a topic that will be discussed in Chapter 7. But for many years much of the grain that filled government contracts was raised in the three adjoining northern counties of Taos, Mora, and San Miguel. Representing 39 percent of the territory's population and containing 37 percent of its farms, these counties together produced in 1870 almost 34 percent of New Mexico's corn, 60 percent of its wheat, and 74 percent of its oats. Many of these farms were small affairs. Roughly 76 percent were under fifty acres and 48 percent were under twenty acres. Yet the largest farms in New Mexico were located in the north. Mora County claimed three farms of 1,000 acres or more and San Miguel County claimed one.[113] Farming throughout New Mexico depended upon irrigation, utilizing waters from a few principal streams and rivers. Moderate mountain winter snowfalls usually guaranteed good crops in the fall. Too much snow brought destructive spring floods, and too little snow resulted in partial or total crop failures. Crops were also damaged by early frosts, too much or too little rainfall, and visitations by grasshoppers.

The typical farmer in these northern counties was a Hispano who cultivated a small plot of land using a simple wooden plow—perhaps tipped with steel—drawn by a yoke of oxen. Threshing was done by mules, goats, or horses trampling the grain on the ground. Anglo observers frequently criticized such elementary farming practices and urged adoption of "American" tools and techniques. Governor Henry Connelly told legislators in 1865 that agriculture, as practiced in the territory, was "very little in advance of the dark ages."[114] Fourteen years later, Governor Lew Wallace made a similar observation, stating in his annual report: "Agriculture in New Mexico is yet in its primitive condition. The wooden plow of the Mexican fathers holds preference with the majority of farmers."[115] Lawrence G. Murphy, in describing farms near Fort Stanton in 1866, noted that it took three years of labor using a Mexican plow "to produce a crop equal to that raised in the first year on land broken up with an American plow."[116]

But newcomers also praised farmers for practicing modern agriculture. James F. Meline, who traveled through New Mexico in 1866, noted that Salvador Armijo of Albuquerque was probably the only Hispanic farmer who spread manure on his fields. Armijo also was among the first to use steel plows and to plant improved seed corn.[117] A second visitor, William A. Bell, who arrived with a railroad surveying expedition in 1867, described William Kroenig's property six miles south of Fort Union as a model farm. Kroenig stored run-off rain water from nearby mountains in two or three earthen reservoirs, from which he irrigated more than 2,000 acres.[118] He was the largest grain producer in Mora County in 1870, raising 2,800 bushels of wheat, 6,786 bushels of corn, 9,100 bushels of oats, and 40 bushels of rye. Kroenig was also among the first to utilize a threshing machine in New Mexico, paying an employee in 1867 $75 per month and board to run it.[119]

Both Armijo and Kroenig planted improved seed corn, and both sold their grain to the army or to other government contractors. Military officials strongly encouraged farmers to plant "American" seed corn (described as soft white or yellow corn) rather than "Mexican" corn (described as blue Mexican, hard, black Mexican, or common Mexican flint corn). The chief quartermaster for the district occasionally distributed American seed to local farmers, and contractors who offered grain raised from this seed gained preference on opening of bids. American corn also commanded a higher price, sometimes as much as a cent more per pound.[120] Some officers objected to feeding Mexican corn to animals. Major Andrew J. Alexander, who commanded Fort Garland in 1873, claimed that animals could not chew Mexican corn and instead swallowed it whole, "which produces colic and frequent death."[121] But at least one officer agreed with Abraham Staab, who argued that the Mexican corn raised near Fort Stanton was a "good nutritious article and by many preferred to American seed corn."[122] The debate became academic in the 1880s as more and more forage contracts were filled from the east.

The vast majority of contractors who furnished grain to New Mexico's army posts were merchants rather than farmers. During a ten-year period beginning in 1866, for example, the army signed contracts with twenty-six men for delivery of corn and oats at the Fort Union Depot. Eighteen of the twenty-four who have been identified operated mercantile establishments in Mora, Las Vegas, Santa Fe, and other northern villages. One contractor, Robert Allen, operated a Las Vegas gristmill, and another, Benjamin C. Cutler, owned a Las Vegas hotel. Several of the merchants, including Henry Korte of Mora and John Dold of Las Vegas, held contracts in more than one year, and others, like Abraham Staab of Santa Fe and Frank Chapman of Las Vegas, delivered grain to more than one post.[123] Merchants in northern towns who contracted to supply posts in southern New Mexico customarily purchased grain from producers living in the vicinity of the posts. Santa Fe merchant Adolph Seligman, who as bondsman for a failing contractor undertook to fill the Fort Stanton contract in 1875, employed an agent to procure corn from local farmers. Likewise, to fill contracts at Fort Craig in 1877, Santa Fe merchants Abraham Staab, Bernard Seligman, and the Spiegelberg brothers bought corn in small quantities from local producers. As scarcely any oats were raised in the vicinity of Fort Craig, however, contractor Charles Ilfeld, a Las Vegas merchant, shipped that commodity from the north to complete his contract.[124]

Merchants like Charles Ilfeld secured much of their grain through their retail stores, selling merchandise and receiving in payment corn, wheat, oats, and bran, as well as hay, lumber, cattle, and sheep, which they could then sell to the army or to other government contractors. Ilfeld's supply area, which may have been typical for other merchants, was immense, extending from Roswell and Albuquerque in the south to Mora and Cimarron in the north.[125] For the larger merchants like Louis Huning of Los Lunas government contracts were a boon. From 1871 through 1875, Huning signed sixteen contracts with the quartermaster's department for delivery of corn and bran, amounting to more than $59,000. He held additional contracts to deliver "as needed" beans, cornmeal, flour, and salt to posts in both Arizona and New Mexico.[126] Profits for Huning and other merchants were twofold: from the original sale of merchandise to customers and from the sale of raw materials to the army.[127] Clearly government assistance to private enterprise was significant in territorial New Mexico, for most successful merchants held supply contracts during their careers.

But federal aid reached other segments of society as well. Nineteenth-century business records document how federal money was divided between contractors and their suppliers. For example, in September 1871 Charles Ilfeld agreed to deliver 200,000 pounds of oats at $2.87 per hundred pounds to the Fort Union Depot. Soon after, he paid David J. Wilkins of Mora $1.50 per hundred pounds for oats delivered on the contract. One year later Ilfeld permitted George Berg, who raised large amounts of grain at Golondrinas, to pay his store account with oats at the Fort Union contract price. Some merchants, in fact, took contracts at low prices in order to liquidate outstanding debts with their customers.[128]

The prices paid to farmers fluctuated according to supply and demand. When the army required vast amounts of corn to issue to the Navajos at Bosque Redondo, the chief quartermaster agreed to pay Lawrence G. Murphy 4 cents per pound for corn he delivered at Fort Stanton. He in turn paid local farmers about 3½ cents, although they were unhappy with this amount and threatened to hold their corn while awaiting a rise in prices. Once the emergency was over, prices fell. Murphy received a contract in September 1871 to deliver 900,000 pounds of corn to Fort Stanton at 3½ cents per pound. In July he had been purchasing corn from local farmers at 2½ cents, but by November he was forced to pay 3¼ cents in order to begin delivery on his contract. As more corn came on the market, prices again declined to 2½ cents. During the same year, Charles Ilfeld and John Dold held contracts to deliver corn at Fort Union at $2.17 and $2.18 per hundred pounds. Records do not show what these merchants paid to their suppliers, but they would have received even less than their counterparts in Lincoln County.[129] To small farmers like Ricardo Fajardo, one of L. G. Murphy's suppliers, the military market was a godsend, for it allowed them to exchange their produce for family necessities—coffee, sugar, cloth, thread, candles.

By 1870 the chief quartermaster for the District of New Mexico purchased locally all the forage needed by his department, the amount being regulated by troop strength. In 1870, when 1,598 officers and enlisted men were assigned to the district, the quartermaster awarded contracts to local suppliers for 5,520,000 pounds of corn, which equaled about 15 percent of New Mexico's crop (35,886,088 pounds) and cost more than $136,000. The army also contracted for 305,000 pounds of oats at a cost of about $12,210. In 1877, when troop strength dropped to 884, the army signed contracts for only 1,320,000 pounds of corn, valued at a little more than $32,840; 280,000 pounds of oats, costing about $5,892; and 140,000 pounds of barley, valued at $6,150. In later years, when additional troops were sent to help subdue Apaches, grain consumption increased. With about 1,700 men attached to the district in 1881, the quartermaster contracted for 6,545,236 pounds of corn, amounting to about $161,889, and 2,012,418 pounds of oats, costing more than $63,000.[130] By this date, however, some grain was being imported from Kansas so that it is impossible to know what percent was purchased locally.

Wholesale agricultural prices declined in both national and world markets following the Civil War. In New Mexico the lowest prices paid on grain contracts occurred between 1876 and 1878, when troop strength was also reduced. Joseph B. Collier, a Mora County farmer, agreed in 1876 to deliver 1,105,000 pounds of corn to the Fort Union Depot at a record low of 79 cents per hundred pounds. The following year, another Mora County farmer, Joseph B. Watrous, contracted to deliver 100,000 pounds at the same post for $1.00 per hundred pounds, and in 1878 merchant Henry Korte of Mora agreed to deliver the same quantity for 98 cents per hundred pounds. Higher prices prevailed at other New Mexico posts, ranging in 1878 from $1.43 per hundred pounds at Santa Fe to $5.71 per hundred pounds at Fort Bayard.[131]

Merchants and grain dealers in southern New Mexico found additional markets at military posts in west Texas, where such supplies as were not obtainable locally had to be transported overland from San Antonio, a distance of 690 miles in the case of Fort Bliss. The army encouraged New Mexicans to bid on west Texas contracts in hopes that competition with Texas residents would reduce expenditures.[132] Although few financial records for the west Texas posts have been located, patterns of supply and competition can be identified.

Federal troops reoccupied Fort Bliss in October 1865. The post was located near El Paso, Texas, whose residents—numbering 764 in 1870—used water from the Rio Grande to irrigate their vineyards, orchards, gardens, and wheat and corn fields. Grain and garden produce also were raised in the farming communities below El Paso: at Ysleta with 799 residents, Socorro with 627, and at San Elizario with 1,120 inhabitants.[133] In addition to what was raised in these settlements, Fort Bliss contractors could secure grain harvested across the Rio Grande in Chihuahua, Mexico, as well as in New Mexico's Mesilla Valley.

Starting in 1868, six different suppliers received contracts to deliver corn to Fort Bliss before it was temporarily abandoned in 1877. The highest-priced corn was supplied by William P. Bacon, an El Paso lawyer and district judge, who agreed in October of 1868 to deliver a small amount (550 bushels) at $3.97 per bushel (about 7 cents per pound). Eight months later, William B. Knox, who ran a line of freight wagons between San Antonio and Chihuahua, agreed to deliver "corn as required" at $2.20 per bushel. The brothers Samuel and Joseph Schutz, El Paso merchants, signed a contract in 1870 for "corn as required" at $1.56 per bushel, and in 1876 their nephew Solomon Schutz agreed to deliver 928 bushels at about 95 cents per bushel. Las Cruces merchant Henry Lesinsky supplied more corn than any other contractor, holding contracts in 1871, 1872, and 1873 for a total of 709,420 pounds at prices ranging from about $1.03 to $1.85 per hundred pounds. Lesinsky's partner, Ernest Angerstein, retailed merchandise in El Paso, Mexico (modern Juárez), and the two partners most likely gathered corn from both sides of the river to fill the Fort Bliss contracts.[134]

Following the El Paso Salt War, a bloody economic contest to control salt beds east of El Paso, Fort Bliss was reactivated in 1878.[135] During that year, the government let five contracts for the post's supply of corn. Three Mesilla Valley merchants (Simon M. Blun, Thomas J. Bull, and Charles Lesinsky) agreed to deliver a total of 162,000 pounds; John H. Riley of Lincoln County contracted for 100,000 pounds; and Adolph Krakauer of El Paso would supply 240,000 pounds. Blun and Bull again signed contracts in July of 1879, the former agreeing to supply 150,000 pounds of corn at $2.37 per hundred pounds and the latter to deliver 125,000 pounds at $2.44 per hundred pounds.[136] Corn later became scarce. During the summer the Rio Grande had dried up from Albuquerque to a point below El Paso, Mexico, a problem that would recur as farmers in southern Colorado brought new lands under cultivation and placed greater demands on the river. A drought compounded the hardships. In November Blun reported no corn in his section of the country and asked that his contract be annulled.[137] With the coming of the railroad to El Paso in 1881, contractors no longer had to rely on Rio Grande farmers. Adolph Krakauer imported 120,000 pounds of corn from California that year, which he sold to the government for $2.52 per hunded pounds.[138]

Fort Quitman was located about seventy miles southeast of Fort Bliss near the point where the El Paso–San Antonio road left the valley of the Rio Grande. Reoccupied by federal soldiers in January 1868,

Quitman was permanently abandoned nine years later. A few civilians, mostly Hispanos, resided near the post, supplying it with milk and sometimes butter, eggs, and chickens. The nearest settlements were on the Mexican side of the river at Guadalupe and San Ignacio about twenty-five miles above the post.[139] Four of the seven contractors who supplied Fort Quitman with corn also held Fort Bliss contracts: William P. Bacon, Samuel Schutz, Henry Lesinsky, and George Zwitzers, a resident of El Paso. Lesinsky again supplied more corn than any other contractor, signing contracts in 1871, 1872, and 1873 to deliver a total of 976,934 pounds at prices ranging from $1.19 to $2.25 per hundred pounds. According to Captain Charles Bentzoni, commanding at Fort Quitman in 1875, most of the post's grain supply came "from the interior of Chihuahua."[140]

Possibly the most successful merchant to deliver supplies to Forts Bliss and Quitman was Henry Lesinsky of Las Cruces. From 1871 through 1874, Lesinsky supplied on the average seven military posts a year with commodities ranging from flour to charcoal. Profits from these supply contracts were pumped into other enterprises. In addition to government contracting, Lesinsky also ran a stage line, served as a government mail contractor, operated a Las Cruces flour mill and mercantile establishment, invested in banking, mining, and merchandising in Silver City, and developed the famous Longfellow copper mine in southeastern Arizona. The Las Cruces *Borderer* reported in December 1871 that H. Lesinsky and Company's grain trade had amounted to 1,850,000 pounds that year. "A large share of the grain of southern New Mexico passes through their hands," the writer concluded.[141]

But New Mexicans generally could not compete successfully with Texas entrepreneurs in supplying grain to the more distant west Texas posts. Fort Davis, located about 125 miles southeast of Fort Quitman, was reoccupied by federal troops in July 1867.[142] The government let three contracts that year to supply the post with a total of 29,000 bushels of corn at prices ranging from $3.23 to $3.75 per bushel. The three contractors, Louis Zork, Hardin B. Adams, and Peter Gallagher, were residents of San Antonio. In later months the quartermaster at Fort Selden reported that agents for the Fort Davis contractors were in El Paso and the Mesilla Valley negotiating for corn.[143]

For the next four years most of Fort Davis's corn came from Mexico. Anton F. Wulff of San Antonio, who held contracts to supply

Fort Davis with corn in 1868, 1869, and 1870, employed an agent, William Hagelseib, to buy grain from Mexicans who crossed the Rio Grande with their wagons near Presidio del Norte, about 90 miles southwest of the post. In 1871 Hagelseib himself agreed to deliver 420,000 pounds of corn, and John Burgess, a resident of Presidio del Norte, would deliver 200,000 pounds. The only New Mexican who contracted for corn at Fort Davis was Henry Lesinsky, who in 1872 agreed to deliver 340,000 pounds.[144] At this time the only land under cultivation near the post, according to the commanding officer, was a garden and one small farm where a little corn was raised. About 150 Hispanos and 25 Anglos lived nearby, all dependent upon the army or the overland stage for their support.[145] From 1873 through 1876 Fort Davis's grain contracts were held by residents of the Fort Stockton region, where irrigated farms were producing significant amounts of grain.

The population around Fort Davis remained sparse through the early 1870s even though new areas were opened to cultivation. A group of families led by Samuel Miller and Daniel Murphy settled on Toyah Creek thirty-five miles north of the post in 1871, and they soon began raising grain and herding cattle.[146] According to the post commander, in 1879 about 1,000 men, women, and children, mostly Hispanos, were living on ranches scattered throughout the length of Toyah Valley for a distance of twenty-two miles. He also noted that Limpia Canyon running northeast from Fort Davis was only sparsely settled with cattle and other ranches. A few families were located along the road to Presidio del Norte, and a settlement of 100 Hispanos could be found forty miles north of Presidio on the banks of the Rio Grande. Despite this moderate growth, Fort Davis's grain contracts for 1879 were held by Fort Stockton men, Michael F. Corbett, a former post trader at Fort Stockton, and George M. Frazer, who ranched at Leon Holes.[147] Corbett and Frazer probably secured grain to fill their contracts from the Fort Stockton region. In 1883 the quartermaster at Fort Davis reported that "neither oats or corn are grown in any considerable quantities" in his immediate vicinity.[148]

Fort Stockton was located seventy-five miles northeast of Fort Davis on Comanche Creek, having been established in 1859 to protect the San Antonio–El Paso stage route. Evacuated by Union soldiers at the start of the war, the fort was reoccupied by federal troops in 1867. This was about the same time that Peter Gallagher of San Antonio began buying up property along Comanche Creek, including the land on which the post was located, for the purpose of speculating in irrigable lands and producing supplies for west Texas garrisons. Others soon followed Gallagher to the region. On land acquired late in 1869 and early in 1870, Cesario Torres, Bernardo Torres, and Félis Garza constructed an irrigation system, including ditches and a diversion dam on Comanche Creek, about three quarters of a mile from the post. Eight miles west of the post three large springs known as Leon Holes provided additional water to irrigate farms. By 1870 about 500 settlers, mostly Hispanos, lived on ranches and farms along Comanche and Leon creeks.[149]

Even though local farmers were soon raising substantial amounts of grain, suppliers in the early years obtained most of what was needed to fill Fort Stockton contracts from Mexico. Colonel Edward Hatch, commanding the post in February 1868, complained that the government paid contractors $4 to $10 per bushel for corn, which their agents had purchased from wagon trains coming from Mexico at between $2 and $2.50 per bushel. Contract registers show that two San Antonio residents provided the post with corn in 1867: Gustav Groos, a merchant, who agreed to deliver 7,500 bushels at $3.92 per bushel and Peter Gallagher, who contracted for 10,000 bushels at $4.95 and 15,000 bushels at $4 per bushel. Frederick Groos, Gustav's brother and partner, received a contract in July 1868 to supply 14,000 bushels at $3.80 per bushel.[150] Unrecorded purchases may account for Hatch's apparent exaggeration in price; even so, Gallagher was receiving a good price for his grain, almost 9 cents per pound.

Like their counterparts in New Mexico, military officials stationed at the west Texas posts were critical of this Mexican corn. The post commander at Fort Stockton in November 1868, Captain George H. Gamble, even blamed it for causing the death of eighty of his horses. Autopsies revealed, or so Gamble reported, that the lining of the animals' stomachs had been burnt away, which he attributed to the corn having been soaked in lime and then dried to prevent insects from injuring it.[151] Lieutenant Colonel James H. Carleton, who inspected all four west Texas posts in 1871, described Mexican corn as "hard and flinty," and he recommended that it and all other kinds of grain be well soaked before feeding to animals.[152]

While inspecting Fort Stockton, Carleton interviewed Peter Gallagher, who told him that local farmers had raised 12,000 bushels of corn in 1870 and could double this amount to furnish all the grain required at both Forts Stockton and Davis. But rather than purchase from Stockton farmers, contractors Hardin B. Adams and Edwin Wickes in 1870 had obtained most of the corn to fill their contracts from Mexico, receiving from the government $2.33 per bushel.[153] Between 1872 and 1878, however, Cesario Torres, Peter Gallagher, Joseph Friedlander, George M. Frazer, and Michael F. Corbett, all Fort Stockton area residents, held contracts at much lower prices, which suggests that they were buying grain from local farmers rather than transporting it from Mexico at added expense. Friedlander, for example, contracted for more than 232,000 pounds in 1874 at $1.10 per bushel.[154]

From the foregoing, it is clear that contractors for west Texas posts secured much of their grain during the 1860s and early 1870s from Mexico, while obtaining smaller amounts from settlements on the U.S. side of the Rio Grande. Residents of New Mexico successfully competed for grain contracts to supply Forts Bliss and Quitman; only one New Mexican, Henry Lesinsky, supplied grain to Fort Davis; and none contracted for grain at Fort Stockton, although in 1874 Lesinsky supplied that post with flour. Like Fort Stanton in southern New Mexico, the military market at Fort Stockton attracted a sizable agricultural population, whose irrigated farms would provide grain for both Fort Stockton and Fort Davis. Much of the rest of west Texas between the Pecos and the Rio Grande, however, was devoted to cattle ranching, an industry that profited greatly from the military's presence.

New Mexico contractors also tried to capture the military market in Arizona, but they had limited success. Because forage was so costly in Arizona, the army encouraged competition and actively promoted agricultural expansion. But it refused to pay exorbitant prices for local products. A cardinal rule for supply officers was to purchase wherever prices were the cheapest. To the dismay of Arizona's farmers, this rule was firmly enforced in 1866 when officials in San Francisco repudiated contracts that had called for delivery of Arizona-raised grain at Camp Lincoln for 16 cents per pound and at Prescott for nearly 17 cents per pound.[155] Grain was shipped to these posts from California, an action that angered the editor of the Prescott *Arizona Miner,* who claimed that the California grain would cost even more because of transportation expenses. It was wrong, he said, to go out of the territory to purchase grain for any reason, even if prices were high. He added:

> What are our farmers here for? What in the name of common sense is the object of our Government in sending a military force here? Is it not that our country, known to be valuable, may be developed and produce a revenue to the United States?[156]

The army's action had a salutary effect. In March 1867 local suppliers agreed to furnish grain at Fort Whipple and Camp Lincoln for about 7 cents less than the repudiated contract prices. The Prescott editor applauded these new arrangements, but he advised farmers not to seek extravagant rates in the future and thus force the government to purchase elsewhere.[157] But that is what the army planned to do three years later when farmers in central Arizona threatened to combine to force prices up. To circumvent their scheme, the quartermaster at Fort Whipple quietly sent an agent into New Mexico with instructions to purchase 500,000 pounds of grain as cheaply as possible.[158]

By 1870 Fort Whipple was at the center of the hay- and grain-growing region of central Arizona. The Prescott *Arizona Miner* frequently published articles extolling the excellence of soils found in nearby valleys and recording the acreage that Yavapai County farmers cultivated. Such articles were intended to attract investors and to offset the impression that Arizona was "a dry, useless, barren country."[159] Because agricultural statistics for the county were not reported in the 1870 federal census, the articles are important for gaining an understanding of the role the military played in Arizona's agricultural expansion, even though allowance must be made for journalistic exaggeration.

The same newspaper estimated that 1,600 acres had been cultivated in the county in 1867, mostly planted in corn with only 75 acres planted in wheat and barley. Several varieties of corn had been sown with excellent results, yielding on the average 45 bushels to the acre. Farmers near Prescott preferred American flint corn and the "hard New England variety known as King Philip," while farmers in other localities preferred "the large yellow dent."[160] In September 1868 the paper listed the county's principal growing areas and the settlers who farmed there. Among the most important districts were Lower Granite Creek south of Prescott, where James S. Giles had about 300 acres in corn; Willow Creek northwest from Prescott; Skull Valley, twenty miles southwest from Prescott, where Beach and Smith cultivated 100 acres; Kirkland Valley, four miles south of Skull Valley, where 200 acres were planted in corn and vegetables; Walnut Grove, thirty miles south of Prescott on the Hassayampa; Rio Verde Valley, fifty miles east of Prescott, where 150 acres of corn had been planted; and the new settlement of Phoenix on the Salt River, where 800 acres were planted in corn and 200 acres in sorghum and vegetables. The editor calculated that Yavapai County farmers in 1868 had planted 4,000 acres in corn, which he believed would yield 6,720,000 pounds. The following year, after the county experienced an extended dry spell, he estimated that the harvest would yield no more than 5,000,000 pounds of corn.[161] At least one officer commanding at Fort Whipple in later years was less sanguine about the county's productivity. Lieutenant Colonel John Wilkins, reporting on local resources, wrote in 1878: "The arable land is not very extensive, cultivation is confined to valleys that retain moisture and the rainfall is principally relied upon for success." In dry seasons, he noted, the crop was almost an entire failure, as had happened the previous year when farmers raised scarcely enough grain to supply their own wants.[162]

Yavapai County farmers sold their surplus grain primarily to the government or to government contractors, but none were more dependent upon such sales than farmers in the Rio Verde Valley. Camp Lincoln (soon to be renamed Camp Verde) was established in 1864 in a section of the valley that was about seven miles wide. With its rich alluvial bottom lands, this area became known as "one of the largest and best agricultural districts" in the territory. But so long as Apaches resisted white settlement, few farmers risked their lives to till the good cropland.[163] The Prescott newspaper reported in 1867 that the nine ranchmen who were living in the valley had a total of 200 acres under cultivation, mainly in corn and barley. Although located only fifty miles east of Prescott, an "almost impassable hill" (Grief Hill) west of the post prevented farmers from transporting grain to the Prescott market, where mining and freighting companies competed with the army for locally raised forage. In 1868 soldiers constructed a wagon road over Grief Hill, but apparently it failed to improve matters very much. Two years later enlisted men were at work building a new road to avoid the hill altogether, which was still being called "the great bug bear of this region."[164] Gradually more and more settlers took up farms. Sixty-seven of the 174 residents listed in the 1870 census for Rio Verde were farmers, and most of the others were laborers who probably worked on other men's farms or at the post. Only two were women, both living on farms with their husbands and without any children.[165] By 1878 farmers had constructed a canal six miles in length, and on irrigated land they were raising more barley and corn than was needed by the garrison at Camp Verde, which remained "almost their sole market," according to the post commander.[166]

Some of the best farm lands in Arizona were located in the Santa Cruz, San Pedro, and Sonoita valleys south of the Gila River in Pima County, where tilling the soil required constant vigilance against Apache raiders. According to the local press, white settlement depended "entirely upon the vicinity of a military post, a town, or other source from which protection may be obtained."[167] When this protection was withdrawn, as it was along the Sonoita in August of 1870, farmers abandoned their crops and found safer quarters elsewhere. Farming in the neighborhood of Tubac along the Santa Cruz was so dangerous that armed escorts of one or two men accompanied farmers as they plowed their fields or gathered their corn.[168] Leslie B. Wooster and his wife Trinidad Aguirre, who cultivated land about seven miles above Tubac, were only two of several settlers who were killed during Indian attacks in the early 1870s. A veteran of many raids, Peter Kitchen successfully farmed lands close to the Sonora border, where in 1873 he had 150 acres planted in grain and 20 acres in potatoes. He too had suffered losses: Apaches had driven off his stock, killed and wounded his employees, and murdered his son.[169]

In spite of these dangers, farmers in Pima County raised on their 140 farms during the year ending 1 June 1870, 26,957 bushels of spring wheat, 30,069 bushels of corn, and 54,697 bushels of barley. They produced two crops annually: barley and wheat, sown in November, were harvested in May; and corn, planted in June, was harvested in October.[170] Unfortunately, Pima County farmers faced stiff competition in marketing their crops. A correspondent for the Prescott *Arizona Miner,* who was surveying business opportunities throughout the territory, found conditions in Pima County late in 1869 "discouraging in the extreme. . . . Farmers are unable to sell their grain at a living price, because grain can be laid down at the military posts in that county, from Sonora, for less than it has cost the farmers of Pima County to raise and harvest it." As late as 1877, Estévan Ochoa and Pinckney R. Tully were importing grain from Mexico to fill their contract at Camp Bowie.[171]

Fine agricultural land was also found in the valley of the Gila River to the east of the Pima and Maricopa villages, where the town of Florence was located in 1866. In February of 1869 the Tucson *Weekly Arizonian* reported that the settlement had a thriving population of about 1,000. Most of the settlers had been drawn there by the fertility of the soil, an abundant supply of water, and the security offered by the presence of friendly Indians. On lands irrigated by the Gila, Hispanic and Anglo farmers together raised in 1873 about 4,000,000 pounds of barley and 500,000 pounds of wheat, while the Pimas and Maricopas produced 4,000,000 pounds of wheat and 500,000 pounds of barley.[172]

Although heralded as a promising commercial and farming community, Florence soon was overshadowed by the growing settlement of Phoenix on the Salt River, which by 1875 had become Arizona's principal agricultural region.[173] Farmers in both districts were spurred by the prospect of filling government contracts, and the increasing demand they made on land and water resources was to bring about tragic changes in the fortunes of the Pimas and Maricopas. It is the story of the founding of Phoenix, however, that reveals most clearly the central role that government contracting played in the territory's economic and urban development.

Credit for promoting Arizona's first modern reclamation project is given to John W. Swilling, who was attracted to the Salt River Valley by expanding agricultural markets at Camp McDowell, Fort Whipple, and the nearby mining camps. Late in 1867 a group of miners, merchants, and farmers joined him in financing the Swilling Irrigating and Canal Company. By spring, workers had completed a temporary ditch that tapped the Salt River near the junction of the McDowell–Wickenburg road, allowing Swilling and other settlers to plant several hundred acres in wheat, corn, and barley. The Swilling Company affixed the name of Phoenix to the little farming settlement, as they envisioned a new civilization rising from the prehistoric ruins of the Hohokam, who once had irrigated this region with an extensive network of canals.[174] In 1871 Salt River Valley farmers planted 2,200 acres in barley, 1,200 acres in wheat, and 700 acres in corn. The following year some of their grain was transported to Camp McDowell and to posts in northern Arizona to fill government contracts. Salt River Valley farmers harvested between 4,000,000 and 6,000,000 pounds of barley in 1874, and by 1881 they were producing two thirds of all cereals grown in Arizona, as well as bountiful crops of alfalfa, fruits, and vegetables. Maricopa County, which in 1871 was carved out of southern Yavapai County with Phoenix as its new county seat, well deserved the title "garden-spot of the Territory."[175] Federal expenditures played a key role in this remarkable growth.

But in reclaiming the Gila and Salt River valleys, Florence and Phoenix residents became embroiled in acrimonious dispute with the Pimas and Maricopas over land and water rights. For more than twenty years military officers labored to keep the peace. Frequently they spoke on behalf of the Indians, but their words had little impact in the long run. The army looked upon the Pimas and Maricopas as friends and allies; they had served as regular soldiers and scouts in campaigns against the Apaches, and their crops had sustained both the soldiers and the army animals. For many years they had aided travelers who crossed the continent using the southern route. They had even helped early Phoenix settlers in crossing their animals over the Salt River during high water, earning praise from one settler, who remarked upon their friendliness and stated: "When they are around one can feel a degree of safety not otherwise felt, as they are ever vigilant."[176]

Shortly after this appraisal appeared in the Prescott *Arizona Miner,* Phoenix residents accused the Pimas of entering their gardens and corn fields, eating and destroying melons and vegetables, and carrying away corn to their villages. Farmers on the Gila reported similar occurrences. These complaints surfaced during the early part of September 1868, about the same time that floodwaters on the Gila demolished three Pima villages and wrecked the nearby flour mill and storehouses of William Bichard. According to the local press, Pimas had seen nothing like this flood for more than sixty years.[177] The flood's tragic consequences—homes destroyed, crops washed away—may in part account for this first major disturbance between Pimas and settlers, but the larger issue driving a wedge between them was that of land ownership.

For several centuries, the Pimas had cultivated the rich bottom land of the Gila River Valley. When the United States acquired sovereignty over this area through the Gadsden Purchase of 1853, it promised to recognize land rights established by the former Spanish and Mexican governments. The federal government finally established a twenty-mile reservation for the Pimas and Maricopas in 1859, embracing the lands that the Indians then irrigated. Pima chief Antonio Azul protested the smallness of the reservation, claiming that the two tribes owned the entire Gila River Valley for a distance of one hundred miles. Indeed, Lieutenant Sylvester Mowry, who surveyed the condition of Indian tribes in the Gadsden area in 1857, later submitted evidence that the Mexican government had granted fifty leagues of land (approximately 150 miles) to the Pimas and Maricopas.[178]

At the time that settlers began farming above the Indian villages, the two tribes were self-sufficient, prosperous, and planting more wheat than they had ever done before. Lieutenant Colonel Clarence E. Bennett, who commanded Fort McDowell in 1866, encouraged the tribes to raise more wheat and corn in order to double the amount they had sold in previous years. Bennett assured his superiors that "this would be a direct and great benefit to the Military Service in getting forage for Government animals right when it is wanted, flour for troops and citizens of the Territory, and of course it would greatly benefit the producers."[179] But the Pimas and Maricopas were becoming increasingly resentful of the whites who settled on the upper Gila, cultivating land they considered their own and stealing their water. Lieutenant Colonel Thomas C. Devin, commanding officer in southern Arizona, informed Arizona Superintendent of Indian Affairs George W. Dent in mid-March of 1869 that so much water was being diverted from the Gila at Florence that the Indians would not have enough to irrigate their fields in dry seasons. Both army and Indian Bureau officials recommended that the reservation be enlarged, but no significant changes were made for several years.[180]

During the fall of 1869 Pimas and Maricopas expressed their resentment by riding through fields at Phoenix and on the upper Gila, stealing and destroying the crops. In November, Frederick E. Grossman, agent for the tribes, reported that 300 Indians were making winter quarters in settlers' fields near Florence, subsisting on grain.[181] A headman for the Pimas who was interviewed during these troubles succinctly stated his tribe's grievances: "The Whites have gone above and settled on our land, cut down our Mesquite trees and now they take our water so that we have only a few acequias filled and cannot cultivate all our land."[182]

Pimas continued to raid settlers' fields through the early Seventies, while more and more white farmers diverted water from the Gila. As less water became available downstream, Indians in increasing numbers left the reservation to cultivate lands they claimed on the Salt River and farther upstream on the Gila. In November 1873 an inspector for the Indian Bureau reported that many Indians were destitute. In 1876 the government extended the reservation eastward to include about 9,000 acres the Indians then were cultivating, but this was an empty gesture. The Pimas and Maricopas still lacked sufficient water to remain self-supporting on the reservation. During a drought in 1878, the river on the reserve was dry from May to December, forcing abandonment of almost the entire western end.[183]

The military point of view at this time was best expressed by Colonel Orlando B. Willcox, commanding the Department of Arizona, who believed the tribes had been illegally deprived of water. It is "the duty of the Government," he wrote to division headquarters, "to either restore the waters of the upper Gila to the Indians, or give them other irrigable lands." Responding to the suggestion made by Arizona Governor John C. Frémont in 1879 that the Pimas be removed to the Colorado River, Willcox wrote:

> The Pimas and Maricopas are on different footing from all the Indians in the country, and deserve great consideration from the fact that they have never fired a shot, but on the contrary, rendered frequent assistance as friends and allies and whatever is done, I think you will agree with me, should be done in their best interests, as well as in those of the whites.[184]

The reservation again was enlarged in 1879, which allowed the tribes to expand their irrigated lands during the early 1880s and to appear relatively prosperous. But construction of the Florence Canal eight miles above the town of Florence signaled further disaster. By 1887 this canal was utilizing almost the entire flow of the Gila River to irrigate white people's farms; little or no water reached Pima fields downstream. Thereafter conditions on the reservation continued to deteriorate, and the government in 1895 was forced to issue rations to Indians because of food shortages.[185] The destitution of the Pimas and Maricopas in the mid-Nineties stands in marked contrast to their prosperity thirty years earlier when they assisted General Carleton and the California Column on their overland march to the Rio Grande.

Both Indian and white farmers in Arizona, like their counterparts in New Mexico, depended upon water from a few principal streams and rivers to irrigate crops of corn, beans, barley, and wheat. The two territories, however, exhibited noticeable differences in other farming practices. An Arizona farmer, for example, tilled more land than the typical New Mexico farmer. According to the 1870 census, 67 percent of the farms in Arizona and 7 percent in New Mexico consisted of 100 acres or more, while 85 percent in New Mexico and 6 percent in Arizona were less than 50 acres. A decade later differences in size remained, although the gap had narrowed: 72 percent of Arizona's farms and 20 percent of New Mexico's farms now contained 100 acres or more.[186]

In addition, farmers in Arizona planted more barley than corn in each decade, whereas in New Mexico the opposite occurred. Climate was a controlling factor; barley flourished in the Gila and Salt River valleys where high temperatures prevailed, while corn was better suited to cooler climates.[187] Moreover, army officers in Arizona encouraged the cultivation of barley, claiming that animals would not thrive on corn in hot climates. Major John Green, commanding at Camp Grant in 1870, complained about the quality of forage in a letter to his superior officer: "Why the Quartermaster's Department wants to persecute the animals of the government with their eternal corn, where there is other grain to be had, is more than I can comprehend." At Camps Bowie and Crittenden in 1870 animals were fed barley during the hot months "when corn does not answer so well" and corn during the cooler months.[188]

Farm machinery was more conspicuous in Arizona than in New Mexico. Newspaper editors in both territories heralded the introduction of modern equipment, but Arizona's press more often reported new acquisitions. In April of 1868 the *Arizona Miner* described how a shipment of plows arriving in Prescott from California had been sold even before the plows were unloaded from the wagons. Moses Langley rode twenty-five miles to purchase one and packed it home to Kirkland Valley on a horse.[189] Among the most successful pioneer farmers in central Arizona who used modern machinery were Nathan and Herbert Bowers, government contractors and sutlers at Fort Whipple, who in 1867 bought King Woolsey's ranch on the Agua Fria twenty miles east of Prescott. During the following year, the brothers employed between ten and thirty men on the ranch, operated four team-drawn plows, and planted 400 acres in corn, 20 acres in alfalfa, 20 acres in Hungarian grass, and smaller amounts in melons and vegetables. They also owned a steam-powered corn sheller, which they set up in a Prescott furniture factory to shell corn for local farmers.[190]

During the summer of 1869, Isaac Q. Dickason, who owned a ranch on the lower Agua Fria, received from California a reaper and a thresher, said to be "the first machines of this kind ever brought to this part of the Territory."[191] The first header and thresher appeared in Salt River Valley in 1870; three years later the press reported that headers were plentiful and threshers sufficient to handle the valley's harvest. In addition, reapers were operating in Santa Cruz Valley and at least one threshing machine was reported on its way to the San Pedro Valley in 1873. But mechanization was most apparent in the Salt River Valley, where in 1877 William H. Hancock's steam threshing machine threshed more than 5,000,000 pounds of wheat and barley, a second steam thresher processed more than 3,000,000 pounds, and one horse-powered machine threshed more than 600,000 pounds. Salt Valley farmers paid about three dollars per acre to have their grain harvested and stacked by a header and one twelfth of their crop for threshing.[192]

Primitive farming practices prevailed in many parts of Arizona, however, even after modern equipment was introduced. Farmers in the Santa Cruz Valley during the early 1870s often plowed with wooden plows, harvested with sickles, and utilized horses to thresh grain by tramping on it. The first barley offered for sale in 1873 in Phoenix belonged to Gordon A. Wilson, who had it cut by hand and tramped out by horses. And several farmers on the Gila in 1876 were threshing their grain with animals, although the first threshing machine had been brought to the valley at least by 1870.[193]

As in New Mexico, farmers in Arizona rarely competed for government contracts. Instead, they sold their grain and hay to merchants and grain dealers, who received the majority of the contracts. Two examples illustrate this point. During a ten-year period starting in 1866, the army signed contracts with sixteen different individuals or partnerships for delivery of grain at Tucson Depot and at Camp Lowell. The contractors included one freighter, one physician, a miner, two grain dealers, and the following nine merchants: Charles T. Hayden, John B. Allen, Estévan Ochoa, Isaac Goldberg, William Zeckendorf, Pinckney R. Tully, Frederick L. Austin, Edward N. Fish, and Samuel Drachman. Only three contractors were farmers. Peter Kitchen farmed lands near the Arizona–Sonora border; Charles W. Lewis, a former officer in the California volunteers, farmed near Tubac; and Edward K. Buker was a Salt River Valley farmer. Thirteen contractors supplied grain to Camp McDowell between 1867 and 1876. They included Charles W. Beach, a freighter; Gideon Cornell, carpenter; William Bichard and William B. Hellings, flour mill owners; and eight merchants or mercantile firms—G. F. Hooper and Co., Hooper, Whiting and Co., Aaron Barnett, Charles T. Hayden, Edward N. Fish, John Smith, and Morris and Michael Goldwater. Several merchants, like Hayden, Fish, and Drachman, held contracts in more than one year and delivered grain to more than one post.[194] Among the most successful at obtaining government contracts, however, were the partners Estévan Ochoa and Pinckney R. Tully of Tucson. This firm was never without a government contract between 1865 and 1880. During this period, they supplied grain to Camps Bowie, Goodwin, Grant, Huachuca, Mason, and Tucson Depot, and they held transportation, hay, fuel, and bean contracts as well. The money they received from government contracting contributed to their success in a host of other enterprises: merchandising, freighting, mining, flour milling, operating stage lines, and running sheep and cattle herds.[195]

Contractors frequently sublet their contracts to other merchants or dealers, although officially the government recognized only the original contractor. Such an arrangement was made by Marcus Katz of Tucson who received a contract in December 1877 to supply Camp Thomas with 50,000 pounds of barley at 6 cents per pound, the contract to run through June 1878. He subsequently sublet the contract to Lord and Williams, one of the largest mercantile firms in Tucson, at 5¾ cents per pound. This company in turn purchased the barley from local farmers, paying perhaps as much as 4 cents per pound before new barley came on the market. By late July of 1878, dealers could buy large lots of barley on the Gila for 2 cents per pound.[196] Some merchants, like Albert E. Jacobs of Safford, operated their own reapers and threshing machines, which helped them in gathering grain to fill government contracts. And by threshing for farmers who were indebted to them they were assured of collecting those debts.[197]

As in New Mexico, grain prices in Arizona declined during the 1870s, reaching an unprecedented low in 1874, which was about the time that farmers began raising more grain than local markets could absorb.[198] Between 1866 and 1874, contract prices fell at Tucson from $6.50 per hundred pounds to $2.18 per hundred pounds and at Camp Bowie from $8.25 to $3.23 per hundred pounds. At Camp McDowell contract prices declined from $6.25 per hundred pounds in 1868 to $1.97 per hundred pounds in 1874. An even greater reduction was recorded at Fort Whipple where James Grant received $9.75 per hundred pounds for delivery of 40,000 pounds of corn in 1867, and Jefferson H. Lee received $2.24 per hundred pounds for delivery of 100,000 pounds of corn in 1874. Prices fluctuated in later years, but Forts Lowell, Bowie, and McDowell all recorded their lowest prices of the decade in 1879, with barley costing as little as $2.14, $2.98, and $1.75 per hundred pounds at the respective posts.[199]

Even before prices declined in the 1870s, farmers in central Arizona attempted organization to keep them up. During the summer of 1868, angry growers met in Prescott to organize a Farmers' Association, voting not to accept less than 8½ cents per pound for grain delivered at Fort Whipple on penalty of a $1,000 fine. The farmers claimed that government contractors, by purchasing grain from the producers at prices far below the contract price, were "crippling the resources of the country and oppressing a large proportion of the laboring class." The association later declared the agreement void, however, when several members failed to support it.[200]

Thereafter, Curtis C. Bean, as agent for some of the farmers, accepted the contract to supply Fort Whipple with 500 tons of corn at 8 cents per pound in coin, agreeing to pay local planters 7½ cents per pound. This arrangement also fell through after farmers learned that the military required the grain to be delivered over several months. Many needed cash immediately and could not afford to store their grain for six months before realizing a profit. As a result, most of them sold their grain that season for 3½ cents and 4 cents per pound.[201] They had greater success in 1869 when Bean purchased their grain at what was considered a fair price—5 cents per pound. Bean on his part was to receive from the government between 7 and 7¾ cents per pound for 800,000 pounds of grain delivered at Fort Whipple and 1,000,000 pounds delivered at Camp Toll Gate.[202] In 1872 Phoenix area farmers spoke of forming a Farmers' League to support their demand for 4 cents a pound for new barley, but the league was never organized.[203]

Attempts by Arizona farmers to control prices had failed. But a sympathetic territorial press kept their plight before the reading public. Through the mid-Seventies, newspaper editors were highly critical of speculators, accusing them of taking military contracts at such low figures that producers were becoming impoverished. One editor suggested in 1869 that contractors cast aside their "unwholesome avarice" and form a society for the purpose of regulating the market, thereby assuring farmers just prices for their produce. A second editor recommended in 1873 that companies be formed in farming settlements for the purpose of building warehouses to store farmers' grain and advancing them capital on the grain's value. This scheme was similar to those being implemented in the East that enabled farmers to hold grain off the market until fair prices could be obtained.[204]

The press also encouraged farmers to make themselves less dependent upon the sale of grain by producing more of their own necessities, such as vegetables, milk, butter, cheese, eggs, and meat. When the price level continued to fall, the Tucson press placed some of the blame on New Mexico's grain dealers, who were willing to supply Arizona's posts at prices that "our farmers deem below the cost of production and marketing." Such prices already had impoverished the farming and laboring class in New Mexico, or so the newspaper claimed. By 1874 editors in both territories recognized that the government market was limited, and they urged farmers to diversify—to raise tobacco, sugar, cotton, or more alfalfa as feed for horses and cattle and hogs.[205]

While prices were falling during the 1870s, New Mexicans supplied grain to only three posts in Arizona, all in the eastern part of the territory. Louis and Henry Huning of Los Lunas supplied corn or barley to Camp Apache in 1872, 1873, 1875, 1878, and 1879; Nathan Bibo furnished oats there in 1871; and Numa Reymond of Paraje supplied barley in 1874. Las Cruces merchant Louis Rosenbaum furnished barley to Camp Thomas in 1876 and to Camp Bowie in 1878. The only other New Mexican who held an Arizona grain contract in these years was Simon M. Blun of Las Cruces, who delivered barley to Camp Bowie in 1877. It is clear that New Mexico's penetration into Arizona's grain market was limited, its merchants providing real competition only at Camp Apache. But even here, Arizonans signed more than half the grain contracts awarded in the 1870s.[206]

The largest grain contract issued in Arizona to a New Mexico supplier went to the Hunings, who agreed to deliver 1,000,000 pounds of barley to Camp Apache in 1879. This represented about 15 percent of the barley required by the Department of Arizona that year and cost the army $56,300. To supply the other posts, the department let contracts for an additional 5,675,000 pounds, which equaled almost 50 percent of Arizona's barley crop, and for 1,250,000 pounds of corn, amounting to 64 percent of the territory's crop.[207] Beginning in 1881, California residents began receiving contracts to furnish some of the grain required in the department, although Arizona contractors would continue to supply local grain through the years of this study.

By awarding supply contracts locally, the government encouraged economic development throughout the Southwest, and Anglos and Hispanos alike profited. Although most grain contracts were held by merchants, farmers were encouraged to expand their plantings to help fill the army's requirements. Even Hispanic farmers, who traditionally raised only sufficient amounts to feed their families, increased their harvests. The military market was limited, however, and by 1874 residents in the Southwest were raising all the grain required by the army. Even so, grain contracts continued supplying investment capital for other enterprises. And citizens could also turn to the land for two other commodities the army needed: hay and fuel. Digging mesquite roots and cutting natural grasses, the subject of Chapter 3, proved vital to the army's mission.

(TOP) Wood haulers at Fort Davis, Texas. (Courtesy Fort Davis National Historic Site.)

(BOTTOM) Making hay in Arizona. (Arizona Historical Society Library, Tucson, Arizona.)

(TOP) Threshing wheat with sheep and goats, Galisteo, New Mexico, 1918. (Courtesy Museum of New Mexico, Santa Fe, New Mexico, Neg. No. 12583.)

(BOTTOM) Rear view of Ceran St. Vrain's stone mill, Mora, New Mexico, as it appeared in 1971. (Courtesy Museum of New Mexico, Santa Fe, New Mexico, Neg. No. 53360.)

(TOP) John Becker's store and forage agency, Belen, New Mexico, 1877. (Courtesy Museum of New Mexico, Santa Fe, New Mexico, Neg. No. 66013.)

(BOTTOM) Otero, Sellar and Company, Las Vegas, New Mexico, ca. 1879. (Courtesy Museum of New Mexico, Santa Fe, New Mexico, Neg. No. 9431.)

(TOP) Ox train near Camp Grant, Arizona, 1871. (National Archives.)

(LEFT) Abraham Staab, Santa Fe merchant and military contractor. (Courtesy Museum of New Mexico, Santa Fe, New Mexico, Neg. No. 11040.)

(TOP) Jacob Gross, Las Vegas merchant and military transportation contractor. (Courtesy Museum of New Mexico, Santa Fe, New Mexico, Neg. No. 40177.)

(TOP RIGHT) John Lemon, Mesilla merchant and military contractor. (Rio Grande Historical Collections, New Mexico State University Library, Las Cruces, New Mexico.)

(RIGHT) Estévan Ochoa, Tucson merchant and military transportation contractor. (Arizona Historical Society Library, Tucson, Arizona.)

3

Forage and Fuel

MORE than a century ago large sections of the Southwest provided good pasturage for sheep and cattle; nutritious natural grasses were cut, dried, and fed to work animals as fodder. Government contractors were quick to take advantage of nature's bounty. They harvested its grasses to sell to the army for hay, and they built up herds of cattle to satisfy the army's fresh beef requirements. Nature also supplied materials that contractors sold to the army for fuel—typically mesquite and other local wood but also coal, particularly after the introduction of railroads made large-scale coal mining profitable. In all sections of the Southwest federal money trickled down to a large number of laborers who cut and delivered hay and wood on government contracts. Although problems of supply were similar throughout the region, variations encountered in Arizona and New Mexico merit separate treatment for each territory.

The army's work animals and cavalry horses consumed prodigious amounts of forage. Regulations stipulated that horses receive a daily ration of fourteen pounds of hay and twelve pounds of grain; mules were fed the same amount of hay and nine pounds of grain.[1] In 1870 the chief quartermaster for the District of New Mexico awarded contracts that called for delivery of 4,950 tons of hay. More than a third of this amount—1,800 tons—went to Fort Union, the district's major supply depot, while amounts delivered elsewhere ranged from 85 tons at Fort McRae to 800 tons at Fort Wingate.[2] Besides feed for its own horses and mules, each post required forage for transient animals. In 1872, for example, fifty-five animals were attached permanently to district headquarters in Santa Fe, but during the year the post provided more than 12,500 rations to other animals.[3]

During the 1860s and 1870s southwestern grasslands were largely unfenced and unencumbered by claims of private ownership, yet in most instances only residents with substantial finances harvested natural grasses to fill government contracts. New Mexico's chief quartermaster Herbert M. Enos believed that it required more capital to furnish hay than it did to furnish corn.[4] Contractors typically employed Hispanos to cut and then haul the hay to army posts, and some hired superintendents to oversee their work force. In 1865 John Lemon of Mesilla sent forty men to cut grass near Fort Bowie, where he had agreed to put in 200 tons. William Chamberlin of Silver City employed more than thirty cutters in 1880 while filling a contract at Ojo Caliente. Contractors like Lemon and Chamberlin sometimes had difficulty hiring laborers since hay ripened about the same time that villagers were busy harvesting their own crops.[5]

How much contractors normally paid laborers for cutting hay is not known. When the government purchased hay by the burro load in open market, as it did in Santa Fe in September 1871, hay cutters probably received a fair return for their labor, in this case, 1 cent per pound for bottom hay and 1½ cents per pound for grama hay.[6] Lawrence G. Murphy, who signed contracts in October 1872 to supply Fort Stanton with 150 tons of grama hay at $24.90 per ton and 100 tons of bottom hay at $18 per ton, paid local residents more than 1 cent per pound for grama and slightly less than 1 cent per pound for bottom, still fairly reasonable prices.[7] But when contractors agreed to furnish hay for as little as $8.75 per ton—the price paid to John M. Shaw for hay delivered at Fort Craig in 1872—hay cutters must have received very small wages.[8]

A contractor's greatest expense was for transportation as the hay frequently had to be hauled from distant fields. In 1879 Jesús M. Perea used sixty wagons to haul hay from the vicinity of St. Johns, Arizona, to Fort Wingate, New Mexico, a distance of more than 100 miles.[9] Sometimes suppliers specified the cost of freight in their proposals. In 1868 William B. Stapp and Charles Hopkins, who ranched near Fort Bascom, offered to deliver 150 tons to the post for 4 cents per pound. Since the hay field was located 130 miles from the post above the Red River Crossing on the Cimarron Road, the ranchers calculated 2 cents per pound for freight, ½ cent for baling, and 1½ cents for cutting. Fort Selden's quartermaster reported in 1867 that transportation expenses would more than double the cost of hay. If purchased at the hay grounds eighteen miles east of the post near San Agustín Springs it would cost $15 per ton, but if delivered at the post it would cost $35 per ton.[10]

Bidding on hay contracts was similar to gambling, since success in harvesting the natural grasses depended on ideal weather conditions. Too little summer rainfall stunted growth; late rainfalls ruined hay fields; early frosts, it was believed, destroyed nutritional content; and early snowfalls mired ox-drawn wagons in snowdrifts and mud. The quartermaster's department usually opened bids and awarded contracts in July and August. When summer rains were late, bidders were unable to calculate the probability of securing a good hay crop before submitting their bids. Fort Stanton's quartermaster consequently recommended that contracts not be let until the rainy season was well advanced; otherwise bidders would calculate on a bad season, he implied, and substantially raise their bids to cover this risk. But when the quartermaster's department delayed the letting of contracts until late September or early October, contractors had little time for harvesting hay before the first frosts occurred, and hay in some areas was already past its prime.[11]

Contracts stipulated that suppliers furnish the best hay "that the season produces, free from dirt, sticks, roots, etc. . . . cut when the crop is in the best condition for making hay, and all before heavy frosts."[12] Quartermasters preferred hay that was cut from grama grass, although the army also awarded contracts for "bottom" and "upland prairie" grasses. Contractors delivered several varieties of grama, including grama Navajo, blue grama, and white grama, all of which commanded higher prices than either upland or bottom hay. Prices established at Fort Stanton in 1869 reflect the comparative value of the grasses: $30 per ton for grama, $25 per ton for upland, and $22 per ton for bottom hay.[13] At some posts the difference in contract price between grama and bottom was as much as $10 or even $15 per ton.

Most observers agreed that the best time for cutting grama grass was in September and October. Officers at Fort Stanton reported that bottom grass cut in June and July made a fair quality hay, but it became coarse if allowed to grow until August. Rufus C. Vose, a former officer in the California volunteers who was cutting hay near Fort Sumner in 1868, claimed that bottom grass there should be cut between 15 July and 31 August and that grama grass should be cut in September, October, or November. Contractors who delayed cutting beyond October, however, risked having an early frost "destroy" the grass.[14]

Quartermasters sometimes accepted inferior grasses (such as vega and sacaton) when grama and upland became scarce. In 1865, after floodwaters damaged the grain crop south of the Jornada, the army experimented with feeding government animals another type of forage—mesquite and tornillo beans—to reserve corn and grain for residents. In July the government awarded a contract to Albert H. French and George B. Lyles for delivery of 700,000 pounds of mesquite beans at Franklin, Texas, for 4½ cents per pound. During the following month, the quartermaster's department authorized purchasing mesquite and tornillo beans in open market, paying citizens 4½ cents per pound at Las Cruces, 4 cents at Forts Selden and Craig, and 5½ cents at Fort Cummings. Government officials believed that purchasing beans in open market would give destitute citizens "an opportunity to help themselves."[15] Many families in the Mesilla Valley and at San Elizario, Socorro, and Ysleta, Texas, faced starvation because of the flood, and they subsequently became dependent upon the army for support. During the most critical days, the army distributed among the people large amounts of beef, flour, wheat meal, and other subsistence stores as well as corn and beans for seed.[16]

Residents in these lower Rio Grande villages gratefully received government rations, but animals balked at eating tornillo and mesquite beans. General James H. Carleton, who believed that horses and mules had to acquire a taste for beans, sanguinely wrote his chief quartermaster: "By chopping up the beans fine with hay, or feeding finely chopped beans with bran, or a little grain, the horses and mules soon get to be fond of the beans, and will afterwards eat them without all this trouble."[17] Nonetheless, a large amount of mesquite beans remained on hand at Fort Bliss in December 1866, and New Mexico's chief quartermaster reported them as a total loss to the government. Finally, in April 1867, Dr. George H. Oliver, surgeon at Fort Bliss, recommended that they be destroyed since he considered the decomposing beans "detrimental to the health of the garrison."[18]

Much of the hay fed to government animals was cut with sickles and scythes, a smaller amount with hoes. The sickle proved the best implement for cutting hay near Fort Stockton, Texas, where grass was scattered among chaparral.[19] The army frequently refused to accept hoe-cut hay since during the cutting large amounts of dirt and roots became mixed with the grass. When grass was sparse and difficult to obtain, however, contractors sometimes were forced to cut with hoes.

John A. Miller turned in hoe-cut hay at Fort Cummings in 1880 even though the post's commanding officer condemned it as inferior and almost worthless. Supply officers issued Miller's hay, nonetheless, rather than allow animals to stand on the picket line without forage.[20]

In some areas of New Mexico contractors employed mowing machines to cut natural grasses. Rufus C. Vose in 1867 or 1868 employed one McCormick and three Buckeye mowing machines, as well as two revolving horse rakes, for harvesting hay near Fort Sumner.[21] Richard Hudson in 1873 employed two mowers about twelve miles from Silver City where he was cutting between eight and ten tons of hay a day. A newspaper correspondent who visited Hudson's hay camp reported: "The mowers seem to have gone over nearly all the surrounding country, as it is dotted pretty thickly with the piles of hay, putting one in mind of a meadow in the States, only it is minus the fences."[22] The Buckeye was among the most popular mowing machines in the country, and it may have been the model used by Robert H. Stapleton of Socorro, who was harvesting hay near Fort Tularosa in 1872, and by James J. Dolan of Lincoln, who was cutting with two machines on the Rio Peñasco in 1874.[23]

The quartermaster's department erected platform scales at most western posts for weighing hay delivered in wagons. After the hay was weighed, contractors were required to stack it in such quantities and places as designated by post quartermasters. General Carleton in 1865 issued orders that all public hay should be put in ricks or stacks of not more than 100 tons each. When hay was moist, the army furnished contractors with salt (thirty pounds per ton), which was mixed between layers of hay to preserve its "sweetness" and to keep it from molding. Contractors agreed to complete delivery of hay by a certain date, usually by 15 November or 31 December.[24]

High winds, lightning, and man-made fires created special problems for contractors and quartermasters. Fierce winds sweeping down on Fort Union prevented contractors from stacking freshly cut hay. At Fort Garland canvas used to cover haystacks was "daily torn or blown off" during windstorms, allowing the hay to whirl away "in perfect clouds." Even though Garland's quartermaster attempted to preserve the stacks by placing "heavy logs of wood on each side of the comb of the stack, with rope lashings crossing its crest," large quantities were continually lost.[25]

Sometimes the haystacks were totally or partially consumed by fire. Fort Bascom's haystack in 1868 was totally destroyed after being struck by lightning. In 1875 the haystack at Fort Bayard, containing about 130 tons, caught fire from unknown causes. By applying wet blankets, soldiers checked the fire long enough to pitch off and save about eight tons. When fire broke out in a 250-ton stack at Fort Davis, Texas, in 1881, the entire command turned out to fight the blaze. Working from late afternoon through the following morning, soldiers succeeded in saving about half the hay by cutting down through the middle of the stack and "fighting the fire back with snow and water." They also covered adjacent stacks with wet canvas to prevent the fire from spreading.[26]

The commanding officer at Fort Davis later claimed that the fire was the work of an incendiary, a possibility since haystacks were not routinely guarded by soldiers. A careless smoker also might have started the blaze, although civilians and soldiers were forbidden to smoke around the stacks. When schoolboys began using the quartermaster's hay corral in Santa Fe as a playing ground, the chief quartermaster asked their teacher to intervene, since he knew that the "boys sometimes smoke and . . . are accustomed to carry matches in their pocket."[27] At least one officer believed that haystacks had to be protected from disgruntled suppliers. The quartermaster at Fort Craig in 1867 requested that two sentinels patrol the hay corral at night after he refused to buy any more hay from local citizens; he feared they would set the stacks afire.[28]

Quartermasters also had to be alert for dishonest suppliers. A case of attempted fraud was discovered in 1870 at Fort Selden, where John Lemon and Pedro Serna, respected Doña Ana County businessmen, held contracts to deliver a total of 750 tons of hay. Both contractors had hired Hispanos to cut and haul hay to the post, with the laborers to be paid according to the number of tons they delivered. Army officials later discovered that the workers had mixed stones with the hay before the wagons had been weighed, and when soldiers subsequently overhauled an unfinished haystack, they uncovered about 3,100 pounds of rocks. Soldiers then ran iron rods through finished stacks, but no additional stones were found. Since the army held the contractors legally responsible, Lemon and Serna had to make up the deficiency in hay.[29]

A similar case occurred in 1872 at Fort McRae, where army officials accused three Hispanos who were delivering hay on John M. Shaw's contract of putting between 200 and 300 pounds of "river sand and other foreign matter" in each wagonload of hay. These men were held part of one day in the post guardhouse and then released. The commanding officer at Fort McRae, who forwarded details of the case to his superiors, observed: "The attempt [at fraud] was most bold and glaring and in my belief the Government will not be safe from such fraud until it resorts [to] severe punishment where detected."[30]

The government would have lessened the risk of fraud had it received baled rather than stacked hay. Baled hay had other advantages as well: it was easier to handle and easier to care for, and it involved less waste and kept in better condition than hay stacked in the usual way.[31] But baling the hay added to its cost, and the army, with few exceptions, contracted only for stacked hay until 1880, when contractors began importing baled hay from Kansas.

The army, nonetheless, used machines to bale some of the hay it received in stacks, so that the troops could carry hay more easily on expeditions. During November 1874 civilian employees at Fort Union turned out thirty bales a day by using a hay press on loan from Trinidad Romero of Las Vegas. In the spring the post quartermaster requested authority to hire additional citizens in order to speed the work, for Romero needed the press for baling wool.[32] To avoid using borrowed equipment, Chief Quartermaster John H. Belcher in 1876 requisitioned three Dodge Excelsior Hay Presses and three loopers (devices for tying hay) for the garrisons at Santa Fe, Fort Union, and Fort Selden. Belcher speculated that W. H. Banks and Company, a Chicago firm that manufactured the machines, would sell the presses for $500 each and the loopers for $20 each. Since these machines would enable commands to carry their own forage, he believed that the army would soon abolish civilian forage agencies.[33]

The presses eventually arrived, but few extant records mention them. The Santa Fe *Weekly New Mexican* reported in October 1877 that the Dodge Press located in Santa Fe's military corral produced a bale that was "very small, cylindrical in shape and easily handled." Four animals—either horses or mules—were required to run the machine.[34] A report in 1878 showed that the unskilled laborers who operated Fort Selden's hay press took half an hour to make one bale and another half hour to wire it. Fort Selden's quartermaster was still having difficulty in 1882, although the machine now in use was a P. K. Dederick's Perpetual Baling Press. After laboring two days, workers had completed only one bale. The quartermaster feared that animals would not eat two thirds of the hay being baled as the machine, which packed hay by repeated percussion blows "something like a horizontal pile driver," crushed it into small pieces and coated it with dust.[35] Quartermasters usually recommended that civilians be hired to operate the baling machines because, as one officer stated in 1880, enlisted men were "inexperienced and notoriously awkward with machinery." Nonetheless, soldiers continued to assist in working hay presses, and Fort Union's quartermaster reported in 1883 that they would turn out between 80 and 100 bales a day if conditions were favorable, adding that all work was suspended during rain and high winds.[36]

Some of the best hay grounds in New Mexico were located near Fort Union, Fort Craig, and Santa Fe. Colonel George W. Getty, who commanded the District of New Mexico in 1870, reported that grazing at Fort Union was good throughout the year and that hay delivered there cost between $12 and $15 per ton, which was about 50 percent cheaper, he said, than the average price paid for hay at other posts. He also reported that hay at Fort Craig cost between $14 and $23 per ton.[37] But in fact the army almost always paid less for hay delivered on contract at Fort Craig than at Fort Union. During a ten-year span (1868–77), the average contract price at Fort Craig was $11.43 per ton and at Fort Union, $17.47 per ton; only in 1876 was hay cheaper at Fort Union. Chief Quartermaster John C. McFerran in 1865 had referred to the Fort Craig district as "one of the best and safest for a supply of hay in New Mexico," and these prices give support to his observation. Most hay delivered to the government in Santa Fe was cut near Galisteo, an Hispanic settlement located about twenty miles south of the capital. The contract price at Santa Fe averaged about $25.50 per ton between 1868 and 1877.[38]

Quartermasters preferred receiving hay by contract rather than in open market, since by contract they were more certain of receiving a steady supply. But open market purchases spread government patronage throughout the community, allowing the army to function as a welfare agency. In the eyes of many westerners, the army was the one government agency that could help the destitute and the unemployed survive hard times. Fort Sumner's quartermaster, for example, recommended open market purchases in July 1866 so that soldiers soon to be discharged would find employment cutting hay.[39] The following summer, when floodwaters from the Rio Grande destroyed houses in the Rio Abajo villages of Los Lentes, Pueblitos, Bosque de Belen, and Sabinal, U.S. Assistant Assessor Henry Hilgert of Los Lunas sent this message to Chief Quartermaster Enos:

> A great many persons are looking for relief from you, and hope that you will tender it, by issuing an order that no hay contracts shall be let in the Rio Abajo, and that the posts along the river shall buy hay in open market, in order to better distribute the benefits of such purchases among the destitute, and at the same time to get better and cheaper hay for Government, than can be got by giving out contracts.[40]

The flood damaged property in the Mesilla Valley as well, and Enos directed the quartermaster at Fort Selden to purchase small amounts of hay from several parties to distribute government money as widely as possible. He recognized that for many sufferers the "sale of a few tons of hay to the government would be quite a relief."[41] A severe drought in 1873 stunted crops so badly near Santa Fe that by spring of the following year many of the city's poorer residents were facing starvation. Business leaders petitioned New Mexico's commanding officer for relief on their behalf. What steps the government took to alleviate suffering is not known, though later in the summer the quartermaster's department was purchasing hay in Santa Fe by the cart and burro load.[42]

Despite the superior quality of nearby hay fields, contractors at Santa Fe and Fort Union sometimes had difficulty meeting their obligations. Bad weather, poor roads, and disgruntled competitors were the chief impediments. Neither of the contractors at Fort Union in 1870—Hugo Wedeles of Mora and Alexander Grzelachowski of Las Vegas—completed his contract on time. Grzelachowski's problems are not mentioned in the records, but Wedeles explained in a letter to the chief quartermaster that heavy September rains followed by snowfalls in October had caused delays.[43] Samuel Kayser of Las Vegas, who had subcontracted to deliver 100 tons of hay at Fort Union in late 1874, was able to deliver only 22 tons because of bad roads. During the fall of 1875 Trinidad Romero delivered 1,000 tons of hay at Fort Union, but was unable to deliver the remaining 125 tons because rains had "spoiled a great deal of hay which would otherwise have gone in on the contract."[44] Lack of rainfall during the summers of 1873 and 1874 stunted natural grasses near Santa Fe, forcing contractors there in 1873 to haul hay a distance of 120 miles. J. E. McEachran in October of 1874 simply abandoned his Santa Fe contract, explaining that the drought had destroyed all the local grama grass.[45]

Residents who banded together to corner the hay supply created additional problems. Simon Filger, who in 1871 was awarded a contract to supply Santa Fe with 250 tons of hay, claimed that not less than 150 people living at Galisteo and Agua Fria had agreed in writing not to sell him any hay. He also alleged that members of the combination had stated openly that they would go wherever "I go, for the purpose of cutting hay, and cut it all away from me." Under these circumstances, Filger refused to accept the contract. Thereafter the government purchased hay in open market, holding Filger and his securities responsible for the difference between the contract price and the open market price, which amounted to about $1,250. The chief quartermaster, perhaps suspecting that monied interests were behind the combination, issued orders to purchase hay from "original parties . . . the people who have labored and worked to cut the grass and haul it to this market," rather than from merchants and speculators.[46] Colonel J. Irvin Gregg, commanding the District of New Mexico in August 1874, reported a similar combination at work in the Fort Union district, where parties owning or claiming to own all hay grounds within fifty miles of the post had agreed not to sell hay to the lowest bidder unless the contract was awarded to a member of the combination. The army briefly contemplated removing the cavalry from Fort Union and quartering men and horses at different posts to defeat the ring's attempt to control the price. Samuel Kayser, however, came to the army's assistance by supplying the post with 500 tons of hay at the moderate price of $18.50 per ton.[47]

Elsewhere in New Mexico, contractors struggled to complete contracts under all sorts of adverse conditions. Indians were troublesome and sometimes deadly. Apaches attacked Charles Hopkins's hay camp near Fort Stanton in December 1863, running off some of his work animals. In October 1863 and in January 1864 Apaches raided Martin Amador's hay camp located at the foot of the Organ Mountains about twenty-five miles from Las Cruces. In the first attack Apaches ran off

Amador's livestock valued at $1,600, and in the second they killed three men and destroyed property worth $400. In August 1866 about seventy-five Indians captured a hay team of eight oxen near Fort Selden and killed the Hispanic teamster, though pursuing troops forced the Indians to abandon the cattle. More than a year later, a party of Apaches, whose number was variously estimated at between twenty-five and seventy-five, attacked the quartermaster's herd at Fort Selden, and in the ensuing melee Aguapito Olguín of Doña Ana, who was cutting hay for the post, was shot and killed by a soldier who had mistaken him for an Indian.[48]

During the Victorio War of 1879–80 contractors at Ojo Caliente and Fort Stanton reported they were unable to hire men to cut hay because of the increased danger. William Chamberlin reported in April 1880 that Indians had killed three of his laborers who were cutting hay for Ojo Caliente, causing nearly thirty more men to leave his camp for safer quarters. Alphonse Bourquet, also delivering hay at Ojo Caliente, reported that Indians had driven his hay cutters from the fields on three separate occasions and subsequently they refused to cut any more hay. Nor could Bourquet find freighters who were willing to haul the hay that had already been cut. Patrick Coghlan, a contractor at Fort Stanton, had to haul his hay from El Paso, a distance of 150 miles, because men were unwilling to cut hay near the post. Even then he lost twelve tons of hay and seven wagons, burned by Indians while his hay wagons were passing through the Mescalero reservation.[49]

Lack of rainfall, however, rather than Indian hostilities, accounted for most delays in filling contracts. The quartermaster at Fort Bayard in July 1868 reported that prospects for obtaining hay there were discouraging: "The grass all around is parched and dried up from drought, and fires have been raging for over a month here, burning up the dry grass, trees and vegetation to a great distance."[50] A drought in southern New Mexico stunted the growth of natural grasses to such an extent in 1871 that contractors at Forts Bayard, Stanton, Craig, Selden, and Wingate all experienced difficulty in meeting their obligations. Two contractors, Mariano Barela at Fort Bayard and Peter Ott at Fort Stanton, eventually abandoned their contracts. Willi Spiegelberg, contractor at Fort Wingate, put in only 650 of the 1,000 tons for which he had contracted, some of the hay having been cut fifty miles or more from the post.[51]

Drought again struck sections of New Mexico in 1874 and 1879, making it impossible for some men to fill their contracts.[52] Others who turned in the required amount lost money on the transaction. Jesús M. Perea of Bernalillo, who held a contract in 1879 for delivery of 280 tons of hay at Fort Wingate, said that he was unable to locate any hay within 300 miles of the post: "little or no rain has fallen during this past season . . . every place is dry and no grass is to be found to even deliver a few tons." Perea finally located hay along the Little Colorado in Arizona and completed his contract at a financial loss.[53] Because of the failure of New Mexico's hay crop that year, Chief Quartermaster James J. Dana requested information on the cost of importing hay from Kansas.[54] In later years New Mexicans and Kansans would compete for the privilege of delivering hay to military posts in New Mexico, and army officials would debate the merits of local and Kansas hay.

The army worked hard to assure a steady supply of hay, on occasion establishing pickets at cutting grounds and furnishing arms and ammunition to contractors. The army even furloughed ten enlisted men and issued them scythes to help contractors cut hay on the Rio Mimbres in 1865.[55] But the quartermaster's department continually experienced difficulty getting its hay supply, a point graphically illustrated in the records of Fort Stanton. Army officers frequently praised the Fort Stanton area for its rich soil and nutritious grasses. Sophie Poe, who as a young bride in 1884 lived on a ranch fifteen miles southwest of the post, later recalled that "grama grass grew so thick and high that it almost hid the fat cattle grazing upon it."[56] Despite these accolades, the post quartermaster rarely was assured a steady supply of hay.

During a fifteen-year span, 1865–79, the army awarded contracts to fifteen different individuals or partnerships for supplying hay at Fort Stanton. Lawrence G. Murphy and his partner Emil Fritz agreed to deliver more hay than any other contractor: 450 tons in 1870, 300 tons in 1871, 250 tons in 1872, and 100 tons in 1878. They furnished additional hay through open market purchases, and their close friend, James J. Dolan, who entered the firm of L. G. Murphy and Company in 1874 (the year that Fritz died), contracted to deliver 100 tons in 1873 and 250 tons in 1874. Like Murphy, Fritz, and Dolan, most other contractors who supplied the post with hay were merchants or businessmen, including Willi Spiegelberg and Benjamin Schuster of Santa Fe, Tularosa storekeeper Patrick Coghlan, hotelkeeper Peter Ott of

Doña Ana County, and Lincoln County residents José Montaño, Paul Dowlin, Elisha Dow, and Frank Lesnet. The average contract price for hay at Fort Stanton during these years was about $27 per ton, with bottom hay usually costing between $18 and $25 per ton and grama hay between $23 and $35 per ton.[57]

In at least ten of the fifteen years under consideration the quartermaster at Fort Stanton faced shortages or other difficulties pertaining to the hay supply. Scarcity due to lack of rainfall has already been mentioned. Other problems included grasshopper and prairie dog infestations, burning of the hay supply, combinations working to advance prices, and new settlements encroaching on hay fields.[58] Problems such as these continued to plague the post quartermaster in later years. And on at least one occasion, in 1881, violence threatened to interrupt the hay supply. Local residents reported that "certain parties" threatened to burn the hay of Hispanic hay cutters unless the Hispanos sold hay to John DeLany, the post trader who was bondsman for one of the bidders for the hay contract.[59]

The best-known Lincoln County resident who failed to complete a government hay contract was Pat Garrett, the sheriff who killed Billy the Kid. In early 1884, two years after he had relinquished his sheriff's badge, Garrett turned his attention to ranching. In April he submitted a bid for twenty tons of hay, which the government accepted, but Garrett later failed to execute a contract, telling the post quartermaster that "his hay was spoiled." This occurred about the time that cattlemen in the Texas Panhandle asked Garrett to organize a group of rangers to stop cattle rustling, an occupation Garrett would find infinitely more satisfying than cutting hay. But his inattention to hay contracting raised the ire of army officers. In late July Fort Stanton's quartermaster received this message from department headquarters: "Bids from Garrett offering to furnish Government supplies should not, in the future, be entertained by you."[60]

By mid-1885 most posts in New Mexico were receiving baled hay from Kansas. Only Forts Union and Stanton were supplied with locally grown native hay. And even though farmers in New Mexico were raising significant amounts of alfalfa, the army awarded no contracts to local suppliers for cultivated hay. In contrast contractors were supplying cultivated hay to about one half the posts in Arizona by this date. During most years of this study, however, the military in Arizona relied upon nature's bountiful grasslands for its hay supply.[61]

Early travelers to Arizona were impressed by its abundance of natural grasses. Some foresaw the day when large herds of cattle would fatten on its ranges. Describing conditions he observed in the early 1860s, J. Ross Browne wrote:

> The valley of the Santa Cruz is one of the richest and most beautiful grazing and agricultural regions I have ever seen . . . grass is abundant and luxuriant. We traveled, league after league, through waving fields of grass, from two to four feet high, and this at a season when cattle were dying of starvation all over the middle and southern parts of California.[62]

Newspaper writers often praised Arizona's grasslands in articles similar to the one appearing in the Prescott *Arizona Miner* on 8 February 1868, which stated in part:

> There are several species of excellent grass covering the ground [in Yavapai County]—two or three varieties of grama grass, which are good feed for cattle the year round; this grass grows luxuriantly over most of the valleys, plains, and mesas, or high table lands, through the Territory; while the mountains are (most of them) covered to their tops with bunch grass, which keeps green through the year.[63]

Arizona contractors relied upon these grasslands to fill government contracts, sometimes hiring as many as eighty laborers to harvest the hay crop. Because of transportation costs, sparse population, and the danger of Indian attacks, hay was expensive, especially during the 1860s. In 1866 contractors received $48.50 per ton for hay delivered at Camp Lincoln, $58.50 per ton at Prescott, $60 per ton at Camp McDowell, and $84 per ton at Fort Yuma.[64] Profits also were high. Army Inspector Roger Jones estimated that in 1867 the contractor for Camp Wallen had cleared between $25 and $27 per ton. Jones recommended that soldiers cut hay to save money, but enlisted men rarely were saddled with this task.[65]

The experience of Leslie B. Wooster and John Spring, who were putting in hay at Camp Wallen in 1868, shows some of the pitfalls that made government contracting risky. Certainly they received no profits such as those Jones reported. Camp Wallen was located on the north bank of Babocomari Creek about sixty-five miles southeast of Tucson. Wooster and Spring established their hay camp two miles below the post in Babocomari Valley where grass "was quite abundant." They had purchased mules, wagons, scythes, and camp equipment, and had employed several laborers to help put in the hay, though Spring later recalled that men "who could handle a scythe with any skill or load up a wagon with hay properly" were scarce. The hay cutters, housed in tents close to Babocomari Creek, soon complained of chills, fever, and ague, and some became too ill to work. Since the quartermaster demanded hay, Wooster rode twenty miles to borrow hay cutters from Samuel Wise, who was putting in hay at Camp Crittenden near the source of the Sonoita River east of Tubac. Wooster and Spring had delivered only 50 tons of hay when the government canceled their contract, anticipating the abandonment of Camp Wallen. Contracts carried a clause at this time allowing the government to terminate the agreement at any time, and contractors such as Wooster and Spring suffered heavy losses when this happened.[66]

Contractors working in exposed areas kept constant watch for Indian raiders, but they could not always avert disaster. Indian attacks on contractors' camps or their men and animals were reported at Camp Bowie in 1867, at Camps Lincoln and Crittenden in 1868, and at Camp Grant in 1869 and 1870.[67] The Tucson *Weekly Arizonian* described in detail the attack on Thomas Venable's two hay wagons in 1869 as they passed through a canyon a few miles from Camp Grant. About fifty Apaches fired on the drivers, killing one instantly. The other driver, together with two soldiers detailed as an escort, held the Indians at bay for half an hour, when the second driver was shot in the chest and died instantly. The soldiers, one critically wounded, made their way to the post on foot. Even though a company of cavalry immediately started in pursuit, it was unable to recover the twelve mules captured by the Indians. The newspaper correspondent lamented these deaths and the financial loss of Venable, who after thirteen years spent in accumulating property was "divested in a moment of almost everything he possessed."[68] Even though Apaches posed a constant threat, the army could not place guards in every hay field. A few months after the attack on Venable's wagons, Pinckney R. Tully and Estévan Ochoa were denied an escort for their hay cutters working near Camp Bowie. With only a small number of men at the post, the commanding officer advised them to hire a larger work force that could protect itself.[69]

Some Arizona residents tried to take advantage of the government by turning in substandard hay. In 1873, for example, three Hispanos employed by Mart Maloney, the contractor at Camp McDowell, testified that he had showed them how to place dry, cured hay on the outside and top when stacking hay and to put green, uncured hay in the center. Since the post commander had suspected that Maloney's hay was not well cured, he had it tested by removing a center section from an incomplete stack that had been allowed to stand for nine days. A board of survey then decided that the hay would spoil if it remained stacked for any length of time. Because the garrison needed forage, the post quartermaster was authorized to receive Maloney's hay, deducting two sevenths from its weight as compensation for its greenness.[70]

Indians supplied some of the hay required at frontier garrisons. Arizona pioneer Thomas T. Hunter recalled that Papago Indians in the late 1860s were cutting hay with sickles for use at Camp Crittenden. Nathan Bibo in 1871 filled his contract at Camp Apache with hay that Indian women had cut with butcher knives. The women "came in constant procession," he remembered, "carrying from sixty to one hundred pounds of grass on their backs bundled up with strings of soap weed strapped around their forehead and besides the heavy burden carried their papooses on top of the hay." The post quartermaster at Camp Mojave indicated that same year that suppliers there were purchasing hay from Mojave Indians.[71]

And preceding the infamous Camp Grant Massacre in the spring of 1871, Apaches had been turning in at the post about four tons of hay daily on the contractor's account. General John M. Schofield, commanding the Division of the Pacific, later suggested that greed for contracts helped explain the massacre, in which as many as 150 Indians may have been killed. The evidence, however, does not support his statement. Tucson residents who led the attack had blamed Apaches at Camp Grant for recent depredations and had been motivated by the desire for revenge.[72]

Moreover, it was only after Crook took command of the department in June 1871 that contractors were eliminated as middlemen between Indian suppliers and the government. To enable Indians living on reservations to become self-sufficient, Crook wanted to provide a ready cash market for all they had to sell. Government supply advertisements soon carried this proviso: "The right is reserved by the United

States to purchase during the fiscal year such quantities of Forage and Fuel, as may be offered by, and accepted from Indians at Camps Apache, Grant, Mojave and Verde, and at the San Carlos and Colorado Indian Reservations: awards of contracts will be made for the remaining quantities required."[73] The amount of forage that Indians subsequently furnished is unknown, but in October 1875 Colonel August V. Kautz, Crook's successor, reported that White Mountain Apaches would sell to the quartermaster's department that year more than 100,000 pounds of corn and possibly 300,000 pounds of hay.[74]

When Crook resumed command of the department in 1882, he renewed efforts to aid reservation Indians. Indians at Fort Apache, for example, were paid one cent a pound for grass they brought in and five dollars a cord for wood. Will C. Barnes, a military telegrapher at the post, later recalled that Apaches delivered in one year more than 1,500 tons of hay, cut with sickles and butcher knives, as well as 600 cords of wood. They brought in some of the hay on horses and burros; much of it they carried on their own backs.[75]

One of the best hay-producing regions in Arizona was Williamson Valley, about twenty miles west of Prescott. During the 1860s and 1870s much of the hay used at Fort Whipple and Prescott came from this area. As late as September 1869, no permanent settlements could be found in the twenty-mile-long valley, though hay cutters had established three or four temporary ranches. Lieutenant Colonel Frank Wheaton, commanding the Subdistrict of Northern Arizona with headquarters at Fort Whipple, described the valley in 1870 as "the great hay region of this section," and he estimated that 2,000 tons could be cut there annually.[76]

Despite the superior grasslands in Williamson Valley, the cost of hay delivered at Fort Whipple was high. During a ten-year period ending in 1877, prices ranged from $22.50 per ton in 1874 to $60 per ton in 1877.[77] These prices may reflect a decline in natural grasses. Records show that the natural hay crop near Fort Whipple failed in 1871 and in 1877. During these same years the quality of grazing near Fort Whipple also declined. A report issued in 1870 indicated that grazing was excellent in summer and poor in winter. A decade later grazing was reported as poor the year around, except in a rainy year when it was "fair."[78]

Twelve individuals or firms received contracts to deliver hay at Fort

Whipple during the ten-year span mentioned above. Only three held contracts in more than one year: Herbert Bowers, post trader, agreed to deliver 565 tons in 1868 and 725 tons in 1869; Rufus E. Farrington, a farmer, agreed to deliver 35 tons in 1868 and 50 tons in 1869; and Eli Putenny, a farmer, would deliver "hay as required" in 1872 and 150,000 pounds in 1877. Jeremiah W. Sullivan, a Prescott resident, later recalled details of the contract Bowers held in 1869. Bowers and his partner, Curtis C. Bean, had contracted to deliver 200 tons at $33 per ton and 525 tons at $35 per ton. Subsequently they made a subcontract with a Mr. Zimmerman and a Mr. Johnson to cut hay in Williamson Valley at $11 per ton. Sullivan took a one third interest in the subcontract, agreeing to run two mowing machines, working one team from daylight until 10:00 A.M., the other until 2:00 P.M., and then working until dark with the first team. Heavily armed to guard against Indian attack, Sullivan cut about eighteen tons a day.[79]

The government paid about the same price for hay at Camp Mojave as it did at Fort Whipple. During the 1870s prices at Mojave ranged from a high of about $67 per ton in 1870 to a low of $24.50 per ton in 1873. Paul Breon, the post trader, received more hay contracts than any one else, delivering hay in 1872, 1874, 1878, and 1879.[80] But suppliers encountered special problems at Camp Mojave, where natural hay crops depended upon the overflow of the Colorado River. The country surrounding the post was a desert, except for the river bottom where Mojave Indians in good years raised wheat, corn, beans, and melons. In years of no overflow, crops failed. When this occurred in 1869, the army issued rations to the Mojaves to keep them from starving. In 1871, a year of no overflow, the natural hay crop failed, and men who had submitted low bids before it was known whether an overflow would occur refused to sign the contracts awarded them. The difficulty was well understood by Captain Richard H. Pond, commanding the post in 1872, who reported that in years of overflow hay could be purchased in the Colorado River Valley for $15 and $20 per ton but in years of no overflow it might cost as much as $69 per ton. He recommended that the quartermaster receive two types of bids, "one predicated on an overflow occurring and the other on a contrary result."[81]

Hay was much cheaper at Tucson and Camp Lowell, where during the 1870s prices ranged from $8.74 per ton in 1874 to $15.68 per ton in 1877 and 1878. Natural grasses grew abundantly in Santa Cruz Valley and adjacent areas. The local press claimed that between Tucson and the Rillito, where Lowell was relocated in 1873, natural hay fields yielded several tons to the square mile. Captain Gilbert C. Smith, post quartermaster, reported in 1880 that enough grama and sacaton hay was raised in the vicinity to meet all demands. Tucson merchants—like William Zeckendorf, Pinckney R. Tully, Isaac Goldberg, Hiram S. Stevens, and Samuel H. Drachman—held most of the hay contracts. But Frederick L. Austin, post trader at Camp Grant in 1871, signed more contracts than did anyone else, agreeing to deliver hay "as required" in 1871, 455,000 pounds in 1877, and 255,000 pounds in 1878.[82]

Camps Apache, Goodwin, and Bowie were the only posts in Arizona where New Mexico contractors put in hay. Solomon Bibo held the contract at Camp Apache in 1871, and Numa Reymond, a Paraje merchant, held it in 1874. John Lemon of Mesilla held contracts at Camp Goodwin in 1868 and at Camp Bowie in 1865 and 1868. David Wood of Las Cruces received the contract at Camp Bowie in 1871.[83] Both Lemon and Wood transported hay cutters and equipment to Camp Bowie from Mesilla Valley, a distance of about 160 miles. In years of adequate rainfall Camp Bowie's suppliers obtained enough grass to fill their contracts from nearby hills and valleys, but they sometimes had to cut it with hoes. Quartermasters judged hoe-cut hay inferior, and in 1870 a board of officers deducted 15 percent in weight from the hoe-cut hay that contractor Mortimer R. Platt turned in because of the large amount of roots and dirt it contained. Platt later obtained "a fair quality" hay from San Simón Valley, sixteen miles from the post, which he cut with a mowing machine and scythes.[84]

Two years after David Wood filled his contract with hoe-cut hay, Army Inspector Delos B. Sacket implicated him in a scheme that must have cost the government several hundred dollars. Wood's problems began on 23 July 1871, the day he arrived at Camp Bowie with men and teams to put in hay. Major Andrew W. Evans, commanding the post, hesitated to receive hay from Wood for two reasons: a copy of the contract had not yet arrived, and daily showers made it impossible to put in well-cured hay. Evans rejected several wagonloads of damp hay delivered on 29 July. And even though Evans considered Wood a "respectable man evidently acting in good faith," he would not receive hay until after the rains had stopped late in August.[85] The delay added to Wood's expenses.

Testimony recorded by Colonel Sacket suggests that Wood subsequently broke the law to recoup some of his losses. But since Wood never had an opportunity to face his accusers, only one side of the case can be presented. John Dobbs, an employee of Wood, testified that a Sergeant Shaffer, quartermaster sergeant in charge of weighing hay, frequently credited Wood's account with more hay than was delivered and that the two men then divided the profits. Dobbs believed that Shaffer and Wood had "defrauded the Government out of about one hundred tons of hay." Another civilian claimed that "it was a notorious fact among the enlisted men" that Shaffer and the hay contractor were swindling the government. Civilians and enlisted men alike testified that an immense amount of dirt, gravel, and roots had been mixed with the hay that Wood delivered. Sacket estimated that this debris, left on the ground after the hay had been consumed, weighed more than 150,000 pounds. Because Shaffer had been discharged by the time Sacket recorded these events, the army soon dropped the matter.[86] Later Wood would champion the cause of law and order as deputy sheriff of Doña Ana County.

Contract prices at Camp Bowie during the 1870s ranged from a high of $22 per ton in 1870 to a low of $12.20 per ton in 1875, with hay costing an average of about $16.50 per ton. These prices were only a little higher than those at new Camp Grant, established late in 1872 about twenty miles southwest of modern Safford. During the remainder of the decade, prices at Grant ranged from a high of $19.75 per ton in 1873 to a low of $10.63 per ton in 1875, the average being about $14.25 per ton.[87]

Like his counterpart at Fort Stanton, New Mexico, the quartermaster at Camp Grant rarely was assured a steady supply of hay. Contractors in 1875 and 1876, for example, failed to complete their contracts, even though the quartermaster in each year reported plenty of good grama grass growing in the area. Sidney R. DeLong, a prominent businessman and the first mayor of Tucson, was the failing contractor in 1876. A board of officers later reported that DeLong had sublet his contract to Warner Buck, the post trader, whom the officers described as an "irresponsible person." About 10 November, after putting in less than one fourth of the 2,000,000 pounds of machine or scythe-cut grama called for in the contract, DeLong and Buck abandoned their efforts. The post quartermaster then purchased hay from Miles L. Wood at contract price, though Wood had to deliver hoe-cut hay because of the lateness of the season. Officers who examined this hay said it was withered, dirty, and without nutrition. Thirty percent of its weight consisted of dirt, gravel, and roots. The officers recommended that additional hay be purchased to cover this loss, the cost to be charged to DeLong.[88] Much of the hay delivered at Camp Grant in later years, in fact, was hoe-cut hay, though quartermasters disliked receiving it. Abraham M. Franklin, who delivered hay in 1882, recalled that in good seasons native hay could be cut with mowing machines three to seven miles from the post, but during poor grass years "we cut it with heavy hoes as it was too spotty to use mowers."[89]

The only post that received large amounts of cultivated hay during the 1870s was Camp McDowell, where contractors raised wheat, sorghum, barley, and alfalfa hay on the government farm. John Smith, who rented the farm in 1869, held more contracts than any other supplier, agreeing to deliver hay in nine of the following twelve years. His first contract called for delivery of wheat, barley, or sorghum hay (as required) at $41.75 per ton. In later years prices were considerably lower. Smith's contract in 1875 called for 500,000 pounds of sorghum hay at $19.90 per ton and 100,000 pounds of alfalfa and barley hay at $21 per ton.[90]

During the 1870s farmers in the Salt River Valley increased significantly the amount of land planted in alfalfa—from 16 acres in 1870 to 2,000 acres in 1881.[91] Some of the alfalfa they raised was sold to the army, although it cost more than native hay. When Fort Whipple ran low in hay in 1878, the quartermaster purchased 50,000 pounds of alfalfa from Jefferson H. Lee, who hauled it from the Salt River Valley, a distance of about one hundred miles. A writer for the *Weekly Arizona Miner* estimated that the army paid about $70 per ton for Lee's hay, an estimate supported by the statement of an officer who claimed that alfalfa at Prescott would cost 100 percent more than native hay.[92] The army probably bought small quantities of cultivated hay whenever native grass was unavailable. However, registers indicate that no Arizona post, other than McDowell, was supplied with alfalfa by contract until 1879 when suppliers delivered alfalfa to Yuma and Whipple depots. The following year Forts Whipple, Verde, and McDowell received alfalfa by contract. At Verde the contract price for alfalfa was $10 more per ton than for grama hay, and at Whipple, $15 more. By mid-decade about half the posts in Arizona were receiving cultivated hay.[93]

Extant records for the west Texas posts suggest that contracting for hay in Texas differed little from what has been observed in Arizona and New Mexico. Most west Texas contractors employed Hispanos to cut hay in fields that were distant from the posts. In 1868 a Fort Stockton officer reported that the post's hay supply was cut twenty miles away. And in 1870 the supply for Fort Davis was being cut at a distance of thirty-three miles. Much of the hay was cut with sickles and hoes. Contractors at Fort Stockton in 1870 or 1871 tried using a mowing machine in a field forty miles from the post but with little success. Almost all contractors were Anglos. During a ten-year span, 1867–76, only one Hispano in west Texas contracted for hay—Cesario Torres, who farmed near Fort Stockton. Some contractors, like Anton F. Wulff, were residents of San Antonio and employed agents to supervise the work.[94]

Southwest residents profited from the army's forage requirements in still other ways. In the late 1860s New Mexicans with access to natural hay fields won contracts to pasture government animals during the winter. William Kroenig cared for an unspecified number of animals on his ranch near Fort Union during the winter of 1866–67, receiving for the month of March (the only month for which records are available) about $8,900. His facilities for wintering animals included four stone corrals, excellent pasture, and plenty of water. He charged the army 4¾ cents per pound for corn and 2 cents per pound for hay fed to government animals.[95] Vicente Romero, a wealthy Mora County farmer, received the contract for the following winter. He agreed to take care of from 400 to 700 government animals, receiving 1⅞ cents per pound for corn fed to the animals and 1 cent per pound for hay.[96]

Like many Southwest residents, William Kroenig also served as a government forage agent, providing forage and accommodations to army commands traveling across country. Army officials regarded forage agencies as crucial to operations in the field. Small detachments of troops frequently left posts at a moment's notice, and their success against Indian foes depended upon speed, which in turn depended upon agents furnishing a steady supply of forage for army animals.[97]

Consequently the quartermaster's department arranged with civilians to establish agencies along both well-traveled roads and roads in isolated regions. Common fixtures throughout the Southwest, forage agencies symbolized the military's symbiotic relationship with the local populace.

A quartermaster's circular issued in 1869 for the District of New Mexico listed some of the duties and responsibilities of agents. They were required to furnish forage "in such quantities as may be needed," receiving in exchange 1¾ cents per pound for hay and 4 to 6 cents per pound for corn, oats, and barley, depending on the location of their agencies. They were required to furnish without further remuneration meals to expressmen, corrals for government animals, and fuel to army teamsters and small detachments of soldiers. They also were required to furnish bed and board for officers who stopped at their stations "upon reasonable charges to the said Officers." In addition the army expected agents to promote the general interest of the United States by protecting government property, recovering stolen or stray animals, taking care of sick animals and indisposed government employees, providing soldiers and employees with means for cooking their meals, and circulating supply advertisements for the quartermaster's department. Agents received additional fees for stabling express animals, shoeing horses, repairing wagons, and issuing fuel to troops.[98]

The number and location of agencies in New Mexico varied from year to year depending upon military strategies, but most villages on well-traveled roads claimed at least one. Between Fort Craig and Albuquerque, for example, agencies were established in San Antonio, Socorro, Lemitar, Sabinal, Belen, and Los Lunas. On a less-traveled route from Santa Fe to Fort Stanton via Dow's Trail, agencies could be found at Galisteo, Antelope Spring, Gallinas Spring, and Jicarilla Station. Within the Fort Union District, which stretched from Walsenburg, Colorado, south to Anton Chico, New Mexico, twenty-seven individuals operated agencies in 1877, including Fred Walsen at Walsenburg and Richens L. ("Uncle Dick") Wootton at Raton Summit.[99] Some agencies—called stations in Arizona—became well-known stopping places for civilians and soldiers alike, including stations at Cienega, San Pedro Crossing, and Sulphur Springs on the Tucson–Fort Bowie road. In 1878 at least thirty-four individuals managed stations in the Tucson District and at least forty-two in the Prescott

District, which included Black Mesquite Station on the Prescott–McDowell road operated by noted Indian scout Al Seiber.[100]

Citizens competed fiercely and went to considerable expense to secure agencies by building new corrals and comfortable quarters for men and officers, digging wells to provide water, and stockpiling grain and hay. Because agents invested so much money, Army Inspector Nelson H. Davis in 1867 recommended they be appointed for a definite period of two or three years, with tenure assured as long as they furnished the required accommodations.[101] His recommendation was never adopted, however, and quartermasters continued to appoint and remove agents on the advice of civilians and officers acquainted with local conditions. The system was open to abuse. Citizens not infrequently campaigned to have themselves or a friend replace an incumbent.

The most elaborate campaign mounted against an incumbent may have been the one directed at Washington G. Rifenburg of Trinidad, Colorado. In 1873 Charles G. McLure, an army veteran with thirty years' service as soldier and wagonmaster, sought to replace Rifenburg as agent. McLure solicited a recommendation from his former superior, Major John G. Chandler, depot quartermaster at Fort Leavenworth, who described McLure as a capable and honest man who recently had sold his property in Illinois and moved to Trinidad to be with his aged parents. McLure then sent Lieutenant Colonel Fred Myers, New Mexico's chief quartermaster, a copy of an indictment issued against Rifenburg in 1868 for embezzling a government horse, so that officials "may see what sort of a man he is." Myers's subsequent investigation revealed that McLure was the brother-in-law of D. W. McCormick, who had sworn revenge against Rifenburg for an incident that had occurred during the winter of 1867. Rifenburg then had been running a sawmill six miles above Trinidad, the only one in the region. Because he was filling a large contract for the army, he had refused to sell McCormick lumber. McCormick probably played a role the following spring in securing the indictment. A jury later acquitted Rifenburg when testimony revealed that after he had bought the horse in question, he returned it to the seller, suspecting it belonged to the government.[102]

McLure traveled to Fort Union to press his case with Captain Gilbert C. Smith, depot quartermaster, in whose district the agency was located. But Smith refused to recommend McLure for the agency. Smith was offended both by McLure's resurrecting the old indictment and by his offer to "make things satisfactory" with Smith if McLure obtained the position. McLure also had told Smith's clerk that he would supply the clerk's family and Smith's family with fresh potatoes free of charge should the appointment be made. Rifenburg held the agency for two more years before another agent was appointed in his place late in 1875.[103]

Rifenburg had suspected that McLure wanted the agency because it was making money. But it is no longer possible to determine how much money Rifenburg or any agent cleared annually. Records show that some agents received substantial incomes during months of intense troop activity. John Ward, agent at San Antonio in Socorro County, received more than $300 for forage fed to public animals during June and July of 1872, and Ramón López, agent at San José in the Fort Union District, was paid nearly $500 for forage issued in December 1874 and January 1875. Indeed, these sums were only a small part of the money that the army pumped into the local economy through these agencies. During a seven-month period ending 30 June 1876, the chief quartermaster for the District of New Mexico requested $49,000 for purchasing fuel and forage from government agents.[104]

The rates agents received for feeding government animals depended on their location: those living far from hay and grain markets were paid more money. John S. Chisum, agent at Bosque Grande on the Pecos in 1872, was paid 7 cents per pound for grain, which was 2¾ cents more than other agents in the Fort Union District received. Chisum's agency was more than 100 miles from the nearest agricultural settlement. Warren F. Shedd, agent at San Agustín Springs about twenty miles east of Fort Selden, received one cent more per pound for his grain than agents in the Mesilla Valley. Two other agents in the Fort Selden forage district received higher prices to cover the army's use of water. John Martin, who had gone to considerable expense sinking a well midway on the Jornada del Muerto, and John D. Slocum, who hauled water twenty-two miles from the Rio Grande to his ranch on the Fort Cummings road, were each paid 8 cents per pound for grain and 2½ cents per pound for hay. In contrast, agents in the Mesilla Valley typically received 5 cents for grain and 1¾ cents for hay.[105] Rather than raising forage prices to cover water expenses in

Arizona, the quartermaster's department there paid station keepers between 10 and 25 cents per head for watering government animals.[106]

The highest forage rates paid in Arizona in 1878 were to station keepers on the Prescott–Mojave road, who received 10 cents per pound for grain and 4 or 5 cents per pound for hay. Prices were almost equally high on the Yuma—Maricopa road where station keepers were paid 10 cents for grain and 4 cents for hay.[107] Rarely were prices this high in New Mexico. The average price paid in the Fort Union District in 1881 was 3 cents for grain and 1½ cents for hay; prices were higher in the Fort Stanton district, where agents received 5 cents for grain and 2 cents for hay.[108]

The army considered agencies sufficiently important to investigate carefully any complaints of inadequate compensation. When grain and hay became scarce, the quartermaster frequently increased rates to prevent agents from resigning. James B. Adams, John Ryan, William Crane, and Luther Martin, all agents in the Fort Wingate District, successfully petitioned in March 1869 to have grain prices raised, arguing that grain had become scarce because several government contracts were being filled in their vicinity. Fort Stanton's quartermaster in January 1874 recommended that agents in his district receive higher rates, for hay was becoming more difficult to obtain "on account of the large herds of cattle and sheep that are now grazed." When crops partially failed in the Mesilla Valley in 1877, the district quartermaster increased rates paid to local agents for both grain and hay. And during the fall of 1880 the quartermaster authorized increased payment to Patrick Coghlan at Tularosa since Coghlan had been unable to find any grain at all near his agency and consequently had had to import corn from the States.[109]

In the competition to secure agencies, some men supported their applications by stating they were veterans or soon-to-be-discharged soldiers. Henry Feldwick, in applying for the agency at Ocaté, New Mexico, in August 1867, wrote: "I served in the California Volunteers from the outbreak of the Rebellion until the close, during which time my health became much impaired."[110] Even more men supported their applications by stating they were family men and thus able to provide better accommodations for officers and their families. Moses Sachs of Belen stated succinctly: "Having a family, I can keep the agency more comfortable for others."[111] In fact, the majority of government agents in New Mexico were married men, who in partnership with their wives provided services requested by the quartermaster's department.[112] Government documents testify to the value of "women's work" in maintaining a successful agency. John Sandford, upon resigning as forage agent at San Antonio in the Fort Craig District, recommended as his successor John Ward, stating that Ward, "who is a married man, can, and I know will, give much better accommodation especially to officers who should at any time wish to stop at his agency." Louis Trauer, a single man, resigned his agency in Los Lunas when he found it impossible to employ a cook.[113]

Although some wealthy married agents hired male cooks, most relied upon the culinary talents of their wives to accommodate military officers and government employees. That the army recognized the importance of the women is clearly revealed in a letter written by army surgeon William B. Lyon supporting the application of Martin L. Hickey for the agency at Mesilla. Hickey himself had been unwell for several months, but Lyon believed that Mrs. Hickey—whom he described as "a notable housewife"—would successfully manage the agency. "Mrs. Hickey sets an excellent table," he noted, "and is now possessed of every facility for accommodating guests and animals."[114]

At least a dozen New Mexican women, mostly widows, acquired agencies in their own names. Quartermasters appeared quite willing to designate widows as agents, possibly because they knew from previous business transactions that the women would provide the necessary services. Louisa Frenger, appointed agent in Socorro in 1877 after the death of her husband, received endorsement by the quartermaster at Fort Craig, who wrote: "Mrs. Frenger will no doubt keep as good a station in the future as she has done heretofore and her appointment as agent is strongly recommended."[115]

Agencies were awarded to women whenever their appointments appeared to be advantageous for the army, as in the case involving Martin and Ellen Koslowski, who owned a ranch on the Santa Fe Trail twenty-seven miles from Santa Fe. During the Civil War their house had served as Union headquarters during the Apache Canyon–Glorieta Pass engagements. After the couple separated in 1877, the quartermaster revoked Martin's appointment as forage agent, replacing him with Ellen who retained possession of the house and corrals.[116]

Many widows had some prior business experience before they became agents. At least three had run hotels in partnership with their husbands—Augustina Castillo Davis in Mesilla, Josefa Ortiz de Clark in Plaza Alcalde, and Esther Martin at Aleman on the Jornada del Muerto. Ellen Casey was also an experienced businesswoman by the time she became a forage agent following the death of her husband Robert in 1875. Together they had established a ranch on the Rio Hondo in Lincoln County, and before Robert's death Ellen handled official correspondence relating to his government agency. Ellen ran the agency in her own name for at least four years, and later, in 1884, she contracted to supply Fort Stanton with hay.[117]

For other women, running a forage agency was their first venture into the world of business. This was true for Gertrude Hobby, whose husband A. M. Hobby died soon after being named a forage agent on the Mimbres River in Grant County. In April 1881 Gertrude notified the district quartermaster that she was ready to carry on the business. Somewhat apologetically she added: "Pardon this very unofficial letter as I am ignorant in such matters, this being my first attempt at business of any kind." There was nothing apologetic or timid, however, about the way she managed the agency. She received nearly $300 for forage issued during May and June, but when business subsequently declined she wrote to military authorities claiming that so few military trains stopped at her agency that it did not afford her a living "at the price we pay for our supplies." She therefore boldly requested permission to purchase subsistence stores from the army, a privilege many sought but few received. "As Government agents are we not working for the government," she asked, "and thus would we not be entitled to buy our supplies from the Commissary?" A few months later in May 1882 she inquired about obtaining a government contract, informing the district quartermaster that she had the necessary machines and teams to put in hay at nearby Fort Bayard.[118]

Even though forage agents performed valuable services, at least one officer looked upon them with suspicion. Major Andrew W. Evans, commanding Fort Wingate in 1869, accused agents on the Fort Wingate road of getting soldiers drunk and harboring deserters. Evidence surfaced that confirmed part of his allegations. In June of that year a search party traced several deserters to the house of John Q. Adams, recently-resigned agent at the Albuquerque crossing of the Rio Puerco. Soldiers searched the premises and nearby fields but failed to find the men. Delgadito, a Navajo guide, however, tracked them to a hole in the ground, but as he approached the opening Adams begged him "not to betray those poor fellows." Delgadito turned aside and said nothing of the incident until later. Some deserters eventually were captured near Trinidad, and they had in their possession a diary, in which reference was made to Luther Martin, forage agent at Old Fort Wingate.

> We proceeded to the Old Post to get our breakfast and rest ourselves all day, knowing Mr. Martin very well indeed. He received us very well, told us he would have some breakfast for us in a short time, which he did, and put a good breakfast on the table of eggs and chickens, hot cakes, etc., having had the above we retired to a private room where we could sleep and not be disturbed; he also gave me and H a colt revolver apiece for one carbine, and me and Edwards a suit of clothes for his carbine, we laid there all day, and after having a good supper, and one dozen eggs and 3 loaves of bread, for which he only charged us 2 dollars, to take with us, we thanked him very kindly for all he had done for us, and took our leave. He accompanying us a short distance to show us the trail.

Evans believed that without the aid of government forage agents, deserters from Fort Wingate would never get away.[119]

As an economy measure, the army discontinued forage agencies in New Mexico in 1882. Commands going into the field thereafter were instructed to carry forage with them or obtain it from nearby posts. If this were not possible, officers were allowed to purchase forage from local ranchmen. The same need for economy led to orders being issued in the Department of Arizona in June 1883 for commands to avoid purchasing forage at road stations except when absolutely unavoidable. Forage agents, however, continued to serve the army in Arizona through the years of this study.[120]

The need for economy was a persistent theme in military correspondence, and enlisted men frequently bore the brunt of economy measures that the army adopted. Soldiers stationed at Fort Craig, New Mexico, for instance, wielded pick and shovel to extract coal from nearby mines in an attempt to reduce the post's fuel expenditures by using coal for heating. Grates were placed in quarters, and during the winter of 1865–66 the post was supplied with bituminous coal, which was hauled about twenty-eight miles from coal mines located south and east of Socorro. The experiment gave satisfaction in the short run. In 1866 Army Inspector Nelson H. Davis reported that the coal was of "a fine quality," and Colonel Francisco Abréu, commanding the post, claimed it cost one third less than a comparable supply of wood. The army used both civilians and soldiers to work the mines that year, but early in 1868 it awarded a contract to William V. B. Wardwell, the post trader, for delivery of 10,000 bushels of coal at 68 cents a bushel. In the long run contracting for coal proved uneconomical; the post commander in 1869 claimed that coal delivered at the post would cost twice as much as mesquite or piñon wood. Thereafter wood rather than coal was used for heating and cooking.[121]

The quartermaster's department tried to achieve further savings by using soldiers to produce charcoal for its blacksmiths. At Fort Bascom in 1867, for example, five enlisted men worked fourteen days to produce about 300 bushels.[122] Records suggest, however, that civilians supplied most of the charcoal, either by contract or in open market purchases. Often the men who contracted for charcoal were forage agents, post traders, or local farmers rather than town merchants. Between 1868 and 1882 the army let eleven contracts for delivery of charcoal at Fort Wingate. Six went to William Crane, forage agent at Bacon Springs, who agreed to deliver a total of 6,000 bushels at prices ranging from 25 to 49 cents per bushel. The largest supplier at Fort Stockton was Joseph Friedlander, the post trader, who contracted to deliver a total of 4,400 bushels between 1873 and 1876 at prices ranging from 30 to 54 cents per bushel. John Smith, trader at Camp McDowell, held the most contracts for that post. Between 1875 and 1882 he agreed to deliver a total of 5,500 bushels at between 40 and 70 cents per bushel. Other post traders who supplied charcoal in Arizona were Paul Breon at Camp Mojave, Sidney R. DeLong at Camp Bowie, John Norton at Camp Grant, and Frederick L. Austin at Fort Huachuca.[123]

Between 1866 and 1885 Anglos received more than 90 percent of the charcoal contracts awarded in the Southwest. Only ten contractors were Hispanos; eight lived in New Mexico, and two, Benjamín Velasco and Mariano Samaniego, lived in Arizona. The largest Hispanic contractor was Rafael Tafolla of Socorro County, who supplied

Forts McRae, Tularosa, and Apache with a total of 3,700 bushels between 1872 and 1875 at prices ranging from 18 to 36 cents per bushel.[124]

In its drive to economize, the War Department on 12 November 1867 issued General Order No. 97, which stipulated that "at every post where it is possible, fuel and hay shall be procured by the labor of the troops."[125] As mentioned earlier, enlisted men in the Southwest rarely cut their own hay, but many had been gathering their own wood supply at the war's end. Some post commanders complained that they had too few men to employ any as wood cutters; others argued that contracting for fuel would save wear and tear on army wagons and mules and would cost less than supplying fuel by labor of troops.[126] Their arguments eventually won out, but not before Apaches had made a daring raid on army wood cutters at Fort Stanton. On the afternoon of 3 September 1870 Privates Charles Hoeffer and John McGrath set out from the wood camp with two loaded wagons for the eight-and-one-half-mile drive to the post. The road was deemed "absolutely safe." Citizens traveled it daily unarmed. McGrath and Hoeffer had felt so secure that one had left his carbine and pistol at the post and the other had placed his revolver out of reach. At 3:00 P.M. on the fatal day a mule galloped into the post with an arrow in its body. Within fifteen minutes every available man was mounted and on the road to the wood camp. A short distance from the camp they found Hoeffer and McGrath lying beside the wagons—their bodies filled with arrows and stripped of clothing.[127]

It was not until 1873 that fuel was supplied by contract to all posts in the Southwest, although some had received contract fuel at a much earlier date.[128] The army issued contracts for both hard and soft wood with hardwood costing as much as a dollar more per cord than soft. Contractors delivered several kinds of wood: primarily mesquite roots, oak, piñon, and pine in New Mexico; oak and mesquite roots in west Texas; and mesquite roots, pine, oak, cedar, juniper, and cottonwood in Arizona. Mesquite roots, classified as a hardwood and dug from the ground, generated the most controversy. Its quality as a fuel was not in question. Colonel James H. Carleton, inspecting the fuel supply at Fort Bliss in 1871, called mesquite roots "a most excellent wood indeed." But contractors found it difficult to stack mesquite roots in a manner acceptable to quartermasters. When poorly stacked, large spaces appeared in the pile, and a half cord could be stretched to measure a cord. Carleton described the difficulty this way: "Cording up rams horns, cattle horns, and other of the most crooked imaginable things of all lengths and sizes would give a faint idea of cording mesquite roots."[129] To compensate for spaces, the quartermaster at Fort Cummings in 1868 demanded that each cord delivered by John Lemon contain 192 cubic feet rather than the standard 128 cubic feet. Even though Lemon protested, resenting the implication that he would knowingly cheat the government, the district chief quartermaster upheld the action of his subordinate.[130]

To avoid problems like this, Carleton suggested that a standard weight be established for a cord of mesquite roots, a suggestion the army later adopted. Joseph Friedlander's contract for delivery of mesquite roots at Fort Stockton in 1875 stipulated that a cord weigh 3,000 pounds, and it was this figure the army would most often use to define a cord.[131]

Some of the details and hazards of contracting for fuel are revealed in correspondence relating to the contract awarded to Oscar M. Brown in 1868 to deliver 2,000 cords of wood at Fort Sumner for $9.98 per cord. During the spring of that year Brown and his associate Rufus C. Vose, both veterans of the California volunteers, established a wood camp twenty-eight miles from Fort Sumner on the east side of the Las Vegas Road. Brown hired about twenty wood choppers, furnishing them axes, water barrels, and a grinding stone and paying them $1 for each cord they cut. He also hired an overseer to receive and measure the wood for $60 per month and board.[132]

About ten weeks after the wood choppers began work, Chief Quartermaster Marshall I. Ludington stopped delivery on the contract, a decision having been made to abandon the Bosque Redondo reservation. For this breach of contract, Ludington was willing to make a settlement "on equitable terms." Brown wasted little time in submitting a detailed account of his expenses: $159 for the overseer, $1,376 for the wood choppers, $162 for equipment and miscellaneous expenses, and $1,176 for teamsters and teams used in hauling pine poles and constructing wood racks.

But Brown also charged the government for lost opportunities. Brown had intended to send his ox train of fourteen wagons to Hays City, Kansas, with a load of hides. In order to complete delivery of the wood on time, however, he decided to keep his teams in New Mexico and to freight his hides to Kansas at a cost of $1,380. This amount he now charged to the government. Because his teams then had stood idle fourteen days at the wood camp awaiting clarification of the contract, Brown charged an additional $1,176 for demurrage. He estimated that his total loss on the contract was $5,429.[133]

Despite the risks involved, ambitious men considered fuel contracts good investments. The price that contractors received, however, varied widely from post to post. Within the District of New Mexico contract prices for fuel were lowest at Forts Garland, Wingate, and Stanton. With abundant sources of wood near by, these posts were among the last to receive wood by contract. During a ten-year period starting in 1873, the average price for a cord of wood at Garland was $3.08, at Wingate, $3.46, and at Stanton, $3.24.[134]

Fort Garland was located in the San Luis Valley of Colorado, twenty-three miles west of Sangre de Cristo Pass. Established in 1858 to protect settlers and to restrain Utes and Jicarilla Apaches, the post provided a market for the local populace, which remained predominantly Hispanic through 1883 when the post was abandoned. Kit Carson, who commanded the post in 1866, reported that Hispanos inhabited some twenty nearby villages, where they cultivated fields of corn, oats, wheat, barley, beans, potatoes, and chile.[135] However, all but one of the contracts to supply the post with corn, oats, hay, wood, and charcoal between 1866 and 1883 went to Anglos, who settled in the valley in increasing numbers following the Civil War. Ferdinand Meyer and Joseph Hoffman between them received more than a third of the contracts. Meyer, a merchant in Costilla, supplied corn and oats primarily, but he also held two contracts for delivery of hay and two for piñon wood. Hoffman, a Costilla County farmer, received contracts for hay in six separate years and wood in two. The only Hispanic contractor was Antonio A. Mondragón, another county farmer, who agreed to deliver 250 cords of piñon wood in 1875 at $2.50 per cord.[136]

Fort Wingate's contractors found an abundance of wood on hills about five miles from the post. Román A. Baca of San Mateo was the only contractor between 1873 and 1885 who supplied fuel in more than one year. His contracts called for delivery of 800 cords of piñon at $2.65 per cord in 1876, 900 cords of piñon at $2.75 per cord in 1878, and 1,500 cords of pine at $3.50 per cord in 1881.[137] Baca had founded the village of San Mateo about 1864 at the foot of Mt. Taylor, eighteen miles northeast of modern Grants. He soon became one of Valencia County's largest sheep and land owners. Over the years he supplied Fort Wingate with a variety of commodities: hay, oats, beans, barley, and wood. Like other contractors, Baca sometimes had difficulty completing his contracts. To cite one example, in 1881 construction of the Atlantic and Pacific Railroad west from the Rio Grande placed a premium on labor and teams. Baca had to pay top prices to secure sufficient men and teams to complete his wood contract, which he did at a financial loss. Baca had earned the respect of army supply officers, however, and during this difficulty, the district quartermaster sent the commanding officer at Fort Wingate this message:

> Mr. Baca is a man so well known in New Mexico that I need not say a word as to his fidelity to his agreements or his integrity. He has long served the Government, and always faithfully and well. I therefore bespeak for him your most favorable consideration in all matters he may bring before you, relating to his contract.[138]

About half the fuel contracts for Fort Stanton between 1873 and 1885 were held by merchants James J. Dolan and José Montaño and sawmill owner Frank Lesnet. Twenty-three-year-old William R. Ellis, whose father's store fronted on the main street of Lincoln, held the largest contract, calling for delivery in 1883 of 2,200 cords of soft wood at $2.98 per cord. Other fuel contractors included George W. Nesmith, a sawmill owner; Patrick Coghlan, a Tularosa merchant; and Samuel S. Terrell, a bookkeeper who probably worked for the post trader.[139]

Fuel prices in New Mexico were the highest at Fort Cummings where timber was scarce. Captain Joseph A. Corbett, post quartermaster, stated in 1867 that a small growth of trees could be found twelve miles from the post on a rocky mountain slope, but a road would have to be built so that wagon teams could reach the area. A small amount of mesquite was located about five miles from the post. Contractor Jacob Appelzoller in 1869 hauled his wood from the Florida Mountains twenty miles south of the post. With a force of twenty men and nine eight-mule wagons, it took Appelzoller one and one-half months to furnish 250 cords.[140] Starting in 1868, four contractors would supply the post with fuel before it was temporarily abandoned in 1873: John Lemon in 1868, Appelzoller in 1869, William V. B. Wardwell in 1870 and 1871, and Estanislao Montoya in 1873. Of these, Lemon's wood was the most expensive at $15.50 per cord; Montoya's was the cheapest at $5.10 per cord.[141]

Fort Cummings was reactivated in 1880 during the Victorio War. The renewed Indian hostilities kept fuel prices high, for contractor Samuel P. Carpenter had to pay top wages to lure freighters and teamsters to his wood camp located in a dangerous area some twenty-eight miles from the post. Colonel George P. Buell, commanding troops at the post, authorized purchasing Carpenter's wood at $25 per cord, which was the lowest price offered by anyone in the area. Carpenter claimed that it cost him $21 for each cord he delivered at the post. Even so, the price appeared exorbitant, and Colonel Edward Hatch, commanding the District of New Mexico, later authorized payment of only $15 per cord.[142]

The quartermaster at Cummings had feared that unless high prices were offered for fuel, the post would be reduced to using sagebrush during an overnight visit of a distinguished guest—the President of the United States. The presidential party—which included President and Mrs. Rutherford B. Hayes, their son Rutherford B. Hayes, Jr., Secretary of War Alexander Ramsey, General William T. Sherman, and Rachel Sherman (the general's daughter)—was en route from California to the East Coast. They arrived at Shakespeare, New Mexico, on 25 October on the Southern Pacific Railway and then traveled overland sixty miles to Fort Cummings. After a night at the post they would travel to the end of the track of the Atchison, Topeka and Santa Fe Railway near old Fort McRae, where they would board a train for Santa Fe. Everything possible was done to provide for their safety and comfort—troops were stationed along the road; tents and provisions of all kinds were stockpiled at each stopping point; ambulances and relays were provided for rapid transportation.[143] Considering the occasion, it is not surprising that Carpenter was promised good rates to assure a supply of fuel.

Contractors who furnished fuel for the west Texas posts frequently were merchants or post traders. The most expensive fuel was delivered at Fort Stockton, where the cost of wood averaged $9.30 per cord between 1867 and 1880. Most of it—mesquite—was obtained from eight to twenty miles from the post, although a newspaper correspondent in 1868 reported that Peter Gallagher had hauled his wood a distance of forty miles. Gallagher's wood was also the most expensive, costing the government $16 per cord.[144]

Contract prices in west Texas were lowest at Fort Davis, where a cord of wood cost an average of $5.06 between 1867 and 1880. Among the merchants who supplied the post with wood were two brothers from Virginia, Otis and Whitaker Keesey. Whitaker Keesey had arrived in west Texas in 1867 as civilian baker for Lieutenant Colonel Wesley Merritt. Otis apparently followed his brother west, and both became leading merchants in the Fort Davis community.[145]

Although most west Texas merchants enjoyed good relations with army officers, few generated as much hostility while delivering fuel as Samuel and Joseph Schutz of El Paso. In 1869 the Schutz brothers subcontracted to complete T. A. Washington's wood contract at Fort Bliss, which called for delivery of a year's supply of hardwood at $11 per cord. After Captain Edward Meyer, the post's commanding officer, rejected 300 cords of mesquite roots delivered on the above contract, Samuel allegedly told Meyer: "I will soon show you that it won't do for a little Captain of Infantry to attempt to stand between a monied contractor and the government." Meyer soon left the post on new assignment. Before his replacement arrived, Post Quartermaster James D. Vernay accepted the 300 cords. Later the new commander, Major Henry C. Merriam, heard that the Schutz brothers and attorney Benjamin F. Williams had paid Vernay a total of $500 to receive the wood. Merriam relieved Vernay from duty and questioned one or both of the brothers. But nothing further came of the matter, and the Schutzes continued supplying goods to the army.[146]

Arizona businessmen received the highest prices for fuel at Fort Bowie, a difficult and dangerous garrison to supply. During a span of ten years, 1871–80, the contract price averaged $10.22 per cord. Bowie had been established in 1862 near Apache Pass in the Chiricahua Mountains. By 1871 fuel close to the post had been exhausted, and contractors had to obtain their supply on hills and mountainsides about eight miles away. The area was so rough that wood had to be "snaked" downhill before it could be hauled away in wagons.[147]

The first contract at Bowie was awarded in 1871 to Robert H. Whitney for hardwood "as required" at $9.50 per cord. Whitney and his partner Mariano G. Samaniego were furnished three enlisted men to guard their wood camp. But the 31-year-old Whitney himself was killed a short time later when Apaches attacked an army command in which he was serving as guide. Post trader Sidney R. DeLong received half the fuel contracts awarded at Bowie during the ten years mentioned above. His wood was also the most expensive, priced in 1873 at $12.50 per cord.[148]

Elsewhere in Arizona fuel prices were lowest during the ten-year span at Forts Apache, Lowell, and Whipple, where the average cost for a cord of wood was $3.27, $3.51, and $4.34, respectively. Each had an abundant source of wood near by—mesquite near Lowell, pine within four miles of Whipple, and cedar, juniper, and pine near Apache. William R. Milligan, who had switched his residence from New Mexico to Arizona, delivered the highest-priced fuel at Apache. In 1872 he received $4.45 per cord for an unspecified amount of hardwood.[149]

Frederick L. Austin and Frederick Maish won most fuel contracts at Fort Lowell and were the only contractors there to furnish the supply in more than one year. New York-born Austin was post trader at Camp Grant when the massacre occurred and later served as trader at Lowell. Maish, a native of Pennsylvania, was a well-known cattleman who built Tucson's first hotel in 1875. Between 1870 and 1885 Austin and Maish each furnished fuel in five different years. The largest suppliers of fuel at Fort Whipple were well-known Arizona entrepreneurs: Charles H. Veil, who had arrived in the territory in 1866 as a first lieutenant; post trader George W. Bowers; and Samuel C. Miller, pioneer freighter and Prescott rancher.[150]

Fuel contracting in the Department of Arizona remained basically unchanged through 1885, but significant changes occurred in the District of New Mexico, where the arrival of railroads stimulated coal mining. Increased coal production coupled with cheap transportation allowed several military posts to adopt coal as their primary fuel for cooking and heating. The first contract to supply Fort Union with coal was signed on 24 June 1879, shortly after the Atchison, Topeka and Santa Fe Railway was built over Raton Pass and into New Mexico. The contract called for Trinidad Coal and Mining Company to deliver at La Junta (shortly to be renamed Watrous), a station on the railroad about nine miles from the post, 1,100 tons of bituminous coal for $5 per ton. The following year Albert Stark (representing the Trinidad

Coal Company) agreed to deliver 500 tons of coal at Watrous for $6 per ton, and Gerard D. Koch of Santa Fe agreed to deliver 980 tons of Trinidad coal to Santa Fe, where a branch line had been completed in February 1880, for $7.25 per ton. In recommending the contract with Koch, District Quartermaster James J. Dana estimated that the army would save $4,000 in fuel expenditures during the 1880–81 fiscal year, even though heating and cooking stoves would have to be purchased for the Santa Fe headquarters.[151]

Trinidad coal was not an immediate success. The acting assistant quartermaster at Santa Fe reported in May 1881 that "the coal is of poor quality; it has very little heating power and contains a great deal of slate; it does not answer the purposes of the Quartermaster's Department."[152] The depot quartermaster at Fort Union had formed a similar opinion concerning coal supplied there in 1879. But he judged the coal delivered the following year as superior: "All the officers at this post, and their families, use it, and prefer it to wood."[153] Nonetheless, the only contract issued for coal within the district in 1881 went to John G. Price, who agreed to deliver 1,600 tons at Fort Lewis, Colorado, at $5.75 per ton. This post had been established in 1880 about twelve miles west of Durango and adjacent to the Southern Ute Reservation. The Denver and Rio Grande Railway reached Durango in 1881, and thereafter Fort Lewis received most of its fuel and other supplies in railroad cars.[154]

The coal-mining boom that New Mexico witnessed in the 1880s was linked to railroad expansion. Newly opened mines west of modern Raton supplied the Atchison, Topeka and Santa Fe with coal in 1880, the same year that the railway built a depot at Willow Springs, a stage station that was soon renamed Raton. The Santa Fe also located its division repair shops there, and town boosters envisioned Raton as the Pittsburgh of New Mexico. The Santa Fe and later the Southern Pacific and the Mexican Central railways consumed much of Raton's coal, but the army also used it. In 1882 the Las Vegas firm of Browne and Manzanares agreed to deliver 300 tons of Raton coal at Fort Union for $12.50 per ton. Fort Cummings also received coal that year from contractor Samuel P. Carpenter, who would supply 390 tons of bituminous coal for $14.89 per ton. By this time Carpenter could ship coal to Florida station on the Atchison, Topeka and Santa Fe and then haul it overland five miles to Cummings. In later years mines in the Gallup area furnished Fort Wingate with coal, and the Cerrillos mines southwest of Santa Fe supplied Fort Marcy and district headquarters.[155]

Residents of the Southwest from all walks of life helped supply the quartermaster's department with forage and fuel. Some labored in fields grubbing out mesquite roots and cutting grama grass with hoes and knives and shared with contractors the government's patronage. For many businessmen forage and fuel contracts were major investments, which involved financial risk as well as opportunity for profit. Men and women alike managed government forage agencies, which the army considered essential to its mobility. Other entrepreneurs supplied the commissary department with items in the soldier's ration. The military demand for bread and meat literally created the milling and cattle industries in the Southwest, which is to be the subject of Chapters 4 and 5. Through all these enterprises, civilians came to rely upon federal funds—both to obtain the necessities of life and to finance other economic ventures.

4

Millers and Merchants

When the Army of the West arrived in Santa Fe in August 1846, it created an immediate market for all kinds of supplies. Government wagons filled with subsistence stores had accompanied General Stephen W. Kearny's soldiers on the march across the plains, but even so the army relied upon the local populace for additional amounts of flour, beef, hominy, beans, cornmeal, salt, and other provisions. Though it would take years before New Mexicans provided a significant part of the soldier's ration, the army's need for flour soon sparked a revolution in New Mexico's milling industry.[1]

At the time of the American occupation, New Mexican gristmills were small, primitive affairs powered by water, horse, and mule power, capable of grinding only a few bushels of grain a day. Flour manufactured at these mills was coarse, gritty, and, according to some of Kearny's soldiers, unpalatable. A Missouri volunteer commented: "When we first came here the commissary issued to the companies flour made here, called Taos flour which is half bran, dirt and sand and not fit for a white man to eat."[2] Soon, however, local entrepreneurs constructed larger and more efficient gristmills to take advantage of the new market.

About 1850, Ceran St. Vrain built the first modern gristmills in New Mexico. The one he constructed near the village of Talpa a few miles south of Taos used grain from Taos Valley, the principal wheat-growing area in New Mexico. The second mill was erected at Mora and drew wheat from settlements east of the mountains. Both mills were powered by water. A third modern mill, also operated by water, was erected a short time later at Peralta, eighteen miles south of Albuquerque. The mill was originally owned by William L. Skinner, who died in 1851, and then by Antonio José Otero, a prominent landowner and merchant. The Peralta mill was capable of grinding about

500,000 pounds of flour a year. In 1858 Joseph Hersch, a native of Poland, erected New Mexico's first steam-powered gristmill on the Rio Chiquito in Santa Fe. The local press reported that it could grind "some hundred *fanegas*" of flour a day. The army's demand for flour coupled with the construction of these and other modern mills spurred New Mexico's farmers to more than double their output of wheat between 1850 and 1860.[3]

Once the Confederate threat in the Southwest had ended, construction of new mills continued apace. A second steam-powered mill was erected in 1863 by Franz Huning south of his home in Albuquerque. A native of Hanover, Germany, Huning had arrived in New Mexico in 1849 and was joined a few years later by a younger brother Charles. Together they engaged in freighting, government contracting, ranching, and selling merchandise to valley residents.[4] Two other brothers, Louis and Henry, soon settled at Los Lunas, four miles south of Peralta, where they became well-known merchants. In 1870 Louis and Henry Huning constructed a large gristmill six miles south of Los Lunas, said to be "one of the best full roller process flouring mills in the Territory."[5] The first large modern flouring mill in Arizona was erected in 1869 by William and Nicholas Bichard at Adamsville, three miles west of Florence. Earlier they had purchased Ammi White's mill at the Pima villages, but this mill had been destroyed in the flood of 1868.[6]

The census for 1870 lists thirty-six gristmills in New Mexico and two in Arizona. Four of the New Mexico mills were powered by steam, two by horses, and thirty by water, while in Arizona one was steam-powered, the other water-powered. Less than half the mills furnished flour under contract to the army, though the majority at one time or another probably furnished flour to government contractors.[7] The mill owners mentioned above—St. Vrain, Otero, Hersch, the Hunings, and the Bichards—were typical of contractors who supplied flour to the army. Merchants as well as mill owners, they had the necessary financial resources to take advantage of the military market and to win contracts to supply the army with a wide assortment of commodities.

Merchants were as important in furnishing subsistence stores to the commissary department as they were in supplying grain to the quartermaster's department. Between 1865 and 1880 the main commodities that the commissary department secured under contract in New Mexico and west Texas were flour, beans, salt, and beef, and in Arizona, flour, beans, and beef. Except for beef, furnished for the most part by ranchers and cattle dealers, these food items typically were supplied by merchants and less often by farmers. Federal dollars expended for these supplies helped to finance the pioneer mercantile establishments that dominated much of the southwestern economy. Profit-minded proprietors attempted to furnish the commissary department with all the subsistence stores it required. But no amount of ingenuity on their part could make certain local commodities competitive with items imported from outside the region. In 1865, for example, contracts were awarded in New Mexico for hominy and vinegar, but the army soon discovered it cost less to import these items from the East.[8]

Only a portion of the cornmeal that the army required was supplied by local contractors. Between 1865 and 1875, sixteen individuals or firms received contracts to supply cornmeal within the District of New Mexico, although others furnished this commodity in casual transactions. At least fourteen of the contractors were merchants or mill owners, and some, like Henry Lesinsky and Louis Huning, were both. Between them Lesinsky and Huning held more than one third of the contracts. James (Santiago) L. Hubbell, a prominent land and cattle owner living at Pajarito south of Albuquerque, supplied the most expensive cornmeal under contract, agreeing in 1865 to deliver 10,000 pounds at Fort Craig at 16 cents per pound and 4,000 pounds at Fort Stanton at 18 cents per pound.[9]

Local contractors supplied the west Texas posts with cornmeal between 1870 and 1875, although this commodity may have been purchased in casual transactions in other years as well. Ward B. Blanchard of Ysleta, Texas, supplied more cornmeal under contract than any other west Texas supplier. In 1870 he agreed to furnish the supply needed at Forts Bliss, Davis, and Quitman, and the following year he received the Fort Quitman contract. The last recorded contracts were issued in 1875 to Peter Gallagher, who would furnish cornmeal at Forts Davis and Stockton for 4 cents per pound.[10]

In Arizona contracts for cornmeal were awarded in four consecutive years, starting in 1874. At least five of the seven men receiving the contracts were merchants or mill owners. Charles T. Hayden, who had built a modern gristmill on the Salt River, received more contracts than anyone else, agreeing to supply Fort Whipple and Camps

McDowell and Verde in 1874 and McDowell and Whipple in 1875. Two New Mexicans, Santiago L. Hubbell and Louis Huning, furnished Camp Apache with cornmeal in three separate years.[11]

As steel rails approached the Southwest, it became cheaper to import cornmeal than to purchase it locally. In 1876 Captain Charles P. Eagan, chief commissary for the District of New Mexico, rejected all local bids as being too high and subsequently obtained a supply from Fort Leavenworth. Two years later, while Eagan was serving as chief commissary for the Department of Arizona, he informed his superiors that cornmeal manufactured in Arizona and New Mexico "not being what is technically known as 'kiln dried' does not keep in good condition except for brief periods of time."[12] Thereafter posts in Arizona received most of their cornmeal from San Francisco.

In both Arizona and New Mexico the commissary department tried to stimulate local production of bacon and salt pork, meat items in the soldier's ration issued about two days in every seven.[13] When transported overland from depots at San Francisco or Fort Leavenworth, bacon put up in canvas sacks and salt pork packed in brine-filled barrels were subject to much wastage and spoilage. Supply officers hoped to secure a cheaper and fresher article from local contractors. In neither territory, however, were swine available in adequate numbers, primarily because corn was too expensive to feed to hogs. Then, too, the Indian danger in Arizona discouraged ranchers there from raising all kinds of livestock.

George E. Blake of Doña Ana County may have been the first New Mexican to furnish a large amount of locally cured bacon to the army. In the summer of 1868 Blake informed the commissary department that he could manufacture on his ranch at San Agustín between 25,000 and 75,000 pounds of bacon, which he would deliver at Fort Selden for 33 cents per pound. Captain Charles McClure, the district's chief commissary, hesitated to accept this proposal, for he doubted that New Mexico would ever "become a pork or bacon making country, [even] though the government should encourage it by giving the preference to New Mexican bacon."[14]

Nonetheless, the army purchased 3,600 pounds of Blake's bacon as an experiment. The commissary officer at Fort Selden was instructed to report each month on the quality of the bacon, its acceptability to the troops, and how well it kept in storage. The bacon would be placed in racks labeled "Blake's experimental bacon." Evidently the experiment was a success, because in May 1869 the army purchased from Blake an additional 10,000 pounds at 30 cents a pound delivered at Fort Selden. This was still about 3 cents a pound higher than the current cost of bacon delivered from the East.[15]

McClure also accepted the offer of Lawrence G. Murphy in 1869 to deliver at Fort Stanton 10,000 pounds of homemade bacon and 5,000 pounds of pickled pork at 25 cents per pound, a price slightly less than the cost of these items if freighted from eastern depots. McClure gave detailed instructions to Fort Stanton's commissary officer: "You will carefully inspect each piece of bacon, and each barrel of pork and accept none that is not properly cured or pickled. . . . Clear sides bacon has all the bones removed and should be well smoked, and without any disagreeable odor." Bacon would be placed in racks with cards attached, showing weight when received and amount of shrinkage when issued; a half dozen pieces of the first delivery would be retained to judge its keeping qualities. The barrels of pork should weigh either 100 or 200 pounds each and the pieces of pork no more than 10 or 20 pounds. "Please give especial attention to this matter," McClure wrote, "as the Subsistence Department is desirous of encouraging the production of bacon and pork in this country and upon the result of this experiment further encouragement may be extended."[16]

In January 1870 McClure sought authority to advertise locally for all the bacon required in the district. Because corn had been plentiful the previous two years, New Mexico's farmers were raising more hogs, and McClure now believed that with the army's encouragement they would soon manufacture all the bacon and pork required by the troops.[17] In May the army awarded contracts to eight individuals or firms for delivery of bacon to posts in the district. Charles Keerl and John A. Miller, Fort Bayard merchants, agreed to supply the largest amount: 9,000 pounds at Fort Cummings, 18,000 pounds at Fort Selden, and 27,000 pounds at Fort Bayard, all for 28 cents per pound. Contract prices ranged from 19 cents per pound for bacon that Joseph B. Watrous furnished at Fort Union to 28½ cents per pound for bacon that George E. Blake delivered at Fort McRae. At least half the contractors were merchants. Frank M. Willburn and Thomas L. Stockton, who furnished bacon at Fort Stanton, however, were cattlemen; they also agreed to deliver a total of 39,000 pounds of pork at Forts

Union, Wingate, Bascom, and Stanton at between 20 and 22 cents a pound. The only other contract awarded for pork went to Ferdinand Meyer who agreed to furnish Fort Garland with 3,000 pounds at 16⅞ cents per pound.[18]

The letting of bacon contracts in New Mexico seemingly had little effect on the local swine industry. Three fourths of the bacon delivered on the contracts came from the States, and Keerl and Miller secured theirs from Santo Tomás in Chihuahua. Captain William H. Nash, who replaced McClure as chief commissary, consequently judged the attempt to secure bacon in New Mexico a failure. And even when locally manufactured salt meats were delivered, he noted, "after keeping a short time [they] turned soft and hardly fit for issue."[19] The commissary department, nonetheless, solicited new proposals in 1871. Competition was lively, with between four and seven bidders seeking the bacon contract at each post. When the chief commissary of subsistence for the Division of the Missouri learned of the bids, he ordered them rejected for being too high. But the decision came too late. Nash already had awarded contracts to Las Vegas merchant Frank Chapman to deliver bacon at all posts in the district, except Fort Garland, at between 20 and 26 cents per pound.[20]

The records are silent concerning the quality of Chapman's bacon, but the army appeared willing to continue letting bacon contracts in New Mexico for at least another year. On 21 May 1872 Nash opened bids in his Santa Fe office. Competition again had been keen. Twelve individuals sought the contract at Fort Union; five submitted bids to provision the entire district, including Frank Chapman and his partner Charles W. Kitchen, who offered to furnish 190,000 pounds at 20 cents a pound.[21] Kitchen recently had advised farmers through a notice in the *Weekly New Mexican* that he had on hand "25 pairs of pigs of the Poland China breed," which he would sell cheaply to improve the local variety.[22] But New Mexico was not destined to become a major swine-producing territory. All bids received in 1872 were rejected for being too high, except the one submitted by R. E. Cullen for provisioning Fort Stanton at 15¼ cents per pound.[23] Thereafter no contracts for bacon were awarded in New Mexico, and the troops received their supply from the East. New Mexico's swine population reflected this loss of market, falling from 11,267 in 1870 to 7,857 in 1880, a reduction of about 30 percent.[24]

Arizona's swine population, on the other hand, experienced a dramatic increase, mainly in response to the growth of settlement rather than to the military market. The census reported 720 hogs in Arizona in 1870 and 3,819 in 1880.[25] The military awarded bacon contracts only in 1874 and pork contracts in 1874 and 1876, though each commodity was obtained at other times in private transactions.[26] Cotsworth P. Head, a Prescott merchant, received all the bacon contracts, agreeing to supply Camps Apache, Lowell, McDowell, Mojave, and Verde with a total of 50,000 pounds for 23 cents per pound. Head furnished locally cured bacon only at Lowell, McDowell, and Verde; the bacon for Apache and Mojave came from St. Louis and San Francisco.[27] William B. Hellings, a Phoenix merchant and mill owner, signed contracts in 1874 to deliver a small amount of mess pork at McDowell, Verde, and Whipple at between 26 and 29 cents per pound.

In 1875 the commissary department decided not to advertise locally for salt meats, an action that disappointed many farmers. Army Inspector James A. Hardie reported that because the army previously had received locally prepared meats with favor, farmers "had turned attention to the raising of hogs and the curing of salt meats as a supposed permanent market resource." According to his calculations, territorial bacon could be supplied at Bowie, Grant, Lowell, McDowell, Verde, and Whipple at 19 cents per pound; imported from San Francisco, bacon would cost 21 to 24 cents per pound. Territorial pork in barrels could be supplied at most posts for $50 per barrel, while imported pork would cost from $54 to $58 per barrel. Hardie believed that the quality of locally cured salt meats was high, and he recommended that it be purchased for the troops. Commenting upon bacon cured in the Salt River Valley, he wrote: "The bacon bought was not breakfast bacon specially prepared for dainty appetites; but sides bacon for issue to troops and sale to miners and farmers. Better bacon I have never eaten. And so say the men of the various commands inspected."[28]

Despite Hardie's recommendations, Arizona's commissary department issued only three contracts for salt meats in 1876; they went to the firm of Foster and Veil, who agreed to supply Fort Whipple and Camps Verde and McDowell with mess pork "as needed" at between $44 and $48 per barrel. Thereafter most of the pork and bacon furnished to troops in Arizona was sent from Chicago or San Francisco. Commissary Eagan reported in 1879 that the army no longer purchased territorial salt meats because the supply was limited, prices high, and the quality substandard.[29] For similar reasons the army purchased only a small amount of salt meats locally for posts in west Texas. Until the mid-1870s most of the bacon and pork furnished at Forts Stockton, Davis, Bliss, and Quitman came from New Orleans via San Antonio. With the advance of railroads, however, these garrisons would be supplied from Midwest supply centers.[30]

Although it was not part of the ration, butter manufactured in the more settled regions of the United States was shipped to frontier posts and there made available for purchase by officers' families and enlisted men. The butter deteriorated in transit, however. Packaged in glass jars or in firkins placed in brine-filled half-barrels, it frequently arrived soft, liquid, and rancid. Martha Summerhayes recalled that the butter she purchased at Camp McDowell about 1877, shipped from San Francisco in glass jars, "had melted, and separated into layers of deep white, deep orange and pinkish-purple colors."[31]

Army Inspector Nelson H. Davis in 1872 reported that officers and enlisted men in New Mexico rarely bought the poor quality butter imported from the States. On occasion the district commissary purchased butter in small lots from Colorado and New Mexico suppliers at 45 and 50 cents per pound, which Davis said was "readily taken by officers' families." But soldiers continued to boycott butter shipped from the East. Consequently Commissary Eagan requested authority in 1876 to purchase the entire butter supply within the district, claiming it could be obtained near Fort Union at 40 or 45 cents per pound and near Forts Garland, Bayard, and Stanton at 50 cents per pound.[32]

Captain Frederick F. Whitehead, Eagan's replacement, was less sanguine. Because he was unable to purchase good quality butter near Santa Fe for less than 70 or 75 cents per pound, he ordered 500 pounds from a firm in Junction City, Kansas, for sale to officers' families.[33] In an 1879 letter to Washington, Whitehead noted that improved transportation facilities made it practical to furnish all posts with high quality eastern butter.[34]

Both before and after the arrival of railroads, residents who had established dairy farms on or near military reservations supplied some of the butter that soldiers consumed. These suppliers included Nelson Thayer, a recently discharged California volunteer, who in 1866 operated a small milk ranch on the Fort Union reserve, and Peter

Donnelly, a former soldier and quartermaster's employee, who was living on the Fort Stockton reservation in 1871 where he supported his wife and six children by selling milk and butter to the garrison.[35] A decade later, in 1881, the commanding officer at Fort Stanton gave William DeLany permission to open a dairy on the reservation and then established rates for DeLany's dairy products: one dollar for thirteen quarts of sweet milk or twenty-five quarts of buttermilk, 40 cents per pound for butter in the summer, and 45 cents for butter in the winter.[36]

For several years after the Civil War, butter was both scarce and expensive in Arizona. In 1868 the price of butter in Prescott ranged from $1 to $1.50 per pound, while it cost from 35 to 45 cents per pound in San Francisco. The editor of the Prescott *Weekly Arizona Miner* in October 1869 reported that the town was entirely out of butter. What was needed, he editorialized, was "some enterprising California dairyman" who would relocate his herd in central Arizona.[37]

Officers and enlisted men in Arizona received their butter from San Francisco. Because much of it became unpalatable in transit, the commissary department in 1870 experimented to determine the kind of package that was best suited for shipping this commodity to Arizona. Butter was placed in tins, oak kegs, and crocks, "all packed in wooden cases filled in with salt." Cases were marked "P. butter," "R. butter," and "C. butter," denoting respectively eastern butter sent via the Isthmus of Panama, eastern butter sent via the railroad, and fresh-made California butter. The experimental butter was shipped to each post in Arizona with requests for officers to submit their views upon the quality of butter and packaging. Only the report of Captain Charles W. Foster, commissary officer at Fort Whipple, has been located. He judged that all the butter had arrived in good condition. Because it had been shipped through Yuma during cool months, the kind of packaging had made little difference. But he believed that butter packed in crocks would withstand the extreme summer heat better than butter packed in kegs or tins.[38]

Before the advent of railroads the army never solved the problem of delivering palatable butter to the Arizona garrisons. When the commissary at Camp Bowie complained that six cases sent there from California in 1877 had arrived "unfit for use," the chief commissary for the Department of California flatly stated: "The article is too delicate for transportation in quantities over so long a route in a hot climate."[39] Limited amounts were purchased locally. Captain Charles P. Eagan reported in August 1877 that butter could be purchased near Camp Apache and Fort Whipple at 75 cents per pound, near Camp Verde at between 75 cents and $1, and in the vicinity of Camps Grant and Lowell at $1 per pound.[40] In subsequent months officers throughout the territory reported their inability to secure a sufficient supply at reasonably low rates. Consequently the commissary general of subsistence in 1878 arranged to have eastern butter sent to San Francisco for distribution in Arizona. In this way he hoped to ensure a fresher and better article than previously supplied from California.[41] Because the quantity of locally manufactured butter could not meet demands, the commissary department continued to import butter from outside the territory during the years of this study.

The army also imported most of the salt it required in Arizona. Between 1865 and 1885 the commissary department awarded only ten salt contracts to territorial residents, though it probably made additional purchases in casual transactions. Five of the seven men holding the contracts were merchants, including Charles T. Hayden, who agreed in 1867 to supply the Tucson depot with 16,298 pounds of fine salt at 8½ cents per pound and 9,175 pounds of coarse salt at 6½ cents per pound. The only salt contract held by a Hispano was issued in 1869 to Pedro Aguirre, then a resident of Altar, Sonora, who charged almost 7 cents per pound for the 9,000 pounds of fine salt he would deliver at Tucson. The last recorded salt contract was signed in 1872 by Cotsworth P. Head for 80 pounds of fine salt furnished at Prescott at the enormous price of 25 cents per pound.[42] Because there were few workable salt beds in Arizona, contractors probably imported most of the salt they sold to the military.

In contrast much of the salt supplied to posts in New Mexico and west Texas was furnished under contract and obtained from local salt lakes. The unrefined salt that contractors usually delivered received mixed ratings from officers. Captain Charles McClure, chief commissary for the District of New Mexico, reported in 1868 that contract salt had not met his expectations because of its dark color. Nonetheless, he described it as being pure and of good quality. A year later McClure praised the salt that contractor Joseph A. DeHague delivered and requested authority to allow DeHague to put in an additional amount. "So much rain fell during the spring," McClure wrote, "that the salt is whiter and better this season than it has been for years."[43] Marian Russell, the wife of DeHague's partner, recalled that salt was obtained from a salt sink near Tecolote. The partners hired Hispanic laborers who waded into the sink and shoveled salt into windrows. Russell remembered that "the waves lapping over the windrows would clean it nicely. Then it was loaded into the wagons and hauled away."[44]

Some of the salt supplied at Forts Bascom and Sumner was obtained from a lake seventy miles east of Fort Bascom near the north bank of the Canadian River. During the summer of 1864 Captain Edward H. Bergmann and a detachment of New Mexico volunteers loaded six ox-wagons with nearly 24,000 pounds from this source. Two years later Lieutenant Colonel Silas Hunter, 57th Infantry (a black regiment), led a salt-gathering expedition to the lake and returned to Fort Bascom with about 11,000 pounds. The lake then was about 200 yards long, 100 yards wide, and 3 feet deep. The colonel judged that the lake held an abundance of salt, though approaching cold weather would prevent men from working in the water. Hunter had boiled a gallon of lake water as an experiment and obtained excellent results: "one quart of beautiful white salt equal in appearance and grain to the finest quality of Lake Salt."[45] In 1873 Captain Samuel B. Young, commanding a summer camp on the Canadian, described salt from the lake as having "a pure fine composition and good taste." His guide Frank DeLisle had told him that "Mexicans" in the past had obtained salt here by wagonloads. Their method, DeLisle explained, was to wade into the middle of the lake, shovel salt from the bottom, and throw it onto a wagon sheet, letting the water drain. The residue was almost pure salt.[46]

The army contracted for both fine and coarse salt, reserving the fine for officers and enlisted men and the coarse for animals. Coarse salt, also used in salting hay, was sometimes ground into a finer article at local gristmills.[47] But unprocessed salt was readily available. The commanding officer at Fort Union noted in 1866 that salt could be purchased almost daily from local residents who brought it into the post (see Table 4).[48] Immediately after the Civil War, most contract salt in New Mexico was delivered at Fort Union and then distributed else-

Table 4
Salt contracts at Fort Union

Date	*Contractor*	*Amount in pounds*	*Type*	*Price in cents per pound*
September 1865	William White	15,000	——	6¾
December 1866	William White	30,000	——	4 97/100
August 1867	A. D. Johnson	25,000	fine	4 9/10
	Julian Sandoval	10,000	coarse	4
August 1868	Juan Gonzales	10,000	fine	3 85/100
July 1869	J. A. DeHague	30,000	fine	2½
		20,000	coarse	2¼
August 1870	Nathan Eldodt	20,000	fine	3 14/100
	Hugo Wedeles	10,000	coarse	1 99/100
January 1872	Sigmund Wedeles	16,500	——	3¼
January 1873	Sigmund Wedeles	10,000	fine	2 95/100
		6,900	coarse	2 30/100
March 1874	Sigmund Wedeles	as required	fine	2¾
		as required	coarse	2
January 1878	Sigmund Wedeles	3,500	fine	2 5/10

Data from Reg. Cont., CGS, RG 192, NA.

where. William White, a merchant in Santa Fe who contracted to deliver a total of 45,000 pounds, secured his supply from a lake said to be 200 miles from the post.[49]

Between 1865 and 1878, the last year for which a contract was issued, eight individuals received contracts to deliver salt at Fort Union. Sigmund Wedeles, a merchant at Mora, held more contracts than anyone else. His brother Hugo, also a merchant, delivered the cheapest salt at 1 99/100 cents per pound. Salt contracts first appear on the registers for Forts Bascom and Craig in 1867, Wingate in 1868, Stanton in 1872, Bayard, Santa Fe, Selden, and Tularosa in 1873, and McRae in 1874. More than half the salt contracts registered in New Mexico were held by four merchants: Henry Lesinsky and Louis Rosenbaum of Las Cruces, Sigmund Wedeles of Mora, and Louis Huning of Los Lunas.[50]

The amount of salt the army purchased within the district declined in later years. In 1869 McClure had reported that all salt issued in the district was furnished locally. But during the fiscal year ending 30 June 1882, only a small quantity was purchased in New Mexico because salt could be supplied more cheaply from Kansas.[51] The last registered salt contracts were issued to Sigmund Wedeles in October 1882 for immediate delivery at Forts Bayard, Bliss, Cummings, Selden, and Stanton, but his supply probably came from the East by railway.

There is reason to believe that some salt contracts for the west Texas posts were not recorded, but, based on those that were, salt was the cheapest at Fort Bliss, where contract prices ranged from 2⅓ to 5 cents per pound. Only El Paso merchant Benjamin Schuster among the eleven men who held contracts at Fort Bliss delivered salt there in more than one year. Two Hispanic residents contracted for a year's supply at both Forts Bliss and Quitman. In 1872 Máximo Aranda of San Elizario, Texas, received 4 cents per pound for salt delivered at Bliss and a cent more per pound for salt delivered at Quitman. The following year Gumecindo Pendragón of Ysleta, Texas, received 2⅞ cents per pound at Bliss and a cent more at the other post (see Table 5).[52] Much of the salt delivered at the two posts came from the great salt deposits near Guadalupe Peak about one hundred miles east of El Paso.

Contractors for garrisons at Forts Davis and Stockton most likely secured their salt from the Pecos salt lake thirty-five miles northeast of Fort Stockton. Fort Stockton area residents Peter Gallagher, Michael F. Corbett, and the brothers Bernardo and Cesario Torres held eight of the nine contracts recorded for the two posts. Cesario Torres delivered the most expensive salt, receiving in 1870 8 cents per pound at Fort Davis and 7 cents per pound at Fort Stockton. Corbett furnished the cheapest salt, receiving 4 cents per pound for the supply he delivered at Fort Stockton in three separate years.[53]

In all parts of the Southwest the army relied upon local suppliers to furnish the mainstay of the soldier's daily ration—the common but nutritious bean. Merchants won most contracts for this commodity, but farmers also benefited from the military market. Prior to American occupation, farmers in New Mexico had raised beans only as a subsistence crop. The army's subsequent demand, however, stimulated crop production. Census reports for New Mexico show that beans and peas (combined in the reports) more than doubled in output between 1850 and 1860.[54]

Table 5
Salt contracts at Fort Bliss

Date	*Contractor*	*Amount in pounds*	*Type*	*Price in cents per pound*
November 1869	W. P. Bacon	as required	fine	5
December 1870	A. H. French	as required	fine	4 45/100
November 1871	H. J. Cuniffe	as required	fine	4
May 1872	Máximo Aranda	as required	fine	4
June 1873	G. Pendragón	as required	fine	2 7/8
June 1874	W. B. Blanchard	as required	fine	2 33/100
August 1878	George Zwitzers	900	fine	3
May 1879	A. Boulet	1,500	——	$4.75 per 100 pounds
October 1879	Solomon Schutz	2,400	——	3 94/100
August 1880	Benjamin Schuster	13,000	——	3 61/100
October 1881	Benjamin Schuster	2,000	——	$2.74 per 100 pounds
October 1882	Sigmund Wedeles	1,580	——	$3.25 per barrel

Data from Reg. Cont., CGS, RG 192, NA.

Contract prices reached a record high in 1866. During that year five individuals or firms agreed to deliver within the District of New Mexico a total of 161,000 pounds at prices ranging from 9 to 14½ cents per pound (see Table 6). Mesilla merchants Joseph Reynolds and James E. Griggs held more contracts than anyone else and received the highest prices. Two contractors, Santa Fe merchant Hyman Rinaldo, a native of Russia, and Santiago L. Hubbell, would each supply two other garrisons. Joseph Hersch, another Santa Fe merchant, agreed to deliver the largest amount. Hersch had signed his contract late in the year following a poor harvest when beans were scarce. McClure reported that Hersch controlled more beans "than any other person probably in New Mexico." In January 1867, McClure increased Hersch's contract by a third, instructing him to deliver an additional 21,666 pounds.[55]

The price of beans fell dramatically in later years. In February 1868, McClure purchased from 2,000 to 6,000 pounds each for Forts Selden,

Table 6
Bean contracts—District of New Mexico, 1866

Post	*Amount in pounds*	*Price in cents per pound*	*Contractor*
Los Pinos	5,000	9	Santiago L. Hubbell
Albuquerque	6,000	10½	William and C. W. Lewis
Fort Bliss	10,000	10¾	Reynolds and Griggs
Fort Cummings	10,000	14½	Reynolds and Griggs
Fort Craig	20,000	11¾	Santiago L. Hubbell
Fort Marcy	5,000	9¼	Hyman Rinaldo
Fort McRae	10,000	14½	Reynolds and Griggs
Fort Selden	10,000	12¼	Reynolds and Griggs
Fort Union	20,000	12¼	Hyman Rinaldo
	65,000	11⁷⁄₁₀	Joseph Hersch

Data from Reg. Cont., CGS, RG 192, NA.

McRae, Bascom, Plummer, Wingate, Stanton, Craig, and Garland at prices ranging from 4¾ to 8⁷³⁄₁₀₀ cents per pound. Although the bidding had been competitive, no contracts were issued because the amounts were too small. But Joseph Hersch received contracts to deliver 20,000 pounds at Fort Union at 5⁷⁴⁄₁₀₀ cents per pound and the same amount at Fort Sumner for a slightly higher price.[56]

McClure solicited new bean proposals in November 1868 for supplying a total of 79,000 pounds to the twelve posts in the district. More than half the twenty-seven individuals who submitted bids were merchants, who would secure their supply from Hispanic farmers.[57] Five individuals shared the ten contracts awarded in December and January, including John Lemon of Mesilla and Benceslado Luna of Los Lunas who each agreed to deliver 19,000 pounds. Jacob Gold, the 16-year-old son of Santa Fe merchant Louis Gold, received the contract to put in 8,000 pounds at Fort Stanton, but only after his father had given written consent.[58] Because the amount required at Forts Garland and McRae was small, the army purchased the supply for these posts without issuing contracts.

In October of 1869 McClure reported that all beans issued during the fiscal year ending 30 June had been furnished within the district, and this probably held true for the next four or five years. During much of this time, Las Cruces entrepreneur Henry Lesinsky seemingly controlled the southern New Mexico bean market. Between 1870 and 1873 he received fifteen of sixteen contracts issued for delivery of this commodity at Forts Bayard, Cummings, Craig, McRae, Selden, and Tularosa. No contractor similarly dominated bean contracts in the north, where only Zadok Staab and Nathan Eldodt held more than one contract during the above four years.[59] Starting in 1875 some posts probably began receiving their bean supply from Fort Leavenworth. In the previous year the district's acting commissary had complained about the quality of the New Mexico bean. He claimed that Mexican beans were "a very poor substitute for the white bean and are a not very desirable article of diet, for continual use."[60] But the main reason that eastern beans came to be substituted for the New Mexico variety was the fact that they were cheaper and more plentiful. The territorial bean crop failed in 1877 and again in 1879, making it impossible to get beans locally at "a reasonable price." The district's supply in 1880 was shipped from Fort Leavenworth, and thereafter only small amounts were purchased in New Mexico.[61]

During the late 1860s the west Texas posts received their supply of beans from San Antonio. Colonels Edward Hatch and Wesley Merritt, commanding officers at Forts Stockton and Davis, deplored this practice as a needless waste of money. Hatch suggested buying beans for his post in Chihuahua or New Mexico. One hundred pounds delivered at Fort Stockton from El Paso or Chihuahua, he advised, would cost $6.50; delivered from New Orleans (via San Antonio), $19.25.[62] Merritt recommended that contracts be let closer to Fort Davis, where Mexican beans recently had been delivered at 5½ cents per pound. Merritt described them as being "very much superior to the white beans furnished by the Commissary Department."[63]

By at least 1870 bean contracts for the west Texas posts were being awarded to local residents. During the ensuing decade prices would range from 4 to 12½ cents per pound. Not surprisingly, many contractors were merchants, including New Mexico resident Henry Lesinsky, who delivered beans at Forts Bliss and Quitman. Railway transportation eventually made it cheaper to import beans, and by 1885 soldiers stationed in west Texas no longer consumed large amounts of Mexican frijoles.[64]

Farmers in Arizona would not raise enough beans to meet the army's requirements until the 1870s, which is not surprising considering the unsettled condition of the countryside. Sherod Hunter's invasion had caused Union sympathizers to abandon their lands for safety in Mexico; other residents had fled to escape fierce Apache raiders. On 19 February 1866 Colonel Charles A. Whittier, on an inspection tour through Arizona, reported: "The people have commenced to return to Tucson, Tubac, and the intervening country since its reoccupation by United States troops, but very little work upon the land appears to have been done."[65] Consequently during the remainder of the decade the army had to import from San Francisco the bean supply for Camps McDowell, Hualpai, Date Creek, Mojave, Verde, and Fort Whipple.[66] It is impossible to estimate what percentage delivered elsewhere came from the bean-growing region of southern Arizona, for contractors there imported large amounts from Sonora. At least three contractors who delivered beans at Tucson late in the decade were merchants who had business dealings in Sonora. These men—Charles T. Hayden, Estévan Ochoa, and Barron M. Jacobs—probably secured beans to fill their contracts from both sides of the border.

Among Arizona's most successful pioneer merchants were the brothers Barron and Lionel Jacobs. They had arrived in Tucson in 1867 to establish a retail store in partnership with their father, Mark I. Jacobs, a prosperous California merchant. Mark, the senior member of the firm, bought merchandise in San Francisco and arranged to have the goods transported to Tucson. Besides selling merchandise, the brothers established an exchange and loan business, exchanging gold coins for greenbacks and making short-term loans at 5 percent interest, twice the rate charged in San Francisco. Like many frontier merchants, Lionel and Barron believed in diversifying their investments, "so as not to have all our eggs in one basket." Government contracting helped them secure the financial success that allowed the Jacobs Company in 1875 to be listed as the fifth largest retail firm in Tucson.[67]

The Jacobs firm received its first supply contracts in October 1869, agreeing to deliver both beans and flour to the army depot in Tucson. The firm contracted to deliver by 1 December 32,000 pounds of beans at a low 3 74/100 cents per pound. Grain contracts also had been awarded at unprecedented low rates, causing dismay for local farmers like Leslie B. Wooster. Wooster had opened a farm that year a few miles above Tubac, where he planted corn and beans intending to sell his crops to government contractors. But because of the low prices Wooster refused to sell.[68] Local farmers may even have joined forces two years later to prevent Tucson merchant William Zeckendorf from filling his contract, which called for a supply of beans at a low 3 24/100 cents per pound. After Zeckendorf failed to furnish the requisite amount, the army purchased beans in open market from merchants Edward N. Fish, Mariano Molina, Isaac Goldberg, and one other supplier, at 15 and 25 cents per pound. In following years the contract price for beans delivered at Tucson and Camp Lowell ranged from 4 9/10 cents per pound paid to Zeckendorf in 1874 to 9 94/100 cents paid to Frederick L. Austin in 1871.[69]

Surviving records of the Jacobs firm reveal some of the complex business transactions that characterized government contracting. In the fall of 1871 William B. Hellings of Phoenix received contracts to put in a total of 54,952 pounds of beans at five posts north of the Gila River at prices ranging from 8 to 11½ cents per pound. Later that year the Jacobs brothers were purchasing beans at 6 cents a pound from Modesto Bórquez of Altar, Sonora, which they subsequently tried to sell to William Bichard of Adamsville for 10 cents a pound. The Jacobs Company needed cash, and in hopes of quickly closing the transaction Barron reminded Bichard that contractors for the northern posts would have to purchase beans in southern Arizona. By controlling part of the supply, Bichard would make a "handsome" profit. That the Jacobs brothers profited from this or a similar transaction is clear from the note Barron sent his father on 18 January 1872, reporting that "our trade with Borquez for Beans at 6¢ was a good one."[70]

When business slowed in the family store during the spring of 1872, Barron concentrated on the military market. In a letter dated 24 April to his father and Lionel, then in San Francisco, Barron wrote: "The way business is now I can see no opening unless we go for and get contracts, as the idea of selling $11 to $40 a day [in merchandise] savors more of the ridiculous than the profitable."[71] He then formed alliances with other entrepreneurs wanting to share the army's business. He joined William F. Scott, James Lee, Charles H. Lord, and Wheeler W. Williams in submitting a bid for the Tucson flour contract, and he united with Charles A. Shibell, Joseph Goldtree, and Oscar Buckalew in seeking the McDowell hay contract.[72] In neither venture were the men successful. But Barron did receive the bean contract at Tucson.

Table 7
Bean contracts—Department of Arizona, 1873

Post	*Amount*	*Price in cents per pound*	*Contractor*
Camp Apache	as required	7 85/100	Santiago L. Hubbell
Camp Bowie	as required	6	Louis Rosenbaum
Camp Date Creek	7,500 pounds	9 25/100	Morris Goldwater
	as required	9 25/100	Sprague and Webb
Camp Grant	as required	6½	Louis Rosenbaum
Camp Hualpai	as required	10 2/10	Morris Goldwater
Camp Lowell	as required	8 7/10	Barron M. Jacobs
Camp McDowell	as required	11 15/100	Morris Goldwater
Camp Verde	as required	8 85/100	Santiago L. Hubbell
Fort Whipple	as required	9 85/100	Santiago L. Hubbell

Data from Reg. Cont., CGS, RG 192, NA.

He and Sabino Otero, who farmed near Tubac, had submitted a joint bid for the supply "as required" at 8 72/100 cents per pound. Otero agreed to raise and deliver the beans at market rates, not to exceed 5 cents per pound; then he and Jacobs would divide the profits.[73]

Between 1871 and 1877 all posts within the Department of Arizona, with the exceptions of Yuma and Mojave, were supplied with beans under contracts issued in the department. Yuma and Mojave received their supply from California. Three New Mexicans—Nathan Bibo, Santiago L. Hubbell, and Louis Huning—received all the bean contracts issued for Camp Apache, and Hubbell held contracts in 1873 for Camp Verde and Fort Whipple as well. A fourth New Mexican, Louis Rosenbaum, obtained more than a half dozen contracts to supply beans at Camps Bowie, Grant, and Thomas.[74] Clearly, a large amount of New Mexico beans was being transported into Arizona, a fact that may have pleased Arizona's chief commissary, Captain Charles P. Eagan, who reported in August 1877 that the New Mexican frijole "is the best bean [and] is preferred by troops" (see Table 7).[75]

But Eagan also recommended that posts in Arizona receive their supply of beans from San Francisco. Under a recently published advertisement, Eagan had received only one proposal for this commodity.[76]

According to the registers, Louis Huning signed the last bean contract issued within the department on 7 December 1877, agreeing to deliver 8,000 pounds at Camp Apache during the first half of the new year. Two years later in his 1879 report, Eagan wrote that although he had repeatedly advertised for beans, they could not be produced in Arizona and delivered at reasonable rates. He therefore was ordering them direct from the depot in San Francisco "as the cheapest and surest manner of supply."[77]

With the coming of steel rails the military would also import flour to supply its Southwest garrisons. The story of the army's quest to secure good quality local flour, however, illustrates the great impact the military had on the region's economy. In response to the military market, farmers grew more wheat, mill owners received scarce investment funds, and local residents reluctantly adopted new technologies.

By the close of the Civil War, local gristmills were manufacturing most of the flour that the army required within the District of New Mexico. But soldiers complained that bread baked from this flour was sour, musty, gritty, and unpalatable. Complaints arose in part because of the inexperienced bakers, primarily enlisted men who received extra duty pay of between 20 and 35 cents per day.[78] On occasion, the army hired civilian bakers, as it did at Fort Quitman in 1869 when the commanding officer there reported that "not a man in the command . . . can perform the duty. Every one that has tried spoiled all the flour that they attempted to bake."[79]

The majority of complaints focused upon the quality of flour that the bakers used. Territorial millers manufactured their flour from wheat threshed on the ground by driving animals over it, a practice that often left the flour dark, dirty, and gritty. Major Hugh Fleming, commanding officer at Fort Garland, was only one of several officers who believed that soldiers deserved a better quality flour, and he requested that his post be furnished with "good Saint Louis flour." In a letter to department headquarters dated 10 March 1868, he graphically described some problems with locally ground flour:

> The wheat is gathered and tramped out on the ground, generally by sheep or goats; then scraped up; winnowed, and sent to the mill. Some calculate that our flour is composed of one third wheat, one third earth, manure, etc. This calculation is not cor-

> rect, but the flour we have is not what troops have a right to expect. Troops should have good flour. I have not seen any really good bread since I have been in this District—it all has grit in it and cannot be ground exceedingly fine.[80]

Two years later officers at Fort Selden reported that about 55,000 pounds of flour that Socorro mill owner Manuel Vigil had supplied were musty, full of weevils, and unfit for issue. Each sack reportedly contained from one to two handfuls of worms. Lieutenant Edmond G. Fechet, the post's commissary officer, attributed the flour's condition to the fact that millers in Socorro used no smut machines but rather "cleanse their wheat by washing it, previous to grinding; they then dry it in the sun, and grind it." According to Fechet, "wheat that has been washed is very difficult to dry; consequently the flour made from this imperfectly dried wheat becomes heated, fermented, and spoiled after a short time."[81]

Officers who served as chief commissary in the district more often praised than condemned the locally ground flour. Captain Charles McClure in 1868 reported to his superiors that excellent flour, free of all grit, was manufactured in the district. Officers and enlisted men alike, he stated, preferred New Mexico flour to States flour. When complaints surfaced, the blame rested with uninformed officers who believed that all New Mexico flour contained grit and consequently accepted inferior flour from contractors.[82] Captains William H. Nash and Charles P. Eagan, who in turn succeeded McClure, defended local flour, believing that it made good, nutritious bread suitable for troops, though neither considered it equal to the best brands of St. Louis, San Francisco, or Leavenworth flour.[83]

The three captains identified four mills as grinding the best flour in New Mexico. Two were in Taos, one owned by Frederick Mueller and Joseph Clothier and the other by David Webster; the third was Henry Lesinsky's mill in Las Cruces; and the fourth belonged to Louis and Henry Huning of Los Lunas. McClure claimed that flour manufactured by Mueller and Clothier was superior to that of any mill in northern New Mexico. He reported that "even in Santa Fe for family use it brings in the market a higher price than any other of home manufacture."[84] Nash judged Webster's flour equally superior, a judgment apparently shared by a citizen baker at Fort Marcy who refused to bake bread unless he was supplied with Taos flour (he did not specify which brand). Nash also praised the flour that Lesinsky manufactured.[85] And Eagan reported in 1876, after Lesinsky no longer furnished flour to the military, that Taos flour and the Hunings' flour were the best in New Mexico.[86]

But even the best flour was sometimes not good enough. In 1869 the post commissary at Fort Craig rejected about 13,000 pounds of David Webster's flour, and in 1873 a board of officers at Fort Stanton rejected a lot of Henry Lesinsky's flour, in both cases the flour being judged inferior to samples the contractors had submitted with their bids. And in 1878 officers at Fort Craig refused to accept about 21,000 pounds of Louis Huning's flour, claiming it had become wet in transit.[87]

The army devised various methods for testing the quality of contract flour, including the simple technique of chewing a small amount to detect grit. A frequently used method, and one McClure considered the fairest, was the baking of bread using contract flour and another brand and then comparing results.[88] Nash incorporated a variation of this method in 1871 while investigating complaints at Fort Union where Vicente St. Vrain and Vicente Romero were delivering flour under contract. St. Vrain's flour gave satisfaction, Romero's caused the complaints. Their mills were less than ten miles apart and drew grain from the same area. St. Vrain, the son of Ceran St. Vrain, probably operated the stone mill his father constructed at Mora, a mill capable of grinding 100 bushels of grain a day. Romero's mill was on his property at La Cueva and could grind 220 bushels a day. After carefully inspecting the St. Vrain and Romero flour, Nash concluded that "one was but little if any better than the other." He then set out to prove his point. He had samples of each baked by two different people not connected with the post bakery, the result being "a fair article of bread from both brands." Next, he placed some of Romero's flour in a sack branded as St. Vrain's flour and placed St. Vrain's flour in a sack marked as Romero's. He sent both to the post bakery to have baked into bread. Later the post commander judged the bread marked St. Vrain's (in reality Romero's flour) as light, sweet, "the best bread . . . seen at the post for issue to troops." The bread marked Romero's (in reality St. Vrain's flour) was judged as inferior. Thereafter Romero's flour was treated as equal in quality to St. Vrain's.[89]

Two years later, in 1873, Nash employed a different test on some of Z. Staab and Company's flour that a board of officers had rejected at Fort Union. Selecting a sack at random, Nash sent part of the flour to a private baker and part to the post baker. The citizen baker produced bread that was white, light, and of excellent quality; the post baker's bread was yellow and "bad looking." Nash concluded that "as both flours came out of one and the same sack . . . the fault is in the Post Baker."[90]

In August of 1865 the commissary department awarded near record prices to New Mexico's flour contractors. They were the highest paid in the district at any time during or after the Civil War, ranging from 11 to $18\frac{95}{100}$ cents per pound.[91] Acting Inspector Major Andrew J. Alexander later suggested collusion between contractors and Commissary William Bell, but Bell's successor, Charles McClure, believed that a poor harvest in 1865 accounted for the prices. McClure noted that even at Fort Union, located in the best grain region in New Mexico, the army had had to pay Lucien B. Maxwell 11 cents per pound and Ceran St. Vrain 15 cents per pound to secure a supply of flour.[92] Despite high prices, the army accumulated such a surplus of flour that in 1866 it issued only one new contract. And in November of that year McClure recommended that some of the surplus, particularly bad and deteriorating flour, be sent to Bosque Redondo for issue to Navajos.[93]

The army awarded new flour contracts in 1867 for a total of 1,550,000 pounds at much reduced rates of $4\frac{7}{10}$ to $8\frac{1}{2}$ cents per pound.[94] Contractors were required to deliver "the best quality of superfine flour, well bolted, and perfectly free from grit, smut and all foreign substance, and put up in strong cotton sacks containing one hundred pounds net each, branded with the name of the contractor and the date of manufacture."[95] The largest contract went to Vicente Romero for delivery of 500,000 pounds at Fort Union, and the second largest to David Webster, who would put in 250,000 pounds at the same post. William H. Moore received the most contracts, agreeing to deliver a total of 230,000 pounds at Forts Bayard, Cummings, Selden, and Stanton. Moore and his partner William Kroenig operated a water-powered mill near La Junta (now Watrous), capable of grinding 100 bushels a day. Commissary McClure described the flour that Moore later delivered as being "of very superior quality."[96]

Between 1867 and 1879 flour delivered under contract in the District of New Mexico was least expensive at Forts Marcy and Union, where the average price was about $4^{59}/_{100}$ and $4^{90}/_{100}$ cents per pound, respectively.[97] Most contractors for the two posts were mill owners. But a few, like Abraham Staab and Louis Clark, were merchants who obtained flour from the larger mills. In November 1871, for example, Staab offered to furnish Fort Marcy with 10,000 pounds of Webster's Taos flour, whose superiority over other brands was so well established that Staab did not think it necessary to submit a sample.[98] Taos flour was transported both to Santa Fe and to Fort Union by burros. David Webster told military officials in October 1867 that he would start fifty pack animals at a time across the mountains carrying flour to Fort Union. He would send an outfit this size, attended by eight or ten Hispanic packers, two or three times a week and sometimes daily until the contract was completed. Because Apaches had been raiding in the area, Webster requested that soldiers accompany the packers who usually were without weapons.[99]

Flour delivered under contract (between 1867 and 1879) was most expensive at Forts McRae, Bascom, and Cummings, where average prices were about $6^{4}/_{100}$, $6^{23}/_{100}$, and $6^{56}/_{100}$ cents per pound, respectively. None of the posts was close to a major gristmill, and transportation costs added significantly to the cost. For the same reason, prices were nearly as high at Forts Wingate and Bayard: $5^{94}/_{100}$ cents per pound at Wingate and $5^{99}/_{100}$ cents per pound at Bayard.[100] David Webster, who supplied flour at Fort Wingate in 1871 at $6.71 per 100 pounds, paid freighters $2.50 per 100 pounds for transporting flour from Taos to the post. And freight expenses could be even higher for other posts. When the wheat crop failed south of the Jornada in 1868, the bids at Forts Cummings and Bayard were so high—between 11 and 12¼ cents per pound—that McClure rejected them and shipped flour from Union depot. The contract price at Fort Union was $5 per 100 pounds, but freight expenses raised the cost of flour delivered at Fort Cummings to $9.10 per sack and at Fort Bayard to about $9.56 per sack.[101]

Mesilla Valley residents received most of the flour contracts at Forts Bayard, Cummings, and McRae, and Henry Lesinsky and Louis Rosenbaum held more than anyone else. Both contractors also furnished flour at Forts Bliss, Quitman, and Selden, and Lesinsky held contracts at Forts Craig, Davis, Stanton, Stockton, and Tularosa as well. Contract registers lend support to the statement appearing in an 1872 issue of the Las Cruces *Borderer* that Lesinsky's Las Cruces Flouring Mills "grinds the most of the flour used in this section by the military as well as citizens."[102]

Lesinsky had purchased the steam-powered mill in January 1870, paying Numa Granjean and Arnold Audetal $8,000. The mill was located on the main road leading from Las Cruces to Mesilla and was then capable of grinding only forty-two bushels a day. To fill his government contracts, Lesinsky purchased wheat from grain dealers in the Mesilla Valley, El Paso, Texas, and in villages farther south. Sometime in 1874 he leased the mill to Swiss-born Jacob Schaublin and soon stopped bidding on government contracts, concentrating his energies and resources on developing the Longfellow Mine in Arizona.[103] Rosenbaum, who signed his first flour contract in 1870, took in wheat from the surrounding countryside and then had it ground at one of a handful of flour mills operating in the Mesilla Valley.[104]

The Hunings of Los Lunas, rather than Lesinsky or Rosenbaum, however, became the premier flour contractors in New Mexico, receiving more flour contracts after the Civil War than any other residents. Louis Huning, the older of the two brothers and in whose name most contracts were issued, started bidding on flour contracts in both New Mexico and Arizona soon after establishing his mill in 1870. Success came in 1872 when he was awarded his first contract at Camp Apache in Arizona. In 1874 he signed his first flour contract in the District of New Mexico, agreeing to furnish the amount needed at Fort Tularosa.[105] Clearly, Huning had needed time to establish his reputation as a superior miller. His bids in 1872 for supplying flour at Forts Wingate and Craig had been lower than those of Abraham Staab, who received the contracts. Army Inspector Nelson H. Davis called attention to this fact, stating that "L. Huning's flour was reported to me to be equal to any made in New Mexico, if not superior."[106]

Between 1872 and 1883 Huning contracted to deliver flour at Forts Craig, Marcy, McRae, Stanton, Tularosa, and Wingate in New Mexico and at Forts Apache, Verde, and Whipple in Arizona. He was the only individual who delivered flour under contract at Fort Wingate from May 1875 through 1879 and at Fort Apache from 1878 through mid-1884. It is not possible to know exactly what these contracts meant in actual dollars, as many were for the "supply needed" rather than for an actual amount. None of the five contracts Huning held for New Mexico posts in 1876, for example, specified an amount. In 1878 he contracted to deliver a total of 240,000 pounds in New Mexico, which would cost the army almost $11,000. His contracts at Fort Apache called for delivery of a total of 685,000 pounds at a total cost to the government of $41,978.[107]

Huning's chief competitor for the flour contract at Fort Marcy was Z. Staab and Company, a Santa Fe firm composed of the brothers Zadoc and Abraham. The Staabs had emigrated from Germany in the 1850s and then settled in Santa Fe, where they found employment in the mercantile establishment of the Spiegelbergs. Zadoc and Abraham established their own firm in 1859, and within two decades Z. Staab and Company had become one of the largest stores in Santa Fe. Sometime after the Civil War, the exact date is unknown, Zadoc moved to New York as the resident buyer while Abraham remained in Santa Fe to run the business.[108] During the 1870s the brothers won contracts to supply the army with a variety of commodities—salt, beans, cornmeal, and flour, as well as corn, oats, and bran. Their first flour contracts were awarded in January 1872 to supply Forts Craig, Garland, Marcy, Stanton, Union, and Wingate with a total of 420,000 pounds, for which they would receive about $22,550. The following year they held contracts at Marcy and Union. In 1874 contracts for Garland, Marcy, and Union were issued in the names of Staab and Alexander Gusdorf, the Staabs' nephew, who had emigrated to the United States to work for his uncles. The Staabs purchased flour from Taos millers to fill their first contracts. But sometime in the early 1870s, at least by September 1875, the brothers owned or had obtained a financial interest in one of the Taos mills. During the 1870s Gusdorf lived in Taos and managed the firm's flour business.[109]

Money from government contracts aided the Staabs in expanding their commercial enterprise from a small retail store carrying a stock of $20,000 to a mammoth wholesale-retail business with a stock worth $300,000 in 1882. The firm sold goods throughout New Mexico, southern Colorado, eastern Arizona, west Texas, and into Mexico. Staab money helped build the magnificent Palace Hotel in Santa Fe and provide the city with its first gaslights.[110]

Although one of the strongest firms in New Mexico, Z. Staab and Company did not monopolize military supply contracts. In 1875 the firm lost the Fort Marcy flour contract to Louis Huning. And Huning would win six of eight flour contracts issued there between 1875 and 1879. A grasshopper infestation in Taos Valley in 1877 gave Huning an edge over Taos millers in that year and probably the year following. Frederick Mueller told the Santa Fe press in August that the entire Taos Valley wheat crop had been destroyed and that little flour would be sent to Santa Fe. Another correspondent reported that farmers had not harvested enough wheat "to seed the land for next Spring's planting," a great loss to citizens who usually sold wheat to the two large Taos flouring mills.[111]

From all indications the Staabs and the Hunings enjoyed harmonious relations with the commissary department. But some other contractors, including Santa Fe merchants and mill owners Joseph Hersch and Louis Gold, earned reputations as troublemakers and were prohibited from bidding on government contracts. For Hersch and Gold, the partnership of federal government and private enterprise brought more adversity than prosperity.

Prior to the Civil War Joseph Hersch had been one of four gristmill owners who had supplied the army with the bulk of the flour it consumed in New Mexico. In 1859 he had received a contract for 320,000 pounds, but an army inspector later classified his flour as "indifferent—or bad." Subsequently Secretary of War John B. Floyd ordered that flour of this quality not be accepted.[112] According to Captain John P. Hatch, commissary officer at Santa Fe, the commissary general of subsistence also directed that no further contracts be given to Hersch. Hatch felt justified, therefore, in rejecting the bids that Lipman Meyer submitted in June 1861 for supplying flour at Forts Garland, Marcy, Albuquerque, and Stanton, even though he was the lowest bidder. Meyer, a Santa Fe merchant, was manufacturing his flour at a mill formerly occupied by Hersch, and Hatch concluded that Meyer's bid was a cover for Hersch, who intended to provide the flour.[113]

Despite Hatch's claims, Hersch was awarded contracts in the 1860s for furnishing flour at Los Pinos, Bascom, Marcy, and Sumner. Some of the flour he delivered, however, was gritty, and when he and the firm of Mueller and Clothier offered in September 1868 to supply flour at Fort Bascom at the same price, Commissary McClure awarded the contract to Mueller and Clothier, believing theirs the superior flour.[114] Complaints about Hersch's flour again surfaced early in 1869. McClure subsequently instructed the commissary officer at Fort Marcy to warn Hersch that flour would be purchased elsewhere at his expense unless he supplied a higher quality flour.[115] Meanwhile, Hersch was becoming entangled in a bitter controversy that involved the army, Louis Gold, and his son Abraham.

A native of Poland, 33-year-old Louis Gold reported having personal assets in 1860 of $8,000, which would place him among the merchants in Santa Fe of modest means. During the war years he received contracts to supply Fort Marcy and Albuquerque with flour, but he failed to complete the Albuquerque contract. A few years later, in 1868, his 21-year-old son Abraham won contracts to put in flour at Forts Marcy, Stanton, and Wingate, but officers soon complained that the flour Abraham delivered was inferior. In February 1869, McClure received permission to exclude both Abraham and his father from bidding on army subsistence supplies. "It is not the first instance," he explained, "in which the Commissary Department has experienced trouble and annoyance from the failure on the part of the Golds to strictly comply with their contracts."[116]

Abraham Gold was still legally responsible for filling his contracts. But the one at Fort Wingate, calling for delivery of 225,000 pounds of flour at 6 81/100 cents per pound, caused the trouble. Joseph Hersch, the subcontractor, failed to keep the post supplied with flour, and in early April 1869 the garrison ran out. McClure doubted that either Gold or Hersch would complete the contract, and consequently he withheld payment on flour that already had been delivered. Learning of this action, Hersch tried to salvage some of his investment. Convinced that Gold would never pay him for flour already delivered, Hersch sent word to the freighter en route to Wingate with a new shipment to deliver the flour to the post trader rather than to the commissary. When the freighter arrived, however, the post commissary seized Hersch's flour, claiming that it had been shipped to satisfy Gold's contract. McClure suspected that Hersch had wanted the post trader to hold onto the flour so that it could be sold later to the army at a greatly increased price.[117]

Hersch protested the government's action, claiming that the flour was his and that his arrangement with Gold as subcontractor had been "entirely a private matter . . . not understood and entirely unrecognized by the Government."[118] But McClure would neither return the flour nor pay for it until outstanding charges against the contract were settled. In June McClure received authority to purchase 150,000 pounds of Taos flour from Stephen B. Elkins, who agreed to deliver it at Fort Wingate at $8.25 per 100 pounds. The difference between this price and the contract price would be charged to the contractor. By this time Abraham Gold and one of his bondsmen had left the country, and the other bondsman was bankrupt. Neither Hersch nor the Golds again held contracts with the commissary department, though they may have furnished supplies and services to other contractors.[119]

By the mid-1870s a new controversy was brewing over the quality of even the best New Mexico flour. The first salvo was fired by William H. Gardner, surgeon at Fort Union, who submitted a detailed report on 16 February 1876 comparing the quality of Taos and Leavenworth flour. Gardner found that Taos flour was of a much lower grade, coarser, and contained much more dirt and foreign matter. He concluded that "the bread made from the Leavenworth Flour is white, well raised, sweet and nutritious, and will keep without turning sour for four or five days. The bread made from Taos Flour is of a dirty yellowish tinge, is not so light and spongy, has not the same sweet pleasant taste and readily turns sour if kept more than forty-eight hours." Major James F. Wade, commanding at Fort Union, forwarded Gardner's report to District Commander Edward Hatch, with a recommendation that Leavenworth flour be furnished at his post.[120]

District Commissary of Subsistence Charles P. Eagan was quick to defend the local product. He noted that Taos flour gave universal satisfaction among the troops and that by purchasing territorial flour the army encouraged local industry and avoided the risk of having eastern flour damaged in transit. Eagan warned his superiors that the subject of substituting eastern flour for the local article was a delicate one "apt to raise a storm," and during the spring and early summer he mounted a campaign in defense of territorial flour. He forwarded samples of local flour to General Robert Macfeely, commissary general of subsistence, and to Captain Jeremiah H. Gilman, chief commissary of subsistence for the Department of the Missouri, requesting that the flour be examined and graded. He told Gilman that bread made of Taos flour was served at his own table and that the same flour was used by the wealthiest New Mexicans who could afford a superior flour if they desired it. He told Macfeely that it would cost the government more than $25,000 to supply the district with flour from Leavenworth, while the best grades of New Mexico flour would cost $8,000 less. He believed the savings justified purchasing local flour. Furthermore, he had encouraged millers to import threshing machines, which would eliminate impurities in locally ground flour. It was possible, Eagan wrote, that if the government refused to purchase New Mexico flour for the coming fiscal year, more threshing machines would be imported and "good clean flour [would be furnished] in future at lower figures than it can be laid down from the East, at the posts."[121]

Though Colonel Hatch supported the importation of eastern flour, Macfeely ruled in May that New Mexico flour would be purchased if the quality was equal to Huning's and Rosenbaum's.[122] This ruling gave the two men a decided advantage. In June Louis Huning and his brother contracted to deliver flour at Forts Craig, Marcy, McRae, Stanton, and Wingate, and Rosenbaum agreed to supply Forts Bayard and Selden. However, extension of the railroad to within 110 miles of Fort Union and 25 miles of Fort Garland allowed these posts to be supplied with eastern flour.[123]

General Macfeely in his 1876 annual report expressed hope that the introduction of threshing machines would allow residents to manufacture a superior grade flour so that in the following year all flour required in the District of New Mexico would be purchased from local millers. Captain Frederick F. Whitehead, who replaced Eagan as commissary, sent a copy of this report to Rosenbaum, Huning, and Mueller and Clothier with the additional information that Z. Staab and Brother had already introduced a modern threshing machine, and he wanted to know what steps they had taken to improve the quality of their flour. Whitehead also ended many complaints about the bread ration by issuing orders, sanctioned by Macfeely, to mix one fourth eastern flour with three fourths New Mexican flour in the baking of bread.[124]

In the following months the army continued its efforts to induce change among New Mexico's farmers and millers. In June 1878, Macfeely instructed Whitehead to accept proposals only for flour manufactured from machine threshed wheat. However, Whitehead doubted that anyone would submit bids under these conditions. He reported that Hispanic farmers objected to the use of threshing machines because they reduced the number of laborers needed for threshing and by thoroughly cleaning the wheat reduced the weight of the quantity to be sold. In addition, merchants and dealers were unwilling to pay a higher price for better cleaned grain. Macfeely consequently issued new instructions to receive bids from all suppliers, giving preference to flour ground from machine threshed wheat.[125] The army's campaign to effect change seemingly was successful. Records suggest that at least some flour delivered in 1878 by Louis Rosenbaum and Louis Huning, who between them held most of the contracts that year, was ground from machine threshed wheat.[126]

Changes and improvements came slowly—too slowly for the army. Starting in 1879 the commissary department began shipping more eastern flour to New Mexico for issue to the troops. By the end of the year States flour was being issued at Forts Union, Garland, Lewis, and Marcy and comprised half the ration at all other posts. Whitehead reported that the increased allowance of States flour was a decided improvement, "as it seems impossible to manufacture a good grade of flour in this Territory."[127] Captain Charles A. Woodruff, who replaced Whitehead as district commissary, stated in his 1882 annual report that only a small amount of flour had been purchased locally during the preceding fiscal year, as flour was more cheaply supplied from Kansas.[128] The millers of New Mexico had lost their biggest customer, and enlisted men now were consuming bread baked from Kansas flour. This change, however, failed to stop complaints about the flour ration. Captain John B. Guthrie, post treasurer at Fort Cummings, informed his commanding officer in December of 1882 that the flour on hand, "XXX Fogarty," manufactured in Junction City, Kansas, was of inferior quality and that "the different methods of making good bread from it prove unsuccessful." He requested that the post be furnished with a superior brand, "XXX Hoffman and Son," milled at Enterprise, Kansas.[129] The names of the millers had changed, but the nature of the soldiers' complaints remained much the same.

By the 1880s west Texas posts were also receiving flour manufactured outside the Southwest, for railroads allowed the importation of a finer and cheaper flour than was available locally. At a much earlier date, and prior to the arrival of railroads, Forts Davis and Stockton had received their supply from New Orleans by way of San Antonio. This was expensive flour. Colonel Edward Hatch, commanding at Stockton, reported in 1868 that a barrel of flour shipped from New Orleans cost about $36; transported from El Paso, about $21. Colonel Wesley Merritt, commanding at Davis, claimed that the reason the commissary department purchased flour in the East was because the quartermaster's department bore the expense of transporting it west, enabling the commissary department to report a savings. Merritt thought this was "a poor economy."[130]

Among the first to hold flour contracts at Forts Davis and Stockton were Hardin B. Adams and Edwin D. Wickes, who in December 1870 agreed to furnish the supply needed at Fort Davis for $8\frac{23}{100}$ cents per pound and in January 1871 contracted for the supply at Fort Stockton for $9\frac{23}{100}$ cents per pound. Well on their way to becoming the dominant freighting firm between San Antonio and the west Texas posts, Adams and Wickes probably obtained their flour along the Rio Grande, possibly in Presidio del Norte, Mexico, or in San Elizario and El Paso, Texas.[131] Colonel James H. Carleton, in his 1871 inspection report, stated that the previous contractor at Fort Stockton, S. S. Brown, had obtained his flour from Presidio del Norte, about 150 miles southwest of the post, and had charged the army $11\frac{3}{10}$ cents per pound. On the same inspection tour, Carleton found that Fort Quitman received a superior quality flour ground at the "Army Mills" at San Elizario.[132]

Carleton also reported on the quality of bread issued at the west Texas posts. He recorded that even though Fort Stockton's bakehouse was dilapidated and its oven almost worthless, the bread was good and wholesome. A sergeant and two privates did the baking at Fort Davis, baking one batch a day except on Saturday when two were baked. The capacity of the oven was 560 loaves; the bread was of excellent quality. Two soldiers baked excellent bread at Fort Quitman, though the bakehouse was old and would soon fall unless supported by buttresses. The bakehouse at Fort Bliss was also in need of repair. Henry Curley, a civilian, baked for the post, receiving two sacks of flour a month for his wages. Carleton, however, directed that Curley's services be terminated and that soldiers resume baking their own bread.[133]

Between 1870 and 1880 seven individuals or firms contracted to deliver flour at both Forts Davis and Stockton, where the average contract price was $6\frac{11}{100}$ cents and $6\frac{3}{10}$ cents per pound, respectively. Besides Adams and Wickes, contractors included Las Cruces mill owner Henry Lesinsky, who furnished flour at Davis in 1873 and 1874 and at Stockton in 1874, and El Paso merchant Joseph Sender, who supplied flour manufactured in El Paso to both posts in 1876 and 1877. Daniel Murphy, who farmed on Toyah Creek thirty-five miles north of Fort Davis, supplied the posts in 1875 with flour he had manufactured on or near his property. It is not known where Edward Fenlon, a freight contractor, and James L. Millspaugh, post trader at Fort Concho, obtained flour they supplied in 1878 and 1879. But William Heuermann, who also supplied Forts McKavett and Concho, bought his flour in St. Louis in 1879, shipping it by rail to Austin and then freighting it to the west Texas posts. Heuermann also furnished flour in 1880, the last year in which flour contracts for Davis and Stockton appear in the registers.[134] Thereafter the army supplied the posts with eastern and northern flour.

The price of flour delivered under contract at Fort Bliss between 1868 and 1882 ranged from 3⁶⁄₁₀ cents per pound in 1879 to 7³⁄₁₀ cents per pound in 1870; the average price was about 4⁹⁶⁄₁₀₀ cents. Most contractors were El Paso merchants or post traders, but flour was also supplied by Las Cruces residents Louis Rosenbaum in 1871 and Henry Lesinsky in 1873. Moritz Lowenstein, postmaster and merchant at Ysleta, Texas, furnished flour in 1879, and Sigmund Wedeles, now selling goods out of Santa Fe, New Mexico, supplied 26,000 pounds late in 1882.[135]

Wedeles probably imported his flour by railway, and Rosenbaum and Lesinsky undoubtedly furnished flour manufactured in the Mesilla Valley. The El Paso contractors may have secured some of their flour from Hart's Mill (El Molino) located at the head of the rapids on the Rio Grande near the center of modern downtown El Paso. Simeon Hart, a native of New York, had marched to New Mexico with the Missouri Mounted Volunteers during the Mexican War. In 1849 he and his bride Jesusita Siqueiros settled on the east bank of the Rio Grande, where Simeon built a house and a small mill—he did not finish his large mill until 1854.[136] Before the Civil War, Hart supplied large quantities of flour to the U.S. army, but much of it was ground at his father-in-law's mill at Santa Cruz, Chihuahua. A supporter of the Southern Confederacy, Hart left the area with Sibley's retreating troops, whereupon Union officials seized his property. In December

1865 El Molino was sold at public auction to Nathan Webb and Henry J. Cuniffe for $3,000.[137]

Hart secured a pardon from President Andrew Johnson in November 1865, but it was not until 1868 that the Supreme Court overturned the El Paso confiscation cases. Even then Hart was unable to resume his profitable milling business because of a lengthy lawsuit that Unionist William W. Mills brought against him for personal damages alleged to have been inflicted during the early stages of the war.[138] Meanwhile the mill suffered from neglect. A traveler writing to the Las Cruces *Borderer* in April 1871 noted: "We jogged along slowly by Hart's Mills, now in a venerable and dilapidated condition."[139] Shortly before this notice appeared, Louis Cardis, an Italian who had arrived in El Paso toward the end of the war, announced in the pages of the same newspaper that he had "come into possession of the Flouring Mill at the head of the Rapids in the Rio Grande at El Paso del Norte (Texas)." He had equipped the mill with steam machinery, and he promised government contractors that he would grind flour as cheaply as any other mill in the country. "The established reputation of this mill for turning out the very first quality of flour will, we hope, guarantee us a full share of patronage."[140] Exactly when the mill was restored to its former working capacity remains unknown.

None of the sources mention El Paso as having more than one gristmill in the 1870s. Perhaps some flour was imported from Mexico, for with a population of only 736 in 1880, the town conducted a surprising amount of business "with the frontier and interior towns of Northern Mexico."[141] Twenty miles below El Paso at San Elizario, a town of about 900, a large steam gristmill ground wheat from both sides of the Rio Grande, and Fort Bliss contractors probably secured some of their flour from this mill. In 1877 Charles E. Ellis, a former soldier in Carleton's California Column, purchased the mill from J. W. Campbell. By the end of the year, both Ellis and Louis Cardis, the one-time operator of Hart's Mill, were dead, killed during the opening skirmishes of the El Paso Salt War.[142]

Although Carleton reported in 1871 that Fort Quitman's flour came from San Elizario, he failed to record the contractor's name, and it is not listed in the registers. During a five-year period, starting in 1872, the average price of flour delivered under contract at Fort Quitman was 4 96/100 cents per pound. Contract prices were the highest in 1872, when Louis Rosenbaum received $5\frac{95}{100}$ cents per pound, and the next highest in 1873, when Henry Lesinsky received $5\frac{20}{100}$ cents per pound. The three remaining contractors were El Paso residents Joseph Sender, Joseph Schutz, and Adolph Krakauer.[143]

Farmers and merchants in Arizona benefited as much from the army's call for flour as their counterparts in New Mexico and west Texas. Following the Civil War, however, only a few gristmills manufactured flour in Arizona, and they proved incapable of meeting the home demand. Ammi M. White's mill at the Pima villages supplied some of the flour used at Camp McDowell and at garrisons south of the Gila. Sometime in 1867 William Bichard and his younger brother Nicholas, both English emigrants, purchased White's mill and operated it under the name of W. Bichard and Company. Army Inspector Roger Jones rated their flour as superior to that ground in Tucson or in Sonora, and in August of that year William and Nicholas signed contracts to furnish the army with 100,000 pounds (delivered at the mill) at $6\frac{7}{10}$ cents and 199,470 pounds at $5\frac{95}{100}$ cents per pound in coin.[144]

The great flood on the Gila in September 1868 destroyed the Bichards' mill, storehouses, and 28,000 pounds of army flour stored therein. Major Andrew J. Alexander, then commanding at Camp McDowell, blamed the Bichards for the army's loss, for the brothers had neglected to send the flour forward as ordered. Thereafter the army stopped receiving flour at local mills and required contractors to make deliveries directly to each post.[145] In 1869 the Bichards constructed a new gristmill at Adamsville, three miles west of modern Florence. Exactly when the Bichards' "Pioneer Mills" started grinding is not known, but in June the brothers signed contracts to deliver 100,000 pounds of flour at Camp Grant and 300,000 pounds at Camp McDowell at $9\frac{1}{4}$ cents per pound in coin.[146] The Bichards' flour received mixed ratings. Officers at Camp McDowell complained that the flour was inferior, but officers at Camp Grant expressed satisfaction, reporting that it was "free from all foreign substances and pleasant and agreeable to the taste, making as good an article of bread as could be made from California flour."[147]

During the 1860s much of the flour issued to troops south of the Gila came from Mexico. In June of 1865 Charles T. Hayden contracted to supply 21,000 pounds of Sonora flour at Fort Goodwin and the same amount at Fort Bowie, all at 17 cents per pound in coin. The only other flour contract issued in Arizona that year was awarded to Amos F. Garrison, post trader at Fort Mason, who probably imported Sonora flour to fill a large contract at Tubac.[148] Between 1867 and 1870 Thomas M. Yerkes, a merchant at Tubac, contracted to supply Tucson depot with a total of 823,750 pounds of flour at prices ranging from $5\frac{94}{100}$ to $8\frac{74}{100}$ cents per pound in coin. Most likely he obtained flour from across the border to fill at least part of his government contracts. The local press noted in April 1869 that Apaches had attacked one of his wagon trains on its return from Sonora, capturing forty-eight of his mules.[149] And almost certainly during the summer of 1869 James P. T. Carter, ex-secretary of the territory, furnished flour ground at his mill in Altar, Sonora, to fill a contract at Tucson depot. The 45-year-old Carter had agreed to put in 135,000 pounds at $8\frac{41}{100}$ cents per pound in coin by 1 July. He died later in September, and his widow was soon offering for sale a two thirds interest in the Altar steam-powered mill.[150]

Flour manufactured in Sonora had a reputation for being badly bolted. Millers there washed their wheat prior to grinding, and according to one commissary officer the moisture retained in the wheat "causes insects to be developed in the flour in a few months." "The strangest thing is," he added, "that the flour retains its sweetness even after worms and weevils are in it."[151] A better quality flour was ground at Joseph Pierson's mill at Terrenate, Sonora, about 120 miles from Tucson. By 1870, and possibly even earlier, Pierson's mill was supplying some of the flour issued to troops in southern Arizona. Army Inspector Charles A. Whittier had visited Terrenate in 1868 on an overland trip from Tucson to Guaymas. "In Terenate [*sic*]," he later reported, "we find a decaying, idle town, the only life in the place comes from the flouring mill of Joseph Pearson [*sic*], French by birth but living in Sonora as many do, as an American citizen." Pierson's water-powered mill was capable of grinding 60 *fanegas* of wheat per day, and, Whittier noted, "his flour bears a high reputation."[152]

Until 1872 garrisons in central Arizona received most of their flour from San Francisco at tremendous cost to the government. The Prescott *Arizona Miner* reported in 1867 that flour rarely cost less than $20 per hundred pounds at that place. "If there is any one thing our people need more than another," the editor claimed, "that thing is a good flouring mill."[153] By the end of the year, at least two small mills were processing some of the local wheat crop: Herbert and Nathan Bowers's mill on Granite Creek a short distance above Prescott, and Timothy Lamberson's mill at Walnut Grove twenty-five miles to the south. About 1 October 1868 the Bowers's mill burned to the ground, and a few weeks later the press reported that Prescott was entirely without flour.[154] But the brothers soon began building another mill on the Agua Fria, eighteen miles from Prescott, by importing machinery and a miller from San Francisco to oversee operations. In October 1869 the press reported that the Bowers's mill was grinding corn and soon would start grinding wheat. Some Prescott merchants at this time were also stocking their shelves with New Mexico flour.[155] In later years mill owners near Florence and Phoenix would hold most of the government contracts at Fort Whipple.

In 1873 nine steam- and water-powered gristmills in Arizona manufactured an estimated 3,786,400 pounds of flour, still not enough to meet the home demand.[156] The army alone required at least one third of this amount, although the military represented only about 18 percent of the population. Several mills that supplied the army with flour were relatively new, with their owners hoping to tap the growing market. The enterprising Bichards built two mills—one at Adamsville in 1869 and the other at Phoenix in 1871, both at the center of important grain regions. On 4 July 1871 the Phoenix mill reportedly produced the first flour ground in the Salt River Valley, but in September the mill was destroyed by fire. W. Bichard and Company never rebuilt, though for a short time the firm maintained a store and flour depot at Phoenix.[157]

The Bichards faced strong competition from William B. Hellings, who in partnership with his brother Edward and two others erected the Salt River Flouring Mill near Phoenix in 1871. The Prescott *Arizona Miner* called the Hellings Mill "the finest flouring mill this side of San Francisco," put up at a cost of nearly $70,000. Powered by steam, the mill could manufacture 30,000 pounds of flour daily.[158] The first large flour contracts held by W. B. Hellings and Company were for supplying posts north of the Gila, including Date Creek, Hualpai, Verde, and Whipple, during the first half of 1872. The manner in which the firm obtained these contracts signaled to the Bichards that a formidable rival had emerged. These posts previously had been supplied from San Francisco, but in 1871 Lieutenant Colonel Marcus D. L.

Simpson, chief commissary officer for the Division of the Pacific, decided that the posts might be furnished with flour as cheaply or perhaps at a lower price by local contract. When permission to advertise for supplies was delayed, Simpson decided to accept the proposal Nicholas Bichard had submitted in June for supplying flour to the above four posts. Simpson apparently felt justified in taking this action because Bichard's prices were less than the cost of importing flour from San Francisco and because it would assure the posts a steady supply of flour. In explaining his decision, Simpson later wrote: "Mr. Bichard was a man of large means, owned two flouring mills in Arizona, and manufactured flour of excellent quality, especially from his mill at Adamsville."[159]

But Simpson's action was premature. Without knowledge of this arrangement with Bichard, the commissary officer at Tucson distributed posters calling for proposals for flour and other supplies required throughout the department with 12 August designated as the day for opening bids. Hellings soon learned of Bichard's contracts. Armed with information about the prices Bichard was promised, Hellings submitted bids for supplying the central posts that were ½ to 1½ cents lower. Hellings also wrote to the Secretary of War, claiming that the arrangement with Bichard would result in "great loss to the Government" and asserting that his own flour was superior to Bichard's. Despite Simpson's protest, the commissary department canceled the contracts with Bichard and awarded new contracts to Hellings.[160]

Hellings would soon face stiff competition from yet another mill owner, Charles T. Hayden, who began erecting a mill at Hayden's Ferry eight miles east of Phoenix shortly after locating there in 1871. Hayden operated a store and a ferry in addition to the water-driven gristmill, and the growing settlement that gathered about his establishment acquired the name of Tempe. Hayden's mill began producing flour in 1874, the same year that he received contracts for supplying Camp Verde and Fort Whipple. He would later hold flour contracts at Camps Huachuca, Lowell, and McDowell and additional contracts at Verde and Whipple.[161]

Two other mill owners competing for government contracts were James Lee and William F. Scott, who built Tucson's first steam-powered mill in 1870. During the previous decade the partners had operated a water-powered mill a short distance from town. In 1874 Lee and Scott sold the steam flour mill, known as the Eagle Mill, to Tucson merchant Edward N. Fish. Fish operated the mill until 1888, supplying flour to Camps Bowie, Grant, Lowell, Thomas, and Huachuca.[162]

Between 1869 and 1885 flour delivered in Arizona under contract was the least expensive at Camp McDowell, located close to the Salt River Valley, the principal wheat growing district in the territory. Contract prices at McDowell were the highest in 1869 at $9\frac{25}{100}$ cents per pound and the lowest in 1885 at $2\frac{27}{100}$ cents per pound. The average price was about $4\frac{63}{100}$ cents. Most contracts went to mill owners, including William and Nicholas Bichard, Charles T. Hayden, William B. Hellings, Charles H. Veil (who succeeded Hellings in 1876 as sole proprietor of the Salt River Flouring Mill), and John Y. T. Smith, sutler at McDowell and after 1876 part-owner of the newly erected Phoenix Flour Mills.[163] Michael Goldwater, Polish-born grandfather of former U.S. Senator Barry Goldwater, was among a handful of merchants holding flour contracts at McDowell who were not also mill owners. Michael and his brother Joseph had opened a store at the Colorado River mining town of La Paz in the 1860s. In 1872 Michael and his son Morris moved to Phoenix, where they established a branch of the mercantile firm of J. Goldwater and Brother. Another merchant supplying the post with flour was German-born Charles Goldman, who settled in Phoenix in 1878 and managed a dry goods and general merchandise store with his brother Leo. Neither Goldman nor the Goldwaters dealt heavily in flour. According to the registers the Goldwaters held government flour contracts only in 1874 when they agreed to supply Camps McDowell and Mojave, and Goldman held contracts only in 1884 when he contracted to supply Forts Grant, McDowell, Thomas, Lowell, and Verde.[164]

Contract prices for flour delivered at Camp Lowell during the above years were almost as low as prices at McDowell, ranging from $8\frac{41}{100}$ cents per pound in 1869 to $2\frac{55}{100}$ cents in 1885, with the average about 5 cents a pound. Through the mid-1870s merchants rather than mill owners furnished most of Lowell's flour, much of it coming from Sonora. The Tucson *Arizona Citizen* reported in September 1874 that local mills could not grind enough flour for the market and that merchants were receiving shipments from Terrenate in Sonora and from the Bichards' mill at Sanford (formerly Adamsville).[165]

The Jacobs Company in Tucson served as forwarding agent for

Joseph Pierson's mill in Terrenate. The brothers used Pierson's flour to fill contracts to Camp Lowell in 1869, 1871, and 1873, and they also sold his flour to other contractors. As mentioned earlier, Barron Jacobs tried without success to capture the Camp Lowell flour contract in 1872. His maneuverings to obtain the contract and his apparent willingness to work against the successful bidder reflect the business climate of his era and the acquisitiveness of territorial merchants. Jacobs had planned to secure most of the flour to fill the contract from Pierson and a smaller amount from mill owners in Altar, Sonora. In April Barron told his father that he was sending his bid to Prescott with Tucson merchant Isaac Goldberg, paying him an extra $50 so that he (Goldberg) would not interfere with Barron's plans. But in May Jacobs withdrew this bid and joined mill owners James Lee and William F. Scott and merchants Charles H. Lord and Wheeler W. Williams in submitting a joint bid at a higher figure. In reporting this action to his father, Jacobs said that the three companies had "formed an alliance offensive and defensive on the flour question." Even though Williams had been confident of receiving the contract, it was awarded to Bichard and Company, whose bid of $5.15 per 100 pounds was $1.80 lower than the combined bid mentioned above. Bichard also received flour contracts that year to supply Date Creek, Grant, Hualpai, McDowell, Verde, and Whipple. Jacobs believed that Bichard's bids had been too low and indicated in a letter to his father on 25 May that disgruntled competitors sought revenge: "Fish and Co., and others are going for Bichard and Co., on account of their taking the contracts so low." And Jacobs planned to join the attack. He instructed Pierson to buy up all the wheat he could find as flour soon would command high prices. "We must try to play [Bichard] out with flour," he wrote, ". . . it makes no difference whether we have a contract or not."[166]

The records do not reveal whether William Bichard had difficulty filling his contracts, but within a year of receiving them the 32-year-old Englishman was dead. His brother Nicholas carried on the business for a few more years, filling contracts at Camps Date Creek, Grant, Lowell, and Fort Whipple. Sometime after signing his last contract, in June 1876, Nicholas apparently relocated to San Francisco where other members of his family resided.[167] Thereafter three other mill owners dominated flour contracting at Camp Lowell. Twelve of nineteen contracts issued there between 1875 and 1885 were held by

Edward N. Fish, Charles T. Hayden, and John Y. T. Smith. Hayden's flour delivered in 1879 was judged equal to the "best California flour," but flour that freighter Aaron Barnett supplied in 1880 was so inferior that the post baker had to mix it with a better quality flour to make "passable bread."[168]

Flour furnished at Camp Bowie under contract between 1871 and 1885 was almost as cheap as that delivered at Camp Lowell. Contract prices ranged from 6 95/100 cents per pound in 1871 to 3 1/10 cents in 1885, with the average being about 5 1/5 cents per pound. Merchants from the Mesilla Valley in New Mexico (about 160 miles east of Bowie) supplied most of the flour between 1871 and 1879. Louis Rosenbaum and his partner Simon M. Blun held contracts in three of these years and Charles Lesinsky in four. H. Lesinsky and Company had opened a branch store in Tucson during the late 1860s, but it had closed in February 1873. Charles probably filled his contracts with flour ground at Jacob Schaublin's mill in Las Cruces rather than with Sonora or Arizona flour.[169]

Among Arizona residents holding flour contracts at Bowie in later years were the brothers Adolph and Isadore E. Solomon, emigrants from Poland. In 1876 Isadore Solomon and his wife Anna had settled in Pueblo Viejo, forty miles southwest of Clifton, where Isadore produced charcoal for his wife's cousin, Henry Lesinsky, for use at the Clifton smelter. Isadore's business enterprises soon expanded to include merchandising, milling, farming, and cattle raising. He constructed a gristmill at Pueblo Viejo (renamed Solomonville in his honor) sometime after March of 1881, but even before this he and his brother Adolph were engaged in wheat and flour transactions. In September 1879 when flour was scarce along the Rio Grande, Isadore was buying large quantities of wheat in Arizona for grinding into flour, which he then shipped to Silver City, Clifton, and Globe. And in November of that year, Adolph Solomon contracted to deliver 30,000 pounds of flour each at Forts Bowie and Thomas and 155,000 pounds at Fort Grant. Isadore won flour contracts at Fort Thomas in 1882 and at Forts Bowie and Thomas in 1885. The Solomons undoubtedly secured their wheat from farmers who had taken up land in the upper Gila River Valley of Arizona.[170]

Flour delivered under contract in Arizona between 1871 and 1885 was the most expensive at Forts Apache, Verde, and Whipple, where the average price was about $6\frac{55}{100}$, $6\frac{18}{100}$, and $5\frac{98}{100}$ cents per pound, respectively. These posts were far from major wheat-producing regions, and transportation costs added to the cost. New Mexicans dominated flour contracting at Fort Apache, receiving every contract issued between June 1872 and November 1883, the last year a contract appears in the registers. Louis and Henry Huning held most of the contracts, but Charles Lesinsky of Las Cruces furnished flour in 1873 and 1874 and Michaelis Fischer of Socorro in 1876 and 1877.[171] During Victorio's War flour prices at Apache rose from $5\frac{45}{100}$ cents per pound in November 1879 to 7½ cents in March 1880. Higher prices were warranted. A few days after the Hunings received a new contract in 1880, Apaches attacked their wagon train about eighty miles from Los Lunas en route to Fort Apache, killing some of the teamsters. The cargo of flour had to be abandoned.[172]

The awarding of the Camp Verde flour contract to the Hunings in March 1873 served as a warning to Arizona mill owners that outsiders might also capture the government's patronage there. Consequently William B. Hellings, who had held the contract in early 1872, employed ten men that summer to build a road from Phoenix to Verde. A newspaper correspondent, writing in the 27 September issue of the Tucson *Arizona Citizen,* explained the situation: "The people of New Mexico are making a strong effort to secure the trade and furnishing of supplies for Verde and unless this road be completed, it is quite likely they will be successful against all competitors, save by those who grow produce in Verde Valley." The writer predicted that the new road would reduce the cost of freighting to Verde from the grain region by 1½ cents per pound of freight.[173]

The Hunings, in fact, were the only New Mexicans to win a flour contract at Camp Verde, although others supplied it with beef, beans, and cornmeal. Charles T. Hayden held more flour contracts at the post than anyone else, furnishing that commodity in 1874, 1875, 1878, and 1881. Flour was also supplied by Nicholas and William Bichard, Charles H. Veil, Charles Goldman, John Y. T. Smith, Frank E. Jordan, and Martin W. Kales, a Prescott banker. Most of these contractors also furnished flour at Fort Whipple, where Hayden and Smith were the most successful. Each held contracts in five of the fifteen years under consideration.[174]

The Southern Pacific Railroad entered Tucson in March of 1880, and with it came opportunity to receive a finer quality flour. Complaints had already surfaced about the quality of Arizona flour. In commenting upon Edward N. Fish's flour, a commissary officer at Camp Bowie had stated in 1878 that troopers could eat it if absolutely necessary. "Most of the citizens of the Territory," he added, "subsist on articles of the ration of inferior quality."[175] Commissary Charles P. Eagan believed that Arizona flour was inferior to New Mexico flour and that both articles were inferior to California flour. In August 1880 he requested authority to advertise for flour in the Los Angeles *Star* and in the San Francisco *Alta*, hoping that favorable railroad rates would elicit competitive bids.[176] A year later Eagan reported that completion of the railroad had resulted in increased competition and in a reduction of flour prices.[177] Still, between 1880 and 1885, most contracts to furnish flour at Arizona posts went to residents of either Arizona or New Mexico.

Only two California contractors have been identified, although undoubtedly there were others. In 1883 Walter S. Maxwell of Los Angeles agreed to deliver 40,000 pounds of flour at Fort Bowie at 4 cents per pound and 70,000 pounds at Fort Grant at 3 98/100 cents per pound.[178] The following year bids for supplying flour at Apache, Bowie, Huachuca, Thomas, and San Carlos were rejected as too high, and flour was shipped to these posts from the San Francisco military depot.[179] Although two Arizona mill owners, John Y. T. Smith and Isidore E. Solomon, dominated flour contracting in 1885, receiving between them contracts for supplying nine Arizona posts, only one resident was awarded a flour contract in 1886. This was Charles M. Strauss, who agreed to supply Fort Lowell with 20,000 pounds. For the remainder of the decade, Los Angeles resident Isaac N. Van Nuys contracted to deliver between 100,000 and 120,000 pounds of flour annually at Hackberry, Arizona, a town located on the railroad line a few miles above Kingman.[180] During these same years the army must have imported additional quantities from outside the territory, for only a few other flour contracts appear in the registers for the Arizona garrisons.

The arrival of railroads in the Southwest allowed the army to import most of the items in the soldier's ration that two or three decades earlier it had encouraged local residents to supply. The military demand for bread literally created the flour-milling industry in the Southwest. Even after the military market vanished, however, local mills continued for several years to process wheat from local farmers and provide flour for the home market. And even though railroads helped curtail the government market for certain commodities, railroads also encouraged the growth of a local cattle industry that became tied to the national economy. As early as 1867 Commissary Charles McClure foresaw some of the changes railroads would bring to the local economy. "Should the Pacific Railroad pass through New Mexico," he wrote to superiors in Washington, "grain can be brought here so cheaply that the staples of this country will, in a few years, be cattle and sheep and wool and wine."[181] Moreover, it was the military market and the protection that troops extended to cattlemen that allowed the western cattle industry to expand while the North and South were still recovering from a devastating civil war.

5

Ranchers and Contractors

SPANISH explorers introduced cattle into the Southwest in the sixteenth century, but frequent and disastrous Indian depredations prevented the development of large-scale cattle ranching in either the Spanish or Mexican period. After American occupation, U.S. military posts provided both a market and protection for cattle herds. Local residents, however, could not immediately satisfy the army's demand for beef, and eastern cattle had to be driven west to supply the garrisons. By the close of the Civil War government contractors were bringing in Texas cattle. But ruinous competition, Indian raids, bad weather, and harsh terrain often prevented contractors from realizing a profit. Nonetheless, the government market for beef sparked interest in stocking Southwest ranges and led to rapid development of the cattle industry in both Arizona and New Mexico.

At the start of the North–South conflict, contractors in New Mexico were supplying local beef to most garrisons in the department.[1] But the Confederate invasion and General Carleton's subsequent war against the Navajos placed such heavy demands on local beef herds that government contractors soon had to import eastern cattle to supply the garrisons. In September 1864, for example, Charles S. Hinckley journeyed to Kansas to purchase cattle for seven posts, having agreed to supply a total of 4,000 head at 18 cents per pound. The following year James M. Kerr signed a contract to deliver 2,500 head at Fort Union at 11 36/100 cents per pound. Most of his cattle would also come from Kansas.[2]

Contractors soon discovered an untapped reservoir closer to home—the wild and neglected longhorns of Texas. Left unattended during the Civil War, the Texas herds had greatly multiplied. Ranchers and investors alike soon recognized that high profits would be made if Texas cattle were driven to beef-starved eastern markets. Within a short time, cattlemen had blazed several trails north out of Texas to intersect railroad tracks pushing west. Among the earliest and most dangerous was the Pecos Trail, later called the Goodnight–Loving Trail, which followed the Middle Concho River to its headwaters and then moved west across almost eighty miles of waterless plains to Horsehead Crossing on the Pecos River. From there, cattle were trailed northward and entered New Mexico near Pope's Crossing. The trail led up the Pecos River Valley to the Canadian River Valley and entered Colorado through Trinchera Pass or Raton Pass.[3]

Although the trail was named for Charles Goodnight and Oliver Loving, pioneer Texas cattlemen who drove cattle over the trail in 1866, government contractors had brought Texas cattle into New Mexico the previous summer over the same route. Thomas L. Roberts, a former captain in the California Column, had signed a contract in July 1865 to deliver 100 head of cattle at Camp Nichols, a new post that Colonel Kit Carson was establishing about 130 miles east of Fort Union. In September Roberts received additional contracts to supply fresh beef at Forts Sumner and Stanton at 14 and 18 cents per pound, respectively.[4] By this date James Patterson, Roberts's partner, had twice successfully brought cattle into New Mexico from Texas, and on at least one drive the army had furnished him a military escort. General Carleton instructed officers at Fort Union early in September to loan Patterson a dozen Sharps carbines, for he was about to return to Texas to get another lot of cattle and would have to travel through Indian territory. Carleton wanted to encourage the introduction of Texas cattle into New Mexico, correctly predicting that the cattle Patterson was bringing in were only the beginning "of a great and profitable trade."[5] The trailing of Texas cattle up the Pecos by Roberts and Patterson preceded by several months the more famous drive north to the railroad terminal at Sedalia, Missouri, by Texas cattlemen in 1866, the year often associated with the start of the open-range cattle industry.

News of Roberts's and Patterson's success must have percolated rapidly through the military-contractor community, for competition to supply the government with cheap Texas cattle soon intensified. Entrepreneurs surely realized that contracting was as much a gamble as prospecting the hills for gold. But supply contracts were among the few investments available at the war's end, and visions of federal dollars excited the gambling instinct in many a westerner. The twelve military posts comprising the District of New Mexico would require substantial amounts of fresh beef. But the big market was the Navajo reservation at Bosque Redondo where more than 8,000 captives received government rations.

Roberts and Patterson felt the full impact of the heightened competition, for in 1866 they lost the Bosque Redondo beef contract to another risk-taking capitalist. In May of that year Jesse Connell of Leavenworth, Kansas, agreed to deliver 6,000 head of cattle in monthly installments to the Navajos at a then-record low of 5 4/10 cents per pound. But Connell would lose money on the contract. When he was unable to make the first installment of 500 head, Chief Commissary Captain William H. Bell purchased 447 head from Roberts and Patterson at 15 cents per pound, charging the difference between this and the contract price to Connell. James M. Kerr, who was then in New Mexico serving as Connell's agent, complained that the purchase of so many cattle and at such a high price was unjust to Connell, who was expected to arrive any day with Texas longhorns.[6]

Captain Bell later explained that because Roberts and Patterson held the only cattle for sale near the reservation, he had to buy from them on their own terms. Carleton defended Bell's actions. The general had learned that Connell probably would fail on his contract and consequently had impressed upon Bell the need to keep at least sixty days' supply of meat on hand for the Navajos. "The beef must be there," Carleton warned, and he instructed Bell to buy it where best he could. Carleton had little sympathy for the Leavenworth contractor. Not only had Connell failed to deliver cattle, but he also had assailed the character of Fort Sumner's officers. Carleton also implied that Connell had recklessly underbid local contractors—"good, careful, reliable business men . . . who counted all the costs of an enterprise of this sort, and proposed accordingly."[7]

Some of the cattle that Roberts and Patterson sold to the army belonged to the first herd that Goodnight and Loving drove up the Pecos in 1866. The men had started near Fort Belknap, Texas, on 6 June with 2,000 head. By 4 July they had sold Roberts and Patterson more than 800 steers for 8 cents per pound and were grazing the remaining cattle north of Fort Sumner near a creek called Las Carretas. Shortly thereafter Loving drove the cattle on into Colorado, where he sold them to pioneer rancher John W. Iliff while Goodnight returned to Texas for another herd.[8]

Meanwhile, in August of 1866, Patterson received contracts to supply all the fresh beef needed during the coming year at Forts Sumner, Union, Stanton, and Bascom at 10, 10¼, 11¾, and 18 cents per pound, respectively. He and Roberts also sold beef on the hoof to Captain Bell at 8½ cents on Connell's unfilled contract. Because Connell was making no attempt to fulfill his agreement, Bell soon advertised for new proposals to furnish 5,000 head of cattle to the Navajos.[9]

Roberts and Patterson again were underbid. In December the army awarded a contract to Andy M. Adams, who agreed to deliver the cattle within ten months at $5\frac{19}{100}$ cents per pound. Patterson's proposal to furnish the Navajos with fresh beef (rather than beef on the hoof) at 7½ cents per pound was rejected. Bell's successor, Captain Charles McClure, estimated that the army would save more than $33,000 by accepting Adams's proposal rather than Patterson's.[10] But Adams would not realize a profit, due in part to the machinations of Roberts and Patterson. That government contracting involved financial risk, cutthroat competition, and financial ruin for some contractors is clearly documented in extant correspondence relating to Adams's unsuccessful efforts to fill his contract.

Adams's early life is obscure, but by 1866 he had acquired sufficient business acumen to secure as a bondsman Lucien B. Maxwell, one of New Mexico's wealthiest ranchers. And it was to Maxwell's ranch at Cimarron that McClure sent word notifying Adams that he had won the beef contract. McClure instructed Adams to deliver 900 head at Fort Sumner by 20 December, 600 head by 20 January 1867, 500 head by 20 February, and 500 head by 10 March.[11] Adams planned to purchase sufficient cattle to meet the first installments from drovers along the Pecos, thus allowing him time to gather cattle in Texas to fill the bulk of the contract. But Adams's low bid had aroused resentment, and the cattle dealers refused to sell. Commissary McClure concluded that the cattle dealers wanted to break Adams's contract and force the government to buy cattle at a higher price. According to McClure, 7,000 head of cattle were being held on the Pecos that winter near Fort Sumner: Loving and Goodnight held 3,000; H. W. Cresswell, 2,000; and Joe Curtis and John Dawson, 2,000.[12]

In January 1867 Adams arranged to purchase 700 steers from William W. Mills of El Paso, Texas, but before this herd reached Fort Sumner Patterson met Mills's drover and bought the cattle himself. Still, Adams or his agent Joseph A. LaRue managed to supply 711 head in January, but failed to furnish any the following month.[13]

Both McClure and Carleton were anxious to break the combination working against Adams. But foremost in Carleton's mind was the need to feed the captive Navajos; he instructed McClure to purchase cattle in open market whenever the contractor failed to deliver. In an attempt to outwit the cattle dealers, McClure suggested that Lucien B. Maxwell intervene or that LaRue employ some person unknown to the combination to make the necessary purchases.[14] Late in February McClure published a new advertisement calling on the public to notify his office of the number of cattle each individual held and their price. In a letter to the commissary officer at Fort Sumner dated 21 February, McClure wrote that he intended to help Adams complete the contract. "It is the object of this office," McClure stated, "to control the speculators, and not to let them ever dictate. By so doing the market price of everything will be kept at reasonable figures. This can be done in a country where the Government is the chief purchaser."[15]

Heading the combination against Adams were Thomas L. Roberts and James Patterson. On 4 March they submitted the following propositions for supplying the Navajos: they would deliver 3,000 head of cattle on the hoof at 8½ cents per pound, 4,000 on the hoof at 8 cents per pound, 5,000 on the hoof at 7½ cents per pound, or 5,000 head on the block (fresh meat) at 8 cents per pound. On the same date Lieutenant Robert McDonald, the commissary officer at Fort Sumner, reported that the combination (he mentioned Loving, Roberts, Patterson, Cresswell, and Dawson) refused to sell less than 2,000 head and were demanding 8½ cents per pound.[16] Convinced that these demands were unfair, Carleton suggested that the Navajos be placed on short rations and that all the sheep and cows on hand be killed to gain time for Maxwell to deliver cattle on Adams's contract. But the agent Maxwell sent to buy from the cattle dealers accomplished nothing.[17]

In March Vicente García of Galisteo offered to deliver at Fort Sumner from 100 to 200 head of cattle at 7 cents per pound. This proposal was quickly accepted, even though his cattle were broken-down and lacking in beef. McClure later reported that the purchase of García's cattle (132 head) had helped reduce the cost of the combination's cattle by one cent a pound. By mid-April the commissary department had purchased 1,427 from Thomas L. Roberts at 7½ cents per pound and at a total cost of $51,979.[18]

Meanwhile, Adams was in Texas arranging to drive longhorns to New Mexico, but misfortune dogged his every step. In May his partner, a Mr. Dillon, delivered 282 head at Sumner, all that remained of a herd of 1,000 after Indians had stampeded it and killed six of Dillon's herders. After losing three separate herds that summer to Kiowa and Comanche raiders, Adams asked to be released from his contract.[19] In forwarding this request to General Amos B. Eaton, commisary general of subsistence, McClure wrote:

> Mr. Adams is at present a ruined man and being unable to pay for the cattle purchased by him, has involved many others in his ruin. He has worked faithfully to fulfill his contract, and has done all in his power to deal honestly by the U.S. and as he has been singularly unfortunate, is entitled more favor in this instance, than is usually extended to contractors.[20]

But Adams was not released from his contract, and the commissary department continued purchasing cattle in open market. In July the army bought 759 head from Thomas Patterson, James's brother, at 6 cents per pound and in August 802 head from Texas cattleman John Hittson at 5½ cents per pound. By the time the contract was terminated in November, Adams owed the government $21,502, the difference between the contract price and the open market purchases.[21]

The federal troops that reoccupied Fort Stockton on 7 July 1867 arrived too late to help Andy Adams avert financial disaster. But the post would offer protection to other cattlemen, for it was located only thirty-five miles southwest of Horsehead Crossing, said to be the most dangerous point on the road leading to Fort Sumner and New Mexico. Patrols got under way almost immediately after the post was reactivated. In late July soldiers discovered signs of a large Indian war party ten miles north of the post going east in the direction of the Pecos River. By rapid marching Captain George Gamble and about fifty black troopers reached Horsehead Crossing in time to prevent the capture of two large herds being driven to Fort Sumner. Colonel Edward Hatch, the post's commanding officer, was confident that his troops could protect all herds if notified of probable time they would reach the crossing.[22]

John S. Chisum, later known as the "Cattle King of the Pecos," was among the cattlemen who successfully drove cattle into New Mexico shortly after Fort Stockton was regarrisoned. His herd of 1,000 head reached Bosque Grande, forty miles below Sumner, in August 1867, and that fall and winter he sold his cattle to "interested parties," most likely to government contractors. Chisum then entered an agreement with Charles Goodnight to deliver cattle from Texas to Bosque Grande, with Goodnight's herders driving what was not sold in New Mexico to Colorado or Kansas for sale.[23]

The army lacked sufficient manpower to protect all herds being driven up the Pecos Trail. During the summer of 1868 Chisum lost a herd of 1,100 head to Comanche raiders shortly after the same band of Indians had captured cattle from three other parties.[24] The army also lacked means to suppress the growing trade in stolen livestock. In reporting on Gamble's scout in 1867, Hatch described part of the problem: "The Indians have not only subsisted on cattle during the last year, but have also sold cattle on the Mexican line, so great has been their success in capturing droves and stealing from the settlements."[25]

Indians sold a good deal of the stolen cattle to the Comancheros, residents of New Mexico who had been traveling to the plains to trade with Comanches since the Spanish era. In the early years Comancheros traded biscuits, flour, cornmeal, and other food items for horses, mules, buffalo robes, and meat. During the 1860s the character of the trade changed, and New Mexicans started bartering powder, lead, guns, and whiskey for cattle stolen in Texas. Army officers attempted to restrict the commerce, but their efforts were ineffective, partly because official policy seemingly changed from day to day.[26]

Following Kit Carson's campaign against the Comanches in November 1864, Carleton ordered the arrest of all parties attempting to trade with these Indians. But a few months later, after a Comanche war leader made a peaceful visit to Fort Bascom, Carleton relaxed his trade restrictions, allowing traders with passes countersigned at his headquarters to travel to the plains for trade. Carleton even authorized Marcus Goldbaum, beef contractor at Ford Marcy, to go into Comanche country the summer of 1866 for the specific purpose of trading with the Indians for cattle, though Goldbaum was forbidden to carry with him for sale to the Indians "any arms, powder, lead, percussion caps, or any intoxicating liquor or spirits."[27]

Carleton must have felt uncomfortable with this decision because he soon reversed his stand. On 18 December 1866, undoubtedly with Carleton's approval, Captain McClure informed contractor Andy Adams that "no cattle bought from the Comanche Indians will ever be received by the Commissary Department of the District of New Mexico on contract, by purchase or otherwise."[28] But Carleton continued to doubt that the military had the legal right to restrain New Mexicans from buying cattle from the Indians, even though the Comanchero market encouraged Indians to raid in Texas. Still, early in 1867, after General Winfield S. Hancock advised him that military intervention was proper, Carleton issued a new ban against the Comanchero trade. Despite the army's persistent efforts to eliminate this illicit trade, the Comancheros sustained a thriving commerce for another decade.[29]

Although Andy Adams had lost heavily on his contract and Comanches threatened life and property, competition to sell Texas cattle to the government soon resulted in record low prices. In June 1868 Santa Fe contractors Charles Probst and August Kirchner agreed to furnish cattle for the Navajos during their exodus from Bosque Redondo at 5½ cents per pound, and about six or seven months later James Patterson received a contract to supply the Navajos with 500 or 600 head at 4 87/100 cents per pound.[30] These prices were in stark contrast to the cost of beef four years earlier when Charles S. Hinckley had received 18 cents per pound for beef on the hoof and 22 cents per pound for beef on the block. Contract prices for both categories of beef had remained high in 1865, but the price of beef on the hoof had fallen dramatically the following year.[31] The fact that the army continued paying high prices for beef on the block caught the eye of Fort Bascom's new commander, Major Andrew J. Alexander, and led to one of the most controversial military inspections in New Mexico's history. It also raises an unanswered question about Carleton's honesty in approving military contracts.

Upon taking command at Fort Bascom in January 1867, Alexander called attention to the high price the goverment paid for beef and requested that the post's contract be annulled. He subsequently wrote to his friend, General Cyrus B. Comstock, who was aide-de-camp to General Ulysses S. Grant, suggesting that "an honest inspector" be sent "to look into the matter of beef contracts . . . in which there has been considerable fraud." Shortly thereafter, Grant instructed Alexander to make a special inspection of military matters in New Mexico, reporting his findings directly to himself. In carrying out this task, Alexander apparently reported every innuendo and rumor that came his way. He had no experience and little understanding of the difficulty in procuring supplies in the West. When he discovered that high prices were paid to contractors, he imputed fraud without conducting detailed investigations.[32]

General Winfield S. Hancock, who commanded the Department of the Missouri, forwarded copies of Alexander's reports to Colonel George W. Getty, then commanding the District of New Mexico, and instructed him to make searching inquiries into Alexander's charges. Hancock expressed reservations, however, about Alexander's statements. He saw nothing out of the ordinary, for example, in high prices paid for beef, as prices fluctuated depending on the state of Indian hostilities and the difficulty in obtaining cattle. Colonel Getty and his staff subsequently searched files, reviewed orders, assembled price lists, and solicited testimony, all of which tended to exonerate the maligned officers in their official conduct.[33]

Captain Charles McClure, who replaced William H. Bell as chief commissary of subsistence on 3 November 1866, found satisfactory reasons for the high price of beef at all military posts in the district except at Fort Bascom. Alexander had claimed, for example, that the 12¼ cents paid to Abraham Cutler for delivering fresh meat at Albuquerque was extravagant. Not so, McClure replied. Only a small amount of beef was required for this garrison. Bidders consequently submitted high bids, for they would have the same expense in hiring butchers, herders, and other laborers as if they had furnished ten times the amount.[34] Nor did McClure believe that the 18¾ cents paid to James H. Whittington for supplying fresh meat at Fort Wingate exorbitant. "On account of the Indians being troublesome near Fort Wingate," he explained, "the prices ranged high, and no person would bid low, when he was in constant danger of losing his herd by capture, and the profits on a contract for a whole year, even when the price was eighteen and three-quarter cents, would be lost by the Indians stealing twenty beef cattle, as the amount of beef required was and is small."[35] And McClure easily explained Alexander's inquiry as to why Bell had awarded a contract to James Patterson for furnishing the garrison at Fort Sumner with beef for 10 cents per pound when McClure a few months later awarded a contract to Andy Adams for supplying Navajos with beef at 5 19/100 cents per pound. The contract Bell awarded was for beef on the block, which always commanded a higher price than on the hoof "as the expense of butchering, herding, etc. falls upon the contractor, as well as the decrease in weight of cattle during the winter months." Then, too, McClure stated, the large number of cattle required on the hoof enabled men to bid at low figures, which even he had found "surprisingly moderate."[36]

Even before Alexander took command at Fort Bascom, McClure had suggested to his superiors that the beef contract at Bascom be annulled. Bell had awarded a contract in August 1866 to James Patterson to deliver beef on the block at 18 cents per pound. In December McClure had learned that Patterson sublet his contract at one half the contract price and that other parties were willing to take the contract at 9 cents per pound. Shortly after Alexander voiced his complaints, the commissary general instructed McClure to terminate the Bascom contract and advertise for new proposals. Despite Patterson's involvement in the combination against Adams, McClure considered Patterson "a responsible man" and allowed him to compete for the new contract. Since Patterson's bid of 8 cents per pound was the lowest, McClure awarded him a new twelve-month contract, commencing 11 March 1867.[37]

Left unanswered was the question of why Bell had awarded the first contract to Patterson at such a high price when William B. Stapp and Charles Hopkins had been willing to take it at a lower figure. Bell's civilian clerk later stated that Stapp's and Hopkins's bid had been rejected because it had not conformed to the terms of the advertisement—the names of the sureties had not appeared on the bid itself but had been attached to it. McClure thought this slight informality would not have been sufficient reason for rejecting the bid. And since Carleton had disapproved the Stapp and Hopkins bid, McClure believed he was the only one who could explain the matter. But Carleton had been relieved of command in March 1867 and had left the territory before he could be questioned.[38] Because correspondence bearing on Carleton's response has not been located, it is impossible to judge whether he improperly intervened to assure Patterson the contract.

During his tour of the district, Alexander inspected troops stationed at Lucien B. Maxwell's ranch at Cimarron, site of the Moache Ute and Jicarilla Apache agency. Even though biased and inaccurate, the report Alexander subsequently submitted raises important questions about the federal government's role in the accumulation of personal fortunes. Clearly Maxwell valued government contracts; how much they contributed to his total wealth is not so clear. The Indian department had established an agency on Maxwell's ranch in 1861 at the suggestion of Kit Carson, the Indians' former agent, and there it distributed food to keep the Indians from raiding.[39] Maxwell was soon the agency's principal supplier of beef and grain. When lack of funds forced the Indian department to stop distributing food in mid-1866, the Indians became angry and restless. To avoid a calamity, Carleton authorized Maxwell to issue a daily ration of meal and meat to the Indians and ordered a company of mounted troops to remain at the agency to maintain order and supervise the issue of provisions.[40] Although the situation was less tense when Alexander visited Maxwell's ranch, the army was still feeding the Indians. In his report on the Maxwell agency Alexander described Maxwell, who even then was one of New Mexico's best-known residents, as rolling in wealth from having the Indians on his property. He also claimed that Carleton had ordered the Indians there "to personally aggrandize Mr. Maxwell."[41] Carleton, of course, had nothing to do with the agency's location, but without doubt Maxwell profited from his contracts to feed the Indians.

The amount Maxwell received from the Indian department is not known, but from September 1866 through August 1867 the army paid him $33,383 for supplying beef and wheat meal to the Indians. During this time he also supplied soldiers garrisoned at his ranch with $1,902 worth of fresh beef.[42] For the first seven months he had received 8 cents per pound for both the beef and meal and thereafter 6 cents for the meal and 8 cents for beef. Captain McClure, in reporting on Maxwell's contract, noted that the government had incurred no additional expense for herding, renting storehouses, or butchering, since Maxwell delivered meal and cattle as they were needed and the Indians did their own butchering.[43]

Despite McClure's favorable report, Maxwell's contract (which had not been made through competitive bidding) was terminated on 31 December 1867, and new proposals for feeding the Indians were solicited. McClure encouraged Maxwell to submit new bids since he had "given great satisfaction" in carrying out his former contract.[44] But new contracts for corn and wheat meal were awarded to Ceran St. Vrain, the lowest bidder, who would supply corn at $2\frac{98}{100}$ cents per pound and wheat meal at $3\frac{85}{100}$ cents per pound. Maxwell had submitted the lowest bid for beef, but because he would deliver all or none of the Indian supplies the beef contract was awarded to Moritz Loewenstein of Las Vegas at $5\frac{9}{10}$ cents per pound. With Maxwell no longer furnishing storage facilities, McClure ordered troops to build a log storehouse at the agency and requested permission to employ a herder at $35 per month and three Indians as assistants at $10 each.[45]

Maxwell again fed the Indians in 1869, receiving 7 cents per ration of one-half pound of corn, one-half pound of wheat meal, one-half pound of fresh beef, and one-fifth ounce of salt. A year later he offered to furnish supplies for $7\frac{1}{2}$ cents per ration, but the contract (now in the hands of the Indian department) was awarded to Frank Willburn and Thomas L. Stockton, who bid $6\frac{43}{100}$ cents.[46] By this date Maxwell was making plans to sell his estate and would soon leave Cimarron for his new residence at Fort Sumner. Exactly how much government contracting contributed to Maxwell's financial success remains a mystery. Like other entrepreneurs, Maxwell took advantage of economic opportunities as they arose. By the time the Cimarron Indian Agency was established, he was already a wealthy rancher, and in 1870 he reaped a fortune by selling his famous land grant. Maxwell's biographer believes that he profited handsomely by furnishing supplies for soldiers and Indians. Even so, having Indians on his property was a mixed financial blessing. Kit Carson testified in 1867 that Maxwell distributed annually to the Indians without reimbursement at least $5,000 worth of supplies. And Colonel George Getty reported in 1869 that Maxwell lost from $10,000 to $20,000 a year because of Indians—either in damages to his stock and property or in the presents he freely distributed.[47]

As more and more Texas longhorns were trailed up the Pecos, the average price of fresh beef delivered on contracts in the District of New Mexico fell from $12\frac{1}{10}$ cents per pound for the fiscal year ending 30 June 1867 to $9\frac{4}{10}$ and $7\frac{6}{10}$ cents per pound for fiscal years ending 30 June 1868 and 1869, respectively.[48] Contracts issued after 1866 (both on the hoof and on the block) were for the supply needed rather than for a specified amount. Captain McClure's statement in December 1867 that troops afforded a market for 6,000 to 7,000 Texas cattle annually gives some indication of the number the army required.[49]

While serving as the district's chief commissary, Charles McClure instituted several changes in contracting in hopes of reducing government expenses. In July 1867 he requested that bidders for fresh beef contracts state at what price they would also furnish mutton twice a week. He believed that issuing mutton as part of the meat ration would be more economical at small posts and at posts with insufficient grass for grazing cattle. A single steer often furnished more meat than a small garrison could consume before it spoiled. To reduce loss in the summer McClure also issued instructions to slaughter cattle late in the afternoon, allowing the meat to hang over night in a cool place, and to issue the meat early the next morning. What remained would be salted and issued a day or two later. But without refrigeration, spoilage remained a problem.[50]

How many garrisons were furnished mutton in 1867 is uncertain. In reporting on the beef contracts awarded in August, McClure told General Eaton in Washington: "A clause has been put in each contract for furnishing mutton, not oftener than twice per week, . . . in lieu of fresh beef on the block, only when the price of mutton was the same as the lowest bid for beef, or lower." New Mexico had more sheep than any other territory in the West, and its sheep had a reputation for yielding tasty mutton. Even so, the army awarded no mutton contracts in 1868 because beef was less expensive. In later years the army infrequently contracted for mutton, which caused the post surgeon at Fort Union in 1875 to lament: "We are in a country where Mutton is one of the commonest articles of food . . . yet it is impossible to get anything but beef." He recommended that mutton be furnished at least twice a week.[51]

The fresh beef contracts that McClure issued on 14 August 1867 required contractors to furnish upon request as many cattle on the hoof as would be needed by moving troops and scouting parties. This eliminated post herds and the expense of extra-duty pay for herders.[52] Contracts also stipulated that fresh beef "must be of a good marketable quality, delivered in equal proportion of fore and hind quarter meat (necks, shanks, and kidney tallow excluded)." Prior to 1867 contractors delivered fresh beef in quarters, and the army bore the expense of employing a butcher at each post to cut the meat into smaller portions. Successful bidders in 1867 agreed to place their own butchers under orders of post commissary officers and to cut up and issue the beef in quantities desired. Contractors also were assigned rooms at the posts wherein they would issue meat to troops and keep a meat market for sale of mutton, fresh pork, veal, and sausages to officers, employees, and company messes.[53]

To improve the quality of beef, McClure required contractors to feed their cattle corn during February, March, and April when grazing was poor. He hoped this would "do away with so many complaints made during those months, when no better cattle can be purchased in the country."[54] The new ruling seemingly had little effect. James Patterson, who received contracts in 1868 to supply beef at all posts in New Mexico, stated that his cattle were Texas cattle and would not eat corn. He was allowed to substitute hay, but soldiers complained about Patterson's beef nonetheless. A board of survey convened at Fort Union in April 1869 found that his beef was "blue and tough, and mostly bone and gristle, and entirely unfit for issue."[55] In later years complaints about contractors' beef ranged from poor and inferior to tough, bony, unwholesome, tainted, sour, destitute of fat, and unfit for issue. Carlos Carvello, post surgeon at Fort Union, reported in March 1879 that the contractor's meat was so bad that he had removed it from his table. Carvello added with some sarcasm that the contractor was too busy furnishing ties to the railroad, then building through the territory, to attend properly to his contract.[56]

Despite reforms that McClure instituted, the meat ration was often a focus of controversy. If sufficiently rancorous, such squabbles threatened to derail the military–civilian partnership that was so important to the efficiency of the army and to the local economy. Complaints about the quality of contract beef, for example, endangered a contractor's livelihood. If he failed to provide a better quality beef, the post commissary officer might stop purchasing from him altogether. When this happened to Prussian-born Marcus Goldbaum, contractor at Fort Bayard, he fought back by accusing officers there of official misconduct. Bayard had been established in August 1866 to protect the Pinos Altos mining district against Apaches. Goldbaum, the post's first beef contractor, had agreed to supply fresh beef for twelve months, commencing 1 November, at 11 85/100 cents per pound. Within a few weeks, officers complained about the quality of Goldbaum's beef, and when he failed to substitute better meat, Lieutenant Charles Porter, the post's commissary officer, suspended the contract. Porter claimed that Goldbaum had good beef in his herd but found more profit in selling it in Pinos Altos. Army Inspector Nelson H. Davis later agreed that Goldbaum's beef had been inferior, though he thought Goldbaum was capable of supplying a better article. Upon receiving Davis's report, General Carleton recommended to department headquarters that Goldbaum's contract be annulled.[57]

On his own behalf, Goldbaum submitted to Davis a four-page letter and a sworn affidavit listing his grievances against the military. He believed that the contract had been suspended for pecuniary reasons. While he remained contractor, officers had little opportunity to sell government beef to citizens for profit. Goldbaum claimed that the commissary sergeant had told the first sergeants of the companies: "I want you to find fault with the beef so we can run off Goldbaum." Goldbaum also reported having overheard the government butcher say: "Damn the contractor, that contract takes $150 per month out of my pocket and we must and will soon play him out." Goldbaum swore that his beef was of good quality and that civilians as well as officers had purchased the same beef that the commissary officer had rejected.[58]

The conflict was resolved in April 1867 when officials in the Department of the Missouri ruled that Goldbaum's contract would not be annulled, but that the government herd at Bayard should be consumed before Goldbaum was called upon to deliver more cattle. If Goldbaum then failed to furnish good beef, it would be purchased in open market with the difference in price charged to the contractor.[59] No evidence has been located to suggest that Davis or any other officer thoroughly investigated Goldbaum's allegations. Military personnel may have sold government provisions illegally, as Goldbaum stated, but it is unlikely that the post's entire officer corps misappropriated government property, as he implied. Goldbaum's knowledge of English was imperfect, and this undoubtedly complicated business transactions with the military.[60]

To what extent language hampered communications between contractors and army supply officers can only be surmised. The large number of European emigrants and Hispanic *ricos* who established successful businesses in the Southwest is one indication that language barriers were not insurmountable. Nevertheless, numerous misunderstandings must have occurred, leading, as in the case involving Charles Probst and August Kirchner, to bitter recriminations.

It is not known when Probst and Kirchner emigrated from their native Germany. Kirchner served five years in the U.S. army and after being discharged in 1859 established a butcher shop in Santa Fe. Probst then or soon after became his partner. They were ambitious young men and soon gained the support of the Santa Fe Jewish community. For at least fifteen years Probst and Kirchner ran the principal meat market in the city. They also dominated the local military market, holding the contract to deliver fresh beef at Fort Marcy in eighteen separate years between 1861 and 1885.[61]

In July 1871 Probst and Kirchner received contracts to furnish all the fresh beef required within the District of New Mexico (except Fort Garland) at a uniform rate of 9 73/100 cents per pound. In previous years they had brought cattle in from Texas, and in April they were grazing between 1,200 and 1,500 head on their ranch near Tierra Amarilla. Probst and Kirchner subcontracted with Emil Fritz and Lawrence G. Murphy of Lincoln County to supply the garrisons at Forts Stanton, Selden, Cummings, and Bayard. They undoubtedly supplied the northern posts from their own herd.[62]

There is little in the records to suggest that either partner engaged in dishonest transactions or attempted to swindle the goverment in filling their contracts. A major scandal appeared to be brewing, however, when Commissary William H. Nash informed Colonel Gordon Granger, commanding the district, that on the morning of 11 May 1872 Charles Probst had offered Nash a $1,000 bribe to obtain new beef contracts. On receiving this information, Granger immediately called the men in for an interview, wherein Probst denied any wrongdoing. What Probst had stated was that if he received the contracts he would clear from $500 to $1,000 per month. Nash, admitting that he had been angry and preoccupied on the morning in question, concluded that he had misunderstood Probst's "broken English." By the end of the month, Nash was referring to Probst and Kirchner as "the best contractors we have in the District."[63]

Probst and Kirchner received new contracts in June 1872 to furnish fresh beef at Forts Marcy and Wingate for 9 99/100 cents per pound. In 1873 they again received contracts to supply beef at the same posts at slightly lower rates, and on 31 March 1874 they signed yet another contract to furnish Fort Marcy with fresh beef at 5 97/100 cents per pound and mutton at 8 cents per pound.[64] But soon after Captain Charles P. Eagan assumed duties as the district's chief commissary on 1

December 1874, he initiated changes that quickly led to bitter conflict with the Santa Fe contractors. Eagan had learned from local merchants that Probst and Kirchner "had been running the Commissary Department and had their own way so long that they . . . evidently thought they could do as they pleased." Eagan learned also that the firm had established a monopoly of the beef trade in Santa Fe, refused to sell beef to people who patronized another shop, and had "suppressed or bought out any and all who have competed with them in this market." According to "undoubted sources," Probst had even boasted that he would teach Eagan a lesson and have him sent away.[65]

All this came to Eagan's ears during the early days of his tenure. It is not surprising, then, that the new chief commissary took steps to assert his authority. First, he called Probst to his office where he suggested that the firm sell beef to officers at contract price. Probst refused. Because officers took only choice cuts, they were charged the same as citizens, about two cents per pound more than the contract price. A few days after this conversation took place, Eagan instructed Probst and Kirchner to issue all beef required by officers and enlisted men from the commissary storehouse instead of from their meat market where they customarily had issued the beef. This change was necessary, Eagan later explained, because he had "no confidence in Probst and Kirchner nor in their scales."[66]

What followed is not clear from the records. Eagan claimed that the firm refused to deliver beef in smaller quantities than quarters. Consequently on 21 December Eagan labeled Probst and Kirchner failing contractors and started purchasing beef in open market. The contractors claimed that they had always been ready to abide by the terms of their contract. However, they had refused to fill an order for five pounds of beef that army surgeon Joseph P. Wright had submitted after the firm had made its daily delivery to the commissary storehouse. On 8 January 1875 Commissary General of Subsistence Alexander E. Shiras instructed Eagan by telegram to permit Probst and Kirchner to fill their contract. Friends of the contractors may have been responsible for this quick reversal in policy. Both Stephen B. Elkins, New Mexico's delegate to Congress, and William Breeden, New Mexico's attorney general, wrote letters on their behalf. The firm continued to win government beef contracts in later years, though Probst withdrew from the partnership in 1879, leaving Kirchner to carry on alone.[67]

The army could not have performed its mission without the aid of entrepreneurs like Probst and Kirchner who willingly risked the pitfalls inherent in government contracting to collect federal dollars. None performed greater service, nor better understood the risks involved, than James Patterson, described in 1868 as New Mexico's largest cattle dealer. Patterson's exploits are not well-known, and important details about his life remain hidden. Born in Ohio in 1833, he moved to New Mexico some time before the Civil War. One researcher believes he may have driven Texas longhorns to New Mexico as early as 1864. Certainly he made several drives in 1865. Patterson may well deserve recognition as "the founder of the Pecos Trail."[68] He also played a role in the founding of an eastern New Mexico ranching community. The small adobe trading post he established near the confluence of the Pecos and Hondo Rivers in 1868 or 1869 may have been the earliest building on the site of modern Roswell. Van C. Smith, recognized as one of the founders of Roswell, purchased this property from Patterson early in 1870. Patterson later acquired forty acres about six miles south of Roswell at South Spring River. He was living there in December 1874 when he traded the land and improvements to John S. Chisum for $2,500 worth of cattle.[69]

Having pioneered the route to Bosque Redondo, it is not surprising that Patterson and his partner tried to control the government market there. It is noteworthy that high-ranking military officers held Patterson in high esteem even after he combined with other cattle dealers against Andy Adams. When Patterson experienced difficulty in supplying Fort Wingate with beef in 1870, Commissary McClure advised the post's commissary officer: "Mr. Patterson having filled all his former contracts with honesty and in good faith is entitled to some consideration."[70]

In previous years Patterson had been awarded several beef contracts, including those signed in August 1868 for supplying all posts in New Mexico with beef on the block at 8¼ cents per pound and beef on the hoof at 7 cents per pound.[71] To fill these contracts, Patterson trailed cattle up the Pecos to Bosque Grande, whence they were driven to their final destination. He employed agents or subcontractors to supervise deliveries, some being more trustworthy than others. Charles M. Hubbell, his agent at Fort Bascom, suddenly left for Texas after he made an unsuccessful attempt to steal cattle from the post herd.[72] And like so many other contractors, Patterson had difficulty delivering high quality beef throughout the year. During the first five months of 1869, supply officers at five different posts complained of receiving inferior beef.[73]

Because of limited grazing and the danger of Indians running off the contractor's herd, Fort Cummings was among the most difficult posts in New Mexico to supply with fresh beef. Contract prices consequently were high, averaging almost 11½ cents per pound between 1866 and 1873. Even so, profits were small at this one company post, and few men were willing to bid for the contract. James Patterson and John S. Crouch, a former officer in the California volunteers, were two of the contractors who experienced difficulty in filling their contracts. In the fall of 1868 Patterson had employed Crouch to oversee business transactions at both Forts Cummings and Selden. A herd of Patterson's cattle was grazed near Fort Selden, whence cattle periodically were dispatched to Fort Cummings. Frequently the supply at Cummings was exhausted before a new lot arrived, leaving the garrison without beef. When this happened in April of 1869 Patterson's butcher himself left for the Rio Grande looking for cattle but returned after six days with only six worthless steers.[74] Later that year Crouch held the Cummings contract in his own name, but in May 1870 he had to suspend beef deliveries after Indians ran off his herd on the Rio Mimbres. The post's commissary officer then began purchasing beef in open market at 15 and 20 cents per pound. By mid-July, when cattle were scarce along the Rio Grande, he had to pay 25 cents per pound to secure a supply. By the end of August when the contract expired, Crouch owed the government $527 for beef purchased in open market.[75] Crouch soon left government contracting to pursue a career in politics. Patterson moved his operations to Arizona. In August 1869, while still a resident of New Mexico, he subcontracted with Henry C. Hooker to supply Arizona posts with beef cattle, and in 1875 Patterson was awarded contracts to deliver all beef required within the Department of Arizona.[76]

Many contractors, including Patterson and Crouch, are best described as cattle dealers rather than as ranchers. They kept few cattle themselves and bought what they needed to fill contracts from other ranchers or from drovers arriving from Texas. This description applies equally well to Frank Willburn and Thomas L. Stockton, who in August 1870 received contracts to furnish all posts in New Mexico with beef at a uniform price of 8⅓ cents per pound. Willburn had trailed cattle up the Pecos in 1866 or 1867. He and Stockton later obtained a contract to feed Ute Indians, and in the spring of 1870 they reportedly were building large corrals and branding pens on the Rio Hondo near modern Roswell to facilitate their cattle operations. A newspaper reporter commented that in the past year (1869) Willburn and Company had purchased at this site cattle valued at $200,000. Each partner listed his occupation in the 1870 census as "cattle dealer."[77]

But other contractors were authentic ranchers, though they ranched on a modest scale. Robert V. Newsham, for example, started a stock ranch on the Mimbres in Grant County in 1873 with several hundred head of cattle. By May of the following year he reportedly had 1,500 head grazing in the Mimbres Valley. He held contracts to furnish Fort Bayard with beef only in 1874 and 1875, and in 1881 he sold his ranch to Thomas Lyons and Angus Campbell, future kingpins of the vast L–C cattle empire.[78] Even more modest were the ranching operations of Rafael Chávez of Valencia County, who claimed ownership in 1870 to fifty head of cattle. Chávez was among the wealthiest Hispanos living near Fort Wingate and had frequent business transactions with the post, including beef contracts in 1877 and 1879.[79] Stock ranchers who supplied Fort Union with beef typically owned substantial property, but never acquired reputations as cattle barons. William B. Tipton, who held the contract in 1872, claimed in the 1870 census to own 1,000 head of cattle as well as a miscellany of other livestock, which he valued at $19,620. Eugenio Romero, contractor in 1874, dealt more in sheep than in cattle. He claimed ownership in 1870 to 2,500 head of sheep and only 24 head of cattle, his total livestock valued at $7,085. Joseph B. Watrous, beef contractor in 1876 and 1877, together with his father in 1870 had owned 756 head of cattle, 124 work oxen, and 200 milk cattle, altogether valued at $18,750.[80] These men, part of New Mexico's expanding ranching industry, shared the military market with dozens of other suppliers.

Because the army consistently awarded contracts through competitive bidding, no individual or firm monopolized beef contracts in New Mexico, with the exception of Probst and Kirchner, who regularly received the Fort Marcy contract (see Table 8). As an established firm, Probst and Kirchner could absorb many of the incidental expenses

Table 8
Beef contracts—District of New Mexico, 1876

Post	*Price in cents per pound on the block*	*Price in cents per pound on the hoof*	*Contractor*
Fort Bayard	8½	5	James J. Dolan
Fort Craig	9	——	Francis H. Creamer
Fort Garland	5³⁄₇	——	R. Williams
Fort Marcy	5⁹⁵⁄₁₀₀	4	Probst and Kirchner
Fort McRae	8½	——	José Pino y Baca
Fort Selden	8¾	——	R. Yeamans
Fort Stanton	7¹⁷⁄₁₀₀	5	James J. Dolan
Fort Union	5½	5	Joseph B. Watrous
Fort Wingate	7³³⁄₁₀₀	——	Rómulo Martínez
	——	5	John H. Riley

Data from Reg. Beef Cont., CGS, RG 192, NA.

that occurred in provisioning a small post. Then, too, profits at this post were limited, and few contractors bothered to submit bids.[81]

Competition was the rule, then, rather than the exception, and this principle held true even in Lincoln County, contrary to traditional wisdom that has L. G. Murphy and Company monopolizing government contracts. During a twenty-year period, 1865–1885, Lawrence G. Murphy or his associates James J. Dolan and John H. Riley held the Fort Stanton beef contract for only five and one-fourth years. In at least four other years members of the firm served as agents for other contractors.[82] Murphy and Company had more success controlling purchases for the Mescalero Indian Agency. Between June 1871, when Mescaleros began gathering near Fort Stanton, and November 1872, the company was the sole supplier for the Indians, furnishing beef, cornmeal, beans, and other articles amounting to $57,596. The Indian department instituted competitive bidding for the agency in October 1872, which allowed other firms to share the government's patronage. Even so, Murphy and his associates frequently served as agents for other Indian contractors, including William Rosenthal, who held the beef contract for three successive years (1875–1878).[83] Because the number of Mescaleros receiving daily rations greatly exceeded the number of soldiers at Fort Stanton, contractors would have preferred supplying beef to the agency where profits were greater.[84]

Whether working for the army or the Indian department, suppliers continued to rely on Texas cattle for filling contracts. During the 1870s increasingly large numbers were driven up the Pecos. Hugh M. Beckwith, who had a ranch at Seven Rivers about twelve miles north of present Carlsbad, New Mexico, reported that 90,000 head had been tallied as they passed near his place in 1872 and 63,000 the following year. John S. Chisum, with headquarters at his South Spring Ranch, held the largest number on the Pecos, possibly as many as 60,000 head in 1875. By this date small ranchers had begun to settle on Chisum's range south of the Hondo.[85] And as the Pecos Valley filled with cattle, Fort Stanton contractors had little difficulty filling their contracts. John N. Copeland, contractor in 1874, for example, kept no herd of his own, but rather purchased cattle as they were needed from different parties. Murphy himself was more a cattle dealer than a rancher. To fill subcontracts in 1871, he purchased cattle from Lucien B. Maxwell, Willburn and Stockton, and John S. Chisum; he then herded the cattle on the Fort Stanton military reservation, holding from 60 to as many as 300 head at one time.[86]

As the Pecos Valley filled with cattle, more and more cattlemen suffered losses from Indian raiders and white rustlers. Madison Tucker of Erath County, Texas, lost his herd of 2,500 after an estimated 300 Indians attacked his party on 6 July 1872 a few miles above Horsehead Crossing.[87] Many outfits were poorly armed. A party of seven men driving 650 cattle near Horsehead Crossing when Tucker's herd was captured were armed with only a rifle and a shotgun. Another party of fifteen or twenty men in charge of about 1,000 cattle had in addition to their pistols "only two or three worn out muskets." The army supplied mounted escorts whenever possible, but, as Captain Francis S. Dodge, commanding officer at Fort Stockton, reported in 1873 so many herds were being driven up the Pecos "it would take half a regiment to furnish the required details."[88] Most raids were made by Kiowas and Comanches, but renegade Apaches also left the Mescalero reservation to steal cattle and horses. Ranchers on the Pecos who suffered heavy losses held all Mescaleros responsible. In apparent retaliation, an estimated thirty citizens attacked Mescaleros camped near Fort

Stanton on the night of 31 December 1874 and drove off sixty Indian horses. Major David R. Clendenin, commanding the post, reported that in the previous two months more than 200 head of stock had been stolen from Indians located on the reservation.[89]

Cattle rustling plagued southern New Mexico for more than a decade after Chisum located at South Spring River. In 1876 reports of Chisum's losses to rustlers were becoming more frequent. Rumor had it that Chisum cattle carrying altered brands were being turned in on the Fort Stanton beef contract. The chief suspect was contractor James J. Dolan, who kept a cow camp near Seven Rivers.[90] Stock thieving proliferated in the following year, adding fuel to a major conflagration then building in Lincoln County. After John Tunstall was killed on his Rio Feliz ranch on 18 February 1878, law and order virtually disappeared, allowing rustlers and desperados to perpetrate outrages with little fear of reprisals. As violence escalated, John H. Riley, who succeeded Dolan as beef contractor at Fort Stanton, began to fear for his life and requested that a military escort accompany his cattle from near Seven Rivers to the post.[91] About the same time, in April 1878, Dolan announced in the local press that Jas. J. Dolan and Company, in which Riley was a partner, was suspending operations because conditions within the county made it unsafe to continue. Dolan informed Colonel Nathan A. M. Dudley, then commanding at Fort Stanton, that the firm had placed its business in the hands of attorney Thomas B. Catron for final settlement. Catron's young brother-in-law, Edgar A. Walz, was already in Lincoln to wind up the affairs of the Dolan firm, which included filling contracts at Stanton and at the Mescalero agency. Despite Dolan's announcement, the army awarded a new contract to John H. Riley on 11 June 1878 for supplying Fort Stanton with beef. But Riley soon fled the county, leaving Walz to fill the contract.[92]

Dolan was not the only Fort Stanton contractor suspected of dealing in stolen cattle. Early in 1881 cowboy detective Charlie Siringo, while on a mission to recover cattle stolen from the LX Ranch in the Texas Panhandle, discovered fresh hides with the LX brand in contractor Patrick Coghlan's slaughter pens at Fort Stanton. Coghlan was suspected of purchasing stolen cattle from William Bonney and other desperados. At Mesilla in April 1882 Coghlan was forced to stand trial under eleven indictments for purchasing stolen cattle. Coghlan got off lightly, considering the evidence. He was allowed to enter a plea of guilty to one count of illegally purchasing cattle and was fined $150.[93]

After the close of the Civil War the cattle industry in New Mexico experienced steady growth. Stimulated by the windfall market at Bosque Redondo and other military installations, it would reach boom proportions during the first half of the 1880s. The number of cattle on New Mexico's ranges rose from 21,343 head in 1870 to 137,314 in 1880. Five years later the superintendent of the census reported 543,705 head in the territory, though local newspapers claimed a million more.[94] During the 1870s New Mexico's cattlemen had concentrated on local and regional markets with government contracting a major avenue to wealth. The average price the army paid for its beef delivered on contract held steady the last four years of the decade at about 6½ or 6¾ cents per pound.[95] But as the new decade got under way and outside investors poured more and more money into large cattle corporations, beef prices rose. The great demand for beef in eastern markets and in mining and railroad camps forced the army to pay on the average 11¼ cents per pound for beef delivered on contract in New Mexico during the fiscal year 1883–84.[96]

The boom in the cattle industry affected the army in still other ways. Some military reservations literally were overrun with cattle. Numerous bands of stock roamed unrestrained on the Fort Union reservation breaking down fences and contaminating the post's source of water—a spring a half-mile from the post. Stray cattle became such a nuisance that the commanding officer ordered them impounded, permitting owners to reclaim their animals for a small fee.[97]

Problems were similar on the Fort Stanton reservation. In 1885 the Angus Cattle Company had 10,000 cattle grazing near the post. The animals contaminated the Rio Bonito, the garrison's chief source of water. During that summer drinking and cooking water used by officers, employees, and the post bakery was obtained from a spring four miles below the post. But because of a shortage in water wagons, companies had to take water from the post acequias running from the Bonito. Some men became ill within minutes of drinking the water. Others constructed primitive filtering systems using barrels, sand, and charcoal, though the commanding officer reported them unsafe. The cattle also monopolized grazing grounds intended for cavalry horses, broke down fences around company gardens, destroyed banks of acequias running into the post, and damaged the fence surrounding the hay yard. An enlisted man who tried to keep the cattle out of a company garden by shooting them with birdshot was arrested by civil authorities and charged with "malicious shooting of cattle."[98]

With the advent of railroads, even the manner of receiving fresh beef changed. In 1884 Samuel P. Carpenter received a contract to deliver fresh beef by express daily at the Selden railroad station. Captain Arthur MacArthur, commanding officer at Fort Selden and father of General Douglas MacArthur, predicted that this system of delivery from a railroad car would result "in confusion and misunderstandings." Captain Charles A. Woodruff, the district chief commissary, assured MacArthur that Selden had been satisfactorily supplied in this manner in 1880 and 1881. When the Rio Grande flooded that summer, however, trains stopped running, and MacArthur had to purchase beef in open market.[99]

Arizona was even less capable than New Mexico of meeting the army's beef requirements at the close of the Civil War. Indians had driven almost everyone away from the ranches. When General John S. Mason assumed command of the District of Arizona in 1865, there were virtually no beef cattle in the country. The three individuals or firms who signed beef contracts in July of that year had to bring cattle in from California and Sonora. Contractor Hiram S. Stevens, a long-time resident of Arizona, delivered Sonora cattle at Tubac, Fort Goodwin, and Fort Bowie for 15 cents per pound in currency, and Banning and Company, a California freighting firm, delivered California cattle at Calabasas, Tucson, Goodwin, and Bowie for 9$\frac{7}{10}$ cents per pound in coin. Joseph K. Hooper, a member of the mercantile firm of George F. Hooper and Company, most likely delivered California cattle at Forts Whipple, Mojave, McDowell, and Maricopa Wells, for which he received 8 cents per pound in coin.[100]

For the remainder of the decade, contractors continued to bring in cattle from Sonora and California. But the terrible drought that afflicted California between 1862 and 1865 caused Arizona contractors to seek new sources of supply in Texas and New Mexico. Hugh L. Hinds, former captain in the California volunteers, and his partner Henry C. Hooker may have been the first contractors after the war to bring Texas cattle into the territory. In March 1867 Hinds agreed to deliver by 1 May 500 head of Texas cattle at Camp Goodwin, located about sixty miles west of the New Mexico–Arizona border, for 8 cents per pound in coin. And on 13 May he signed a contract to deliver an unspecified amount of beef at Tucson at almost 7 cents per pound in coin. On contracts issued in Hooker's name, the firm supplied Sonora and Texas cattle to almost all military posts in Arizona during a three-year period, starting in mid-1868.[101]

Most beef contracts issued in Arizona in the 1860s called for delivery of cattle on the hoof rather than on the block. Apache resistance to white settlement made it too dangerous for contractors to keep a large herd on hand at isolated military posts. General Mason realized soon after taking command that the only way to supply posts in his district was to provide contractors with escorts through Indian country and to receive enough cattle at one time to justify a contractor in making the venture.[102] Hooker's contract at Fort Whipple in 1868, for example, called for delivery of 275 head of cattle by 30 June, 275 head by 30 November, 400 head by 30 March 1869, and 200 sheep by 30 March.[103] The risks were still enormous. In January 1869, Apaches stole 200 head of cattle belonging to Hinds on the Rio Grande near Fort Selden. That same year Apaches killed Hooker's herder at a cattle camp at Sacaton and escaped with forty head of cattle. In 1870 another man in Hooker's employ was killed at Williamson Valley and 300 sheep captured. Hooker estimated that the firm had lost to Apaches at "the Fresnal Rancho" during 1869 and 1870 more than 400 head of cattle, which he valued at $18,000.[104]

The cattle that Hinds and Hooker brought from the Rio Grande were often in such jaded condition when delivered that they could not recuperate at once and lost weight instead of gaining. Cattle delivered in 1869 lost so much weight that Army Inspector Roger Jones concluded "the Government has paid for many thousands pounds of beef that it never received." He described Hooker's cattle at Camp Grant and Camp Crittenden as being "in very poor condition," and reports from other posts led him to believe that Hooker's cattle should have been rejected "in every case."[105] But the commanding officer at Camp Mojave reported on 23 April 1869 that the 141 head of beef cattle received from Hooker on the previous day "were in good condition . . . much better than any I have seen delivered before, at any military post." These cattle probably had wintered in Arizona, and they were among the largest and the strongest in Hooker's herd, carefully selected for the drive across the barren countryside.[106]

Because cattle were delivered on the hoof, military posts in Arizona built up sizable herds that enlisted men had to guard and care for. In December of 1866 Fort Whipple was employing a permanent detail of eight or nine soldiers to herd more than 200 head of cattle and nearly 100 horses and mules. A year later the commissary officer at Camp McDowell complained of having insufficient men to care for about 300 cattle then at the post.[107] Detailed instructions were issued for herding military stock. At Tubac in 1867, for example, Post Order No. 6 specified that company horses would be sent out to graze at 7:00 A.M. in charge of a noncommissioned officer and nine privates. Three men would be constantly mounted, moving around the herd to keep horses from straying and to watch for Indians. Horses would be watered at midday and returned to the post at 5:30 P.M. Beef cattle would graze between 6:00 A.M. and 6:00 P.M. near the horses. Two herders on horseback would move round the cattle. Members of the guard were forbidden to leave the grazing site during the day; they could either carry their dinner or it would be sent out by company cooks. An officer visited the herd daily to see that these instructions were carried out.[108]

Beef cattle were fed hay and corn fodder during the winter when grazing was poor. But even so, cattle lost weight in colder months and some died from exposure.[109] Consequently troops often had very poor beef in the winter and early spring—so poor that the beef ration sometimes was increased from one and one-fourth to one and three-fourths pounds.[110] Enlisted men usually butchered their own cattle, but Lieutenant James Calhoun, commissary officer at Camp Grant, employed a civilian butcher when he discovered that "no enlisted man at the post possessed the slightest knowledge of butchering cattle, or . . . of issuing beef without wastage after it is butchered."[111]

Troops also had difficulty handling the wild cattle turned in by contractors. Some of the California cattle were so difficult to drive that an officer at Camp Mojave reported his inability to send cattle to a distant outpost. Another said that cattle were so wild that "to obtain one is like success to the hunter of wild game." In 1867 or 1868 cattle being driven to Camp Lincoln stampeded several times and finally attacked the herders—a detail of eleven soldiers on foot.[112] On occasion Hispanos experienced with the lasso were hired to assist in herding.

But it was usually a soldier who shot the steer to be butchered. John Spring, stationed at Camp Wallen in 1866, later recalled the difficulties this sometimes entailed. Cattle at his post were penned in a flimsy corral constructed of mesquite branches. The post butcher, a remarkably poor marksman, often missed his target, the bullet merely wounding the animal selected for slaughter or striking another standing nearby. The commotion that resulted inevitably led to a stampede with the cattle breaking through the frail corral and soldiers riding in pursuit. When Spring later became commissary sergeant, he regularly shot the cattle himself, with few mishaps. But he long remembered the time his bullet passed harmlessly through the skin of a heifer he was trying to kill and struck a large bull. The larger animal scattered men bathing in a nearby creek and demolished a kitchen tent before galloping across the parade ground and disappearing in the direction of the Whetstone Mountains.[113]

To stop the enormous weight loss among cattle after their delivery by contractors, the army experimented with hiring civilians to care for them. For about twelve months, starting in August 1867, Herbert and Nathan Bowers herded government cattle on their ranch on the Agua Fria. Their basic fee was $1.50 per head per month and an additional monthly fee of $7.50 per head for feeding the animals corn fodder from 1 November to 30 April.[114] The experiment was not repeated. In July 1868 Lieutenant Colonel Thomas C. Devin, commanding the Subdistrict of Prescott, established a government stock ranch about twenty miles from Fort Whipple where water was plentiful and grazing "limitless." A strong corral 450 by 150 feet was constructed, and one noncommissioned officer and ten enlisted men detailed to guard the stock. Cattle failed to thrive at this location, however, in part because the soldiers were negligent. Captain Charles Hobart, who inspected the ranch in May 1869, discovered that cattle were not turned out to graze until several hours after sunrise and were returned to the corral about two hours before sunset. Only one man stayed with the cattle during the dinner hour, and no regular guard was kept at night. Hobart thought this method of herding unsafe and inefficient and recommended several changes. By March of the following year, the post commissary officer, Captain Charles W. Foster, had returned the cattle to Fort Whipple to better supervise their grazing.[115]

When contractors delivered beef cattle at army posts, the cattle were kept in the post corral without food or water for twelve hours before weighing. A contractor was paid according to the net weight of the stock he delivered (the steer's weight when dressed). Army regulations specified that the net weight be obtained by deducting from a steer's live weight 45 percent when the animal weighed 1,300 pounds or more and 50 percent when weighing less than that. If no scales were available for weighing live cattle, one or more average steers were selected, killed, and dressed. The average net weight of these was used as the average net weight of the herd.[116]

Weighing live cattle was a slow and tedious process. An entire day was consumed in weighing sixty-seven of the one hundred cattle Hooker delivered on 5 April 1869 at Camp Date Creek. It was then agreed to estimate the weight of the remaining thirty-three. A few weeks later Hinds delivered seventy-five head of cattle at Camp Bowie. Because there were no large scales at the post, the herd was divided into three lots—large, medium, and small—and an average steer was selected from each lot and killed. When dressed they weighed 352, 249, and 239 pounds, respectively, making the average net weight 280 pounds.[117]

Commissary officers in Arizona complained that the method of calculating net weight was inappropriate for their region where small Texas and Sonora cattle rarely yielded in beef 50 percent of their live weight. Consequently the army was paying for beef it never received, except on paper, and post commissary officers had to account for the losses. Captain Reuben F. Bernard, commanding at Camp Bowie in 1870, suggested that 55 percent be deducted from cattle weighing between 400 and 800 pounds and at least 60 percent from cattle weighing between 200 and 400 pounds. This advice was incorporated in the contract that Simon Kelley signed in April of the following year. The sixth clause stipulated that 60 percent should be deducted from the live weight of any Texas cattle turned in on the contract.[118]

Small cattle were preferred at one-company posts where beef from large animals often spoiled before it was all consumed. Contracts sometimes stipulated that cattle should not weigh more than 300 pounds after butchering. But even Texas and Sonora cattle often weighed more. The net weight of cattle slaughtered at Camp Bowie in a single month ranged from 275 to 426 pounds. And when John G. Capron delivered some large Sonora cattle at Fort Whipple in 1867, a board of officers rejected them because of their size and poor condition. Some of Capron's cattle would have weighed more than 550 pounds when dressed.[119]

Officers frequently requested that mutton alone be issued in summer months to prevent losses from spoilage. And perhaps because of the searing summer heat, mutton contracts were issued in Arizona in more years than in New Mexico, even though the latter produced more sheep. As in New Mexico, mutton in Arizona was often more expensive than beef, but the contracts Hooker signed in the late 1860s stipulated the same price for each. During the fiscal year ending 30 June 1869 Hooker contracted to deliver a total of 1,200 sheep to Camps Goodwin, McPherson, Wallen, Crittenden, Bowie, Grant, McDowell, and Fort Whipple for 8¼ cents per pound, the same price he received for beef. In later years the army usually issued contracts for mutton only when it cost the same as beef. But contracts awarded to Maish and Jacobs in 1874 and to James Patterson in 1875 for supplying all posts in Arizona with beef and mutton allowed about two cents a pound more for mutton.[120]

With the contracts issued to Henry C. Hooker in 1869, the army started receiving more of its beef on the block rather than on the hoof. Hooker agreed to deliver at Camp Goodwin, Camp McDowell, Fort Whipple, and at all three or more company posts fresh beef and mutton for 15½ cents per pound. At all other posts, with the exception of Tucson and Yuma Depot, he would deliver cattle and sheep at 9 cents per pound.[121] To carry out his contracts Hooker would have to employ butchers and construct corrals at the posts he supplied with fresh meat. He also would experience greater risk since the cattle remained longer in his possession. Indeed, he lost cattle on the night of 28 October 1870 when Apaches broke into his corral at Camp McDowell and ran off eleven head.[122]

When Hooker's contracts ended 30 June 1871, the firm of Hinds and Hooker dissolved. Hinds may already have left the territory, for a July issue of the Tucson *Arizona Citizen* reported that he was then in Louisville, Kentucky, and had no desire to return to Arizona. Hooker himself soon left for California, but would shortly return to engage in the stock business.[123]

Meanwhile, Simon Kelley of San Francisco contracted to deliver fresh beef at all military posts in Arizona, except Yuma Depot, for the fiscal year ending 30 June 1872 at 15¼ cents per pound. His contract required that he deliver cattle on the hoof only when needed to accompany scouting parties or to issue to Indians held at army posts as prisoners of war.[124]

But Kelley had difficulty implementing his contract. During the first weeks of July 1871 he failed to supply any beef at Camps Mojave and Bowie. His agent finally appeared at Bowie on 16 July with twenty-five or thirty head of cattle, but he then had to borrow from the commissary department everything he needed for issuing beef: a corral, slaughter-house, butcher shop, scales, weights, butcher knives. The agent soon departed, leaving behind a butcher and the subagent, Ira O. Tuttle, who would supervise operations. Three days later Tuttle hired a Hispano to herd the cattle, mounting him on a mule borrowed from the quartermaster's department and arming him with a carbine borrowed from the troops. The herd was sent to Bear Spring Valley, a mile away and separated from the post by a range of hills. That afternoon Indians killed the herder and drove the cattle away. Indians no doubt killed the butcher also, for he had left the post that same afternoon looking for stray animals and never returned. Within a few weeks the post commissary officer was forced to purchase beef in open market at 30 cents per pound. Indians again tried to steal the contractor's cattle about six months later while the herder was having his breakfast, but they succeeded only in stampeding the herd.[125]

Kelley relied upon agents and subagents to fill his contract, and he may never have set foot in Arizona while it was in force. Though the government recognized only the contractor in business transactions, several parties actually shared the economic risk. For example, Kelley's agent for supplying beef at Camp Mojave was Bishop Goodrich, who turned this responsibility over to James R. Porter. Porter then contracted with a Mr. Hollister to furnish the beef. A disagreement about the method of delivery led the post commissary officer to purchase in open market 4,425 pounds of fresh beef at 30 cents per pound, amounting to $1,327.50. The difference between this and the contract price, about $652, was deducted from vouchers issued to Kelley. Porter protested the open market purchases, and even though an army inspector questioned their necessity, the department commissary of subsistence told Porter he must look to the contractor to recover any losses he might have incurred in the transaction.[126]

Henry C. Hooker returned to Arizona sometime in 1872 to take charge of the new contract issued to George W. Grayson of Oakland, California, for supplying fresh beef at all Arizona posts for the fiscal year commencing 1 July 1872. The army agreed to pay Grayson 16 cents per pound for beef furnished the troops and 9 cents per pound for beef furnished Indians. The price differential reflected the fact that Indians would receive parts of the slaughtered animals not issued to troops (necks, shanks, etc.).[127]

To fill Grayson's contract, Hooker and other suppliers brought in cattle from Sonora and New Mexico at considerable risk from marauding Apaches. In late May 1872 Apaches had killed a herder and captured 2,000 head of sheep within a mile and a half of Prescott. John G. Campbell and James Baker, who sold Texas cattle to government contractors, said in July of that year that the Indian threat was too great for them to build up a large herd in Arizona. Lamenting the necessity of importing cattle, the editor of the *Arizona Citizen* observed: "Millions of cattle could easily be grown and fatted on the grass of Arizona every year, without any shelter, but for hostile Indians." He then correctly predicted (despite the use of a racial epithet) that "General Crook will soon have so thoroughly punished the savages as to make it safe to engage in the stock business generally."[128]

General Crook's successful campaign against hostile Apaches, commenced in November 1872, encouraged Arizona residents to import cattle from Sonora, California, Colorado, and New Mexico to stock their own ranges.[129] The most successful ranch established at this time was Henry C. Hooker's Sierra Bonita in Sulphur Springs Valley, northwest of modern Willcox. This was one of the best grass sections in southeastern Arizona, where knee-high grama stretched for miles. Hooker located the ranch in February 1873 in partnership with James M. Barney—a member of the mercantile firm of William B. Hooper and Co. The ranch was stocked mainly with Texas cattle to fill the Grayson contract and a contract Barney received from the Indian department for supplying beef to reservation Indians during the fiscal year starting 1 July 1873.[130]

Meanwhile, cattle and sheep in large numbers continued to be driven in from California, New Mexico, and Sonora. Several stock dealers competed for the army contract issued in 1873, including Hooker, who offered to supply the entire department with beef on the block at the uniform price of 19 cents per pound. The winning bid, however, was submitted by Thomas Ewing and Charles E. Curtiss, who agreed to deliver fresh beef at all posts for 13 87/100 cents per pound.[131] Ewing and Curtiss, both natives of Ohio, were stock dealers rather than ranchers, and they traveled the countryside purchasing cattle to meet their obligations. Their agent James Stinson of Las Animas,

Colorado, brought cattle in from New Mexico. Ewing and Curtiss also subcontracted with a Tucson butcher, probably Frederick Maish, to supply Camp Lowell from his butcher shop.[132]

Pennsylvania-born Frederick Maish came to Arizona in 1869 at the age of 34. He and Thomas Driscoll soon brought 400 head of cattle from Sonora to graze on the Santa Cruz below Tucson. In the 1870 census Driscoll listed his occupation as laborer; Maish listed his as butcher with assets of $5,750. In January 1874 Maish and merchant Lionel M. Jacobs together won contracts to deliver fresh beef and mutton to all posts in Arizona, the beef at 12 39/100 cents per pound and the mutton at 14 89/100 cents per pound. By June of that year they reportedly held on Maish's ranch 2,000 head of cattle and 3,000 head of sheep, which they would butcher to provision most of the southern posts.[133] For supplying the other Arizona garrisons, Maish and Jacobs signed three subcontracts: one with José M. Redondo, a Yuma County rancher who agreed to supply Fort Yuma and Yuma Depot, receiving 10 cents per pound for fresh beef and 12 cents per pound for mutton; a second with Cotsworth P. Head and Jake Marks of Prescott, who would supply at contract price all the northern posts, including Fort Whipple and Camps Mojave, Verde, and McDowell, paying Maish and Jacobs a profit of 1 39/100 cents per pound for beef and 1 87/100 cents per pound for mutton; and a third with Miles L. Wood, who agreed to supply Camp Bowie with beef and mutton at the contract price while agreeing to purchase all his cattle from Maish at 7 cents per pound. Extant records suggest that Maish and Jacobs cleared about $45 a month on Redondo's subcontract, with Redondo absorbing all costs of butchering and issuing to troops.[134]

With each passing month, the number of stock grazing on Arizona's lush grasslands multiplied. An end-of-the-year progress report in the Tucson *Arizona Citizen* claimed that "many thousands of cattle and sheep" had been brought into the territory in 1873. J. P. Fuller, Tully and Ochoa, José M. Redondo, Governor Anson P. K. Safford, and Larkin W. Carr and Brother were among those who introduced stock in 1874.[135] Colonel August V. Kautz, commanding the Department of Arizona, noted in his annual report for 1875 that the territory's population had "materially increased in the past year. [Residents] have felt safe to bring in stock, and numerous herds of cattle and flocks of sheep have come into the country during the past summer."[136]

Government advertisements requesting bids for fresh meat give an indication of the amount of stock the army needed. It was estimated that for the year commencing 1 July 1872 soldiers in Arizona would require 2,000 cattle and 1,000 wethers and for the year commencing 1 July 1874, 1,600 cattle and 2,000 wethers. But even with the increase in stock raising, Governor Safford would report in 1877 that the supply of beef within the territory was not sufficient to meet the demand of 30,000 citizens, two regiments of U.S. troops, and 4,500 Indians who were being fed at government agencies.[137] Consequently prices remained high. The average price the army paid for beef delivered on contract in Arizona for the fiscal years ending 30 June 1872, 1873, and 1878 was $12\frac{68}{100}$ cents, $13\frac{87}{100}$ cents, and $13\frac{29}{100}$ cents per pound, respectively. During the Seventies, the army paid the least amount for its beef in the fiscal year ending 30 June 1876, when James Patterson supplied all posts in the department with fresh beef at $9\frac{73}{100}$ cents per pound and mutton at $11\frac{87}{100}$ cents per pound.[138] Patterson's was the last contract the army would issue for furnishing meat to the entire department. Thereafter dozens of Arizona residents received contracts to supply beef and mutton at the several posts.

For a time it appeared that Henry C. Hooker would monopolize army beef contracts. The Sierra Bonita ten miles below Camp Grant was prospering, and Hooker imported Kentucky bulls to upgrade his stock. In 1876 he won contracts to supply beef at Camps Apache, Bowie, Grant, McDowell, Mojave, and Lowell (see Table 9). The following year he contracted to supply Bowie, Grant, Lowell, and Thomas. By 1881 he was recognized as the largest cattle raiser in southeastern Arizona.[139] But Hooker would not dominate military contracting, though he may well have furnished cattle to other contractors.

During a ten-year period beginning in 1876, the only individual who monopolized beef contracts at any Arizona post was Henry Lambert. He held contracts at Fort Mojave in six fiscal years, and his business partners Frederick Schimpf and Augustus A. Spear held contracts in three additional years. These men reportedly had the only cattle ranches between Fort Mojave and the Needles, twenty-five miles below the post, and consequently their beef was expensive, ranging from 22 cents per pound for Spear's in 1878 to 30 cents per pound for Lambert's between 1883 and 1885. According to Captain Charles P. Eagan, the department's commissary of subsistence, these prices were justified because of the sterile countryside, excessive heat, smallness of the garrison, and absence of a local market.[140]

Table 9
Fresh meat contracts—Department of Arizona, 1876

Post	*Price in cents per pound Beef*	*Price in cents per pound Mutton*	*Contractor*
Camp Apache	12½	——	Henry C. Hooker
Camp Bowie	10	10	Henry C. Hooker
Camp Grant	10	10	Henry C. Hooker
Camp Lowell	10	10	Henry C. Hooker
Camp McDowell	12½	——	Henry C. Hooker
Camp Mojave	18	——	Henry C. Hooker
Camp Thomas	15	15	James Patterson
Camp Verde	10	10	C. T. Rogers and Co.
Fort Whipple	8½	10	C. T. Rogers and Co.
Fort Yuma	12½	13½	José M. Redondo

Data from Reg. Beef Cont., CGS, RG 192, NA.

The only other firm to receive contracts in at least half the ten-year span was C. T. Rogers and Company at Fort Whipple. A native of Maine, Charles T. Rogers had established a meat shop in Prescott in 1874. Either then or shortly thereafter Orlando Allen joined the firm, and the two men established a stock ranch in Big Chino Valley. The firm received its first contract in April 1876, when it contracted to supply the post with fresh beef at 8½ cents per pound and mutton at 10 cents per pound. Rogers and Allen then won the Fort Whipple beef contract in each of the succeeding four years.[141]

Only a handful of Hispanos won military beef contracts in Arizona. And without doubt José María Redondo was the most successful. A native of Sonora, Redondo began building his famed San Ysidro ranch near modern Yuma early in the 1860s. He was the first to irrigate farmlands in this area and would later raise large quantities of hay, wheat, barley, oats, and corn. From May 1869 through 1872, and again in 1876, Redondo held contracts to supply Fort Yuma and Yuma Depot with fresh beef, no doubt securing his cattle from both Sonora and California.[142] When Richard J. Hinton visited San Ysidro in 1877, Redondo held only 500 head of cattle, which were grazing on the Colorado and Gila river bottom. However, a later student of the Arizona cattle industry, Bert Haskett, states that the Redondo herds in the late Seventies "numbered into the thousands." Lyman A. Smith, who operated a market in Yuma, succeeded Redondo as the Fort Yuma beef contractor in 1877, almost a year before Redondo's untimely death at age 49.[143]

The completion of two transcontinental railroads through Arizona in the 1880s allowed Arizona capitalists to import thousands of cattle from Texas, Mexico, and Utah so that by mid-decade the ranges were fully stocked. Captain Eagan reported a slight rise in price for beef both in 1881 and 1882 because of the increased demand made by laborers building railroads and by the heavy migration to the Tombstone mining district.[144] But in 1885 production of beef cattle outstripped the local market, and ranchers were forced to seek outlets in the East. For the army this expansion meant record low contract prices, about 9 cents per pound for fresh beef, in the fiscal year commencing 1 July 1886.[145] Thereafter Arizona ranchers concentrated on breeding quality herds, shipping marketable cattle either to the Midwest or the Pacific Coast.

For twenty years following the close of the Civil War, the army paid less for its beef in Texas than in any other state or territory in the Southwest. Home of the ubiquitous longhorn, the average contract price for fresh beef in Texas was usually less than 6 cents per pound.[146] Prices at Forts Bliss and Quitman were much higher than this, however, reflecting their isolation from good range land. The average price at Fort Bliss was a little more than 11 cents per pound and at Fort Quitman, a little less than 11 cents per pound. Contractors for these posts apparently ran few cattle themselves, but rather bought them from stockmen as they were needed. Two of the largest contractors at Fort Bliss were Charles H. Mahle, who furnished beef in three years (1873–75) and the Mundy Brothers, owners of an El Paso meat market, who supplied beef in five of six years between 1880 and 1885. Only John Kennedy won contracts at Quitman in more than one year. He agreed to furnish beef in 1871 and in 1872.[147]

Prices were lower at Forts Stockton and Davis, where average contract prices were about 6¼ and 6⅓ cents per pound, respectively. Several contractors who supplied Fort Stockton with beef operated nearby ranches—George M. Frazer, Michael F. Corbett, Joseph Heid, Francis Rooney, James Johnson. Frazer, one of the largest contractors at Stock-

Table 10
Fresh beef contracts—West Texas, 1876

Post	*Price in cents per pound*	*Contractor*
Fort Bliss	6 49/100	George Zwitzers
Fort Davis	5 7/10	Charles H. Mahle
Fort Quitman	9 1/2	Charles H. Mahle
Fort Stockton	7 9/10	John O'Brennan

Data from Reg. Beef Cont., CGS, RG 192, NA.

ton, also became a major supplier at Fort Davis, holding contracts there in 1874, 1875, and 1879. A chief competitor for the Davis contract was Charles H. Mahle, who had established a ranch near Fort Davis sometime in the 1870s. Mahle won beef contracts in three successive years, starting in 1876 (see Table 10). During part of this time he brought in cattle from Presidio del Norte to help fill the contracts.[148] Another strong competitor was George M. Gaither who came from Kentucky about 1880. He held the contract at Davis in 1881, 1882, and 1884. Some time later he became a meat wholesaler in El Paso.[149]

During the 1880s stockmen moved more and more cattle into the trans–Pecos country, establishing ranches wherever sufficient water was available. Presidio County felt the full impact of the boom, with newcomers and their herds arriving annually. A witness to this growth, Captain Charles W. Williams, quartermaster at Fort Davis, complained in 1883 that the surrounding country was "rapidly filling up with cattlemen, who will not permit hay to be cut from their lands."[150] But even as cattle companies multiplied, the army was forced to pay higher prices for its beef because of increased demand outside the state. The spring drive in 1881, for example, amounted to 250,000 head. The following year, in 1882, the army paid 1 1/2 cents per pound more for its beef at Fort Davis than it had in 1880 and almost 3 1/2 cents per pound more at Fort Stockton. At mid-decade the average contract price for Texas garrisons was 8 22/100 cents per pound.[151] By this date, however, Indian raiders no longer harassed Pecos cattlemen, and Fort

Stockton was abandoned in 1886. Fort Davis held on for a few more years with soldiers engaged primarily in routine garrison maintenance. The post was finally abandoned in 1891.[152]

Cattle ranching throughout the Southwest had received a major boost from the military. The army's demand for fresh beef encouraged western entrepreneurs to risk the hazards associated with government contracting to supply posts with cheap Mexican and Texas cattle. With assured markets at military posts and Indian agencies, as well as increasingly effective protection that the troops offered, more and more Texas cattle entered New Mexico and Arizona via the Pecos Trail. Prices fell as competition rose. But enlisted men everywhere complained about the poor quality of government beef.

Still, it was the military market that initiated the great expansion in the cattle industry following the Civil War. Army contracting laid the foundation for the cattle empires of Henry C. Hooker in Arizona and John S. Chisum in New Mexico. Other ranchers gambled that profits could be made by turning cattle loose on lush grasslands found throughout the Southwest. During the 1870s settlement was rapid south of the Gila River in Arizona and in the Pecos Valley in New Mexico because of the cattle business. With the coming of railroads in the 1880s new grazing lands in more remote parts of each territory were stocked with cattle. During the early part of the decade boom conditions existed everywhere in the Southwest as foreign capital poured into the range cattle industry. By the time the Indian wars were over and the military presence was in decline, cattle raising was found in almost every part of the region, and local ranchers were shipping cattle to markets in the Midwest.

For more than twenty years, then, the army was a reliable consumer of southwestern beef. During these same years the army relied upon local residents to provide services and materials needed to build and maintain frontier garrisons. The army could not function efficiently either in the field or in garrison without a skilled civilian workforce. The nature of the services civilians provided and their importance to the military and to the local economy will be examined in Chapter 6.

6

Forts and Employees

In March 1873 25-year-old Frances Boyd arrived at Fort Bayard, New Mexico, with her 4-year-old daughter, infant son, and officer-husband, Lieutenant Orsemus B. Boyd. For one long year, they lived in a log cabin with dirt floor and dirt roof, "a perfectly miserable hut," according to Fannie. To keep the cabin from being destroyed during summer storms, they opened the back door and allowed rain water to sweep through, running out the front door. The roof also leaked, and the Boyds took refuge under umbrellas until storms subsided.[1]

Accommodations at Fort Bayard were by no means unique. Long after Civil War guns had fallen silent in the East, officers and enlisted men in the West were forced to live in wretched quarters. The nation's highest-ranking officer, General William T. Sherman, reported from Washington, D.C., in 1869 that in some places "the huts in which our troops are forced to live are . . . inferior to what horses usually have in this city."[2] That quarters remained dismal for so long is surprising, for officers held a common belief that comfortable housing was necessary for troop morale and efficiency. But a parsimonious Congress that annually chipped away at army appropriations delayed construction of better accommodations. The quartermaster's department built and repaired most quarters, storehouses, and other military buildings from funds that Congress annually set apart for constructing temporary structures and for repairing buildings at established posts. Erna Risch has pointed out that "since most western military posts were themselves temporary," the War Department assumed that soldiers would construct their own quarters from whatever materials were at hand. Construction of permanent buildings required approval of the Secretary of War and special appropriations from Congress.[3] Whether permanent or temporary, the establishment of military installations usually resulted in jobs for civilians and in contracts for local building materials. Funds rarely were sufficient, however, to construct well-built and comfortable quarters.

During the Civil War, when the primary goal was to defeat the Southern Confederacy, construction problems were compounded by the need for haste and the lack of trained engineers. Consequently many posts in the Southwest were shoddily constructed and soon in need of repair; some had to be totally rebuilt in later years. New Mexicans profited immensely from the army's housing requirements, not only by supplying materials and labor for building new and repairing old posts, but also by leasing land on which military installations were constructed and renting buildings for housing men and supplies in settled communities.

The most expensive property leased in the Department of New Mexico was the rancho known as Los Pinos belonging to Civil War Governor Henry Connelly and his wife Dolores, located eighteen miles south of Albuquerque. The post established there in May of 1862 served briefly as a supply depot, then as a staging area for the Navajo campaign, and finally as a forwarding post for Navajo prisoners en route to Bosque Redondo. The government agreed to pay the lessors $5,500 annually for five years, reserving the right to renew the lease for an additional five years. The army also obtained the right to cut and use wood and brush growing on the premises; permanent improvements made by the quartermaster's department would revert to the owners at the end of the lease. Buildings already on the property included a church; a store; the Connellys' ten-room residence, probably used for staff offices and officers' quarters; a dwelling house, which the army used for a hospital; two rows of buildings, used as company quarters; one building 16 feet by 50 feet, used for shops; four small buildings, three used as stables and one as a carpenter's shop; and a small gristmill. No new quarters were erected, even though at one time about 500 enlisted men were camped on the premises. The most imposing structure the army constructed was a large storehouse, built at a cost of $6,000 but left uncompleted when the depot was discontinued.[4]

The army also leased the site on which Fort Craig was built, a tract of land belonging to the heirs of Pedro Armendaris located on the west bank of the Rio Grande about 110 miles south of Albuquerque. It is doubtful that the owners ever received adequate compensation for the rental of this property. On 28 May 1854 the army signed a lease for five years and then renewed it in 1859 for an additional five years. The lease stipulated that the owners would receive only a nominal yearly rent and that the property and all its improvements would revert to the owners at the end of the lease. When the second lease expired in 1864 Manuel Armendaris of Chihuahua, Mexico, representing the owners, sought a more equitable rental fee. Because Armendaris and the quartermaster's department were unable to agree on an amount, a commission was appointed and the rent fixed at $2,000 per year.[5]

The new lease was dated 27 January 1865, but its approval became entangled in a bureaucratic nightmare. The quartermaster general sent the lease to the commissioner of the general land office, who ruled that land on which Fort Craig was located belonged to the United States and that consequently the army had the right to occupy it without paying rent. In fact, Congress had confirmed title to the land to the heirs of Pedro Armendaris in 1860, and the Secretary of Interior subsequently overruled the land commissioner's opinion. Nonetheless, the War Department delayed so long in paying rent that the Armendaris heirs appealed to the Mexican government for redress. The case was then placed before the United States Department of State, and after a year's consideration the State Department decided that the Armendaris claim was just. Still, payment was delayed when the third auditor in the Treasury Department announced that Fort Craig was not located on the Armendaris land grant.[6]

Even though the third auditor's opinion was overruled, delays in paying the rent continued. Finally John S. Watts, attorney for the Armendaris heirs, placed the case before the Joint Claims Commission of the United States and Mexico, which on 1 May 1871 entered an award of $14,200 in favor of the claimants, "the court adjudging that this sum was to pay for the use of and absolute appropriation of the land, considering the title of the claimants as tolled and vested in the United States." Meanwhile, in October 1870, William A. Bell of London, England, purchased title to the land from the Armendaris heirs. Representing both Bell and the Armendaris heirs, Watts in July 1871 appealed to Colonel Gordon Granger, commanding the District of New Mexico, for back payment of rents.[7] The War Department considered the case closed, however. Secretary of War William W. Belknap declared that the United States had gained possession of the

Fort Craig reservation upon the ruling issued by the Joint Claims Commission. Still, the United States delayed paying any awards that the commission had made in Mexico's favor until January 1878 when Mexico made its first payment on claims awarded to the United States. It is doubtful that Bell ever received any money for the Fort Craig property. When he sought possession of the site after the army abandoned it in 1885, Secretary of War Robert T. Lincoln advised him to seek compensation from the government of Mexico.[8]

It is not surprising that the government paid only nominal rent for land on which some western posts were established, for the army's presence enhanced the value of adjacent land that otherwise would have excited little or no interest whatsoever. When General James H. Carleton discovered that the site selected for Fort Bascom on the Canadian River belonged to John S. Watts, Carleton persuaded the attorney to accept a nominal rent of one cent per annum, reminding him that "the presence of a military force at that point will increase the value of your land more than one hundred percent; and you can well afford to give four square miles to the United States." Carleton threatened to build the post further downstream if Watts demanded greater compensation.[9]

Watts was one of the many lawyers handling land grant claims in New Mexico who received payment for their services in land. By representing claimants to the Montoya Grant, Watts eventually acquired ownership to a vast tract of land that included the Fort Bascom site. In a similar manner he acquired part ownership of the so-called "Jornada del Muerto Grant," stretching almost ninety miles along the Rio Grande from the Mesa del Contadera in the north to Roblero in the south. Though Surveyor General William Pelham had disallowed the grant in 1859, Watts and a co-owner Joab Houghton on 21 March 1865 deeded to the United States for ten dollars a portion of this land near the spring called "Ojo del Muerto" where Fort McRae was being built. It is probable that Watts and Houghton wanted the army's goodwill in their struggle to win confirmation of the grant, but they struggled in vain. In 1871 the Supreme Court declared the grant invalid.[10]

The government also leased the site on which Fort Bliss, Texas, was established in 1854, paying its owner James W. Magoffin $2,760 annually. At the start of the Civil War, Captain Isaac V. D. Reeve abandoned the post on orders of General David E. Twiggs, commander of

Union forces in Texas, and set out with his command for the Texas coast. Shortly before his departure, Reeve had relinquished the post and delivered its keys to Magoffin, an ardent secessionist who was then residing on the site. Confederate soldiers occupied Fort Bliss in July 1861, but a year later they, too, abandoned the post after failing to attach New Mexico to the Confederacy. Before departing with the retreating Texans, Magoffin moved his wife and his personal effects to El Paso, Mexico, and instructed his agents to remove the doors, windows, cases, *vigas* (rafters), and other valuable woodwork from the abandoned buildings. These items were later conveyed to El Paso, Mexico, and sold for benefit of Mrs. Magoffin. When the California volunteers under General Carleton arrived a few weeks after Magoffin's departure, they found the post untenable "and little better than a pile of ruins."[11]

For nearly three years Fort Bliss remained unoccupied, and Union troops were garrisoned in the town of Franklin (now El Paso), Texas. When Captain Andrew W. Evans inspected the post at Franklin in May of 1863 he reported that the three companies of the garrison were quartered in two large buildings, one formerly used by the overland mail company.[12] The quartermaster's department rented other buildings to house officers and supplies, paying William W. Mills about $105 for the rental of five houses, Anthony B. Rohman, $39 for renting three houses, and Nepumoceno Sanbrano, $6 for one house.[13]

In June 1865, the army started rebuilding Fort Bliss with civilian laborers, infusing even more money into the local economy. The civilian payroll that month amounted to almost $800 and in July it would exceed $1,500. The carpenters, masons, painters, laborers, and other employees working there in October received more than $2,500.[14] But even as repairs were under way, the days of Fort Bliss at the Magoffin site were numbered. That summer the current of the Rio Grande washed away part of the bank above the post and threatened to destroy public buildings. To deflect the current to the other side of the river, the post quartermaster received permission to build a wing dam with civilian labor. In late August Army Inspector Nelson H. Davis concluded after examining the river bank that the Rio Grande would pose little danger to the post once the dam was built.[15]

Very little was accomplished that fall, however, to protect the post. The following spring the Rio Grande carried away about two hundred yards of river bank, leaving a portion of Fort Bliss projecting into the bed of the river. Even though the army subsequently constructed two wing dams above the post, the river was not tamed. As floodwaters rose in May 1867 soldiers feverishly worked to remove supplies and woodwork from buildings about to be washed away. On 1 June the commanding officer reported that more than half the newly constructed quarters and all the corrals and storehouses had been swept away by the river. The men stored the rescued supplies in their own quarters and thereafter slept in tents.[16]

The army abandoned the Magoffin site in March 1868 and moved into quarters on Concordia Ranch, about a mile below the former post and on a bluff sixteen feet above the level of the river. Buildings at Concordia included the residence of Hugh Stephenson, the former owner of the ranch, and the homes of two of Stephenson's married daughters, Adelaide Zabriskie and Benancia French.[17] In 1859 Stephenson had deeded Concordia to his six children, and in August 1867 Benancia's husband Albert offered to sell it to the government for $60,000 or to lease it for $3,000 annually. The army occupied the ranch almost two years before Adelaide's husband James, on behalf of the owners, negotiated a lease for $2,500 per annum.[18]

The army rented other buildings in both Mesilla and Las Cruces to accommodate the California volunteers when they entered southern New Mexico. Although it is no longer possible to know how many private dwellings were rented or the total amount paid in rental fees, records indicate that buildings owned by Mesilla residents Thomas Massie, Alexander Duval, and Cenobia M. Angerstein were rented for $20 each per month.[19] An inspection report of the post at Las Cruces, dated 20 May 1863, clearly shows the prominent position that troops assumed in small western communities and the crude quarters they often inhabited. Some of the quarters were without bunks, and the men slept on the floor. The three companies that garrisoned Las Cruces, a village of about 800 residents, numbered 215 enlisted men. They were quartered:

> in large adobe buildings upon the main street, which is broad & long; the town regularly laid out, with a plaza, opposite the church, upon which parades & drills are held, about the middle of the place. The town is built just out of the bottom lands, upon

> quite level ground at the foot of the gravel hills, and which was originally covered with mesquite bushes, and the roads are very sandy & apt to be very dusty with the slightest breeze. There are no shade trees upon the principal streets. The flag-staff from Fort Fillmore has been removed to this place, and planted in the plaza, nearly in front of the church. The quarters are but rude & uncomfortable, with earthen floors, and frequently without window-frames or sashes. The warmth of the climate makes this rather an advantage in point of ventilation.[20]

When Carleton assumed command in Santa Fe in September 1862, military posts within his jurisdiction were in various stages of decay. Officers and enlisted men at Fort Garland were housed in well-built adobe quarters; only limited repairs were required to make troops comfortable. Laundresses were then living in tents, but new quarters were being built, for several women were "in a condition that will not permit their living in tents this fall and winter."[21] Officers' quarters at Fort Marcy in Santa Fe had been destroyed by fire during Sibley's occupation, and windows and doors of enlisted men's quarters had been ripped out. When Union soldiers regarrisoned Santa Fe, they were lodged in rented buildings. Carleton soon had the men making repairs, however, and by mid-January 1863 enlisted men were housed in government quarters, although the army continued to rent quarters for some officers and offices for department headquarters.[22] Not only did Carleton undertake the renovation of older posts like Garland, Marcy, and Bliss, but he also ordered construction of six new posts in New Mexico and oversaw the relocation of Fort Union. Born of military necessity, the gigantic public works program that Carleton initiated provided jobs to hundreds of civilians and profits for general contractors.

Among the most dilapidated posts in need of repair was Fort Stanton, set afire when Union troops evacuated it and then further damaged by Confederate soldiers and Apache Indians. When Colonel Christopher Carson and the New Mexico volunteers regained Stanton in October 1862, they found it "a mass of ruins, all the roofs, floors, doors and windows burnt, even the walls much damaged." Within a week soldiers had roofed and made serviceable a commissary and quartermaster's storeroom, but Carson expressed little hope of seeing his men "under better shelter this winter, than the Sibley tent affords."[23]

By mid-June 1863 only eight of the stone buildings had been made inhabitable, and all leaked badly during heavy summer rains. Once recognized as the best built post in New Mexico, Fort Stanton now was a sorry sight. "All the stone walls were more or less blackened by smoke," Captain Andrew W. Evans wrote following an inspection in June, "and everywhere were masses of earth and rubbish, which no attempt had been made to clear away."[24] More than a year later the quarters, storehouses, and corrals remained in dilapidated condition. Major Herbert M. Enos, who inspected the post in October 1864, recommended that it be repaired or abandoned entirely. Even though Carleton authorized $3,000 for repairs, his attempt to refurbish Fort Stanton with limited funds and the labor of enlisted men met with little success.[25]

No significant improvements occurred at Fort Stanton until 1868 when more than one hundred civilians were employed in rebuilding the post. Twenty thousand dollars had been allocated for new stone quarters and for repairing the old. New Mexico's chief quartermaster, Major Marshal I. Ludington, awarded contracts in July to Lawrence G. Murphy for 200,000 feet of lumber at $46.95 per thousand feet and to Henry C. Harrison for 500,000 shingles at $8.98 per thousand.[26] The number of civilians authorized for employment rose from sixteen in mid-1864 to 102 in September 1868, including one master mechanic at $100 a month, eighteen masons and two plasterers at $3 a day, eighteen carpenters and one painter at $2.75 a day, and forty-five laborers at $30 a month. The civilian payroll amounted to $4,639 in October and reached $5,323 in December. And from 1 May to 30 November 1868 the army expended for labor and building materials a total of $40,450, with more than half going to civilian laborers.[27]

The cost of Stanton's renovation startled General Phil Sheridan, then commanding the Department of the Missouri, and in January 1869 he issued orders to reduce expenditures. Consequently almost one third of the civilians employed at Stanton were discharged. This reduction along with rumors of further discharges made it impossible to keep a full workforce. Many employees left for more permanent work, and the post quartermaster, Lieutenant William Gerlach, could not find replacements. "To come to a place hundreds of miles from the lines of travel," Gerlach explained, "at the risk of their lives, with the prospect of being ordered discharged at any day, is no inducement for good workmen."[28] Construction was further delayed in May when summer rains arrived early, forcing adobe contractor William Floresheim to halt work. The entire renovation program was suspended on 30 June when all but three civilian employees were discharged as mandated by department orders severely restricting civilian employment.[29]

When Lieutenant Colonel August V. Kautz took command of Fort Stanton in October 1869, new stone quarters for officers and enlisted men were only partially constructed, and the sick remained quartered in ruins of the former hospital. Kautz was determined to rectify the situation. Arguing that the government would suffer great financial loss if accumulated building materials were allowed to rot, Kautz was permitted to resume construction in the spring. Under his direction quarters were completed and other facilities built or repaired.[30] Frances A. Boyd spent one happy year at Fort Stanton in the early 1870s living in one of the newly constructed residences. Years later she recalled that the houses built of stone "were very comfortable. Each had two rooms, with a detached kitchen and dining-room about fifteen feet in the rear." Trees had been planted around the parade ground and provided "delightful shade in front of our quarters."[31]

A short time after leaving Fort Stanton the Boyds moved into "clean, sweet, fresh, quarters" at Fort Union, a post totally rebuilt after the outbreak of the Civil War.[32] The first Fort Union, established by Colonel Edwin V. Sumner in 1851, was located at the base of a mesa near the junction of the Mountain and Cimarron branches of the Santa Fe Trail. Soldiers had constructed the post with unseasoned pine logs that had rapidly decayed. By 1857 at least one barracks had been pulled down to prevent it from collapsing.[33]

Before turning command over to Carleton in September 1862, General Edward R. S. Canby had selected a site northeast of the old log fort for building a new post and quartermaster's depot. By November adobe walls of a new large storeroom had been erected, and Carleton issued orders to have it roofed before winter weather damaged the unfinished structure. In the spring the Santa Fe *Gazette* carried a notice that ten or twelve carpenters and fifty laborers could find work at Fort Union; the carpenters would receive $50 per month and the laborers $25. A year later, in April 1864, both wages and the number of civilians working at the post had increased, the twenty carpenters authorized for employment receiving $65 per month and the hundred laborers, $30 per month.[34]

Since Fort Union would serve as the general depot of supplies for

New Mexico, increasingly large numbers of civilians were employed there to expedite construction and to transfer goods to outlying forts. The number of civilians included in the depot's budget rose from 209 in February 1863 to 419 in April 1864. Similarly the monthly civilian payroll rose during this time from $6,310 to $15,570.[35] But New Mexico's chief quartermaster John C. McFerran had difficulty finding skilled mechanics to work at the depot, and in the spring of 1865 he requested that carpenters, tinners, wheelwrights, plasterers, and other artisans be sent from Fort Leavenworth, the government providing transportation and a monthly salary of $85 per month—$20 more than current employees received. The army also furloughed soldiers with mechanical skills to work at the depot, paying them mechanics' wages minus their soldiers' pay. When a large contingent of mechanics arrived from the east in September, old employees complained about the disparity in wages. Dissatisfaction soon evaporated, for Carleton ordered wages raised to the level of the newcomers, a move later approved at department headquarters.[36]

The new buildings at Fort Union were constructed of adobe brick, the walls standing on stone foundations and coated with plaster. All the main structures had tin roofs, except the hospital which was shingled. Some of the lumber required for woodwork and for corrals and stables was purchased from local residents, even though the depot had a sawmill in operation by September 1864. James T. Johnson of Mora County received a contract in 1866 to deliver 400,000 feet of lumber at $45 per thousand feet. Prices declined as construction neared completion, with Pedro Valdes of Sapello receiving $26 per thousand feet in the spring of 1867 and Frank Weber of Golondrinas receiving $20.74 per thousand feet two years later.[37]

The military installation at Fort Union, consisting of a four-company post, quartermaster's depot, and arsenal, was the largest in New Mexico, and, according to Inspector Evans, the most luxurious.[38] But Lydia Lane found little comfort in the newly constructed quarters she occupied early in 1867 with her husband, Major William B. Lane, the post's commanding officer. She later recalled:

> The house we occupied, built for the commanding officer, consisted of eight rooms, four on each side of an unnecessarily wide hall for that dusty, windy country. They were built of adobe, and

> plastered inside and out, and one story high, with a deep porch in front of the house. There was not a closet nor a shelf in the house, and, until some were put up in the dining room and kitchen, the china, as it was unpacked, was placed upon the floor. After great exertion and delay the quartermaster managed to have some plain pine shelves made for us, which, though not ornamental, answered the purpose.[39]

By mid-1868 construction at Fort Union was finished, and the number of civilian employees fell from 407 in January to about 265 in June.[40] Like other posts built of adobe, however, Fort Union constantly needed repairs, assuring a market for local building materials and employment for mechanics and laborers.

One month after assuming command in New Mexico, Carleton ordered the establishment of a new post at Bosque Redondo on the Pecos River, 110 miles southeast of Las Vegas. Located in the midst of good grazing lands, the new post of Fort Sumner would offer protection to ranchers and farmers and prevent incursions from Kiowa and Comanche raiders from the Texas plains and from Mescalero Apaches from the south. It would also stand guard over thousands of captive Indians once Carleton concluded his campaigns against the Mescaleros and Navajos.

Troops at the new post were sheltered in tents and crude jacales the winter of 1862–63. Even though Carleton was determined to have the men in substantial shelters by the following winter, construction of new adobe quarters progressed slowly.[41] Not until July 1863 did the army contract with Joseph A. LaRue of Las Vegas for building materials: 500,000 adobes at $15 per thousand, 75,000 feet of lumber at $75 per thousand feet, and 200 joists at $7 each.[42] A hospital, storehouses, and corrals were to be built first, and civilian laborers were hired to speed construction. Winter arrived before the work was completed, and in November Quartermaster McFerran instructed Captain Prince G. D. Morton, the post quartermaster, to complete the houses already up and to secure more lumber and *vigas* to finish the work in the spring. Window and door frames, doors, and sashes would be made during the winter.[43]

A large civilian workforce was hired in the spring to help build the post. The May 1864 roster listed twelve carpenters, six masons, nine adobe layers, and forty-three laborers. In addition, several enlisted men were given furloughs to work as masons, carpenters, and plasterers. And the army contracted with Valentine S. Shelby of Santa Fe to construct two storehouses and a hospital for Indians at a cost of $18,000.[44]

Construction continued to provide employment for civilians almost until the day the post was abandoned. During the summer of 1867 the department commander authorized construction of one set of company quarters, a residence for the commanding officer, additions to six officers' quarters to house servants, and a commissary and quartermaster's storehouse. To complete these structures the army contracted with James T. Johnson to supply 115,000 feet of lumber at $57.50 per thousand feet, employed Indians to manufacture adobes, and hired other civilians as carpenters and masons. The November roster shows that a total of ninety-nine civilians were employed at the post: a superintendent at $100 per month, two carpenters and two first-class masons at $75 per month, nine second-class masons at $55 per month, five third-class masons at $35 per month, two guides (one at $3 a day and the other at $60 per month), one interpreter at $60 per month, and seventy-seven Indian laborers at $6 each per month.[45] Built at great expense to the government, Fort Sumner was one of the largest military installations in New Mexico. Its facilities included seven officers' quarters (five rooms each), six enlisted men's quarters (capacity for one hundred men each), one hospital of twenty-four beds, three stables (capacity for one hundred horses each), four large storehouses, two large grain houses, and one large bakehouse. The post was abandoned in 1868, and the army sold these buildings in 1870 to Lucien B. Maxwell for a mere $5,000.[46]

While the army channeled construction funds into Forts Union and Sumner, Carleton saved on expenses by having enlisted men build Fort Bascom, a new post established in August 1863 on the southern bank of the Canadian River. Soldiers had camped near the site during the winter of 1862–63, living in crude huts made of cottonwood branches and covered with leaves and dirt, "some under and some above ground," according to Captain Edward H. Bergmann, the post's commanding officer. First called Camp Easton and then Fort Bascom in honor of Captain George N. Bascom, killed in the Battle of Valverde, the post was intended for a four-company post.[47] Construction of permanent facilities began in August. Because Carleton wanted men and supplies properly sheltered before winter set in, he authorized Chief Quartermaster McFerran to contract with civilians for adobe brick and to employ twenty citizen laborers to assist soldiers in building the post. It is unlikely, however, that many structures were erected that year, for in November the low stage of the river prevented adobe-makers from getting water for mixing adobe.[48]

Heavy summer rains interrupted building operations in each of the following two years. On 26 June 1864 rain "fell in sheets" for nearly two hours, destroying a considerable amount of adobe brick, softening walls of unfinished structures, and forcing men to leave their quarters (probably tents or jacales) for higher ground. Three days later the river suddenly overflowed its banks and drove the entire garrison to nearby hills.[49] Summer rains in 1865 destroyed 60,000 adobes that troops recently had made themselves and damaged walls then being built. Nonetheless, Bergmann happily reported on 11 August 1865 that newly constructed buildings—three sets of company quarters and commissary and quartermaster's storehouses—had not leaked, "although the rain poured down for days unceasingly." The last set of company quarters was then ready to be roofed, and Bergmann anticipated completing them the following week. Bergmann also reported having sent ten wagons and fifteen enlisted men to the mountains near Chaperito to secure building material, for the bosque near the post had been cleared of suitable timber.[50]

When Major Andrew J. Alexander assumed command at Fort Bascom in January 1867, he found the newly constructed quarters in wretched condition—"not fit for dogs to live in." "Every building in the place leaks badly," Alexander reported to Washington, "and the [quartermaster's] storehouse is in such a condition that I consider it in great danger of falling down the first heavy rain." Using enlisted men to build the post, he implied, had led to faulty construction.[51] Though his wife Eveline recorded in her diary that quarters were "warm and comfortable," major repairs were undertaken the following year with materials supplied by local contractors.[52] In the spring of 1868 contracts were awarded to William H. Ayres of Albuquerque for 250,000 adobes at $11.50 per thousand, H. D. Gorham of Apache Springs for 610 pine *vigas* at $3.74 each, and Pedro Valdes for 84,000 feet of lumber at $48.98 per thousand feet.[53]

About two years later Chief Quartermaster Marshal I. Ludington secured approval for further repairs, wrongly predicting that Bascom would be "occupied as a military post for a number of years." In March 1870, less than ten months before the post was abandoned, Ludington contracted with James Patterson to remove dirt roofs covering the commissary and quartermaster's storehouses and to replace them with shingle roofs.[54] About the time these repairs were finished, Frank C. Ogden was hired to reroof and repair the hospital and enlisted men's barracks and to build a new magazine at a total cost not to exceed $3,050. But in September, before Ogden had completed his work, the district commander ordered all repairs suspended, and the post was evacuated in December.[55]

The building of Fort Wingate near Gallo Spring, a few miles south of modern Grants, New Mexico, further illustrates the waste that occurred in constructing western military posts. Lieutenant Colonel J. Francisco Chaves, commanding four companies of New Mexico volunteers, established the post 22 October 1862 on a site selected by a board of officers. Almost immediately troops were employed in digging an acequia, cutting and hauling timber, and erecting storehouses, corrals, and quarters. Chaves also sent a party to recover 25,000 feet of lumber stored in the bakehouse at old Fort Lyon, abandoned the previous year because of the Confederate invasion. By the time Chaves's men reached the fort, Indians had set fire to the bakehouse, destroying its contents. But the trip was not a total loss. By unroofing the buildings that had not been burned, the men secured enough lumber to fill four wagons.[56] Like Fort Lyon, the buildings at Fort Wingate would have earthen roofs and adobe walls. The post would also be enclosed by an eight-foot stockade of logs. Lieutenant Allen L. Anderson, 5th Infantry and Acting Engineer Officer, estimated it would cost $45,000 to construct Fort Wingate, using both civilians and enlisted men as laborers.[57]

A letter appearing in the Albuquerque *Rio Abajo Weekly Press* on 23 June 1863 told of the army's progress in building the post. A contract had been awarded to William Pool and Hugh McBride for making 380,000 adobes for officers' and enlisted men's quarters. Twenty-five citizens and about seventy-five extra-duty men were working as carpenters, millwrights, masons, and timber cutters. Praising the newly built commissary building as "one of the best" in the territory, the writer believed it would "stand as a monument of the mechanical ingenuity and architectural skill of the Natives of the country." A dance had been held to christen the building, attended by "Senoras and Senoritas of the Post, and from the neighboring villages of Ceboyeta and Cubero."[58]

The optimism expressed in this newspaper account was not shared by Major Ethan W. Eaton, who assumed command of Fort Wingate in January 1864. At this time enlisted men were being housed in wornout tents that collapsed during windstorms. Eaton allowed one company to occupy the unfinished hospital and another to occupy a barracks having only a temporary roof. Quarters for officers had not yet been started, and the walls of other company quarters remained unfinished. The quartermaster's storehouse still lacked floors, thereby endangering supplies from dampness. In fact, Eaton had quickly become disenchanted with the Gallo Spring site because of the ground's moisture. With ground water less than two feet beneath the surface, adobe walls soon became damp and unsafe, for they had not been built on stone foundations. Eaton found that he could run a stick entirely through the thick adobe walls some six or eight inches above the ground because of their dampness. As early as 7 April, Eaton suggested that it would be cheaper to move the post to higher ground than to continue building at the present location.[59]

Despite Eaton's recommendation, the army continued building at the Gallo Spring site. During the summer of 1865 soldiers were manufacturing 5,000 adobes a day, and three civilian masons were laying them into walls. Soldiers obtained saw logs about six miles from the post and then used a sawmill powered by twenty-four mules to manufacture lumber. Chief Quartermaster McFerran anticipated that the post would be completed by fall.[60]

But the results were disheartening. A board of officers reported in August 1867 that most buildings at Fort Wingate were in poor condition and some in danger of falling down. The post had been "built in an alkali swamp—on the wet ground—without any stone foundation," and alkali was "eating away the base of the walls." If summer rains continued, the guardhouse and at least one company barracks would surely come tumbling down. Walls of some corrals had already been taken down and rebuilt. Other buildings showing little damage from alkali had been "wretchedly constructed" and leaked badly. Captain Edmond Butler, commanding Fort Wingate, recommended that a new post be built of stone about two thousand yards southwest of the present site. He also suggested that it be built by contract "as then it would be built by skilled labor and under professional or skilled direction." Butler was not the first nor the last officer to bewail the wastefulness of military building. "Structures by unskilled labor," he admonished, "and under unskilled direction are a double expense to the Government; it pays to put them up, and then has to pay to take them down."[61]

By the end of the year military officers in New Mexico were giving serious consideration to rebuilding Fort Wingate at a different location. On 4 January 1868 Chief Quartermaster Ludington forwarded to district headquarters plans and estimates for a new post, and in April he reported to officials in the Department of the Missouri on the dilapidated condition of the old one. He admonished his superiors, much in the manner of Major Eaton four years earlier, that it would be as cheap to build a new post as to repair the old.[62] Meanwhile, plans were under way that would affect the future of the Navajos and the relocation of Fort Wingate. In mid-1867 Congress had authorized formation of the Peace Commission to smooth relations between settlers and western Indians and to pacify hostile tribes. That winter General William T. Sherman, the best-known dignitary on the commission, received reports that convinced him the Navajos would have to be removed from Bosque Redondo. In May 1868, after touring the Indian reservation, Sherman ordered that the Navajos be removed to their homeland, and he suggested that Fort Wingate be moved closer to the new Indian reservation.[63] Sherman's recommendations gave military officials an excuse for abandoning a post that was falling down and saved them the embarrassment of admitting to costly mistakes in its construction. In June troops left the crumbling post behind and reoccupied the site of old Fort Lyon, now designated Fort Wingate, located sixty miles northwest of the Gallo Spring site.

Carleton established four other posts in New Mexico while serving as commanding officer: Forts McRae and Cummings in 1863, Fort Selden in 1865, and Fort Bayard in 1866. Troops who garrisoned these posts initially lived in primitive shelters and spent much of their time constructing permanent buildings. But construction went slowly. Eighteen months after McRae was established, officers and enlisted men were still quartered in Sibley tents, though the post quartermaster described them, each with a fireplace and chimney built of stone, as "quite comfortable to the occupants."[64] Log cabins at Fort Bayard offered scarcely more comfort than tents. Lieutenant Frederick E. Phelps penned this description of the post in 1871:

> The locality was all that could be desired; the Post everything undesirable. Huts of logs and round stones, with flat dirt roofs that in summer leaked and brought down rivulets of liquid mud . . . low, dark, and uncomfortable.[65]

In building frontier posts military officials had to decide whether civilians or soldiers would manufacture the adobe bricks used in construction. Since adobe-making was a time-honored craft in New Mexico, army officials during the war saw nothing wrong in employing Hispanic soldiers to make adobes. The commanding officer at Fort Garland described Hispanic volunteers as being "very perfect in the art of making adobes."[66] But even at posts like Fort Bascom, where soldiers were employed in manufacturing adobes, civilians contracted to supply the bulk of the amount needed. And as soon as Hispanos were mustered out of the army, officers spoke against employing soldiers as adobemakers. Chief Quartermaster Enos expressed a common view when he wrote his superiors on 25 July 1867: "I do not think it right to require enlisted men to work in mud up to their knees in making adobes."[67] Thereafter contractors and civilian laborers would furnish most adobes the army required.

Typically contractors transported adobemakers to the post where construction was under way. William Hoberg had fifty men at work at Fort Union in 1874, making nearly 10,000 adobes a day. He had agreed to make and lay adobes in a corral wall for the modest sum of $9.95 per thousand, but he lost money on the contract.[68] Suppliers rarely received less than $10 per thousand for making adobes alone, and John Lemon and John L. May, who supplied adobes at Fort Cummings during the early stages of its construction, received $22 and $19.50 per thousand, respectively.[69]

Officers overseeing the building and repairing of frontier posts often lamented the lack of skilled craftsmen among enlisted men. "There is not a single man in this command," Lieutenant William Gerlach wrote from Fort Stanton in 1869, "capable of doing mason or carpenter work as required in the [quartermaster's department]." Nor could the commanding officer at Fort Garland find a soldier capable of running a sawmill.[70] More than a decade later, in September 1882, Captain James M. Marshall, New Mexico's chief quartermaster, warned against using enlisted men in the sawmill at Fort Lewis, a post then under construction, for "men unaccustomed to such work" endangered lives and damaged equipment.[71]

Whether built by civilians or soldiers, most military buildings were far from being rainproof. Those covered with dirt roofs usually leaked during the rainy season despite constant repairs. Officers complained that the porous earthen roofs made quarters dangerous as well as uncomfortable. Officers' quarters completed in Santa Fe in 1871 leaked so badly during rains the following summer that the ceilings fell. Officers often used rubber blankets or tarpaulins to cover records and supplies in leaky buildings. Even tin roofs at Fort Union failed to keep out the rain. An indignant officer there reported to his superiors in July 1873 that he had had to pitch a tent in his bedroom during recent rains.[72] Alice Baldwin left a vivid description of how she coped with the flat, leaky roofs at New Fort Wingate.

> On one occasion the "heavens broke loose," and a deluge was upon us. There was not a dry spot in the house. The rains descended and the floods came in a turbid torrent. The ceiling in my bedroom was converted into a waterfall. These leaks were no ordinary affairs, but a shower to the right of us and to the left of us. In bed, surrounded by basins, pails, tubs, cups, and dripping pans—any available container to catch the water—I lay motionless, not daring to move for fear of upsetting the various receptacles. The room was too small to move the bed to a dryer place. The baby was snug and dry in her cradle, over which an umbrella was suspended. So passed the night![73]

Since leaky roofs endangered supplies, quartermasters were especially interested in having well-built, rainproof warehouses. Those constructed at Fort Craig during and just after the Civil War stood the ravages of time better than most other military buildings. They were described in 1868 as being "substantially built of adobe and partly underground," having dirt roofs with floors of cement made of ashes, sand, and *jaspe* (a homemade calcimine). When the post was temporarily abandoned in 1878, it was stripped of everything having value until only the walls were left. But Chief Quartermaster James G. C. Lee reported in 1881 that even though the rest of the post was dilapidated, the storehouses were "excellent," needing only slight repairs.[74]

Quartermasters also wanted warehouses that were rodent-, theft-, and fire-proof, but most were not. Mice damaged all kinds of supplies stored in government warehouses—clothing, corn, flour, cheeses, even laundry starch. Officers tried to exterminate them by laying traps and spreading poison but without noticeable success. Mice caused so much damage that Quartermaster General Montgomery C. Meigs issued special instructions to keep rat terriers and cats around storage areas.[75] Soldiers also caused losses, stealing clothing, food, and other supplies to sell to civilians, a topic to be discussed more fully in a later chapter. To reduce thefts, all posts in 1868 were required to construct a room within the commissary storehouse where small packages would be kept under lock and key. Sentinels regularly patrolled around warehouses, and in later years iron grates and bars were placed in warehouse windows. But the dilapidated condition of some buildings made it easy for soldiers to carry away supplies. At Fort Stanton in 1868 "parties unknown" removed enough dirt from a roof to enter the quartermaster's storehouse and make away with twelve pairs of trousers, eight shirts, four pairs of boots, and other articles of clothing.[76]

Although most buildings were constructed either of adobe or stone, enough lumber was used to make fire a constant hazard. Fire precautions were primitive, however. At Fort Union Depot in 1867 water-filled tanks were placed in corrals and between warehouses, and buckets of water were placed in warehouses. Six night watchmen were on duty at different locations, and one large double-decker hand fire engine was on the post. But the quartermaster claimed the engine would be of little use in case of a fire "as it is large, cumbersome, and so hard upon those working it that they would be completely exhausted in a short time." Sometime later the depot acquired a steam fire engine, but when fire broke out in the cornhouse on 27 June 1874 no one was on hand who could operate the machine. The fire spread in all directions, destroying several worthless frame buildings and the depot's entire supply of grain, valued at $20,000. Fully one hour elapsed before soldiers got the fire engine working and were able to check the fire. A board of officers later concluded that had the "steam fire engine been promptly at work under the direction of a competent engineer the fire could have been quickly subdued, and a large portion of the public property saved."[77]

Among the most serious problems that quartermasters faced in constructing frontier posts were labor shortages, limited appropriations, and bureaucratic reversals—all of which delayed construction and added to the cost. Nowhere are these problems better illustrated than in the building of New Fort Wingate at Ojo del Oso (Bear Spring), site of the abandoned Fort Lyon. The new post was established in June 1868 by troops accompanying Navajos returning to their homeland and by companies transferred from Old Fort Wingate. During the fall, soldiers with the help of civilian laborers hastily erected temporary log quarters before winter arrived.[78] At least seven civilians were employed as carpenters and blacksmiths, each receiving $2.75 per day plus a ration, and at least sixty-six Navajos were hired to make adobes—their pay 25 cents each per day and a ration. By December a steam sawmill and shingle machine had arrived from Fort Union, and a civilian engineer had been hired at $100 a month to operate the machines. Also employed was a sawyer, who received $80 per month, and six teamsters, who would receive $25 per month to haul logs. But the work force was too small to keep the sawmill running at full capacity.[79]

The temporary buildings were constructed on a circular plan approved by the post's commanding officer. Even before construction began, however, Chief Quartermaster Ludington had objected to the plan, arguing that the post would be expensive to build and difficult to defend. By the time permanent quarters were constructed, Ludington's arguments had won out, and a new site on a square plan was marked, covering part of the same ground as the temporary buildings.[80]

During the winter of 1868–69 Ludington helped prepare plans and estimates for constructing a four-company post. To keep the total cost as small as possible, estimates were based upon employing cheap Indian labor for burning lime, quarrying stone, and manufacturing lumber, shingles, laths, and adobes. Quartermaster General Meigs and Secretary of War Belknap subsequently approved the final estimate of $37,550 and directed that the work be done by contract if it could not be performed by troops. Construction of permanent buildings would not begin until the President of the United States officially set aside the military reservation.[81]

Meanwhile, Navajos were employed to manufacture adobes and perform other labor at twenty-five cents a day. Having acquired some familiarity with the white man's economic system at Bosque Redondo, these men stopped working on 29 March 1869, demanding wages of seventy-five cents a day. The post quartermaster immediately discharged the strikers and had no difficulty employing other Navajos at the old rate.[82] During the month of June 115 civilians worked at the post, including 69 Navajos. But construction came to an abrupt halt at the end of the month, when, following department orders, all but 8 civilian workers were let go. Unable to work the sawmill with this force, officials decided to contract for lumber, granting the contractor the use of the government sawmill and of government transportation in hauling logs. In January 1870 a contract was awarded to Thomas Dunbar, the lowest bidder, who agreed to supply lumber at $14 per thousand feet and shingles and laths at $7 per thousand. These materials would cost more than Ludington had allowed in the original estimate.[83]

Although officials had hoped to have permanent quarters built the summer of 1869, only one or two additional temporary buildings were constructed that year. President Grant officially established the military reservation on 18 February 1870, and shortly thereafter Ludington solicited proposals for constructing the post. Six individuals or firms submitted proposals, including James Patterson, whose bid of $154,979 was the highest, and John and Michael McGee, a Santa Fe architectural firm, whose bid of $82,421 was the lowest.[84] General John Pope, commanding the Department of the Missouri, subsequently disapproved all the bids for being too high, and on 13 May he directed Captain Augustus G. Robinson, then chief quartermaster for the district, to shelter troops and supplies at Wingate "as it can be best done with the sum authorized by the Secretary of War, the total expense not to exceed $37,550." Pope also informed Robinson that he could employ as many civilians as necessary to build the post, their wages to be paid from the above appropriation.[85]

Since all the proposed buildings could not be constructed for the sum allotted, Robinson gave instructions that a quartermaster's storehouse be built first, then corrals and stables, followed by enlisted men's quarters. Possibly as many as twenty skilled mechanics and one hundred Navajos helped erect the buildings—made of adobes with stone foundations, shingle roofs, and brick chimneys. By the end of September the appropriation was nearly exhausted, and the post quartermaster estimated that an additional $67,134 was needed to complete the post.[86]

While officials in New Mexico pleaded for more funds, General Pope was suggesting major shifts within the Department of the Missouri, which, if implemented, would eliminate small, isolated posts like Wingate. Completion of the Kansas Pacific Railroad to Denver in August 1870 allowed men and supplies to be transported across the plains rapidly and cheaply. In his annual report to Secretary of War Belknap dated 31 October, Pope argued that it was more economical to concentrate troops at a few large central posts on or near the railroad than to scatter them among numerous small posts remote from sources of supply. He recommended that garrisons from ten posts east of the Rockies and between the Platte and Canadian Rivers be concentrated at Fort Hays, Kansas, and a new post to be established on the railroad near River Bend or Cedar Point in Kansas. In the District of New Mexico he suggested that troops be concentrated at Forts Union, Garland, Selden, and a new post on the Mimbres River to replace Forts Bayard and Cummings. He believed that Wingate, Craig, and Stanton were "wholly unnecessary." With the Navajos at peace, Wingate no longer served a good purpose; Craig was on private land and was a needless expense; and Stanton protected only a few hundred settlers who depended upon trade with the post for their livelihood. "It seems rather absurd," Pope remarked in regard to Fort Stanton, "that a military post, once established, must be forever kept up for the protection of a few settlers who live by trading with it."[87]

Despite Pope's recommendations, which he reiterated in later years, the army continued to spend money building and repairing posts in New Mexico that Pope had labeled as useless. In the summer of 1871 Fort Wingate received an appropriation of $20,000, and work was begun on officers' quarters and a hospital. About 200 Navajos were employed making adobes at 50 cents a day. An additional $17,500 was allotted in 1872 for building quarters, stables, and corrals. Of the 28 Navajos on the payroll in December of that year, 2 were hired at $1.50 per day (probably as adobe layers), 3 at $1, and the balance at 50 cents per day.[88] Repairs and construction occurred almost yearly thereafter, and in 1881 Chief Quartermaster Lee described Fort Wingate as "the best post in the District." Like his predecessors, Lee recognized the importance of cheap Indian labor. When $10,000 was appropriated for erecting new buildings at Wingate, Lee instructed the post quartermaster to employ all the Indians he could "as it will enable you to accomplish so much more with your allotment."[89]

Forts Bayard and Stanton, targeted for abandonment by Pope in 1870, continued functioning as military posts into the 1890s, and thereafter each was converted into a government hospital. Construction of permanent quarters at Fort Bayard got under way soon after Belknap gave his approval in July of 1870. Chief Quartermaster Ludington estimated it would cost $45,823 to complete the buildings.[90] By 1877 thirteen sets of officers' quarters had been completed as well as barracks to accommodate four companies. But Chief Quartermaster Lee painted a dismal picture of Bayard in June 1881, attributing its run-down appearance to hasty construction by untrained soldiers. "The men's quarters are wretched," he reported, "the officers' quarters barely inhabitable. . . . All the roofs are bad and leaky, many of the floors are worn out. The corrals and stables scarcely deserve the name. Extensive repairs are imperatively demanded."[91]

Permanent quarters to accommodate two companies at Fort Stanton had been completed by January 1872, but additions and repairs required further expenditures. Paul Dowlin, owner of a local sawmill, contracted in 1876 to build twelve sets of laundresses quarters for about $3,632. His brother Will received a contract the following year to construct a hospital for $7,335.[92] The army expended about $1,346 on repairs in 1878, and the following year it authorized an additional $8,095 for building one double set of officers' quarters and some shops and corrals. The army had anticipated building the officers' quarters by contract, but because the lowest bid was almost double the amount allotted, the structure was built with materials and labor procured by the quartermaster's department.[93] The most important improvement built in later years at Stanton was a modern water system consisting of two reservoirs, a steam pump and boiler, and iron pipes to carry water to the buildings. Built at a cost of about $5,000, the system was completed in 1885 or 1886.[94]

Over the years the army made several improvements at military posts that made serving on the frontier less onerous than it might otherwise have been. Officers and enlisted men alike benefited from the construction of modern water systems, bathing facilities, ice houses, schoolhouses, and meeting rooms. But shoddy construction detracted from the army's efforts. Even in the territorial capital of Santa Fe, where the army was the government's most visible symbol of authority, military buildings were in a permanent state of disrepair. And in attempting to remedy this condition, the army infused a significant amount of money into the local economy.

Fort Marcy had been established in Santa Fe in 1846. But the post was deactivated in 1867 and the garrison withdrawn. Military buildings there continued to be used in conjunction with district headquarters, however, as quarters for escorts, guards, the regimental band, and for storehouses. From 1867 to 1875 (when the post was reactivated), the military establishment in Santa Fe was designated as the Post at Santa Fe, New Mexico, but civilians and military personnel still informally referred to it as Fort Marcy.[95]

For several years after the Confederates evacuated Santa Fe, the army rented quarters for officers attached to district headquarters.[96] To eliminate this expense, the Secretary of War in 1870 authorized an expenditure of $31,812 to build eight new buildings—one building for offices of district headquarters, one for the district commanding officer, and six officers' quarters. Work began in the fall. By mid-April of 1871 the quartermaster's department had overspent its allocation, and the buildings were still not finished. Almost all the money spent to this date, $36,586, had been paid to civilians: $22,349 to mechanics, teamsters, and other laborers and the rest to thirty-one local residents who furnished adobes, lime, lumber, and other supplies.[97] Quartermaster General Meigs was not pleased when he received a request for additional money, and he called the situation in Santa Fe "as grave a case of mismanagement as has come to my notice." Funds nonetheless were forthcoming, and by the end of the year six buildings had been completed. The two remaining, including the headquarters building, would not be finished until mid-1872. The total cost of construction must have amounted to about $48,000.[98]

Lieutenant Colonel Fred Myers, the district's chief quartermaster, inspected these new adobe buildings in June 1872 and found serious defects in their construction. Because green lumber had been used, the woodwork had shrunk, causing large cracks to appear in the plaster. The shingled roofs leaked, and brick fireplaces smoked. And because of the imperfect construction of chimneys and fireplaces, every building occupied during the winter had been on fire—some, several times. Enlisted men continued to live in old adobe structures with dirt roofs and floors, all badly in need of repair. The hospital, which Mexican troops had used prior to American occupation, consisted of several adobe buildings built in the form of a square with an interior courtyard. According to Myers, its condition was "very bad."[99] By the end of the year Belknap had authorized an additional expenditure of $3,327, most of which went to repair the newly constructed buildings, although $900 was spent repairing roofs on soldiers' quarters.[100]

A new military hospital was finally authorized in 1875, and contracts were awarded in June for furnishing supplies and labor. John T. Mullaley agreed to do the mason and adobe work, furnishing all the brick and adobes required, for $1,800; John Dalton would furnish lumber, costing $649; Ysidro Torres would receive $210 for supplying 600 bushels of lime; and Stephen Lacassagne agreed to do the plastering and painting for $1,130. In November the firm of Henry Schultz and Samuel E. Blair contracted to do the carpentry for $1,875. With shingles and laths purchased in open market, officials estimated the entire building would cost $6,160. Also in that year the quartermaster's department requested an additional $10,147 for general repairs, giving priority to replacing with shingle "the present very old and wretched earth roofs" that covered many of the military structures in Santa Fe.[101]

Despite repairs and new construction, Chief Quartermaster John H. Belcher reported in 1878 that military buildings in Santa Fe were "all more or less dilapidated; all the roofs leak, and much of the plastering has fallen, and all the quarters and offices need painting and whitewashing."[102] Major James G. C. Lee mounted a campaign late in 1880 to introduce gas lights into the headquarters building and officers' residences. A local gas company was willing to install pipes and other fixtures for about one third the actual cost to secure the government's patronage. Lee argued that gas lights would allow officers and clerks to work at night, completing special reports. Moreover, gas lights were essential to the comfort of officers and their families stationed far from enjoyments found in more populated places.[103] Even though his first request was rejected, Lee continued his attack, reminding his superiors that the government had a reputation to uphold among the people of New Mexico:

> The Head Quarters offices are in the very heart of the town, adjoining the main plaza, and adjacent to the Governor's Office and residence and many of the large business houses, all brilliantly lighted with gas. The heads of these establishments, their clerks and accountants, sit under the blaze of bright gas jets, and do their work easily and pleasantly, while we, officers of "the best Government the world has ever saw," are required to do our work by the weak and flickering light of a "tallow dip."[104]

Lee's arguments were persuasive. Early in 1881 permission was granted to introduce gas into the buildings, and within a year gas lamps were lighting the military grounds as well.[105]

The building of frontier forts in west Texas was as difficult an undertaking as it was in New Mexico. Lumber had to be transported long distances and laborers recruited from distant areas. Forts Davis and Stockton, both badly damaged during the Civil War, had to be completely rebuilt following reoccupation by federal troops in 1867—at great expense to the government. During the fiscal year ending 30 June 1868, the army expended $81,978 for labor and building materials at Fort Stockton and $83,108 at Fort Davis, labor accounting for more than four fifths of the total cost. One hundred and thirty civilians were employed at Fort Stockton in December 1867 and 107 at Fort Davis. Masons and carpenters composed the bulk of the employees, but wheelwrights, blacksmiths, quarrymen, limeburners, teamsters, cooks, and laborers were also employed.[106]

Quarters at Fort Stockton were to be built of adobe brick with stone foundations, but construction frequently was delayed for lack of lumber. Two steam sawmills that arrived in August 1867 proved totally useless since the nearest suitable timber for sawing was 100 miles away. Lumber consequently was hauled more than 300 miles from Uvalde.[107] Some soldiers were still living in tents as late as November 1869. According to the surgeon general's report issued late in the following year three adobe barracks had been completed, each eighty by twenty-four feet long with a thatched roof, and five adobe officers' quarters, each with a shingle roof, front porch, boarded floor, and inside walls that were plastered and whitewashed. Lieutenant Colonel Wesley Merritt, the post's commanding officer, claimed in 1872 that enlisted men had built their own quarters, but it is unclear whether soldiers or civilians actually made the adobes.[108]

Private citizens claimed ownership of land upon which both Davis and Stockton were built, and the United States government was forced to obtain leases. On 17 September 1873 the army entered into a one-year lease with Peter Gallagher, owner of the Fort Stockton site, agreeing to pay an annual fee of $800—this amount was still being paid in 1878.[109] The Fort Davis site was leased both before and after the Civil War from John James of San Antonio, who on 29 November 1867 executed a new fifty-year lease at $900 per year.[110] Timber for building Fort Davis was obtained in Limpia Canyon, about twenty-five miles from the post, but by mid-1869 at least one of the pineries located there was nearly exhausted. In later years lumber was purchased at Blazer's Mill in Lincoln County, New Mexico, and hauled by government wagons more than 350 miles to the post.[111] Most buildings were constructed of adobe bricks. Four of ten officers' quarters inspected in 1872, however, were built of stone quarried within a mile and a half of the post.[112] Buildings at both Davis and Stockton suffered damage in later years from natural and man-made causes. An early morning fire in November 1872 completely destroyed a lieutenant's quarters at Fort Stockton, the fire apparently caused by coals rolling onto the floor from a fireplace. A violent windstorm at Fort Davis in March of 1876 caused ceilings in officers' quarters to fall, tore away the tin roof of the new hospital, and toppled corral walls.[113]

When federal troops reoccupied Fort Quitman early in 1868, they found the post almost in total ruin. Doors, windows, and all available woodwork had either been burned or torn out and carried away. Even roofs of some quarters were gone, leaving only the adobe walls standing.[114] Unlike Forts Davis and Stockton, Quitman was not rebuilt, though repairs were authorized to make the buildings inhabitable. But pity the poor soldiers stationed there, for even with repairs the quarters were dismal. In 1870 Major Albert P. Morrow, commanding the post, reported that the buildings were no longer tenable. During a heavy July rainstorm roofs of two buildings collapsed and others were in danger of falling. Morrow could find no dry place for writing his reports: "the offices are all dripping and filled with mud, washed down from the roof and walls." Mud covered the floor of his own residence, ruining his carpets and furniture. Morrow reminded his superiors that plans and estimates for building new quarters had been submitted and numerous reports had been made showing the condition of the post. But to no avail. Morrow's disgust for the situation is clearly apparent in his report:

> This post, is in my opinion, a disgrace to the Government, and a gross injustice to troops, who are required to keep up to the standard of a soldier in Garrison; and to officers who are responsible for property they have no means of protecting. Last night the soldiers had to remain awake all night, as it was simply impossible to sleep with rain and mud pouring down upon them, and expecting every moment that their houses would fall and crush them. If these men mounted guard this morning with soiled clothes and rusty guns, whose fault is it?[115]

Sometime within the next twelve months, the army invited bids for building a new post about three and one-half miles farther up the river. When it became apparent that the proposed four-company post would cost an estimated $249,500, the project was abandoned. In February 1872 Morrow submitted another critical report. "The condition of [Fort Quitman] is about as bad as it can be," he wrote, "the roofs are all leaky and some of them are actually dangerous in the rainy season." Nonetheless, sufficient repairs were made in following months so that the surgeon general could report in 1875 that quarters for both officers and enlisted men were "quite comfortable."[116]

Seventy miles north of Fort Quitman, the post established at Concordia Ranch in 1868 was falling into disrepair. The quartermaster reported on 29 August 1876 that buildings at Fort Bliss (as the post had been redesignated) were old and unsafe and that troops risked their lives in entering some of them.[117] Partly as an economy measure, the post was abandoned in January 1877, much to the dismay of local citizens who feared that desperados would overrun their communities. Indeed, the tragedy known as the El Paso Salt War might have been averted had soldiers been garrisoning the post. As it turned out, at least sixteen people were killed later that year in the deadly battle to control salt beds east of El Paso.[118]

In December Colonel Edward Hatch, commanding the District of

New Mexico, marched with troops to the troubled area, and the disturbance came to an end. Because of the bloodshed, however, Fort Bliss was reestablished—but not at the old location. Hatch reported that the Concordia property was unfit for occupancy: roofs had caved in, doors and windows been removed—the place was in "a generally dilapidated condition." Even though James A. Zabriskie offered to sell or lease the old post on reasonable terms, the army chose instead to rent buildings in El Paso to quarter troops. By mid-1878 the post of Fort Bliss consisted of a dozen or more houses on different streets, for which the army paid a rental fee of $2,500 per year.[119] No one was satisfied with this arrangement. The buildings in most cases were small, badly built, and impossible to secure against theft. And with troops scattered throughout town, discipline was hard to maintain.[120]

This arrangement was only temporary, however, for sometime in 1878 plans and estimates for a new four-company post had been forwarded to Washington for approval. On 4 February 1879 Congress appropriated $40,000 to build a new post at El Paso.[121] Upon learning of the appropriation, Joseph Magoffin, an El Paso civic leader and son of ex-Confederate James Magoffin, offered to sell to the government "Magoffinsville," the old Fort Bliss site, for $5,000 or to lease it for ten years at $300 per year. Hatch, however, preferred the old Hart's Mill property, about a mile above the town, for the heirs of Simeon Hart were willing to sell it for the nominal sum of $100.[122] The army hoped to build the post by contract, but none of the bidders came anywhere close to the amount appropriated, the lowest bid being about $160,000. The quartermaster's department consequently purchased building materials from individual suppliers and employed M. J. Cavanaugh of Las Vegas at $150 per month as a master mechanic to supervise the construction crew, which included civilian carpenters and masons hired at $3 per day and soldiers assigned to extra duty.[123]

Construction of new Fort Bliss on the Hart property got under way in the summer of 1880, but was slowed by delays in securing lumber from northern New Mexico. The Las Vegas firm of Lockhart and Company furnished most of the lumber, charging $18 per thousand feet for sheathing and $20.50 per thousand feet for dimension lumber loaded onto railroad cars at Las Vegas. It was then shipped to the end of the track at San Marcial, New Mexico, where it was loaded into government wagons for the final 160-mile haul to Fort Bliss. Heavy snowstorms in November, however, interrupted the supply, for wagon teams were unable to haul lumber from the firm's sawmill in the mountains.[124] Some El Paso residents furnished other building materials. Antonio Hart, son of the deceased mill owner, supplied adobes at $12 per thousand, and J. B. Stout supplied brick at the same price.[125]

In the long run, railway transportation reduced army expenditures, but in 1880 railroad construction in New Mexico greatly advanced the price of lumber, labor, and other supplies. Consequently, in order to stay within the appropriation, the army scaled down its plans for Fort Bliss, limiting it to a two-company post. By the end of 1880, troops had moved out of rented buildings and were being quartered in tents at Fort Bliss, where construction of barracks and officers' quarters was still under way. A report dated 21 April 1881 stated that no public buildings had yet been completed, except the bakehouse, and the garrison probably remained in tents throughout the summer. By May of the following year, however, everyone was living in new but unfinished quarters.[126]

Soldiers in Arizona were no better housed than those serving in Texas and New Mexico. Crude and unhealthy housing simply added to the discomfort and danger of soldiering in a territory where summer temperatures soared above the 100-degree mark. Primitive accommodations contributed to illness and to the number of desertions and deaths recorded among troops. More than one soldier in Arizona was crushed to death beneath a dirt roof or adobe structure that suddenly collapsed. At newly established posts enlisted men often constructed their own quarters—sometimes nothing more than hovels made of whatever materials were at hand. This was standard military policy. An officer in San Francisco writing in 1866 to Colonel Charles A. Lovell in Arizona expressed the army's position in these words:

> The troops wherever sent have always soon made themselves comfortable by their officers direction and by their own labor and hutted themselves in the same way that prospecting miners have done and are continually doing by the use of either stone, wood, adobes, poles placed upright and filled in with clay, turf, sods, reeds, willows, etc.[127]

Even though officers in the field bemoaned the lack of suitable quarters, housing in Arizona remained wretched for several years. In 1872, a year after assuming command of the Department of Arizona, General George Crook observed that quarters in his department were "unfit for the occupation of animals, much less the troops of a civilized nation."[128]

The worst quarters in Arizona were those erected at Fort Bowie soon after it was established in July 1862. Located in the Chiricahua Mountains in the heart of Apache country, the post was intended to protect travelers on the road leading through Apache Pass and to guard a nearby spring, the only dependable water supply in the area. The men were housed first in tents, but these were soon torn into shreds by strong winds sweeping through the pass. Soldiers then constructed makeshift shelters of stone walls built against excavations in the hillside and covered with dirt, reeds, and grass. Lieutenant Colonel Nelson H. Davis, who inspected the post in 1864, thought the quarters resembled "cells for prisoners rather than houses."[129] Conditions had not improved materially two years later when Captain Jonathan B. Hager commanded the post. Recent construction included a one-room adobe house for the commanding officer, two one-room log houses, a leaky commissary building built against the hillside, and a quartermaster's storehouse made of canvas. Hager condemned enlisted men's quarters as being "nothing more than holes in the bank, damp, dark, unhealthy, and unfit for a dog to occupy."[130]

In 1868 Camp Bowie was relocated on ten acres of level ground to the east of the old post, where within a year enlisted men assisted by civilian laborers had constructed several "very good adobe buildings." Civilians also had furnished the adobe bricks at $22.50 per thousand. The total cost of the buildings, consisting of two sets of officers' quarters, one company barracks, one subsistence storehouse, one quartermaster's storehouse, and stables and corrals, was an estimated $4,368—a trifling sum compared to the amount spent rebuilding Forts Davis and Stockton in Texas.[131] Construction continued at Bowie in following months, though delays occasionally occurred, as in the fall of 1870 when three of the post's best adobe layers deserted after payday. By 1873, however, three additional officers' quarters, one new enlisted men's barracks, and a hospital had been completed, all made of adobe with earthen roofs.[132] A storm that struck the post early in 1874 badly damaged some of these buildings and rendered the hospital untenable. Nonetheless, Colonel August Kautz, then commanding in Arizona, reported late in 1875 that Bowie was in "excellent condition."[133]

Poor housing and stifling summer heat made Fort Mojave among the least desirable posts in Arizona. Established in 1859 to protect emigrant travel to California, the post was abandoned in 1861 and its garrison sent to Los Angeles where secessionist sentiment was strong.[134] When Captain J. Ives Fitch returned with two companies of California volunteers in May 1863, he found buildings at the post "in a tolerable state of preservation," though stripped of doors, windows, and all other woodwork. The volunteers soon constructed new quarters of "upright logs thickly chinked with clay."[135] Officers were still living in these "stockade buildings" at the end of the decade when Acting Assistant Surgeon F. S. Stirling reported on the health of the garrison. The buildings were much dilapidated at this time, and troops were engaged in building new quarters, with local residents furnishing adobes and shingles. Stirling reported that enlisted men had occupied one of the new buildings, though summer nights were so hot that no one slept inside—"the whole garrison [sleeps] on the open plain, endeavoring to catch the faintest breeze."[136]

In 1869 the Mojave military reservation was enlarged to include Mojave City, a collection of about seven ramshackle buildings from which locals had been dispensing liquor illegally to soldiers. The owners were soon ordered off the reservation for violating military regulations, though they received compensation for their buildings. One or two of these adobe shacks were torn down so that the adobes could be used in other structures, and the rest were used as quarters for laundresses and married soldiers.[137]

Construction at Mojave continued apace, with local residents usually furnishing building materials and soldiers and Indians providing labor. In 1875 the post consisted of two adobe barracks, one having a shingle roof, the other a dirt roof, and both having dirt floors; officers' quarters, comprising two adobe buildings and one old stockade building; two shingle roofed adobe storehouses; a post bakery; a guardhouse; and an adobe hospital that cost an estimated $4,000.[138] In later years the post would experience more than its share of natural disasters that added to the cost of construction and to the sum of human suffering. For example, in late 1876 or early 1877 a fire destroyed officers' quarters, which led the army to contract with local resident Benjamin H. Spear to erect adobe walls for new quarters at $30 per 1,000 adobes. Then on 16 August 1878 a severe thunderstorm damaged roofs of several buildings, including the commissary and the post surgeon's quarters.[139] Two years later, on 22 August 1880, four enlisted men were killed and seven others injured when their barracks collapsed during a tremendous rain and windstorm. Other buildings sustained damage as well—roofs were torn off, window panes broken, walls were cracked, and chimneys blown down. The garrison subsequently moved into an unoccupied barracks, but in August 1881 it, too, was badly damaged in a ferocious rainstorm. The men were then compelled to live in "out buildings" scattered about the post. Plans for building new quarters were submitted to higher authorities, but their estimated cost of $4,614 could not be spared from the military's appropriation. The garrison was still living in makeshift quarters nineteen months later when thirty enlisted men signed a letter protesting the unsafe condition of their quarters, which they felt compelled to evacuate during windstorms.[140]

In contrast to Mojave, officers and enlisted men found Fort Whipple the most desirable post in Arizona because of its mild climate and delightful mountain scenery. For officers and their families, the post also offered a lively social life. Ellen McGowan Biddle, who lived there in the late 1870s with her husband Major James Biddle, recalled years later that Fort Whipple had been "a very gay post, with an entertainment of some kind almost every day and evening."[141] By the time the Biddles arrived, however, the post had undergone major renovation and no longer resembled its earlier and more primitive form.

Fort Whipple had been established originally on 21 December 1863 in the Chino Valley at Del Rio Spring near the Verde River. The only structures erected at this site were four log buildings: a blacksmith's shop, hospital, and quartermaster and commissary storehouses with canvas roofs. In a few instances soldiers had constructed stone and mud foundations for their tents. On 18 May 1864 the post was relocated on the left bank of Granite Creek, about twenty-one miles southwest of the first site and one mile northeast of Prescott, the center of a new gold-mining district.[142] At first the post consisted of a rectangular log stockade, the wall of which formed the outer wall of quarters built within. Quarters were nothing more than log huts—the logs set upright in the ground, crevices filled with mud, and the roofs shingled. Civilians had been employed to help soldiers cut the necessary logs at a rate of fifty cents per log.[143]

By the time Colonel J. Irvin Gregg took command of the District of Prescott in April 1867, quarters at Fort Whipple were "unfit even for the housing of animals," and Gregg warned that attempts to make them habitable would be "unjustified extravagance."[144] But five years would elapse before new quarters were constructed. In the meantime, local citizens were hired and supplies purchased for making repairs on existing structures.[145] The price the army paid for lumber dropped significantly after John A. Rush and his partners erected a sawmill five miles southeast of Prescott in the fall of 1868. This firm provided competition for Albert O. Noyes and George Curtis who had the only other sawmill in the area. According to Lieutenant Colonel Frank Wheaton, commanding the District of Upper Arizona, fair quality lumber cost $50 in gold per thousand feet in 1869, half of what it had cost the previous year, and common, "merchantable" lumber cost $20, the lowest rate ever recorded in Arizona. Rush, however, appeared eager to engage in other enterprises, for he sold his sawmill to the army later that year for $8,000. Wheaton considered this a bargain and predicted the sawmill would pay for itself in less than a year.[146]

The Government Saw Mill, as it was called, was soon supplying lumber to Camp McDowell and to all posts in northern Arizona. Philip Richardson was hired at $200 per month in currency to supervise its operation, and in July 1870 the sawmill payroll carried the names of nine citizens and five soldiers on extra duty.[147] But despite the availability of more lumber for repairs, criticism of the old buildings at Whipple mounted. On 11 January 1871 post quartermaster Charles W. Foster informed officials in California that public property was endangered because of the "miserable character of the storehouses." He thought it was miraculous that fire had not already destroyed the old wooden buildings or that thieves had not made away with supplies, for the walls of the storehouses were so decayed that logs could be pushed in or taken out with very little effort.[148]

Within a month Foster's fears in part were realized. Late on the evening of 20 January a fire broke out in the commanding officer's residence, which was entirely consumed before the fire was extinguished. To prevent the fire from spreading, soldiers had cut away sheds and the palisade that separated the burning building from the storehouses. They also covered the storehouses with wet blankets and kept them saturated with water throughout the night.[149] Though unharmed by the fire, the storehouses remained insecure. And on the night of 4 February an unknown number of soldiers gained entrance to the commissary storehouse and made away with large amounts of sugar, salt, coffee, bacon, and lemon extract.[150]

A second fire that destroyed corrals, stables, and mechanics shops at Fort Whipple on 27 April 1872 ushered in a building program in which the remaining log buildings were torn down and replaced by frame structures. The government sawmill, which had been relocated on Groom Creek about seven miles from Fort Whipple, furnished lumber and shingles, but it, too, was destroyed by fire the night of 21 December. Both civilians and enlisted men provided the labor in rebuilding the post.[151]

By mid-1873 more than half the projected buildings had been completed, but construction slowed in later months because of transportation and manpower shortages. Consequently in November 1874 the army hired M. V. Davis to construct the post hospital (the army furnishing supplies) for $3,000, and William Z. Wilson to complete other buildings for $4,175.[152] By the first of the year 1875, major construction at Whipple had almost ended. Eight sets of officers' quarters had been completed, as well as two enlisted men's barracks, a large frame building (divided into twelve apartments) for married soldiers, an adobe commissary storehouse, a quartermaster's storehouse, blacksmith and carpenter shops, and the hospital.[153] A new department headquarters building was completed in December 1876, described by Colonel Kautz as "a very substantial and commodious building" with adobe walls. But in January 1881 this building was destroyed by fire that had broken out in a privy connected to it by a covered walkway. Fire also damaged one set of officers' quarters in 1875 and destroyed three sets in 1878.[154]

In 1877 work began on a new residence for the department commander, and its construction would involve Colonel Kautz in heated controversy. On 15 June Secretary of War George W. McCrary authorized expenditure of $3,574 for building the new quarters, which were to have pisé walls, brick chimneys, and shingle roof. Kautz subsequently altered and enlarged the original plans, which added to the cost. On 26 June the quartermaster's department, with Kautz's approval, awarded contracts worth more than $5,700 to three Prescott residents. Charles F. Cate contracted to supply bricks at $11.50 per thousand; James G. Wiley would supply dressed lumber at $45 per thousand feet, rough lumber at $19, and shingles at $5.50 per thousand; and Peter B. Brannen agreed to construct the residence for $3,000. By the time Kautz left the territory in March 1878 about $6,200 had been expended for material and labor, and the post quartermaster was asking for more money to finish the job.[155] Kautz later was reprimanded for altering plans without authorization—there was even some talk of court-martial. But Kautz recently had been acquitted in one court-martial trial in which he had been accused of breaching the military code of conduct, and the matter of a second trial was allowed to drop.[156]

With an additional expenditure of about $2,800, the commanding officer's residence was completed in August 1878. It was probably the most spacious house then to be found at any military post in the Southwest. The two-story frame building had a shingle roof and rested on a stone foundation. The walls were formed by filling in between the studs with adobes and were lathed and plastered on both sides. There was a ten-foot-wide piazza on three sides of the building. On the ground floor were the parlor, sitting room, and dining room (each with a large bay window), two spacious halls, and three bedrooms with two bathrooms. The second story contained five bedrooms with space for large closets. Also on the ground floor and adjacent to the dining room were the pantry, kitchen, and laundry. Stairs from the pantry led down to a cellar, and a cistern was located under the porch. A modern plumbing system carried water to both floors and conducted waste to a cesspool located several feet from the house. Army Inspector Edmund Schriver dryly remarked that officers would find this an expensive home to furnish.[157]

Two hundred and fifty-five miles south of Fort Whipple, soldiers at Tucson lived in tents and makeshift shelters into the mid-1870s. The post at Tucson had been established in May 1862 when California volunteers under Lieutenant Colonel Joseph R. West occupied the town. Army headquarters were established in a house owned by Palatine Robinson, a southern sympathizer, and several other buildings were commandeered as well. The post was abandoned in September 1864, reactivated in 1865, and declared a permanent military post the following year under the name of Camp Lowell.[158]

A supply depot was established at Tucson shortly after General John Mason assumed command in Arizona in 1865. To house both men and supplies, Mason instructed his troops to take possession of vacant buildings known to belong to southern sympathizers. By the end of the year the army occupied thirteen houses. Only four, however, had been owned or partly owned by Confederate sympathizers. Another was considered government property. The remaining buildings were rented from local residents at monthly rates of from $10 to $100.[159] Major Murray Davis, who inspected the post in January 1866, criticized this arrangement, for the scattering of soldiers undermined discipline. But Davis also pointed out that the total monthly rent of $278 was much less than it would cost to build new quarters.[160]

Later that year soldiers cleared mesquite from a tract adjoining the town and then settled into tents over which they built brush *ramadas* for protection from the scorching sun[161] The quartermaster's department continued to rent buildings in town for the depot and hospital at a cost of $400 a month in 1867. Army Inspector Roger Jones suggested using this amount for permanent buildings. But supply routes through southern Arizona were in a state of flux, and officials in California would not commit funds for construction until a decision was made about a permanent location for the supply depot.[162] In the meantime, complaints of inadequate housing multiplied. Captain William H. Brown, commanding at Lowell in May 1869, said that tents offered no protection "whatever from the hot sun in summer season, or the cold in winter," and he requested permission to erect adobe quarters. Improvements at the post then included an adobe guardhouse, two kitchens, a bakery, and magazine.[163] In October the army advertised for building materials at Camp Lowell: 170,000 adobes, 892 *vigas*, 40,000 feet of lumber, 262,000 shingles, and 1,000 bushels of lime. The Tucson *Weekly Arizonan* later reported that the lumber contract was awarded to Alphonse Lazard who would furnish the required amount at the rate of $99.80 per thousand feet. Lazard had recently established a steam sawmill in the Santa Rita Mountains, about sixty miles from town, and the press applauded his success in reducing the cost of lumber. According to the newspaper, most timber used in building Tucson had been brought from New Mexico and sold to consumers for $200 per thousand feet.[164]

But the adobe barracks that Captain Brown requested were never built even though their construction had been authorized. Major Milton Cogswell, commanding at Lowell, reported in April 1870 that he would not build until a final decision was made about the depot's location. He recommended that both post and depot be moved to Rillito Creek northeast of Tucson where wood, water, and grass were abundant.[165] But it was not until 1873, after Crook's successful Indian campaign, that Lowell was moved seven miles out of town on the Rillito.

A cavalry unit had been stationed on this site for at least a year, and by 20 March 1873 the infantry had moved out to begin clearing away brush and cactus for the new post.[166] For the next several months the men were employed at building quarters—sometimes working ten and twelve hours a day. In June the army hired about fifty civilians to make adobes at one dollar a day plus rations. Because work had started so late, very little of the $17,237 allocated for construction had been expended by the end of the fiscal year. On 19 June 1873 an indignant Lieutenant Colonel Eugene A. Carr, commanding the post, requested sufficient funds during the next fiscal year so that soldiers would not have to build their own quarters. He said that most men in his command had never lived in proper quarters since their arrival in Arizona; they either had lived in tents or without any shelter whatsoever. "At other places where the duties are less arduous, and the climate more salubrious," he complained, "large sums . . . are allotted for the building of comfortable quarters; and I sincerely hope that more consideration may be shown to the troops serving in this most uncomfortable region, which is far worse than India, where the British troops are provided by the Government with good quarters." He estimated that an additional $75,000 would be required to house comfortably the proposed garrison of two companies of cavalry and two of infantry.[167]

When Colonel Delos B. Sacket inspected the post in mid-June of that year, he found the entire command quartered in tents over which brush shades had been erected. Only one set of officers' quarters was then being built, although plans called for fifteen sets of officers' quarters and four sets of company quarters. The army rented an adobe building in Tucson for a hospital at $95 per month and another adobe building consisting of twenty rooms at $400 per month for the supply depot. In addition, quarters were rented for the paymaster, post quartermaster, army surgeon, and commissary officer, as well as one room for the paymaster's office, and a corral for the use of transient public animals. In sum, the army injected into the Tucson economy about $700 a month in rental fees and $760 a month in wages to civilians working at the depot.[168]

On 25 September 1873 Major Marshall I. Ludington, then assigned to the quartermaster general's office in Washington, reported that the Secretary of War had authorized a total expenditure of $34,000 for building Camp Lowell and a new post (Camp Grant) near the mouth of San Carlos River, by labor of troops if possible.[169] In fact, most of the work at Lowell was performed by civilians. In October the Tucson firm of Lord and Williams received a contract to furnish adobes and lay them into walls at the rate of $30 per thousand. This firm then hired David Dunham as subcontractor, and Dunham supervised construction of the hospital, quartermaster and commissary storehouses, and the quartermaster's corral. Though Dunham lost about $640 as subcontractor, the army subsequently contracted directly with him to erect most of the remaining buildings, all built of adobe with dirt floors and dirt roofs.

In mid-1875 Army Inspector James A. Hardie reported that Dunham had received a total of $15,790 for labor and materials supplied in building the post. Almost all the buildings for a three-company post had been completed or were nearing completion, and Hardie believed their total cost (excluding the cost of hardware, doors, and windows sent from San Francisco) would amount to $20,000.[170] According to the inspector, the post had been finished with dispatch and economy, though Meigs later complained that soldiers should have manufactured the adobes—unpleasant work to be sure but "such as any able bodied man, with very little instruction can perform."[171] By this date Tucson depot had been evacuated, and its stores removed to Camp Lowell. The only residence the army continued to rent in Tucson was the paymaster's.[172]

In contrast to Camp Lowell, construction began at Fort McDowell near present-day Phoenix soon after the post was established in September of 1865. Officials intended McDowell to be "the largest and most solidly built" post in Arizona with work performed by soldiers at little cost to the government. By the first of December California volunteers under Lieutenant Colonel Clarence E. Bennett had manufactured 100,000 adobes and were clearing the site with hoes and axes. Though the men spent that first winter living in tents and holes dug in the hillside, by 25 January 1866 they had completed a hospital, the commanding officer's house, and walls of four sets of officers' quarters and three company barracks. Murray Davis, who inspected the post in

January, was delighted to find "an edifice constructed for the service of the government without enriching a contractor." But construction stopped in February when the men were put to work clearing land for the post farm.[173]

Soldiers resumed construction in February 1867, but the earthen roofs they attached to the long-standing adobe walls failed to keep out rain. In March the post surgeon complained that rain seeping into men's quarters was responsible for a rash of bronchial attacks that swelled the sick list.[174] Even more disturbing was a report the post quartermaster issued in August: "Adobe buildings occupied as quarters by both officers and men at this Post are in very bad condition, some of them in my opinion being unsafe." Later, during four days of heavy rain the latter part of December, soldiers had to vacate their quarters, and one building collapsed.[175]

Soon after assuming command of the garrison in the spring of 1868, Major Andrew J. Alexander made a thorough inspection of the post. In a letter dated 2 June he described his own house as being "very comfortable," but he was appalled by the condition of most other buildings. Beams and rafters made of willow and cottonwood had rapidly decayed, and props were being used to prevent roofs from falling. During rainstorms quarters became unfit for habitation. A board of officers concluded on 11 July that no amount of repair would make the buildings "either comfortable or secure." By October 1869 only one company barracks was in use; three others had been torn down because of their dilapidated condition; and four of the five companies garrisoning the post were housed in tents.[176]

It was not until 1872 that new construction got under way, with contractor Thomas McIntyre erecting the adobe buildings. The total cost of rebuilding McDowell is not known. When Colonel Sacket inspected the post in June 1873, he reported that more than $10,000 was due the adobe contractor. Sacket also reported that only about half the building program had been completed. New officers' quarters had been built, one new company barracks finished, another was being completed, and the walls of a third were about eight feet above the ground. The last mentioned structure—never finished—was demolished in 1874 after part of the walls fell. Still needed at the post, Sacket reported, were a hospital, laundresses' quarters, and a stable.[177]

In mid-1873 the chief quartermaster for the department requested an additional $18,751 for repairs at McDowell. The post probably received a part of this amount, allowing the building program to continue. John Smith, the post trader, agreed to furnish skilled labor to construct cavalry stables in 1875 for $900, and the following year new laundresses' quarters were built.[178] Colonel Kautz, who inspected Camp McDowell in November 1875, found the new barracks, storehouses, and water system all in need of repair. Since 1873 water had been pumped from Verde River into a reservoir and conveyed by pipes into the buildings. The men's quarters were still in poor condition in 1878—without floors, porches, or ceilings.[179] Meant to be the best-built post in the territory, McDowell never reached this exalted status before the military abandoned it in 1890.

The government made many different arrangements for building western military posts. Most were built by a combination of soldier and civilian labor. Often the army contracted for construction of individual buildings or corrals. And in some cases the army hired a master builder to superintend the building of an entire post. All of these methods were used during construction of old and new Fort Grant. The history of the first Fort Grant, established in October 1865 on the east bank of the San Pedro River about fifty-eight miles north of Tucson, is one of misfortune and tragedy. After summer floods in 1866 swept away twenty of its twenty-six buildings, the post was relocated on the nearby site of old Fort Breckinridge.[180] Sickness raged through the garrison that fall, frightening away Hispanic laborers who had been hired to erect adobe buildings. Of ninety-seven men stationed there in September, thirty-four were on sick report.[181] For several years the men lived in shelter tents or jacales while they labored to erect permanent quarters. But scouting expeditions and a shortage of draft animals slowed construction. Lieutenant Colonel Thomas C. Devin, commanding the District of Tucson, expressed hope in 1868 that when new quarters were finished the health of the post would improve.[182] But while soldiers continued their arduous labors, malaria struck them down in alarming numbers. The post surgeon reported treating 1,735 cases of "malarial fevers" in 1868 and 561 cases in 1869 when the command was only about two thirds its former size. Army Inspector Roger Jones described Camp Grant as "the unhealthiest post south of the Gila." During the two years preceding his inspection in May 1869, seventeen soldiers had died there from sickness.[183]

Even though soldiers performed a great deal of work in building the post, Camp Grant in 1870 must have appeared dismal to incoming troops. Lieutenant John G. Bourke, who arrived in March with the Third Cavalry, called it "the most thoroughly God-forsaken post of all." In his classic account, *On the Border with Crook,* Bourke recalled:

> There were three kinds of quarters at Old Camp Grant, and he who was reckless enough to make a choice of one passed the rest of his existence while at the post in growling at the better luck of the comrades who had selected either of the others. There was the adobe house, built originally for the kitchens of the post [Fort Breckinridge] at the date of its first establishment, some time in 1857; there were the "jacal" sheds, built of upright logs, chinked with mud and roofed with smaller branches and more mud; and the tents, long since "condemned" and forgotten by the quartermaster to whom they had originally been invoiced.[184]

The most recent addition to the post was a large adobe storehouse that contractor Newton Israel completed in December 1869. Israel, the post trader, had submitted the lowest bid of $5,790. But Major John V. DuBois, the post's commanding officer, called the storehouse that Israel built "a miserable affair." DuBois claimed that the best interests of the government had been sacrificed to "false economy"—*vigas* were weak, doors and windows insecure, walls too thin.[185] But Israel did not live long enough to enjoy the fruits of his labors.

On a Saturday afternoon, 28 May 1870—the day after DuBois penned these remarks—Apaches attacked two wagon teams carrying provisions to Israel's ranch. Israel was killed by the first volley from their bows and rifles; Kennedy, his partner, was mortally wounded. The rest of the poorly armed party—nineteen men and two women—made their escape. On Saturday evening the mail carrier between Tucson and Florence found Kennedy by the roadside—barely alive with an arrow protruding from his chest. A mile beyond were the mutilated remains of Israel.[186] Eleven months later on 30 April 1871 Arivaipa Apaches living near Camp Grant felt the full fury of Tucson residents seeking revenge for these and similar murders. The Camp Grant Massacre, "one of the worst blots in the history of American civilization," according to John G. Bourke, forms a tragic concluding chapter in the history of old Camp Grant.[187]

The post was abandoned in March 1873, largely because of "malarial fevers," and its name given to a new installation being established at the foot of Mount Graham, about 110 miles northeast of Tucson. Late in April the new post commander, Captain William H. Brown, arranged with Warner Buck, the post trader, to superintend the building of new Camp Grant. Exactly how much Buck profited from this arrangement is a mystery. According to Buck, a master-builder by profession, he assumed total responsibility for "furnishing plans, materials, labor, time, and money to pay for everything required and bought for the buildings." With an initial appropriation of $17,762, soon expanded to $25,000, the post was built of both stone and adobe.[188] A pinery was opened seven miles up Mount Graham, where a soldier-engineer operated a government sawmill. In return for use of his planing machine and shingle saw, Buck received a percentage of the shingles and lumber planed at the mill. In addition, Buck and at least three other citizens received payment in lumber for furnishing ox teams for hauling logs.[189]

Buck and the post quartermaster soon overspent the allocation. By October 1874 construction bills amounted to $44,562, and the department quartermaster was requesting additional money to complete the post. Before authorizing any more funds, Quartermaster General Meigs ordered an investigation, which revealed that Buck, in his capacity as superintendent and chief purchasing agent, had received a total of $42,700 for materials and labor. Buck testified that an unusually wet season was responsible for one third of the cost. Heavy rains had destroyed many adobe walls, which had to be rebuilt. Even though Buck enjoyed "a good reputation in Arizona for truthfulness and honesty," Army Inspector James A. Hardie believed the government's money had been wasted. Camp Grant, he concluded, was a badly built post. The hospital, company quarters, and storehouse were built of an adobe "of no cohesive quality," and mortar used in building stone quarters for officers was already disintegrating. By September 1875 one set of officers' quarters was already in ruins.[190]

Nonetheless, money to complete the post was soon forthcoming. In May 1876 Peter B. Brannen contracted to build two sets of officers' quarters, quarters for laundresses, an adjutant's office, a mechanic's shop, a magazine, and three cavalry corrals with sheds, as well as finishing one quartermaster's corral—for a total cost of $14,977. The government would furnish brick, lime, doors, glass, nails, and other hardware, as well as transportation for hauling stone and adobes from the place of manufacture to the site of the buildings. The government would also cut and haul logs to the government sawmill, giving Brannen the use of the mill for cutting logs into lumber. He would also command the services of at least six extra-duty soldiers.[191]

Brannen's contract provided that work be completed by 31 October, but construction was delayed several months beyond this, for the army found it impossible to keep him supplied with timber. In March 1877 Brannen sought release from the contract because he could not afford to pay idle workers. A board of officers recommended that this request be granted, deducting $180 from Brannen's final payment to complete the unfinished buildings.[192]

For the remainder of the decade, the army contracted with civilians to construct new and repair old buildings at Camp Grant. Officers' quarters were plastered, additions made to the hospital, bathhouses constructed, and wells dug to provide a more adequate water supply. Despite earlier criticism, officers and men considered Fort Grant among the most desirable posts in the territory in the 1880s.[193]

Building projects at the various military installations infused large amounts of money into local communities both for supplies and for labor. But even when major construction was not under way, army posts enabled workers to pick up jobs. During and just after the Civil War the army employed more civilians in the Southwest than any other agency. Chief Quartermaster John C. McFerran late in 1863 estimated that a total of 850 civilians were working in the quartermaster's department in the Department of New Mexico. Fort Union and Union Depot combined employed the largest number—389; followed by Fort Craig with 147 employees; Las Cruces with 80; Tucson, 47; Santa Fe, 45; Sumner, 43; and nine other posts sharing the remaining employees.[194]

Even more civilians were employed in later months. Captain Charles McClure, chief commissary of subsistence for the District of New Mexico, reported in 1866 that during the month of October 1,397 citizen employees had received rations.[195] Most civilians worked for the quartermaster's department; only a few found employment in the subsistence and medical departments. Civilians having special skills such as a blacksmith, carpenter, mason, or saddler were much in demand, but the army also employed large numbers of teamsters and general laborers. For many western men the military must have provided a certain amount of psychological security. If no better offer came along, they could always find a job driving teams or mixing adobe at an isolated military fort.

Typical of civilian payrolls during the mid-1860s was the January 1866 payroll for Fort Cummings, New Mexico. During that month the post quartermaster employed a clerk at $100 per month, an expressman at $125 per month, a carpenter and a blacksmith at $65 each per month, an assistant wagon master at $50 per month, ten teamsters at $30 each per month, and a cook, laborer, and blacksmith's striker at $30 each per month. All the higher paying jobs were filled by Anglos, while eight of thirteen men receiving $30 in wages were Hispanos. In later months the quartermaster employed a saddler, mason, and a wheelwright, each at $65 per month, and a guide at $45 per month.[196]

Wages were higher in Arizona where labor was scarce. In 1866 and 1867 teamsters, herders, and common laborers received $35 per month. Wages for other categories of workers varied from post to post, but typically guides and expressmen received $75 per month, blacksmiths, saddlers, wheelwrights, and other skilled mechanics between $75 and $100 per month, and clerks between $100 and $150 per month in coin.[197] Wages were not static, however, and early in 1868 salaries in New Mexico were nearly comparable to those in Arizona. Among the 396 civilians employed in the quartermaster's department at Fort Union Depot during February of that year, for example, teamsters, herders, cooks, and common laborers were paid $35 per month, most blacksmiths, carpenters, wheelwrights, and saddlers, $75 per month, and clerks between $100 and $150 per month.[198]

In following months the Department of the Missouri ordered reductions both in pay and numbers employed by the quartermaster's department, causing officers in New Mexico to grumble about the consequences. A reduction that took effect on 1 May 1868 reduced the pay of civilian wagon masters from $75 to $55 per month, most teamsters' pay from $35 to $25 per month, and common laborers' pay from $35 to $30 per month. Blacksmiths, carpenters, and wheelwrights would be paid $2.75 per day, about the same wage they had received when paid by the month.[199] Almost a year later civilian employment at all western posts was drastically cut. Under orders from President Ulysses S. Grant, the quartermaster's department reduced the total number of civilians it employed from 10,494 to 4,000. For a time this reduction seriously curtailed building projects in New Mexico, for the district was allowed only 153 employees after 1 July 1869. Of this number Fort Union Depot was allocated 96 employees, Santa Fe, 30, and the other posts from 1 to 3 employees each.[200] By this date wages in the Department of the Missouri also had been adjusted as a result of the new congressional law mandating an eight-hour workday for civilians working in the army. Blacksmiths, saddlers, carpenters, and wheelwrights now received $2.20 per day, common laborers, $24 per month, teamsters with wagon trains, $25 per month, and depot teamsters, $30 per month.[201]

Very few officers welcomed reduction in the civilian labor force. Major Marshall I. Ludington, New Mexico's chief quartermaster, complained that isolated posts would be without the indispensable skills of a blacksmith or wheelwright. He also pointed out that Union Depot would have only thirty-one teamsters for duty on the road, not enough to distribute troops arriving from Texas to their respective posts in New Mexico. Sometime in August 1869 Ludington was authorized to employ fifty additional civilians until troop movements were completed.[202] Ludington's successors, however, struggled to carry out their labors with even fewer employees. In 1871 the district quartermaster was allowed to hire only 117 civilians. Important posts like Bayard, Craig, and Stanton had to carry on with only one civilian employee each, while Union Depot and headquarters at Santa Fe continued to employ the bulk of the civilians.[203]

Similar reductions occurred in Arizona, but the situation there is less clear, for extant records are not as voluminous as for New Mexico. When General John S. Mason took command in the summer of 1865 he ordered all civilian employees at the various posts to be discharged, retaining in the field "only such as are indispensable"—guides, interpreters, expressmen, teamsters for supply trains, and blacksmiths where none could be found among enlisted men.[204]

These reductions, both in New Mexico and Arizona, were meant to save the army money by having enlisted men perform the labor rather than higher-paid civilians. The problem was, however, that enlisted men notoriously lacked the skills necessary to maintain the army's mobility and efficiency. Officers repeatedly bemoaned the lack of mechanical skill among recruits of all races. Entire commands often were without soldiers who could shoe horses, lay stone or adobe, repair wagons, or sink wells. Nor could soldiers be trusted as teamsters, for they often neglected to take proper care of their animals—"they did not enlist in the Army to drive teams" was a frequent excuse. And quartermasters feared using enlisted men to carry mail or transmit telegraph messages. In the one case, they were liable to desert and carry off mail and other valuables entrusted to them; in the other, they might reveal the nature of confidential business transacted over the lines. For these reasons the army could not get along without reliable civilian help, a fact recognized by officials in the Department of the Pacific who in 1866 suspended orders discharging all civilians and substituted another stipulating that civilian employees be reduced to "the lowest possible number."[205]

How many civilians the army employed in Arizona after the Civil War is not known for certain. During March 1867, 56 civilians worked in the quartermaster's department at Tucson and 7 in the subsistence department, the total monthly payroll amounting to $2,930.[206] Major Roger Jones, who made an inspection tour through Arizona in May and June of that year, recorded that the subsistence department employed between 2 and 6 herders at the various posts. He also reported that the quartermaster's department employed 50 civilians at the Fort Yuma Depot, 37 at Fort Whipple, 13 each at Camps Goodwin and McDowell, 11 at a camp near the Santa Rita Mines, 3 at Camp Wallen, and 2 each at Camps Grant, Bowie, and McPherson.[207] Thus, the quartermaster's department in 1867 employed at least 189 civilians in the District of Arizona, and this figure probably would reach just under 200 if employment data for Camps Mojave and Verde were available.

Early in 1869 officials in the Department of California scrutinized post returns to identify positions that could be eliminated. Major John P. Sherburne, the department's adjutant general, questioned the necessity for employing twenty-three civilians at Fort Whipple at a cost of $1,550 per month, and he ordered the discharge of all workers not "actually necessary."[208] In April the Prescott *Weekly Arizona Miner* reported that all but four or five employees at Whipple had been let go, a foolish move according to the press. "It is a well known fact that one citizen will do as much work as two or more soldiers," the writer editorialized, "for the reason, we presume, that the citizen hires to work, while the soldier enlists to fight." Like most Arizona residents this writer believed that the war against Apaches was retarded whenever soldiers remained in camp building houses, chopping wood, or performing tasks better left to civilians.[209]

Just how many civilians in Arizona were discharged that year is not known. General Edward O. C. Ord, commanding the Department of California, reported in September that when reduction orders arrived from Washington his department found it unnecessary to discharge many workers since reductions had already been made to the maximum allowed. In his annual report Ord stressed the importance of civilian labor, and his words are reproduced here in full because they reflect the thinking of many other officers who served in the West.

> I beg leave to call attention to the fact that the duties of blacksmiths, farriers, carpenters, wheelwrights, teamsters, guides, interpreters, packers, and other skilled laborers, are as necessary in building, wagon-making and repairing, shoeing, transporting freight, and other similar duties for the army, as they are for the business and support of our frontier towns; that our army posts are the nucleus around which such towns collect; and there are not mechanics or skillful laborers in the United States willing to enlist as soldiers and perform such duties for sixteen dollars per month in Greenbacks, when in every village or settlement among the mountains and plains such labor is worth from three to ten dollars in gold per day. The result is, we have not soldiers to do such work, and either civilians must be hired to perform it for the army, or the army posts and expeditions in the Indian Country must be abandoned and the troops concentrated at places where they are not needed.[210]

Despite Ord's eloquent defense for hiring civilian labor, in 1871 the quartermaster's department in Arizona was allowed only 106 civilian employees.[211]

The west Texas posts employed only a fraction of the 519 civilian employees allotted to the quartermaster's department in Texas after 1 July 1869. When Lieutenant Colonel James H. Carleton inspected the four posts early in 1871 he found that the quartermaster at Fort

Bliss had no civilians at all on his payroll. Authority had been granted to hire a blacksmith, but none could be hired from the local community.[212] At Fort Quitman the quartermaster employed six civilians—a clerk at $150 per month and a veterinary surgeon, wheelwright, mason, blacksmith, and saddler at $75 each per month. Carleton recommended the discharge of the veterinary surgeon, arguing that "Veterinary Surgeons were not known in our army before the war, and I presume that not more horses and mules were killed by 'doctoring' them, than now."[213] Fourteen civilians were employed in the quartermaster's department at Fort Davis: three clerks, three blacksmiths, four carpenters, two masons, a veterinary surgeon, and a saddler. Fort Stockton reported only four civilians working for the quartermaster—a veterinary surgeon, blacksmith, carpenter, and a mason.[214]

In later years the number of civilians the army employed in the Southwest fluctuated, depending upon the level of congressional funding and the degree of Indian-white hostilities. For example, in January 1874 Quartermaster General Meigs ordered that expenditures for civilian services be reduced by two thirds to keep his department within its appropriation. Meigs indicated in a letter to Lieutenant Colonel Fred Myers, chief quartermaster in New Mexico, that monthly expenses for wages in his district would have to be reduced from the current figure of $5,789 to $1,929. The new allocation, Myers decided, would pay only for employing twenty-four civilians at Union Depot and twenty at Santa Fe. About sixty-five employees throughout the district would be discharged.[215] Meigs informed Myers's counterpart in Arizona that monthly expenses there for civilian employment would have to be reduced from the current $8,846 to $2,948. To comply with Meigs's instructions General George Crook, then commanding in Arizona, ordered forty-five employees discharged, the reduction being about one third of the entire civilian labor force then allowed his department.[216]

A decade later, in June 1884, while Apaches were still resisting confinement to reservations, the Department of Arizona employed 143 civilians in the quartermaster's department, and the District of New Mexico employed 107. The pay differential in the neighboring territories was significant. In Arizona superintendents of transportation were paid between $100 and $150 per month; most blacksmiths and wheelwrights, $100 each per month; messengers and watchmen,

$60 each per month; and teamsters, $50 per month. In New Mexico blacksmiths, wheelwrights, and the superintendent of transportation in Santa Fe were paid $60 each per month; messengers and watchmen, between $30 and $40 per month; and teamsters, $30 per month. The total monthly civilian payroll in Arizona amounted to $10,426; in New Mexico, $4,801.[217]

By 1884 some employees in New Mexico had experienced several salary cuts. Compensation for the superintendent of transportation in Santa Fe had fallen from $100 per month in 1881 to $80 per month in 1883 and finally to $60 in 1884. The army paid the least the labor market would bear, and in many cases workmen accepted reductions to keep their jobs.[218] Other less fortunate men with several years' service in government employ simply lost their jobs when the War Department decreed cutbacks in employment. Take the case of 62-year-old Samuel Price, who was discharged as watchman at Fort Union Depot in March 1878. Price had served more than twenty years as a soldier in the army, enlisting in 1836 at age 20. From 1862 until his final discharge he had worked at the depot as storekeeper, superintendent of the woodyard, and as watchman. When Colonel Edward Hatch, then commanding in New Mexico, sought special authority to retain Price because of his "long and faithful service," he was told that New Mexico should not overspend its allocation.[219]

Whenever there was a reduction in civilian employment, the workload of enlisted men became heavier. Soldiers were the only sizable work force available in remote areas and could be employed relatively cheaply. So they spent much of their time in fatigue details—policing the garrison, carting water, chopping firewood, building roads, and performing most of the heavy labor in construction and maintenance of forts. This constant labor impaired morale and undermined the army's efforts to mold recruits into a trained fighting force. Army Inspector Nelson H. Davis was only one of several inspectors who reported that drill instruction at frontier posts was quite limited "owing to the large amount of work required of the enlisted men."[220]

Officers commonly believed that soldier labor was a major cause for desertion. Major Andrew J. Alexander, commanding at Camp Toll Gate, Arizona, in 1869, predicted a rash of desertions after the paymaster arrived, for men there had been kept at daily hard labor from morning till night. Some deserters may have been treated similar to the new recruits who arrived at Fort Craig the summer of 1882, four of whom deserted after eleven days of incessant work. Unused to heavy labor, their hands had blistered. There had been little time for relaxation, not even on Sunday, and nothing but a bare floor to sleep on at night. One recruit, a barber in civilian life, told his cousin "he could not stand it," and solely on account of the workload, he deserted.[221]

Enlisted men placed on extra duty as carpenters, masons, blacksmiths, saddlers, or in other skilled crafts were allowed extra pay at the rate of thirty-five cents a day, while soldiers classified as laborers received twenty cents a day extra. In 1884 this sum was increased to fifty cents a day for mechanics, artisans, schoolteachers, and clerks at army, division, and department headquarters, and thirty-five cents a day for teamsters, laborers, and other workers.[222] At some western garrisons more than one fourth of the rank and file were assigned extra duty. In 1869 at Camp Crittenden, Arizona, where labor was described as "constant and heavy," 58 of the 227 enlisted men performed extra duty—most of them as general laborers, woodchoppers, and teamsters. At Fort Stanton, New Mexico, where 94 enlisted men garrisoned the post that same year, 40 were assigned to extra duty.[223]

During economy drives the army cut back even on the number of its extra-duty men. Early in 1874 the quartermaster's department almost totally stopped their employment to keep within its appropriation.[224] For the fiscal year ending 30 June 1875 records suggest that posts in the Department of Arizona were allowed to employ no more than one sixth of their commands in extra duty, and during part of 1878 no soldier stationed in Arizona received extra-duty pay.[225] On at least two occasions—in 1876 and again in 1883—the District of New Mexico relieved all enlisted men employed on extra duty except for general service clerks at headquarters.[226]

Even with these reductions it is unlikely that the soldier's workload was lightened to any great extent, for his pay may have been suspended but his labor was not. This conclusion is sustained by the case of Private Daniel Ryer serving at Camp Grant in October 1876. He was detailed as a carpenter in the quartermaster's department with the expectation that he would be allowed extra-duty pay. After working more than a month without extra compensation, he asked to be relieved, "not feeling disposed to give what little knowledge of mechanics I have for nothing." The post quartermaster later explained that a reduced allowance for extra-duty services necessitated relieving several men like Ryer from extra duty and detailing them on daily duty, for which they received no extra compensation.[227] Soldiers throughout the West would have agreed with General John Pope, who wrote in 1877 that military posts were "garrisoned by enlisted laborers rather than soldiers."[228]

Among soldiers receiving extra-duty compensation was a special category of clerks and messengers comprising the General Service Corps and attached to district, department, and division headquarters. During most of the 1870s about ten general service men worked in both the District of New Mexico and the Department of Arizona. Enlisted men competed for these positions, for they conferred additional prestige, pay, and privileges. Clerks assigned to headquarters in Prescott, Arizona, in 1872 received, in addition to their regular pay, extra-duty pay of thirty-five cents a day and commutation for fuel, quarters, and rations—all of which amounted to about $68 per month.[229]

Enlisted men known for their efficiency, good penmanship, and superior character were recruited for these positions. They were expected to serve as models for the other enlisted men. Major James P. Martin, assistant adjutant general, Department of Arizona, required his clerks to dress better than other soldiers and to conduct themselves as gentlemen at all times. Private Edward B. Wheeler, upon receiving appointment as a general service clerk in Santa Fe, was advised to be industrious and refrain from dissipation; this would assure him "a good and comfortable position at District Headquarters."[230] Many clerks served for several years, establishing reputations for their competency, honesty, and diligence.

But even with commutation allowances the salaries of general service clerks did not approach those of civilian clerks working in the supply departments, and this caused dissatisfaction. In both Arizona and New Mexico during the 1870s the chief quartermaster and subsistence officers employed at least one first-class civilian clerk in their offices at $150 a month. Additional clerks were hired at lower salaries. For example, in 1879 the chief quartermaster in New Mexico employed one clerk at $150 per month, four clerks at $125, and one at $100 per month.[231] To save money the army attempted to use enlisted men as clerks at the various military posts. In 1866 Quartermaster

General Meigs had forbidden hiring civilian clerks for strictly post duty. But officers serving in the Southwest, particularly at posts garrisoned by black or Hispanic troops, complained that enlisted men were incapable of performing clerical duties.[232]

It is safe to say that civilian clerks were more in demand than any other category of employee. Their value is reflected in the wages they received—between $100 and $150 per month in New Mexico in 1870 compared to the $80 per month that master mechanics received. But the quartermaster's department in New Mexico was limited to only fifteen civilian clerks that year: four each at Santa Fe and Fort Union, two at Fort Wingate, and one each at Forts Bayard, Craig, Garland, Selden, and Stanton.[233] The situation in Arizona was even more dismal. General Order No. 5, issued on 6 June 1870, limited the number of clerks employed by the quartermaster to four: one for each of the supply depots at Yuma, Tucson, Whipple and Wilmington, California.[234]

Officers without the assistance of either soldier or civilian clerks often could not cope with the prodigious amount of paperwork the frontier army required. In many cases the officer was burdened with multiple assignments. At Fort Garland in 1869, for example, Assistant Surgeon Ely McClellan was in charge of the medical, commissary, and quartermaster departments. The commanding officer at Camp Mojave in 1871 simply complained that making out reports kept him confined to his desk instead of in the field attending to proper military duties. Falling behind with his paperwork, the quartermaster at Fort Bliss in 1878 hired a civilian clerk at his own expense, saying that the long hours he worked—often until after midnight—had caused his health and eyesight to deteriorate. In each of these cases, the officer pleaded for permission to hire a civilian clerk.[235] Somewhat different was the case of Captain William H. Nash, chief commissary of subsistence for the District of New Mexico, who in 1870 hired his wife as a records clerk and carried her on the payroll as S. S. Nash. When the district commander questioned the identity of this employee, the captain bristled, saying in effect that the chief commissary would employ whomever he wished. Shortly thereafter the Secretary of War issued a directive prohibiting officers from employing their wives as clerks.[236]

Every post commander valued a good clerk, but even more important for the success of the army's mission was a group of civilians who accompanied troops into the field—guides, scouts, interpreters, and packers. Most often these civilians were hired for specific tasks or expeditions and were not retained permanently on the payroll. An exception was Spanish interpreter and translator Lycurgus D. Fuller, who served a number of years in the quartermaster's department in Santa Fe smoothing transactions with Hispanic contractors.[237]

To carry on negotiations with Indians, however, the army hired interpreters fluent in an Indian language. Some of the best interpreters were former captives. María Méndez, captured by Apaches as a child, served as Army Inspector Nelson H. Davis's interpreter when he met with Victorio and other Mimbres Apache leaders near Fort Cummings in April 1865.[238] Another woman highly regarded as an interpreter is identified in the records as the wife of Pedro Ledesma. Prisoner of the Apaches for fifteen years, she had a thorough knowledge of the country near Camp McDowell and was employed on several scouting missions leaving the post.[239] A handful of former captives became noted Indian guides and scouts—men like Marigildo Grijalba, captured in Sonora by one of Cochise's bands when a boy, Juan Arroyas, nine years an Apache prisoner, and Micky Free (Féliz Martínez), abducted from his stepfather's ranch on Sonoita Creek in 1861.[240]

A half dozen or so other guides and scouts of mixed heritage carved niches for themselves in Arizona's history during the Apache wars. Part–Cherokee Pauline Weaver, reportedly in the area since 1829 as trapper and prospector, served as guide on several occasions before he died at Camp Lincoln in 1867. That same year Irishman Daniel O'Leary, one of Arizona's most experienced scouts, guided one of the three army columns sent into northwestern Arizona to subdue the Hualpais—a tribe he later befriended. Probably the most noted guide and scout was German-born Al Sieber, who arrived in Arizona in 1868 and served the army for the next twenty years.[241] In New Mexico one of the best scouts was Frank DeLisle, probably of French–Canadian heritage, who by 1872 had been employed by the army for twenty-six years. On occasion Emil Fritz, contractor and sutler at Fort Stanton, guided troops into Mescalero country, and Marcus F. Herring, like Fritz a former California volunteer, guided troops from Fort Cummings into Mimbres territory.[242] Dozens of other men in Arizona, New Mexico, and west Texas served the army as guides and scouts.

Despite the many dangers these men encountered, the army tried to obtain their services at the lowest rate possible. Chief Quartermaster

Ludington reported in 1868 that the pay allowed guides in the District of New Mexico ranged from $40 to $75 per month, depending upon circumstances. Usually their pay was about the same as the skilled craftsmen received, but on occasion it was less. Marcus Herring's pay at Fort Cummings in 1867 was only $45 a month, while blacksmiths and wheelwrights were paid $65 a month.[243] In Arizona most guides, blacksmiths, and wheelwrights were paid $75 per month in 1869, but workers in general were beginning to command higher pay. When army officials in 1870 tried limiting wages of guides and mechanics to $85 a month, post quartermasters complained that they could not hire skilled guides for that amount.[244] Three years later, in 1873, wages for guides ranged from $80 a month at Camp Bowie (where the guide was reported as being "not worth his ration as a guide, but can talk Apache") to $125 a month at Camps Apache and Hualpai.[245] Even Al Sieber fell victim to the army's economy measures. His salary as guide fell from $125 a month in 1872 to $100 a month in 1878, though the higher wage was ordered reinstated the following year, provided that Sieber also acted as interpreter.[246]

Civilians also were hired to manage the army pack trains that carried ammunition and rations on military expeditions. Enlisted men would not or could not learn to pack properly. Major Andrew J. Alexander, commanding at Camp McDowell in 1868, believed that packing was an art acquired only by years of practice. "Any one who knows the delays consequent upon traveling with thirty or forty pack mules with the best packers," he claimed, "will recognize the utter impossibility of moving at all without a proportion of experienced packers."[247] Lieutenant James Calhoun, the post quartermaster at Camp Grant, echoed this sentiment when he requested permission in 1869 to employ three Hispanic packers. He insisted that enlisted men were "nearly useless" as packers and that frequent stopping to repack animals was often the cause for the failure of scouting parties.[248]

One packer was usually allowed for every five or six mules, and a packmaster supervised the entire train. In 1869 the chief packer at Fort Cummings, Charles Harcourt, received $3 a day, and four assistant packers (three Hispanos and one Anglo) received $1.25 a day.[249] In Arizona the salaries were higher. Three Hispanic packers at Camp Mojave received $50 each a month in coin that year, and the chief packer at Camp Toll Gate, a man named Francisco, was paid $100 a month in currency.[250] Similar wages were probably paid Hadji Ali, one of the best-known characters in the West, who was employed as packmaster at Camp McDowell in 1869. A Syrian by birth, he came to the United States in 1857 with a shipload of camels and crossed Arizona with Beale's Expedition. He served as an army scout for several years and was known throughout the territory as Hi Jolly. While with the army at Tucson in 1880 he became a naturalized citizen, adopting the name Philip Tedro.[251]

Much of Crook's success in his 1872–73 campaign resulted from the care he lavished upon pack trains. Robert M. Utley has pointed out that under Crook's supervision mule transportation reached "the highest state of perfection in the history of the U.S. Army," and this allowed his troops to match the Indian's mobility. Crook carefully selected the pack mules, eliminating those that were too large, too small, or too obstreperous, and he hired the most experienced packers.[252] Though he was not the only officer to recognize packing as a trade that required experts, Crook was the most successful in organizing efficient and superbly mobile pack trains.

A second factor in Crook's success was the use of Indian auxiliaries. For his winter offensive Crook sent nine separate commands, each accompanied by a detachment of Indian scouts, to crisscross Apache country north of Camp Grant in Tonto Basin. By applying constant pressure, never giving the enemy time for rest, Crook broke Indian resistance. But without the Indian scouts, mostly Apaches themselves, the troops never would have found the hostiles' camps. Crook recognized their contributions, and in a report dated 30 June 1873 he singled out for special praise ten Apache scouts who had performed gallantly during the various campaigns.[253]

The army continued using pack trains and Indian auxiliaries in its later attempts to crush hostile bands of Apaches. The massive expedition that Colonel George P. Buell led into Chihuahua from New Mexico in September 1880 involved more than 350 soldiers, at least two Apache scout companies, and a large number of civilian teamsters, packers, and scouts. Jack Crawford, known as the "poet scout" because of his verses, preceded Buell into Mexico with instructions to locate the elusive Victorio and persuade him to surrender. Nothing fruitful came of Buell's expedition, for the Mexican commander, Colonel Joaquín Terrazas, had second thoughts about allowing American troops to advance into Mexican territory and requested the American officer to withdraw. Shortly after Buell began his return to American soil, he learned by courier that Terrazas's men had cornered Victorio and in the ensuing fight had killed the Apache leader and sixty of his warriors. Though barren of results, Buell's expedition was costly for the United States government. At least 210 names appeared on the roster of civilian employees hired for the campaign, and their total pay amounted to more than $20,000.[254]

About a year after Victorio's defeat, Chiricahua Apaches fled the San Carlos reservation in Arizona and returned to their old haunts in the Sierra Madre of Mexico. General Crook was ordered back to the territory, and after Chato's band had plundered its way through Arizona and New Mexico in March 1883, Crook crossed the border into Mexico to capture the hostile Apaches. Crook's Sierra Madre Expedition consisted of about 42 cavalrymen, 193 Apache scouts, 5 civilian interpreters and scouts, and 76 packers in charge of 350 mules carrying ammunition and rations. Arizonans condemned Crook for relying upon Apaches to capture Apaches, but the Indian scouts led Crook safely into the Chiricahua stronghold where he convinced Geronimo and other Apache leaders to return to San Carlos with their people.[255] Crook's amazing success vindicated his reliance upon Indian auxiliaries and reconfirmed the value of pack trains in providing maximum mobility.

The army continued using civilian packers during the remainder of the Apache wars, but it did so with an eye to economy. Packers hired for Buell's expedition in 1880 had been paid $60 a month, but in 1882 men employed as packers in New Mexico were receiving only $50 per month. By the close of 1885 the army was attempting to secure their services for $42 a month.[256]

The army's drive to economize led to the official disappearance of a military institution inherited from the British—the company laundress. In the mid-1860s the army allotted four washerwomen to each company, furnishing the women with transportation, rations, lodging, fuel, and medical attention. The women also received cash payments from the individuals for whom they washed, with laundry rates being established by post councils of administration. Rates therefore varied from post to post. Laundresses at Fort Goodwin, Arizona, in 1866 received one dollar a month for each enlisted man, $2.50 a month for a single officer, $3.50 per month for a family of two, $5 per month for a family of four, and $7 per month for a family of six, with officers furnishing soap and starch. Prices were higher at Fort Union, New Mexico, where in 1870 officers paid $5 a month for themselves and $3 a month for each additional member of the family, though the monthly rate for enlisted men was only a dollar.[257]

The army had difficulty recruiting laundresses at remote frontier posts, and consequently some women washed for large numbers of soldiers. At Fort Wingate, New Mexico, in 1871 one laundress washed for seventy-five men, and another—with the help of two Navajo women—washed for more than ninety men. Because there were so few laundresses at the post, company commanders authorized José Castillo, a "good washer and ironer," to establish a laundry, and he in turn hired both Hispanos and Indians to work for him.[258]

Occasionally laundresses balked at washing for officers, though the nature of their complaints can only be surmised—overwork, underpay, and too much criticism. When laundresses at Camp McDowell refused to wash for officers and their families in 1872, their rations were temporarily stopped.[259] Later in the decade, between 1876 and 1878, the army solicited testimony on the feasibility of retaining company laundresses. The War Department concluded that thousands of dollars in rations, fuel, and quarters would be saved by dispensing with this time-honored institution and relying instead on post laundries. As an institution recognized by the War Department, company laundresses passed out of existence with General Order No. 37 of 1878, which denied women the right to travel with troops as laundresses. Still, women continued to find work at frontier garrisons washing dirty laundry, though their right to room and board was now in jeopardy.[260]

Chinese laundrymen made their appearance at frontier posts in both Arizona and New Mexico in the 1880s, a decade that witnessed increased violence against orientals in the West. Two Chinese laundrymen at Fort McDowell, in fact, narrowly escaped being killed in 1881 after wrongfully being accused of theft. Hoping to secure confessions, civilian employees placed ropes around the necks of the accused and lifted them up several times, allowing them "to swing slowly down while a man held on to the other end of the rope." "In this manner," the post commander reported, "they were most harshly and badly treated for the frequency of the act added to the force with which the rope was held, strangled the men severely." The victims instituted civil suit at Phoenix against their assailants, but after the leader was acquitted, the cases against the others were dropped.[261]

In addition to laundry work, women found employment at military posts as hospital matrons and received rations and modest salaries in exchange for washing linens and performing other maintenance tasks. Typically matrons came from the ranks of soldiers' wives or wives of civilian employees, but widows also were hired. Left with a large family to support upon the death of her officer-husband in 1867, Jane Shaw worked at least two years as hospital matron at Fort Marcy in Santa Fe.[262] The army also hired women, on at least one occasion in 1868, to whitewash the post hospital at Fort Marcy—a time-honored craft among Hispanic women. Other Hispanic women worked for officers' families as seamstresses, cooks, and servants, though many families imported either white or black domestics from eastern and southern states.[263]

Civilians performed a wide variety of other services for the army, from teaching Indians and soldiers irrigation techniques to operating ferry boats, carrying mail, surveying military reservations, and removing bodies from old battlefields and abandoned military posts. Some civilians established businesses on the posts themselves—as post traders, tailors, shoemakers, news vendors, and restaurateurs. During the 1870s Adolph Griesinger, a former sergeant in the Third Cavalry, opened a restaurant and beer saloon at Fort Union, and Thomas Parsons in Santa Fe and a Mr. Wood at Camp Grant, Arizona, operated mess halls for officers and employees of the quartermaster's department.[264]

Many civilian employees were, like Griesinger, former enlisted men; indeed at the close of the Civil War, the quartermaster's department gave preference in hiring "to discharged soldiers having good characters."[265] A decade later veterans continued to cite their service records in seeking jobs with the army, and other applicants referred to previous employment with the quartermaster's department. Still others cited poor health as a reason for wanting to work at western posts. Applicants sometimes solicited reference letters from former teachers, military commanders, and hometown acquaintances. Even the quartermaster general himself, General Rufus Ingalls, wrote a letter in 1882 requesting employment in Santa Fe for the son of the late Colonel Fred Myers, former chief quartermaster for the District of New Mexico.[266]

Some civilians spent most of their adult lives working for the army. Joseph P. Verbeskey is a case in point. A soldier during the Mexican War, Verbeskey subsequently was employed for thirty-three years in Kentucky and New Mexico as wagon and forage master. He resigned his position in 1884 at Fort Stanton following difficulties with the post tailor's family. William B. Moores, who arrived in New Mexico with the California Column, worked at least nine years as clerk in the quartermaster's office in Santa Fe before losing his position in 1879 following a reduction in the clerical force. Probably many workers had records similar to Stephen Lacassagne, a plasterer and painter who worked two years in the quartermaster's department in Santa Fe and then sought employment elsewhere, presumably because he had completed the work for which he had been hired.[267] There is no way of knowing just how long the average teamster and unskilled laborer worked for the army, but records suggest a rapid turnover. Of the thirty-seven teamsters employed at Fort Union Depot in June of 1870, twenty-four had been working there less than six months. Only five had been employed two or more years. Pat McManamon, the teamster with the most seniority, had started work on 1 December 1866. Some men moved from post to post as jobs became available. Thomas Estle worked three and a quarter years as wagon master at Fort Wingate, where his pay was $55 a month, and then was hired as cook at Fort Union Depot, where he was paid only $25 a month.[268]

A report of civilian employees retained in the quartermaster's department in the District of New Mexico dated 21 May 1870 shows that Anglos dominated the work force. Of the 116 names on the list, only twenty are clearly Hispanic. Casimiro Reyes was employed at Fort Bayard as guide; Pablo Sandoval and Vicente Romero were employed in Santa Fe as teamster and laborer respectively; and seventeen were employed at the Fort Union Depot—a cook, a laborer, two herders, and thirteen teamsters. The other employees probably were Anglos, although a few may have been blacks. Even though Anglos held all but one of the higher paying jobs, Anglos and Hispanos employed in the same capacity as teamsters and laborers received the same wage.[269]

Whether Anglo, Hispano, or black, male or female, civilian employees were subject to military law and discipline, and they uniformly were dismissed from the service for violating army regulations. The major cause for dismissal was drunkenness, but other causes included theft, disobedience of orders, general worthlessness, mistreatment of animals, and contributing to riots. A carpenter and a wheelwright who struck for higher wages at Santa Fe in 1865 were discharged for "showing a spirit of insubordination and setting a bad example to others."[270] The quartermaster's department circulated the names of these and other troublemakers, including employees who deserted their jobs, with instructions they were not to be employed again at any of the posts.

Though quick to discipline, the army was slow in paying its debts. Civilian employees often waited months for their pay. Teamsters at Fort Union, New Mexico, whose pay was several months in arrears had to request special funds to purchase clothing. The commanding officer at Camp Lowell, Arizona, reported in January 1878 that five employees there—some with families—had not been paid in six months and were "suffering great hardships." John Gates, a packer with Buell's expedition into Mexico, waited more than a year for the $390 the army owed him.[271] On occasion civilians advanced payment to discharged employees, anticipating reimbursement from the quartermaster's department. When ten mechanics were discharged at Fort Stanton in November 1871 and the post quartermaster was without funds to pay them, Emil Fritz paid the men a total of $1,471 so they could leave the post to find employment elsewhere.[272]

Civilian employees at military posts were among the workers affected by the congressional law enacted 25 June 1868 mandating an eight-hour workday for laborers and mechanics in federal employ. Labor spokesmen heralded this law as a victory, for the eight-hour day was a cherished goal among labor organizers. For the military, however, the new law as it affected wages was puzzling. Quartermaster General Meigs complained in a letter dated 13 July that the law did not "execute itself."[273] But the government issued no guidelines for its implementation, leaving various branches to deal with it in an uneven fashion.

The confusion that prevailed was nowhere more apparent than at Fort Union, New Mexico, where in October 1868 the depot quartermaster was working his employees eight hours a day for the same pay as employees in the arsenal received for working ten hours. When depot employees heard rumors that the eight-hour day would be accompanied by a reduction in pay, they requested to work ten hours at the old rates. The depot quartermaster asked Meigs to clarify the law. Meigs, however, failed to state unequivocally whether reduction in wages would accompany the shortened workday. To the depot quartermaster he replied: "It is the duty of all disbursing and employing officers to pay no more wages for a day's work than may be necessary to secure the services of good workmen. . . . If at Fort Union men can be hired at lower rates it is the duty of the officer to take advantage of these lower rates."[274]

Several months elapsed before the rumors at Fort Union became established fact. A pay reduction within the Department of the Missouri became official on 24 March 1869 with publication of General Order No. 3, establishing new rates of pay for certain categories of workers. Blacksmiths, wheelwrights, carpenters, saddlers, and painters would receive $2.20 a day, masons and plasterers, $2.40 a day, a 20 percent reduction from the rates that had been established in General Order No. 6, issued 29 February 1868. But the army allowed extra compensation for men who chose to work overtime. General Order No. 4, dated 26 March 1869, stated:

> On and after the 1st of April next, the workshops and places of labor in this Department will be open ten (10) hours each day, except Sundays. All civil employees who choose may work that number of hours, and will be paid for over-work at the same rate as for the legal day's labor of eight hours.[275]

When two plasterers employed at Fort Dodge, Kansas, went on strike protesting the new pay scale, they were dismissed and denied future employment in the army. Most civilians continued to work ten or even more hours a day, however. The army, in fact, never interpreted the eight-hour law as applying to employees hired by the month, employees like clerks, trainmasters, teamsters, and herders, "whose time may be necessary at any and all hours."[276]

Because so many government departments violated the spirit of the eight-hour law by reducing wages, labor organizers mounted a campaign to restore the law's benefits. Compelled to pay attention to labor demands, President Grant on 19 May 1869 issued an executive proclamation that fixed eight hours as a day's work and directed that "no reduction shall be made in the wages paid by the government by the day to such laborers, workmen, and mechanics, on account of such reduction in the hours of labor."[277] This straightforward directive had no effect whatsoever within the District of New Mexico. When workers at Fort Craig claimed full pay for eight hours' work and additional pay for extra hours, District Chief Quartermaster Ludington denied they had "any good grounds" for their claims. Ludington reasoned that the department commander had authority to fix compensation for civilian employees, which had been done in General Order No. 3, and that General Order No. 4 had fixed eight hours as a legal day's labor. In an incredible statement Ludington told the Fort Craig quartermaster that he could not say whether the reduction in rates of pay made by General Order No. 3 was based upon the reduction of the number of hours to eight. It was "perfectly clear" to Ludington, however, that employees should be paid according to General Orders 3 and 4.[278]

Because so many branches of government ignored Grant's proclamation, the President was forced to repeat it three years later in May 1872. At about the same time Congress appropriated funds to compensate employees for cuts in wages since passage of the eight-hour law. Shortly thereafter the Department of the Missouri rescinded General Orders 3 and 4, issued in 1869, "it having been decided that they were not in accordance with the spirit of the eight-hour law and the proclamation of the President thereon."[279] With the view of fixing new rates of pay, the department quartermaster solicited information in New Mexico concerning prevailing wages paid by other parties than the government. The reports he received were mixed, although almost everywhere mechanics worked a ten-hour day. Blacksmiths and carpenters living near Fort Garland received from $3 to $4 a day, while blacksmiths and wheelwrights residing near Fort Wingate received $3 a day. The quartermaster at Fort Bayard reported that carpenters in his area received from $6 to $8 a day without board, and blacksmiths, $2.50 a day with board. In Santa Fe carpenters, wheelwrights, blacksmiths, and masons were paid $4 a day and received no ration. The quartermaster at Fort Selden simply replied there was "no local standard for the pay of any class of mechanics or workmen."[280]

It was not until March 1873 that the officials approved new rates of pay for workmen in the Department of the Missouri. Employees were to receive for eight hours work the same rates allowed for ten hours work in General Order No. 6 issued in 1868. In other words, most blacksmiths, carpenters, and other mechanics would receive $2.75 per day and a ration, plus additional compensation for overtime.[281]

Meanwhile, post quartermasters started compiling payrolls showing the additional compensation due employees under the congressional law of 18 May 1872. By 31 July 1874 the department quartermaster had approved 531 claims (worth a total of $12,351) of employees who had worked at Santa Fe and at Forts Stanton, Sumner, Selden, Union, and Garland. Among the largest claims was that of Joseph Morgan, an employee at Fort Stanton, for $209. Most claims were more modest—between $5 and $30.[282] Additional claims were approved in later months, but the total number of men entitled to compensation is unknown. Nor is it known how many claims were actually paid, for many former employees could not be located. John H. Belcher, chief quartermaster for the District of New Mexico, reported early in 1877 that he had 460 unclaimed vouchers for payment under the eight-hour law. He believed that all claimants who had been found or who had applied for payment had been paid, and he subsequently returned the unpaid vouchers to Washington.[283]

While quartermasters wrestled with the new regulations, the nation suffered one of the worst depressions in its history. For more than six terrible years, starting in September 1873, unemployment was on the upswing. One student of labor history has written that by the winter of 1877–78 "three million workers were unemployed, and at least one-fifth of the working class was permanently unemployed. Two-fifths worked no more than six or seven months in the year, and less than one-fifth was regularly employed."[284] Unemployment was accompanied by reduced wages and longer hours. New York City workers in the building trades, for example, saw their pay of between $2.50 and $3.00 for an eight-hour day in 1872 reduced to $1.50 to $2.00 for a ten-hour day in 1875.[285] Army expenditures declined during these years, following a trend established at the war's end, but there is little evidence to suggest that the depression triggered radical changes either in the number of civilians the quartermaster's department employed, their wages, or in the hours they worked. New Mexico's allotment of 117 civilian employees in 1871 remained the same in 1875, though as previously noted the army on occasion released employees when its appropriations neared exhaustion. Blacksmiths, wheelwrights, carpenters, and other mechanics who received $2.75 a day in 1873 were receiving $72 a month in 1877 (comparable pay for working 26 days a month).[286] And the eight-hour workday remained intact during most of the depression.

The Supreme Court, however, dealt a heavy blow to the shortened workday when on 13 March 1877 it ruled that the eight-hour law did not prohibit the government from entering contracts with labor fixing a day's work at more than eight hours.[287] Army employees soon felt the impact of this decision. On 13 July Lieutenant Colonel Rufus Saxton, chief quartermaster for the Department of the Missouri, issued a directive that thereafter ten hours would constitute a day's labor for all mechanics employed in the quartermaster's department. Employees would receive no extra compensation for the longer workday. Civilian employees would continue to work a ten-hour day through the years of this study.[288]

The frontier army depended upon civilians for performing much of the skilled labor required in garrison and in the field. Civilians had swelled army payrolls during the Civil War, but even during economy drives the army could not totally dispense with their services. Shoddily constructed military buildings should not be blamed entirely upon inexperienced soldier-laborers since civilians worked along side them in constructing most posts and often had total responsibility for erecting military structures. Inferior construction was often the fault of materials being used. Structures built with unseasoned lumber and adobe brick weathered poorly and needed frequent repairs. Unsatisfactory quarters also resulted from military strategy that scattered the Indian fighting army in small isolated posts that were looked upon as temporary. Even with alterations and enlargement, structures built with an eye to economy required constant work to keep them inhabitable.[289]

Some improvement in housing would come during the 1880s, coinciding with an effort to ameliorate the life of the soldier to reduce desertions. By the end of the decade Congress was appropriating increased sums for building larger and improved posts. The coming of the railroads finally allowed the army to implement the plan that John Pope had suggested in 1870 to abandon small temporary posts and concentrate troops at larger posts near railroad lines.[290] The revolution in transportation would bring other changes to the army and to Southwest residents. But for nearly two decades after Sibley's retreat from New Mexico the army was saddled with the expense of freighting supplies to the Southwest in slow-moving ox and mule wagon trains. The army's quest to reduce transportation expenses and its reliance upon civilian carriers is documented in Chapter 7.

(TOP LEFT) Pinckney R. Tully, partner in the firm of Ochoa, Tully and Company. (Arizona Historical Society Library, Tucson, Arizona.)

(TOP RIGHT) Henry Clay Hooker in the patio of his Sierra Bonita Ranch. (Arizona Historical Society Library, Tucson, Arizona.)

(TOP) Depot and quartermaster's office, Fort Union, New Mexico, 1866. (U.S. Army Signal Corps Collections in the Museum of New Mexico, Santa Fe, New Mexico, Neg. No. 1828.)

(RIGHT) José M. Redondo, Fort Yuma contractor. (Arizona Historical Society Library, Tuscon, Arizona.)

(TOP) Officers' quarters, Fort Union, New Mexico, 1866. (Courtesy Museum of New Mexico, Santa Fe, New Mexico, Neg. No. 1830.)

(BOTTOM) Fort Sumner, New Mexico, ca. 1864–68. (U.S. Army Signal Corps Collections in the Museum of New Mexico, Santa Fe, New Mexico, Neg. No. 28533.)

(TOP) Navajo Indian prisoners constructing building at Fort Sumner, New Mexico, ca. 1864–68. (Courtesy, Museum of New Mexico, Santa Fe, New Mexico, Neg. No. 1816.)

(BOTTOM) Fort Marcy Headquarters Building, Palace Ave. at Lincoln Ave., Santa Fe, New Mexico, 1881. (Photo by Ben Wittick, Courtesy School of American Research Collections in the Museum of New Mexico, Santa Fe, New Mexico, Neg. No. 15844.)

(TOP) Fort Selden, New Mexico, 1867. (Photo by Nicholas Brown, Courtesy Museum of New Mexico, Santa Fe, New Mexico, Neg. No. 1742.)

(BOTTOM) Post trader's store at Fort Stanton, New Mexico. (Rio Grande Historical Collections, New Mexico State University Library, Las Cruces, New Mexico.)

(TOP) Fort Whipple, Arizona, 1871. (National Archives.)

(BOTTOM) Men's quarters, Camp Mojave, Arizona, 1871. (National Archives.)

(TOP) Fort Bowie, Arizona, ca. 1880. (National Archives.)

(BOTTOM) Fort McDowell, Arizona, 1885–1890. (Arizona Historical Society Library, Tucson, Arizona.)

(TOP) Officers and families at Fort Davis, Texas, ca. 1888. (Courtesy Fort Davis National Historic Site.)

(BOTTOM) Fort Verde, Arizona, early 1890s. (Arizona Historical Society Library, Tucson, Arizona.)

7

Freighters and Railroad Agents

On 11 June 1867, Lieutenant Colonel Wesley Merritt with six companies of the 9th Cavalry arrived at Camp Hudson, Texas, about 190 miles west of San Antonio, on his way to reoccupy Fort Davis. To Merritt's dismay, the ration train, sent several days in advance of the main column, was nowhere in sight. Fearing a shortage of rations, Merritt expressed his frustration to the chief quartermaster in San Antonio in these words:

> I cannot impress upon you too urgently the necessity of some means being taken to make the contract trains perform the trip to Davis and interior posts, in a reasonable length of time. At Fort Inge I hear that these trains are frequently a month on the road from San Antonio, and at Clark a corresponding delay is complained of. The fact that the train started from San Antonio on the 20th of last month, has not yet arrived here will give you an idea of the slow and shiftless manner, in which these trains are conducted, especially is this so, when it is well known the train is not within 50 miles of this place.[1]

More than a decade later supply officers serving elsewhere in the Southwest would voice similar complaints, for freighters in both decades struggled with the same problems: long distances, primitive roads, uncertain weather, and the threat of Indian attacks. Captain Frederick F. Whitehead, chief commissary of subsistence for the District of New Mexico, wrote on 26 July 1879:

> As in years past the contract transportation in this District, during the fiscal year ending June 30, 1879, has uniformaly [*sic*] failed to come up to the requirements of the contracts. . . . During the past year several posts in this District ran short in subsis-

tence supplies owing to *unusual* delays in transportation. On one occasion a six months supply of many articles of subsistence stores arrived at Fort Bayard, N.M., one hundred and thirty-six days behind time; on another occasion bacon arrived at Fort Bliss, Texas 87 days overdue.[2]

Some officers wanted the army to operate its own transportation. None spoke more compellingly than Colonel Joseph A. Potter, quartermaster at Fort Leavenworth, who on 15 September 1865 wrote to Quartermaster General Montgomery C. Meigs that "the system of contracting freight is erroneous. . . . The delays, damages, etc. arising from the careless mode of shipment and want of proper care, will be in a great measure avoided by using nothing but government trains."[3] But despite delays and inconveniences, the quartermaster's department had decided long before the Civil War that it was cheaper to hire civilian transportation than to run government wagon trains.[4]

Even so, the army provided some of its own transportation, maintaining teams and wagons for hauling water, hay, fuel, and other supplies and to accompany troops on campaign. A supply depot like Fort Union, New Mexico, was well equipped with wagons and mules. In 1867 the depot had about 250 six-mule teams that were used to supplement contractors' trains in moving supplies within the district. Five years later the depot reported having only 130 wagons and 310 mules, reflecting the army's increased reliance upon contract transportation and its determination to reduce transportation expenses.[5]

Reducing transportation costs was the goal of every supply officer, for the expense of transporting supplies to western garrisons often increased the original purchase price five- or sixfold. Cheap transportation would have to await the coming of railroads, but some reductions were possible by changing routes and modes of supply. Because posts in Arizona were among the most expensive in the Southwest, the army was eager to discover shorter and safer routes for provisioning them. Before the Civil War there were only four military posts in Arizona. Fort Defiance in Navajo country, Fort Breckinridge on the San Pedro River, and Fort Buchanan on the Sonoita received supplies from New Mexico, and Fort Mojave on the Colorado River obtained its supplies from California via ocean and river transportation.[6] As noted in Chapter 1, a new sea and land route opened to southern Arizona in 1861 via the Mexican port of Guaymas. During and after the war this route figured prominently in the army's efforts to reduce the cost of maintaining the ten or more posts that eventually guarded the Arizona countryside.

Shortly after the California Column occupied Tucson in the summer of 1862, General James H. Carleton dispatched Major David Fergusson on a reconnaissance of Sonora. Fergusson's reports of that trip convinced Carleton that the best route for supplying both Arizona and southern New Mexico was through Mexico. Fergusson had learned that a good road connected Guaymas and Tucson, a distance of 350 miles, with an abundance of wood, grass, and water for almost half the route. Two Mexican citizens had offered to transport military supplies from Guaymas to Tucson at rates advantageous to the army.[7]

Fergusson pointed out in a letter to California headquarters that it cost about 3½ cents per pound to ship stores from San Francisco via the Colorado River to Fort Yuma (established in 1850 on the California side of the river) and an additional 12½ cents per pound to haul supplies overland from Yuma by government trains to Tucson, a distance of about 275 miles. Fergusson believed that it would cost only 6 cents per pound to transport supplies from San Francisco to Tucson via Guaymas. Further reductions would be made by freighting from Lobos or Libertad, Mexican ports north of Guaymas and about 213 and 225 miles respectively from Tucson.[8] On a second reconnaissance carried out in October, Fergusson found the roads to Lobos and Libertad to be in good condition; he later claimed that "with the exception of San Diego and San Francisco, California has no harbor comparable to La Libertad."[9] Thereafter Carleton posted letters to government officials, including one dated 8 March 1863 to Secretary of State William H. Seward, urging the United States to purchase a strip of Sonora that would include Lobos and Libertad and a land route to Tucson. Not only would this allow the army to cut its transportation expenses, but it also would provide New Mexico and Arizona with outlets for shipping their valuable ores. Fergusson echoed Carleton's recommendations utilizing words reminiscent of an earlier period of Manifest Destiny: "The Almighty intended that Sonora and Arizona should be *one*, and made the Port of Libertad as a bond of Union." Two and one-half years later, in December 1865, Governor Richard C. McCormick of Arizona called upon the United States to annex part of Sonora to ensure the economic development of his territory.[10]

But Carleton's initial attempts to move military supplies through Mexico met with failure. Even though Governor Ignacio Pesqueira of Sonora seemed willing to allow transit through his state, supplies for Tucson and southern Arizona were shipped via the Colorado River and then transported overland from Fort Yuma. It was not until the summer of 1864 that the army utilized the Guaymas route. In May of that year the U.S. brig *General Jesup* left San Francisco for Guaymas loaded with commissary stores for Arizona. Tucson residents Joseph S. Rogers and George W. Pierce contracted to haul the supplies overland to Tucson for 6½ cents per pound of freight.[11] Delayed en route by heavy rains, the contractors' trains did not reach their destination until 26 July. A board of officers examined the supplies as they were unloaded in Tucson—hundreds of sacks, barrels, and boxes filled with flour, beans, rice, coffee, salt, pork, ham, sugar, vinegar, whiskey, candles, soap, hard bread, mixed vegetables, syrup, and pepper. Captain William Ffrench, who had examined several previous shipments sent via Fort Yuma, claimed never to have seen supplies "that arrived with so little loss or damage, or that could compare in good order with these just received via Guaymas."[12]

Because of the unsettled conditions in Mexico, the army temporarily abandoned the Guaymas route in 1865. French troops overran much of Sonora that year, forcing Governor Pesqueira and his wife to flee to Arizona. Even though the French commander authorized passage of U.S. military supplies, U.S. officials decided not to risk shipping goods through a wartorn country.[13]

When it appeared that Mexican troops would regain control of Sonora the following year, army officers in Arizona mounted a campaign to reestablish the Mexican supply route. Captain Gilbert C. Smith, in charge of the depot at Tucson, pointed out to Quartermaster General Meigs that the total cost of shipping goods from San Francisco via Guaymas in 1864 had amounted to 7¼ cents per pound in coin and that freight shipped via Fort Yuma since then had cost at least 12 cents per pound in coin. Goods were in transit forty-five to sixty-five days via Fort Yuma, but would arrive at Tucson in twenty-five to thirty days via Libertad. A short time later, Lieutenant Colonel Henry D. Wallen, commanding at Tucson, wrote to Senator Cornelius Cole of California explaining the economic advantages of the Libertad route and requesting that Cole use his influence in Washington to authorize an experimental shipment of stores from San Francisco to Libertad.[14]

Even though General Henry W. Halleck, commanding the Military Division of the Pacific, warned against sending supplies through turbulent Mexico, the quartermaster's department contracted with Phineas Banning in April 1867 to deliver government supplies in Tucson via Guaymas at a price of 7 cents per pound for the entire distance from San Francisco.[15] Subsequent events justified Halleck's unease. The collector of customs at Guaymas demanded payment of duties on Banning's cargo, ignoring Governor Pesqueira's orders granting duty-free transit on U.S. army supplies. Mexican officials later said that the customs agent had misinterpreted his instructions. Nonetheless, supplies were detained at Guaymas at least a month or two and did not reach Tucson until December. In light of this trouble, it is not surprising that no one bid on the Guaymas route in 1868 and that supplies consequently were sent to southern Arizona via Fort Yuma.[16]

But the army never lost interest in the cheaper transportation route through Sonora. On 18 March 1868 Captain Charles A. Whittier left Wilmington, California, and traveled overland on an inspection tour of Arizona posts, with special instructions to investigate the cost of moving supplies through Sonora. His report, dated 9 June and written on board the steamship *Montana* en route from Guaymas to San Francisco, gives a detailed picture of the resources of Sonora. Whittier described the road from Tucson to Guaymas as "one of the very best possible" and depicted Guaymas as "a lively little commercial town, with one of the best harbors upon the Pacific." Whittier had located residents in Guaymas and in Hermosillo, the largest town in Sonora, who would transport government freight from Guaymas to Tucson at 4 cents per pound. The captain estimated that the total cost of transporting supplies from San Francisco to Tucson via Guaymas would not exceed $100 per ton (5 cents per pound), which compared favorably with the $175 per ton (8¾ cents per pound) it then cost to send supplies via the Fort Yuma route. He also noted that Mexican teams carried freight more cheaply than teams in the United States because Mexican owners fed their mules no grain and paid their teamsters much less than teamsters were paid across the border.[17]

Whittier was convinced that the Sonora route was superior to the Yuma route, and—like Carleton—he advocated attaching Sonora to the United States. He claimed there was "scarcely a man of property in [Sonora], be he Mexican, English, French or American who would not hail with joy the cession of the state." Men of means wanted law and order, which the Mexican government had failed to provide. He concluded his lengthy report with these words: "May it not be long before the expectations of its best people shall be realized and Sonora with its bright promises, be joined to our Republic—preserving itself and adding to our wealth and influence."[18]

In August 1869 the quartermaster's department contracted with Charles E. Mowry to transport government stores to Tucson by way of Guaymas for 5.44 cents per pound for the entire 2,130 miles from San Francisco. Mowry employed a train of 200 wagons for the Tucson–Guaymas road, and in late December and early January 1870 detachments of 10 and 12 wagons loaded with government supplies began arriving in Tucson.[19]

Despite the fact that moving supplies through Sonora saved the army both time and money, the government decided to abandon the Guaymas route in 1870. For the fiscal year ending 30 June 1871, the quartermaster's department contracted with Hooper, Whiting, and Company to transport freight from San Francisco to Tucson via Yuma for 8⅜ cents per pound for the entire distance of 2,475 miles.[20] A number of factors entered into this decision: political instability in Mexico, bureaucratic red tape, threat of Indian and Mexican bandit attacks, and political pressure from California and Arizona politicians who claimed that "it was to the advantage of the public service" to move government freight through U.S. territory.[21] The editor of the *Weekly Arizonan* had called Mowry's 1869 contract a calamity, for it denied Arizona residents lucrative employment in transporting goods over American soil. Many people in Arizona shared his belief that freighting government supplies over roads within the territory added to the security of those roads and thus encouraged settlement.[22] Sylvester Mowry, brother of Charles Mowry, vigorously protested the change, however. Owner of a valuable silver mine south of Tucson,

Mowry was embittered by the loss of Guaymas as an outlet for his ores. In a letter published in the *Alta California,* he accused Hooper, Whiting, and Company of taking the contract at a ruinously low price simply to destroy the Guaymas route.[23]

A short time later, after Colonel George Stoneman, commanding the Department of Arizona, recommended the removal of the military depot from Tucson and the abandonment of several military posts in southern Arizona, the Tucson *Weekly Arizonan* reversed its stand on the Guaymas route. In a lengthy article appearing 28 January 1871, the press acknowledged that heretofore residents in southern Arizona had been economically dependent upon transactions with the quartermaster's department. With removal of army posts Arizonans would have to take up new enterprises and adopt "the cheapest route by which freight may be brought into the Territory." Clearly the freight route by way of the Colorado River and Fort Yuma had to be abandoned because it was twice as expensive as the Guaymas route. "While government supplies were imported by that route [Colorado River]," the press maintained, "the interests of the settlers along the Gila were a sufficient argument for its continuance." Now economic progress demanded adoption of the Guaymas route. About three months later the press was castigating the army for having abandoned the Guaymas route in the first place![24]

The Guaymas route never again figured prominently in the transportation of government supplies. The Tucson press occasionally mentioned freighters arriving from Guaymas with goods for the government. But until the Southern Pacific Railroad entered the territory in 1878, supplies for military posts in southern Arizona were usually hauled overland from Yuma.

The main supply route for Arizona long remained a land and water route from San Francisco via the Gulf of California. During the 1860s privately owned sailing vessels carried goods from San Francisco around Lower California and up the Gulf to the mouth of the Colorado River, a distance of about 2,000 miles. There stores were reshipped on small steamers that ascended the river to Fort Yuma, about 150 miles upstream, and to Fort Mojave, about 300 miles beyond Yuma. Shipping rates proved fairly stiff. By mid-decade, the army paid about $15 per ton for goods carried from San Francisco to the mouth of the Colorado River and $40 per ton from the mouth of the river to Fort Yuma. Up to 1864, George A. Johnson and Company held a monopoly of river transportation, operating three steamers—the *Cocopah,* the *Colorado,* and the *Mohave.* But the company failed to keep pace with expansion of trade, leaving freight piled up at the mouth of the river and at Arizona City (across from Fort Yuma) before shipment to the interior. Impatience and dissatisfaction on the part of upriver merchants led to the appearance of a rival company on the river in 1864, which in turn led to a reduction in rates. In 1867 the government paid $20 per ton from the mouth of the river to Fort Yuma and $37.50 per ton from Yuma to Mojave. This was one half the former rate and about one half less than the rate charged to civilians.[25]

Johnson regained a monopoly of river transportation in the fall of 1867 when the rival company went out of business. But the lower rates and improved service continued. By 20 December 1869, when Johnson and partners Benjamin M. Hartshorne and Alfred H. Wilcox incorporated as the Colorado Steam Navigation Company, the firm was running four river steamers and a half dozen barges, which were used to tow additional freight behind the steamers. In 1871 the company purchased an ocean-going steamer, the *Newbern,* and soon provided direct service from San Francisco to the mouth of the Colorado River. The following year the company signed a twelve-month contract to transport army supplies from San Francisco to the Colorado River at a rate of $14 per ton, agreeing to make the voyage in ten days by steamer or twenty-four days by sailing vessel. The company purchased a second ocean-going steamer, the *Montana,* in 1873 and reduced the army's shipping rates to $12.50 per ton.[26]

Colonel Delos B. Sacket, who traveled on the *Newbern* en route to Arizona in the spring of 1873, described the ship as being "comfortably arranged for carrying troops and passengers." Cabin passage from San Francisco to the mouth of the Colorado River was $75; deck passage (price per enlisted man) was $40.[27] Martha Summerhayes later penned a classic description of her "never-to-be-forgotten voyage" down the Pacific Coast and up the Gulf of California. She made the trip in August 1874 on board the *Newbern* with her husband and other members of the 8th Infantry en route to new assignments in Arizona. She recalled that the weather was so "insufferably hot" steaming up the Gulf that at night officers and their wives abandoned staterooms to sleep on deck. Odors from rotting provisions permeated the ship, clinging to food placed on the mess table. After a voyage of thirteen days the *Newbern* reached the mouth of the Colorado River. Three days elapsed before the sea was calm enough to transfer troops and baggage to a smaller steamer. Further delays added to their suffering. At least three enlisted men died before the regiment reached Fort Yuma. Summerhayes later recalled her joy upon reaching their destination: "After twenty-three days of heat and glare, and scorching winds, and stale food, Fort Yuma and Mr. Haskell's dining-room seemed like Paradise."[28]

During months of high water, between 1 May and 1 October, steamers made the trip between the mouth of the Colorado River and Fort Yuma in two or three days. For the remainder of the year travel was difficult, and steamers sometimes took ten days to reach Fort Yuma.[29] It was even more difficult to reach Fort Mojave in low water. The commanding officer at Mojave complained in late November 1863 that a steamer carrying badly needed supplies had not yet arrived after being more than twenty days out from Yuma. Thereafter, during the winter and following spring, the garrison was dependent for supplies on overland transportation from San Pedro, California.[30]

One of the best-known entrepreneurs on the Colorado River was Pennsylvania-born Louis J. F. Jaeger. He and several partners established a ferry at the junction of the Gila and Colorado rivers a few months before Fort Yuma was established there in 1850. Jaeger soon bought out his partners and then engaged in a number of enterprises that made him a wealthy man. He hauled supplies for the army, raised cattle, horses, and sheep, operated a mercantile store, and invested in mines. At the start of the Civil War army authorities feared that Confederates might overrun the crossing, and consequently destroyed one of Jaeger's boats anchored twenty miles upriver and moved the others "up to and under the guns of Fort Yuma." The California Column en route to New Mexico crossed the river in Jaeger's boats.[31]

In 1865 the army decided it would be more economical to run its own ferry. Consequently a boat was made in San Francisco, shipped in pieces to Yuma, and there reassembled. It operated only a short time, however, for in March 1866 the ferry was sold to Jaeger at the original cost to the government of $4,587. Jaeger also agreed to ferry army supplies and personnel at one half his charter rates. By this date new military storehouses had almost been completed across the river from Fort

Yuma—site of the Fort Yuma Depot. Since all supplies for southern Arizona would thereafter be delivered at the storehouses, army authorities anticipated a reduction in the amount of ferriage. To their dismay, however, government use of the ferry increased. Ferry bills became "so extravagantly large" that restrictions were imposed on the use of Jaeger's ferry. A small rowboat manned by two soldiers on extraduty was used exclusively for crossing personnel, the "ferryboat being resorted to only for wagons and animals." Still, in some months the army's ferriage bill exceeded $275.[32] In October 1867 the army arranged a new contract with Jaeger, setting rates at $200 a month in coin. Two years later rates were increased to $250 a month.[33] Jaeger's ferry business ended when the Southern Pacific Railroad completed a bridge over the river in 1877.

The government contracted with private firms to freight supplies overland from the Colorado River to interior posts. Supplies destined for Fort Whipple initially landed at La Paz, a mining supply point 150 miles upstream from Fort Yuma, and from there were transported overland about 190 miles to the post. Water was scarce along the overland route. One freighter confided to Army Inspector Nelson H. Davis that he would rather haul freight to Fort Whipple from Leavenworth, Kansas, than from La Paz.[34] A second route opened in 1865 when William H. Hardy built a toll road connecting Prescott with Hardyville, nine miles above Fort Mojave. Competition soon became intense among backers of the two roads. Hardy won a government contract in 1866 to transport supplies from Hardyville to Fort Whipple, a distance of about 160 miles, at 8½ cents per pound of freight. In February of the following year Bernard Cohn agreed to transport 100 tons of government stores to Fort Whipple from La Paz at 8 cents per pound.[35]

Local entrepreneurs and freighters constantly looked for shorter and better routes to attract new customers. Ehrenberg, for example, located in 1867 seven miles below La Paz, replaced the latter town as the main shipping point to mines and posts in central Arizona after the river changed course leaving La Paz without good frontage.[36] But rains turned all these routes into quagmires and slowed movement of supplies. Because of muddy roads in February of 1868 wagon trains on the La Paz road took thirty days to reach Prescott, almost twice as long as it usually took. Later that summer government freight en route to Prescott was left on the desert because of impassable roads. Delays such as these caused critical shortages at the interior posts. On several occasions troops at Camp Lincoln had been "reduced to their last sack of flour," and on one occasion the commissary officer at Fort Whipple had had to borrow coffee from local merchants to send with scouting parties.[37]

The contractor struggling to provision these posts was Virginia-born Dr. Wilson W. Jones, one of the largest freighters on the La Paz–Prescott route. Exactly when Jones arrived in Arizona is not known, but by 1860 he had established a mercantile store at the site of the new gold strikes on the Gila River about fifteen miles east of Fort Yuma. Sometime later in the decade he became partners with Michael and Joseph Goldwater in a freighting firm operating out of La Paz. Wilson W. Jones and Company won its first government freight contract in May 1867, agreeing to transport army supplies from La Paz to Camp Lincoln, Camp McPherson, and Fort Whipple for 4 cents per pound per hundred miles. The firm received additional contracts for hauling stores to Fort Whipple and nearby posts in each of the following three years.[38]

Freighting overland from Fort Yuma to Tucson was as arduous as freighting on the La Paz road. Tucson-bound freighters were faced with heavy sand, scarcity of grass and water, and difficulty in obtaining grain for their work animals. One officer claimed that this route was "almost as impassable at many seasons of the year as the desert of Sahara."[39] General John S. Mason believed that freight contractor Phineas Banning had made no money on his 1865 contract, which called for transporting supplies from Yuma to Tucson for 3 cents per pound per hundred miles. The following year Banning received the contract at 6 cents per pound per hundred miles, a price Mason felt was too high but as cheap as could be done with government trains.[40] In good weather freighters usually made the round trip between Tucson and Yuma in thirty days. But delays often occurred. Drenching rains in early 1874 made roads out of Yuma almost impassable for heavy wagons. One wagon train traveled only five miles in six days and after twenty-three days on the road finally reached Fillibuster Station, just thirty-eight miles east of Yuma.[41]

One of the most successful freighting firms operating out of Tucson belonged to Pinckney R. Tully and Estévan Ochoa. Ochoa had learned the freighting business as a boy when he accompanied his brother's freight train from Chihuahua to Independence, Missouri. His partner Tully had been freighting on the Santa Fe Trail since 1846. Probably the first government freight contract the firm received in Arizona was the one Ochoa signed on 21 December 1863 agreeing to transport military stores from Fort Yuma to Tucson during the coming year. The contract required him to keep in running order thirty good teams and wagons, with an average freighting capacity of not less than 3,500 pounds each. Ochoa would receive 10 cents per pound of freight transported on the first trip to Tucson, 9 cents per pound for the second trip, and 8 cents per pound for the third and all subsequent trips before 1 January 1865. Ochoa would also have the privilege of purchasing grain forage at cost from government depots at either Tucson or the Pima villages.[42]

Probably the next freight contract the firm received was one awarded on 1 June 1868 to transport army supplies from Tucson to posts in southern Arizona for 2.47 cents per pound per hundred miles. Indians made two attacks on the firm's wagon trains while the contract was in force. On 13 July 1868 100 Indians attacked their wagons at Cienega, about twenty miles southeast of Tucson, killing 2 men, wounding 4, and capturing 38 mules. Ten months later a party of 200 Apaches attacked their train carrying government supplies to Camp Grant, killing 3 men, wounding 2, destroying 2 wagons, and capturing 80 mules. The firm's losses in the second attack amounted to about $12,000. In December of 1870 another train of Tully and Ochoa, en route to Camp Goodwin, lost 30 oxen during an Indian attack. At this time the firm was under contract to freight army supplies from Tucson to Camps Crittenden, Bowie, Goodwin, and Grant for 2.39 cents per pound per hundred miles.[43] Despite reversals such as these the firm earned a reputation for honesty and integrity. One officer described Ochoa as a man who "always fills his contracts to the letter, even if he loses by it." Tully and Ochoa won at least three more government freight contracts before the Southern Pacific Railroad brought a halt to their freighting business.[44]

Contractors like Tully and Ochoa found Camp McDowell among the most difficult posts to supply, and their fees often reflected this fact. The firm of Fish and Hellings, for example, in 1873 charged twelve cents more per hundred miles for transporting a hundred pounds of freight 222 miles to McDowell than for hauling supplies 275 miles to Tucson. Wagon trains had to cross both the Gila and Salt rivers en route to McDowell, and during the rainy season the rivers were not fordable. Even with ferries at both crossings it was no small task getting provisions to the garrison. A troubled Major Andrew J. Alexander, commanding officer at McDowell, wrote to his superiors on 18 August 1868: "I desire to call attention to the isolated condition of this post during the rainy season. For nearly a month now, the Salt river has been impassable to a wagon, and all the supplies of flour and grain we have received have been ferried across in a ricketty [*sic*] skiff." Alexander later reported that the most favorable months for supplying the post were January, May, June, September, and October, with trains usually en route sixteen to eighteen days from Yuma.[45]

Arizona historian Henry P. Walker has observed that "freighting was a relatively easy business to get into. All one needed was enough money to buy a wagon and a team of mules or oxen, say $1,000."[46] Luckily for the government, many Arizonans invested money in wagon freighting. As competition increased, freighting rates declined. A dramatic fall in rates occurred at Camp McDowell. In 1866 Louis J. F. Jaeger had hauled freight between Fort Yuma and McDowell for 6 cents per pound per hundred miles; a decade later Estévan Ochoa agreed to do the same for 1.73 cents.[47]

In the fierce competition to obtain government contracts, firms sometimes employed questionable tactics, a fact well illustrated in the letting of the 1871–72 contracts. This was a hotly contested race in which eleven firms or individuals submitted bids. When they were opened on 15 May 1871, it appeared that David Neahr had submitted the lowest bid for transporting supplies from Yuma to posts in southern Arizona and that James M. Barney had handed in the lowest bid for freighting over the Ehrenberg–Prescott route. Barney was a member of the prestigious mercantile firm of Hooper, Whiting and Company, whose main store at Arizona City functioned as a forwarding and commission house. This firm wanted to monopolize government freighting and sought to have Neahr's bid thrown out as a "bogus bid." Neahr, former chief engineer of George A. Johnson and Company, operated a competing commission house in Arizona City. The Hooper firm enlisted the aid of California Senator Cole, who filed papers on their behalf with the government. The company probably also solicited help from the former superintendent of Indian affairs for Arizona, Colonel

George L. Andrews, who soon advised Secretary of War William W. Belknap that Neahr was unreliable, having failed to give bonds and enter into a contract with the Indian Bureau. Quartermaster General Meigs received reports that Neahr's bid for army transportation was "made for the purpose of blackmailing or being bought off; that his guarantors repudiate his use of their names." Meigs preferred awarding the contract to Barney, who represented "a well known firm of good standing." On 14 June 1871 Secretary Belknap issued his decision: "Contract to be made with Barney at the rates of his bid, *provided* the allegations made against Neahr, and his bid are found, on investigation, to be true."[48] The case then was referred to General John M. Schofield, commanding the Military Division of the Pacific, who on 26 July ruled as follows: "The allegations against David Neahr and his bid not having been found on investigation to be true the contract for land transportation in Arizona will be made with him for the posts named in his bid, and with Barney for other posts."[49]

But the controversy was not over. On 30 September Michael Goldwater claimed in a letter to the War Department that he had submitted the lowest bid and deserved the contract. He alleged that Barney's contract had been awarded through political influence. On first glance, Goldwater's bid appears lower than Barney's. Goldwater offered to transport supplies from Fort Yuma to Camps Hualpai, Date Creek, Verde, and Fort Whipple for 2 cents per pound per hundred miles, whereas Barney's bid for transporting supplies to the same posts from Ehrenberg was 2.75 to 2.85 cents per pound per hundred miles. Goldwater apparently planned to move supplies via the Colorado River to Mojave and then freight them overland. Meigs pointed out that the government already had arranged with the steamboat company to move supplies upriver at lower rates. Meigs or one of his officers further noted that Goldwater's bid was not low "when it is considered that transportation by water from Yuma to nearest points on river costs but about 50 cents per 100 pounds per 100 miles." Goldwater had submitted a second bid to carry freight overland from Ehrenberg for 3.20 cents per pound per hundred miles, clearly higher than Barney's bid. In late October Secretary Belknap informed Goldwater that he had "no just ground for complaint."[50]

David Neahr had almost as many problems completing his contract as he had in winning it. Within a few months of signing the contract, his wagon trains had been idled for lack of military escorts, supplies damaged or stolen in transit, and his teams diverted to military sites not covered in the contract. By May 1872 Neahr was even having difficulty securing teams, for the army was moving a large amount of barley from Tucson to Prescott and freighters preferred hauling on that route where the road was better and grass more plentiful. Whether Neahr made any profit on the contract is not known. The chief quartermaster for the Department of Arizona, Major James J. Dana, remarked that Neahr's price of 2 cents per hundred miles "is as low a rate as has usually been obtained in this Department."[51]

Neahr was not the only freighter having problems that year. Because of heavy demands placed on freight contractors, General George Crook reported in September that during the past year contractors had "in some instances failed, or moved so slowly as to render it almost as embarrassing as complete failure." Crook was not unsympathetic to the freighters. Ox and mule transportation was always slow, he wrote, but transportation problems were compounded in Arizona where Indians preyed on the unwary, grass and water were scarce, much of the land a desert, and where the "rays of a burning sun fall with uninterrupted fierceness for months." Crook's successful winter campaign later that year permitted wagon trains to travel in greater safety. As a result Crook would observe in his 1873 annual report that delays in transporting supplies by contract had been greatly reduced.[52]

James M. Barney was among the freighters who profited from the newly established peace. One of Arizona's most successful entrepreneurs, Barney is yet another example of the transplanted easterner whose sky-rocketing business career had its foundation in the military market. The 27-year-old native of New York had arrived at Fort Yuma in 1865 and soon contracted with the army (in partnership with Louis J. F. Jaeger) to transport supplies to Tubac. Thereafter he worked as a civilian employee in the quartermaster's department at Yuma until 1867, when he resigned and bought an interest in George F. Hooper and Company, the predecessor of Hooper, Whiting and Company. The firm dealt heavily in government contracts. In his own name, Barney received the 1872–73 contract for transporting supplies from Yuma Depot to the southern posts, and in June 1874 he contracted to transport company property and baggage of the 8th Infantry moving into Arizona from the Pacific Coast via the Colorado River. In September 1875 he became sole owner of the Hooper firm. When not overseeing government freight contracts and his growing mercantile business, Barney invested in mining properties and soon became principal owner of the famous Silver King Mine. About 1881 Barney turned his full attention to developing the mine and began to close out his other business enterprises.[53]

Barney and other freighters in Arizona charged their highest rates for transporting government supplies to Camp Apache, established in 1870 a few miles north of the Salt River and about 490 miles from the Yuma Depot. The road leading north from the Gila River was wretched. Colonel August Kautz observed after an inspection tour in 1875 that Apache was "almost inaccessible from the West and South by wagon transportation."[54] Consequently the army looked to the east for a safer and more economical supply route through New Mexico, a route made possible by the advance of railroads.

In June 1872 the Denver and Rio Grande Railroad reached Pueblo, Colorado, linking that city with the east via the Kansas Pacific Railroad, which had reached Denver two years earlier. Data from the quartermaster general's office indicated that goods shipped to Apache from Leavenworth via Pueblo would cost $9.52 per hundred pounds and from San Francisco via Yuma, $11.44 per hundred pounds.[55] On 10 May 1873 the quartermaster's department awarded Henry C. Lovell of Topeka, Kansas, the government freight contractor for posts in Kansas, Colorado, and New Mexico, the first contract for transporting stores to Camp Apache from the east. For freighting between Pueblo and Camp Apache, a distance of more than 600 miles, Lovell would receive between 85 cents and $1.10 per hundred pounds per hundred miles, depending on the season of the year. The new route was an immediate success. Writing from Camp Apache on 30 June Army Inspector Delos B. Sacket reported that freight had arrived from Pueblo "in most excellent condition"—not badly broken up and damaged as frequently happened on the route from Yuma Depot. He had learned that the roads through New Mexico were "very good," and he suggested that Camps Bowie and Grant be supplied from the east as well.[56]

For the next several years government stores for Camp Apache were transported from eastern military depots by rail to Colorado and then hauled overland through New Mexico by contract transportation. At the same time Arizona contractors continued to receive contracts for hauling freight between Yuma and Apache. During the 1876–77 fiscal year posts in southern Arizona also received subsistence stores from the East—by railroad to El Moro, Colorado (terminus for the Denver and Rio Grande) and then by wagon to Camp Apache, whence they were sent south. Even though military officials in California argued against this route for supplying posts other than Apache, supplies continued to be funneled through New Mexico—but not without mishap.[57] According to Captain Charles P. Eagan, chief commissary officer in Arizona, the transportation system through New Mexico broke down in 1878. A contract issued in February of that year called for freighter F. F. Struby to move supplies from Fort Garland, Colorado (near the terminus of the western branch of the Denver and Rio Grande), to Camp Apache and posts in southern Arizona. In August Eagan would complain:

> Subsistence stores for Apache are now over four months en route from Chicago and have not yet arrived, and the stores invoiced from Chicago to Bowie, Grant, Lowell and Thomas on March 29th have not arrived and nothing is known of their whereabouts except those for Bowie. It appears that Bowie stores were forty six days from Garland to Santa Fe, showing the ox teams traveled less than 3½ miles per day. At this rate, having passed Santa Fe July 18th they will reach Bowie about the 18th of next November.[58]

Because of these and other delays, the subsistence department temporarily abandoned the New Mexico route and in the fall ordered all stores for posts in Arizona from San Francisco. By late 1879, however, the army was shipping Arizona-bound stores to New Mexico over the Atchison, Topeka and Santa Fe Railroad, which had reached Las Vegas, New Mexico, in July. From Las Vegas supplies for Forts Apache, Bowie, Grant, and Thomas were hauled overland by contract transportation.[59]

Changes in supply routes came rapidly as railroads extended their tracks across New Mexico and Arizona. The Southern Pacific Railroad reached Tucson in March 1880 and crossed the border into New Mexico in September. Only a few weeks earlier, on 30 August, Eagan reported that all posts in Arizona, except Fort Apache, were receiving subsistence stores from San Francisco via the Southern Pacific and then by wagon (in the case of Mojave by steamboat). Almost without exception, stores from San Francisco were being delivered on time and in excellent condition. But transporting supplies to Apache from Las Vegas was, in Eagan's eyes, a complete failure. The commissary officer at Apache attributed the problem to bad roads, describing the 100-mile stretch approaching the post as little better than a well-defined mountain trail. He claimed the road was "almost impassable for wheeled vehicles from about November 1st to March 1st." During the spring one freighter had abandoned his cargo about 70 miles before reaching Apache.[60]

The Southern Pacific connected with the Atchison, Topeka and Santa Fe at Deming, New Mexico, in March 1881, forging a link between California and the Midwest. About the same time the Atlantic and Pacific Railroad, now incorporated in the AT&SF Company, was building west from Albuquerque. By late 1881 it had reached Winslow, Arizona, and then continued along the thirty-fifth parallel reaching the Colorado River opposite Needles, California, in August 1883. A bridge over the river allowed the Atlantic and Pacific to connect with a branch of the Southern Pacific, providing access to Los Angeles and San Francisco. Extension of these lines allowed posts in Arizona to be supplied partly from San Francisco and partly from Chicago, according to local prices and the cost of transportation.[61]

With the building of steel rails across Arizona, local farmers would face stiff competition from California producers. Lieutenant William W. Wotherspoon, acting chief quartermaster for the Department of Arizona, reported that during the fiscal year ending 30 June 1881, Forts Bowie, Grant, Huachuca, and Lowell were supplied with grain, hay, and straw from southern California. Since contract registers indicate that only Arizona residents contracted to supply these posts with grain that year, they must have imported at least part of their supplies from California. During the next fiscal year Walter S. Maxwell and Thomas W. Stackpole, residents of Los Angeles, California, contracted for grain and hay destined for southern Arizona posts. In following years increasingly larger amounts of forage entered the territory from California in railroad cars.[62]

Even with the coming of railroads, territorial freighters were needed to haul supplies between railroad stations and the garrisons. For the fiscal year ending 30 June 1884, for example, Adolph Solomon of Solomonville, Arizona, received contracts to haul army freight from Bowie Station on the Southern Pacific Railroad to Forts Bowie, Grant, Thomas, and to San Carlos. For the fourteen-mile trip between the railroad station and Fort Bowie, Solomon would receive 28 cents per hundred pounds. For the 101-mile haul to San Carlos, the contractor would receive about $1.21 per hundred pounds for the whole distance. Two Arizona contractors were awarded contracts for hauling freight to the northern posts. William S. Head of Fort Verde would haul supplies from Maricopa, a station on the Southern Pacific, to Forts McDowell, Whipple, and Verde. Samuel C. Miller, a pioneer Prescott freighter, would transport supplies from Ash Fork, a station on the Atlantic and Pacific, to Forts Whipple, Verde, McDowell, and Apache. For the 90-mile trip to Apache, Miller would receive $1.80 per hundred pounds; for the 81-mile trip to Fort Verde, $1.60 per hundred pounds; and for the 56-mile trip to Whipple, $1.12 per hundred pounds. Most supplies destined for Whipple and Verde were shipped the shorter route via Ash Fork rather than through Maricopa.[63]

Without doubt railroads greatly benefited the army in Arizona. Military personnel and supplies were transported quickly and cheaply, though some delays in moving freight still were inevitable. Subsistence stores arrived fresher and with less damage and deterioration than before. The large and expensive installation at Yuma Depot would be dismantled, for with railroad transportation its role as a forwarding station had ended. Railroads also encouraged economic development in Arizona, primarily in mining and ranching. But railroads brought reversals to old established mercantile companies like William Zeckendorf, Lord and Williams, and Tully, Ochoa, and Company. These firms were forced to sell goods at a loss to compete with cheaper goods brought in by rail, and they did not survive the competition. Tully and Ochoa closed their doors after selling $100,000 worth of freighting equipment at a loss. The long-distance freighting industry, indeed, went into decline once the Southern Pacific Railroad was built across Arizona. It is an ironic fact that when the first train pulled into Tucson, Estévan Ochoa presented a silver spike to Charles Crocker, the company's president.[64] And even though the army initially encouraged local production of forage and foodstuffs, with the coming of railroads millers and farmers found dwindling markets at military posts.

Many of the problems that impeded transportation in Arizona also occurred in neighboring New Mexico, where transporting supplies across the plains had enriched government contractors since the close of the Mexican War. The most enterprising of the firms carrying goods to New Mexico, the famed Russell, Majors, and Waddell, had signed a two-year contract in March 1855, giving it a monopoly of army freight west of the Missouri River. But the company was virtually bankrupt by the end of the decade. For most of the Civil War the Kansas firm of Irwin, Jackman and Company monopolized the hauling of government freight over the 728 miles between Fort Leavenworth, Kansas, and Fort Union, New Mexico. A contract signed on 30 March 1861 stipulated that the company would receive between $1.30 and $1.50 per hundred pounds of freight per hundred miles, depending on the season of the year.[65]

In time, other wagon-freighting firms won the government's patronage. During the summer of 1864 freighter Andrew Stewart of Steubenville, Ohio, was paid $1.97 per hundred pounds per hundred miles for transporting army supplies to Fort Union. But Indian warfare that year disrupted travel. Colonel John C. McFerran, who crossed the plains in midsummer, reported that wagon trains were camped all along the route unable to proceed because of lack of protection or loss of stock to Indian raiders. "Every tribe that frequents the plains is engaged in daily depredations," he reported, and warned that unless prompt action were taken military supplies would be cut off. The increased danger may account for the higher rates paid the following summer to freighter William S. Shewsbury of Council Grove, Kansas, who received $2.05 per hundred pounds per hundred miles for transporting supplies to Fort Union between 1 May and 30 September 1865.[66]

Shewsbury also benefited from military measures instituted to protect overland travel. In March 1865 General James H. Carleton initiated a system for escorting wagon trains, offering military escorts on the first and fifteenth of each month for merchant trains traveling between Fort Union and Fort Larned, Kansas. The army also increased the number of troops stationed on the Santa Fe Trail. The following year the cost for freighting government stores fell dramatically, partly because of increased security and partly because the Kansas Pacific Railway was advancing west, reducing both time and expense of overland freighting. Between 1 May and 30 September 1866, George W. Howe of Atchison, Kansas, transported supplies to Fort Union for $1.38 per hundred pounds per hundred miles.[67]

In his 1865 report to Secretary of War Edwin M. Stanton, Quartermaster General Montgomery C. Meigs gave some indication of the magnitude of overland freighting. He reported that in July travelers on the stage from Denver to Fort Leavenworth, a distance of 683 miles, had never been out of sight of wagon trains, "belonging either to emigrants or to the merchants who transport supplies for the War Department, for the Indian department, and for the mines and settlers of the central territories." Meigs reported that for the fiscal year ending 30 June 1865 contractors carrying supplies to Fort Union and the interior posts of New Mexico had received $1,439,578, about a third of the total cost of maintaining troops in the territory. Meigs confidently predicted that military expenses would decline with the westward extension of railroads.[68]

Supply officers frequently pointed to another means for reducing expenses—better packaging of commissary supplies to reduce losses incurred in shipping. Chief Quartermaster Herbert M. Enos testified in 1865 that during the previous eight years 75 percent of hams sent to New Mexico in gunny sacks had been condemned. He claimed that the only good hams received had been sent out in boxes. Two years later Chief Commissary Charles McClure recommended packing bread in boxes rather than in barrels to avoid losses. And to prevent spoilage on the long overland haul, he recommended that buckwheat meal, bacon, ham, and dried beef be sent by the last trains of the season—"If in ox-trains not later than the first of September, if in mule trains, by the last of that month."[69]

Other losses occurred through theft. On the long march from Kansas, teamsters pilfered almost everything—for their own use and to sell later at a profit. Army Inspector Andrew W. Evans claimed that canned fruits invariably were opened on the road because freighters found this a cheap means of feeding their hired help. They were also highly prized as antiscorbutics. "Nothing will stop this practice of breaking bulk," Evans suggested, "but making it too expensive for them." General Carleton had recommended in 1865 that freighters be charged three times the cost of the missing article. In later years contractors would pay three times the original cost of the stolen item plus three times the cost of transporting it to New Mexico. Even so, stealing could not be eliminated.[70]

Freighting rates continued to fall as railroads advanced west. On 4 April 1867 John E. Reeside of Montgomery County, Maryland, received a contract to haul freight from Fort Riley, Kansas, or another designated post on the railroad, to Fort Union, New Mexico, for $1.28 per hundred pounds per hundred miles between April and September, when most of the freighting would occur, and $2.34 during the other months. Reeside failed to provide adequate transportation, however, which forced the quartermaster's department to negotiate a special contract with Richard Kitchen of Leavenworth County, Kansas. On 27 June Kitchen agreed to carry from 42,000 to 49,000 pounds of freight to Fort Union at $1.45 per hundred pounds per hundred miles. When it became clear that Reeside could not complete his contract, the army entered into a second special contract with Kitchen and his partner Henry S. Bulkley on 20 July for transporting supplies at the rate of $2.16½ per hundred pounds per hundred miles. Rates varied only slightly the following year. Percival G. Lowe of Leavenworth agreed to carry freight from the railroad in Kansas to Fort Union at $1.29 per hundred pounds per hundred miles between April and October 1868, $1.75 in November, and $2 between December and March.[71]

From the mid-1860s to the end of the decade, the army awarded most contracts for transporting supplies from Fort Union to the interior posts to local residents. The contract awarded Epifanio Aguirre on 1 June 1864 called for him to freight 5,000,000 pounds within the department at $2 per hundred pounds per hundred miles during peak freighting months and $2.25 the other months. The Santa Fe *New Mexican* called Aguirre "the first large Mexican contractor" in the territory; he is a good example of the Hispanic capitalist who tapped into the military's reservoir of federal dollars.

Epifanio's father, Pedro Aguirre, had freighted in Chihuahua before moving his family to Las Cruces, New Mexico, in 1852. Eventually Epifanio and his three brothers became partners in their father's business, freighting on the Santa Fe Trail and in Chihuahua.[72] Epifanio completed the above-mentioned contract by the end of January 1865, and in later months he carried additional freight for the army. For his services between 1 July 1864 and 30 June 1865, he was paid a total of $138,177. There is no way of knowing how much Aguirre cleared on his contracts. The local press reported a sizable loss in December when Navajos stole eighteen of his mules valued at $3,500. Epifanio's bid for freighting army supplies between June and November 1865 was rejected in favor of the one submitted by William H. Moore, sutler at Fort Union, who agreed to transport stores from Fort Union to Forts Sumner and Bascom at $2 per hundred pounds per hundred miles and from Fort Union to the other posts at $2.50. In later years the Aguirre brothers transferred their business interests to Arizona and Sonora. Epifanio was killed in January 1870 near Sasabe, Sonora, when Indians ambushed the stage in which he was riding.[73]

Vicente Romero of La Cueva, New Mexico, was another Hispanic resident who freighted for the army. During the spring of 1866 the quartermaster's department hired six of Romero's teams and wagons to transport enlisted men and baggage to Fort Leavenworth. Romero was asking $8 a day for each wagon.[74] Two years later Romero served as surety on the bid George Berg submitted for transporting army supplies between 1 April 1868 and 31 March 1869. Berg, who farmed near Fort Union, was not an experienced freighter, and his low bid of $1.03½ per hundred pounds per hundred miles raised doubts about his ability to do the work. Because his two bondsmen were known as "reliable men" with combined assets of $100,000, Berg was awarded the contract. Together the three men owned at least one hundred teams.[75]

By the time the contract expired, Berg's inadequacy as a businessman was apparent to military supply officers. In April 1869 a board of survey recommended that Berg forfeit $2,220 for unexplained delays in delivering supplies to Fort Craig. According to the contract, time of delivery by ox teams was not to exceed fourteen days per hundred miles, and a penalty was stipulated of $5 per day for each team exceeding this limit. Three of Berg's teams had arrived at Craig forty-six days late and six had been fifty-one days late.[76] This was only one of several delays that had occurred while Berg was contractor.

The army usually did not assess penalties for delays caused by acts of nature. Consequently contractors gathered as much evidence as possible to justify their tardiness. The testimony that Alexander Grzelachowski submitted in 1869 reveals some of the arrangements contractors made in completing their contracts. Polish-born Grzelachowski had arrived in New Mexico as a young priest with Bishop Jean B. Lamy in 1851. He later relinquished his priestly duties and entered the business world. Toward the end of 1868, while operating a mercantile store in Las Vegas, Grzelachowski submitted the lowest bid of $1.23½ for transporting supplies from Fort Union to the interior posts for the year ending 31 March 1870.[77]

Before awarding the contract, Chief Quartermaster Marshall I. Ludington made diligent inquiry into Grzelachowski's reliability, hoping to avoid the difficulties he had encountered with Berg. In a report to Lieutenant Colonel Langdon C. Easton, chief quartermaster at Fort Leavenworth, Ludington described Grzelachowski as a man of moderate means, having "perhaps fifteen or twenty thousand dollars," who spoke little or no English. He had been engaged in the mercantile business in or near Las Vegas for a number of years and owned teams with which he freighted stores from the States. He and his bondsmen jointly owned seventy-five or eighty wagons and teams.[78]

Once the transportation contract was his, Grzelachowski subcontracted most of the work. He arranged with Juan A. Sarracino of Valencia County to haul supplies from Fort Union to Fort Wingate, a distance of 270 miles, for $1 per hundred pounds per hundred miles. He subcontracted with C. Ramírez y Chávez to transport supplies 345 miles to Fort Selden. When Ramírez y Chávez was thirteen days late in delivering his cargo, Grzelachowski had to submit affidavits to avoid paying a penalty. Testimony showed that the subcontractor had left Fort Union on 10 April 1869 with three loaded wagons. He soon ran into a snowstorm that halted travel for two days. Rain and additional snowstorms caused further delays. When the wagons reached the Rio Grande, high water prevented them from crossing. The owner of a boat refused to carry them across because of strong winds, so they remained on the bank four or five days. At Lemitar the oxen gave out, and Ramírez y Chávez turned over the freight to Cajetano Tafolla, who delivered it to Fort Selden on 11 June. According to Grzelachowski's contract ox teams were required to make the trip to Selden in forty-nine days; his teams had been sixty-two days on the road. Because the delays had been unavoidable, Grzelachowski probably was not assessed a penalty.[79]

In the competition to obtain the transportation contract for the year ending 31 March 1871, Grzelachowski lost out to a more experienced freighter, William H. Moore, whose bid was only slightly lower than Grzelachowski's. For freighting within the District of New Mexico, Moore would receive $1 per hundred pounds per hundred miles from April through September 1870 and $1.25 from October through March. His bondsmen were well-known businessmen: John Dold, Andres Dold, and Marcus Brunswick of Las Vegas and William Kroenig of La Junta.[80]

Though a land of little rainfall, New Mexico boasted a handful of rivers that hindered transportation during stages of high water. Civilians who built bridges and operated ferries provided essential services, and the army compensated them for help in moving men and supplies. Two bridges spanned the Rio Puerco, a stream that rises north of Cuba and flows south more than 145 miles before entering the Rio Grande. During the early 1860s, Tomás Valencia built a small bridge near the mouth of the Puerco seven miles below Sabinal on the road linking Albuquerque with Socorro. Valencia also established a forage agency there at the request of General Carleton, who viewed this as a dangerous location "being one of the outlets and inlets of the hostile Indians."[81]

In later years Valencia clashed with military supply officers. During November and December 1867 Valencia charged the government $1 for each wagon and a lesser amount for each animal crossing over his bridge. Officers considered these rates exorbitant. But Valencia was unwilling to accept less until government wagon trains resumed purchasing forage at his agency. Valencia claimed that for several months government teamsters had made a practice of buying forage either at Belen to the north of him or at Lemitar to the south. Without this additional patronage, Valencia claimed he would have to abandon the place and "find some way of maintaining myself with more ease and with less peril." Reports had it that Valencia's forage and accommodations were inferior, causing the quartermaster at Fort Craig to recommend that the army build its own bridge over the Puerco and appoint a new forage agent to care for it. Later, in 1869, the army paid Valencia a flat rate of $30 per month for use of the bridge during periods of high water.[82]

Santiago L. Hubbell of Pajarito built the second bridge spanning the Puerco, located at the Albuquerque crossing on the Fort Wingate road. His agent was John Q. Adams, who maintained a government forage agency at the bridge. As in Valencia's case, the army considered Hubbell's rates exorbitant: $3 for each wagon and team, $1 for a buggy, 25 cents for a man and a horse, and 5 cents for a man on foot. After lengthy negotiations, Hubbell agreed in 1868 to a flat rate of $30 per month for passage of government transportation.[83]

To transport men and supplies over the Rio Grande, a much wider river, the army relied upon ferryboats. Located at the major river crossings, the boats often were government owned and civilian operated. On at least one occasion the army confiscated a privately owned boat for its own use. Soon after the California Column entered New Mexico, General Carleton ordered Colonel Joseph R. West, commanding troops in Mesilla, to seize a ferryboat that a well-known secessionist had been operating between Mesilla and Las Cruces. The secessionist had fled to Texas, but Mesilla resident Royal Yeamans claimed ownership of the boat. In 1864 Yeamans hired the young attorney Stephen B. Elkins to retrieve his property. The case went badly for the government. It was argued before Judge Joseph G. Knapp, Carleton's sworn enemy, who ruled in favor of the plaintiff. Chagrined by Knapp's decision, army authorities considered an appeal—then offered to buy the boat when Yeamans appeared willing to sell.[84]

During the war years the army used enlisted men to operate both Yeaman's ferry and the government boat at Fort Craig. A tragedy involving the Fort Craig boat may have convinced army officials thereafter to hire experienced ferrymen. On 8 May 1865 several units were attempting to cross the Rio Grande just below Fort Craig on the government ferry. Half of Company F, First New Mexico Infantry, in addition to Lieutenant George H. Pettis, his family, and the company laundresses, crossed successfully on the first trip. On the second crossing water came in over the boat's bow, and the current carried several people overboard. Others became panic-stricken and jumped overboard. Captain Daniel B. Haskell, two civilians, and three men of Company F were drowned in the mishap. A rope eventually was tied to the ailing boat, and it was pulled to shore. Left with only two small boats to cross men and supplies, Lieutenant Colonel Edwin A. Rigg, commanding Fort Craig, ordered the post quartermaster to seize a large boat owned by Andrés Madrid, a post employee, until high water subsided or the government boat was repaired. Rigg also recommended that the army hire a "first class man" to take charge of boats during high water. "Otherwise," he warned, "more lives may be lost."[85]

During the summer of 1866 government-owned ferryboats were operating at Albuquerque, Los Pinos, Fort Craig, and Fort Selden. At this time the army lacked a uniform pay scale for ferrymen. In 1865 Juan Chávez, the chief boatman at Los Pinos, was paid $75 per month. A year later, when Los Pinos was about to be abandoned, civilians in charge of the boat received about $20 each per month. Five citizens were authorized to run the ferry at Albuquerque in 1866—one would receive $45 per month, the others $35 per month, and each would receive one ration a day.[86]

Contracts that Charles B. Tison and Francis H. Creamer signed in 1872 illustrate how government ferryboats were managed in later years. Tison leased the ferry at Fort Selden, and Creamer, the ferry at Fort Craig. Each agreed to ferry across the Rio Grande for one year "all Government teams, escorts, expressmen, employees, and all loose stock belonging to the United States free of charge." They would receive two rations a day—one for themselves and one for an assistant ferryman.[87] By this date, government traffic was slight, for contractors' trains carried nearly all government freight. Profits would come from fees assessed civilian traffic.

When the government crossed men and supplies on privately owned ferries, the standard fee during the first half of the 1870s was $2 a trip. Occasionally the government provided ferrymen with ropes and cables, receiving in return lower ferriage rates than civilians. John Ayers, post trader at Fort McRae, constructed a ferryboat near the post in 1874 with the help of an enlisted man. The quartermaster's department also supplied him with tar, cable, blocks and tackle. The government paid only half the following fares that Ayers charged civilians: $4 for an eight-mule wagon, $3 for a four-mule wagon, $2 for a two-mule wagon, 50 cents for horse and rider, 25 cents for a footman, and 25 cents for each loose animal.[88]

Supply routes and policies changed rapidly with extension of the railroads. By 31 October 1870, when General John Pope penned his annual report, the Kansas Pacific Railroad was completed to Denver. Kit Carson, on its main line in eastern Colorado, was only 280 miles from Fort Union and would serve as the primary transshipment point for New Mexico posts until 1873. Pope envisioned great savings for the government, and among his many recommendations was one to break up the depot at Fort Union. It would be cheaper and easier to haul supplies from Kit Carson direct to each post than to have supplies delivered at Union Depot and then reshipped from there.[89]

In line with Pope's thinking, Lieutenant Colonel Langdon C. Easton, chief quartermaster for the Department of the Missouri, had issued instructions early in October to supply posts in New Mexico directly from the railroad and not from Union Depot.[90] The wagon transportation contract awarded in the spring reflected this change. Contractor Eugene B. Allen of Leavenworth City agreed to transport government supplies from points on the Kansas Pacific Railway to posts in New Mexico, as well as carry supplies from Fort Union to any other post in the interior. By 1876 an official in the Department of the Missouri would observe: "Fort Union is now nearly valueless as a depot of supply for the District of New Mexico, almost all the stores for the District being shipped hence direct to the New Mexican posts. . . ."[91]

The military installation at Fort Union slowly deteriorated as both its military and supply functions declined. Major James G. C. Lee reported in 1881 that only one company of infantry garrisoned the post, hardly sufficient to care for buildings and property. The unoccupied rear two sets of barracks with cavalry stables and sheds were fast going to ruin. Two of the depot's mammoth storehouses, he observed, could be taken down, and there would still remain room for "more property than is likely to be ever stored there again."[92]

Between 1871 and 1879, the year the Atchison, Topeka and Santa Fe steamed into Las Vegas, the majority of contracts for freighting stores to the New Mexico posts were held by Kansas freighters—Eugene B. Allen, Henry C. Lovell, and Edward Fenlon. F. F. Struby, who held contracts in 1878, listed his residence as Garland, Colorado. Contractors usually employed forwarding and commission houses to oversee their business at the new railroad shipping points. The two largest firms handling freight for New Mexico were Otero, Sellar and Company and Chick, Browne and Company. The rival companies moved from place to place as railroads extended their tracks. Miguel A. Otero, former New Mexico territorial delegate to Congress, and John P. Sellar began business in 1867 at Fort Harker, Kansas, then terminus of the Kansas Pacific. The firm moved in rapid succession to

Ellsworth, then Hays City, next to Sheridan, Kansas, and finally to Kit Carson, Colorado. These end-of-the-track railroad towns were filled with gambling houses, dance halls, saloons, and bordellos, catering to a large unattached male work force. In Kit Carson the Otero and Chick companies together employed about one hundred young male clerks.[93]

In 1871 Eugene B. Allen employed Otero, Sellar and Company to assemble and distribute freight from Kit Carson; the following year Allen hired Chick, Browne and Company. Both commission houses moved to Granada, Colorado, in the fall of 1873, about the time the Atchison, Topeka and Santa Fe reached that point in the Arkansas Valley.[94] Whether freight contractor Henry C. Lovell employed either of these firms at Granada that year is unknown, but Lovell appointed as his agent in New Mexico the Las Vegas mercantile firm of Marcus Brunswick and Eugenio Romero. This firm maintained its own agents at Forts Union and Craig to take care of daily problems in moving government freight—including hiring local freighters to haul supplies between military posts.[95]

A handful of Las Vegas merchants in later years managed much of the army's freighting. Marcus Brunswick, long-time Las Vegas resident, served as agent for freight contractor Edward Fenlon for a two-year period ending 30 June 1882. The following year, 1882–83, Brunswick himself held the contract for supplying Fort Stanton from Las Vegas—his rates were $2.50 per hundred pounds for the entire distance of 182 miles. At this time Brunswick was one of the largest military contractors in New Mexico, holding contracts in 1882 to furnish grain to at least seven different posts.[96]

Brunswick's sureties on the Fort Stanton freight contract were Trinidad and Eugenio Romero, members of one of the most prominent Las Vegas pioneer families. Both Romeros became heavily involved in New Mexico politics. Trinidad served two years as probate judge of San Miguel County, was a member of the territorial legislature, and in 1876 was elected New Mexico's delegate to Congress. Eugenio, one of the most powerful Republican leaders in the county, was elected to the territorial legislature, held several county offices, and served as first mayor of the city of Las Vegas from 1882 to 1884. The brothers learned the freighting business from their father Miguel, who in 1851 started freighting between St. Louis and Las Vegas. In 1874 the

Romero brothers, then partners in the mercantile firm of T. Romero and Bro., served as agents for freight contractor Eugene B. Allen.[97]

The Romero brothers also furnished a train of twenty-four six-mule teams in 1874 to accompany the command of Major William R. Price, 8th Cavalry, on the Red River Campaign of 1874–75. This was the campaign that ended Indian warfare on the southern plains. And success came despite the fact that transportation contractors repeatedly failed to move supplies on time. The Romero wagon train, however, "rendered faithful service," according to one officer who was present. But all had not gone smoothly for the Romeros. According to a verbal agreement, their train was required to average about twenty-five miles a day, twice the usual speed demanded of contract trains, and would be in service from thirty to sixty days. The Romeros would receive $10 a day for each team supplied, but were required to furnish supplies for teamsters, herders, and others connected with the train. During the campaign Colonel Nelson A. Miles, commanding a larger column in Texas, took possession of the Romero train and kept it in the field carrying supplies for his command more than a month beyond the terms of the agreement. By the time the train returned to Las Vegas, the army owed the Romero brothers $25,730, an amount some officials judged as excessive. But Chief Quartermaster Meigs believed it was a reasonable expense—considering the "rapid and wearing service" required of animals and equipment.[98]

Trinidad Romero also provided some of the teams needed to move the 8th Cavalry into Texas when it exchanged stations with the 9th during the second half of 1875. A contract signed in November guaranteed Romero $8 per day for each six-mule team and wagon used in transporting property and baggage from Santa Fe to Fort Clark. He would receive the same rate on the return trip if the wagons were loaded and $4 a day if they returned empty. Earlier in the year Santa Fe merchant Abraham Staab had agreed to furnish nine six-mule teams at $6.24 per day for the troop exchange. But when Staab's teams averaged only two miles a day at the start of the trip, Major Andrew J. Alexander impetuously canceled his contract and hired teams at Pajarito at $8 a day.[99]

During the fiscal year ending 30 June 1877 the quartermaster's department reverted to its former practice of transporting all supplies for New Mexico to Fort Union and from there reshipping them to the interior posts. Trinidad Romero was awarded the contracts for freighting between Union and the other posts (see Appendix Table 14). Though Romero had satisfactorily completed his earlier contracts, the army found the old system objectionable and resumed direct shipments at the end of the contract. With an eye to the future the Romero family soon gave up government freighting to concentrate on merchandising. In 1878 Miguel Romero and sons Trinidad, Eugenio, and Hilario founded the Romero Mercantile Company, opening a large two-story store on the plaza in Las Vegas.[100]

The Romeros had anticipated the commercial boom that the Atchison, Topeka and Santa Fe Railroad triggered in Las Vegas in 1879. The railroad and its attendant businesses would pour nearly a million dollars annually into the Las Vegas region.[101] New business houses opened in both the old town and in the new East Las Vegas that sprang up along the railroad tracks. Otero, Sellar and Company relocated near the railroad depot in East Las Vegas soon after the tracks were laid. In 1881 the firm was taken over by Jacob Gross and associates under the name of Gross, Blackwell and Company.[102] Otero died the following year. His son, Miguel A. Otero, Jr., had grown to adulthood working for his father's commission house and would become governor of New Mexico in 1897.

Jacob Gross had started as bookkeeper with Otero, Sellar and Company in 1867 and had followed the firm on its successive moves. Government freighting had continued to be a large part of the company's business. It had served as agent for Eugene B. Allen in 1874–75 and again in 1877–78. Gross received the freight contract in his own name in July 1875—about the time that the commission house moved to La Junta, Colorado, terminal city for both the Santa Fe railway and the Kansas Pacific. In 1876 the Denver and Rio Grande building south from Pueblo reached El Moro, Colorado, five miles north of Trinidad, and the firm relocated there. Gross received the contract that year to haul government supplies from El Moro to Fort Union, a distance of about 100 miles, for $1 per 100 pounds for the entire distance. Two years later Gross held the contract to haul government supplies 328 miles to Fort Stanton from El Moro for $2.85 the entire distance.[103]

Provisions in the contract that Jacob Gross signed in May 1878 illustrate conditions under which contractors carried out their obligations during the 1870s. Gross not only was required to transport government supplies from El Moro to Fort Stanton but also to carry goods to any place located not more than 250 miles from the line of the route. He was required to have an officer or an agent at El Moro, Fort Union, and at Fort Stanton, or at any other place on the route designated by the army. He would receive from five to twenty-five days' notice of the quantity and kind of stores to be transported at any one time.

Gross agreed to provide mule or horse trains whenever the department or district quartermaster judged it necessary for shipping subsistence stores or in other emergencies requiring "expeditious movement of public stores." For this special transportation Gross would receive 25 percent above the regular contract rate. Delivery time was not to exceed eight days per hundred miles for horses and mules or twelve days per hundred miles for ox trains. During the months of November, December, January, February, and March, Gross would not be charged for delays unless a board of survey determined the delay was from causes within his control—or the control of his employees. For delays chargeable to the contractor, Gross would pay $1 per day for every thousand pounds of stores. When trains were detained more than two days by order of an officer, Gross would receive a similar amount. Upon the arrival of the contractor's train at its destination, a board of survey would convene to investigate any deficiencies or damages in the stores transported.[104]

During the 1870s ox trains probably transported more government freight than either mule or horse trains. Oxen were cheaper than mules or horses, pulled better in mud and sand, and did not require grain. Captain Charles P. Eagan, the district's chief commissary officer in 1875, estimated that ox teams brought 90 percent of subsistence stores to New Mexico.[105] When the quartermaster's department requisitioned special mule transportation, it was not always available. Otero, Sellar and Company could hire only ox teams in September 1875 to move subsistence stores from Granada, Colorado, to Camp Apache, Arizona, even though the department quartermaster had considered mule transportation "imperative." In a letter to Eugene B. Allen, dated 15 December 1877, the Otero firm implied that calls for special mule transportation were too frequent, and it requested Allen to get a better understanding of the army's requirements before submitting another proposal to carry government freight.[106]

Even though extension of railroads shortened wagon hauls, delays in receiving supplies at interior posts seemingly were as frequent in the late 1870s as they had been a decade earlier. Bad roads and weather caused most of the trouble, especially for wagons traversing Raton Pass on the New Mexico–Colorado border. During the early years of freighting to Santa Fe, the route through the Raton Mountains was so steep and rough in places that wagons had to be lowered down slopes by rope. Richens L. Wootton made improvements and opened a toll road through the pass in 1866. Although less perilous than before, the pass still remained an obstacle to wagons.[107]

Military correspondence barely suggests the human suffering that occurred when wagons were delayed in the mountains. In notifying district headquarters of probable delays because of storms and bad roads, Captain Frederick F. Whitehead reported simply: "I am advised that [the] train which left El Moro, April 18th [1877], had not passed the Raton Mountains on [May] 3rd."[108] There is more information about the ox train that Otero, Sellar and Company dispatched from El Moro in February 1878. The train was carrying subsistence and quartermaster stores to Fort Stanton, and it was required to make the trip in twenty-seven days. Silvester Chaves and another man with the train later testified that snowstorms had delayed them twenty-two days in Raton Pass and eight days at Anton Chico, about twenty-five miles south of Las Vegas. They were detained three more days near Patos Springs in Lincoln County when they lost their oxen. Snowbound, the freighters had broken into the government cargo. From deficiencies later charged to contractor Eugene B. Allen, it appears that the men helped themselves to warm clothing—boots, trousers, and shirts—and consumed a fair amount of flour, bacon, ham, and prunes. The train was en route eighty-three days, reaching Fort Stanton on 4 May. A board of survey accepted Chaves's testimony but recommended that Allen be charged for twenty-three days' delay. The fee for delays, added to triple the cost of the missing stores, amounted to a hefty $1,665.[109]

Captain Whitehead, who served three years as district chief commissary of subsistence, offered two reasons other than the weather for transportation delays: inferior teams and wagons used on the road and failure of boards of survey to strictly enforce penalties. He believed that railroad construction companies secured the best teams, forcing contractors to employ inferior ones. But officers and enlisted men suffered the consequences. During the fiscal year ending 30 June 1879, several posts ran short of subsistence stores. On one occasion, Whitehead reported, subsistence stores were 136 days late in reaching Fort Bayard; on another occasion bacon arrived at Fort Bliss 87 days overdue.[110]

Marcus Brunswick's business records tell something about the men who carried out transportation contracts. For the year ending 30 June 1882, Brunswick hired independent Hispanic freighters to haul supplies by ox teams from Las Vegas to Fort Stanton. Brunswick paid the vast majority of the freighters 1½ cents per pound of freight for the entire distance of 180 miles. José Lucero, Ramón Ortiz, Eulogio Salas, and four other Hispanic freighters received 2¼ cents for freighting a shipment of bacon by mule teams in February. For some reason contractor Edward Fenlon paid Brunswick, his agent, the full contract rate—$3.25 per hundred pounds. Brunswick's profits probably were more than a cent a pound, for losses in transit had been moderate and all delays excused.[111]

Soon steel rails would crisscross New Mexico, shortening the wagon haul to even isolated posts like Fort Stanton. Military officers were among the nation's most enthusiastic promotors of railroads. Commanding General William T. Sherman in his 1880 annual report attributed the progress of settlement west of the Mississippi to the soldier, the pioneer and "to that new and greatest of civilizers, the railroad." General Philip H. Sheridan, commanding the Division of the Missouri, in his report asserted: "Amongst our strongest allies in the march of civilization upon the frontier, are the various railway companies who are now constructing their new lines with great rapidity."[112]

By the time Sheridan and Sherman penned these remarks, the Denver and Rio Grande had built a western extension across La Veta Pass to the San Luis Valley in Colorado. It had reached Fort Garland in the summer of 1877, and thereafter this post received all supplies by railroad. During the first half of 1878, supplies for most New Mexico posts were shipped via the railroad to Garland and from there transported by ox or mule teams to their destinations. By the end of June 1878 tracks had been extended to Alamosa, thirty miles west of Garland. Wagon transportation contracts awarded in May had designated Alamosa as the transshipping point for Santa Fe and Forts Craig, Wingate, and Bayard. But within a few months the route from Alamosa had been abandoned, "owing to the road being so bad as to be almost impassable." Supplies for New Mexico subsequently were transported from the railroad terminal at Trinidad.[113]

Meanwhile, both the Kansas Pacific and the Atchison, Topeka and Santa Fe railroads had built lines to La Junta, Colorado, in the Arkansas Valley. The Kansas Pacific, first on the scene in 1875, found the line unprofitable and abandoned it in 1878. The Santa Fe railway reached La Junta in February 1876, pushed on to Trinidad the following year, and in 1878 began construction across Raton Pass. Santa Fe construction crews labored long and hard that year, surmounting the summit in December and reaching Las Vegas in July 1879. The entire town turned out on the 4th of July, along with a brass band and a reception committee, to greet the arrival of the first regular passenger train.[114] Townspeople on the railroad's proposed route down the Rio Grande Valley eagerly awaited their turn to celebrate.

For the army, the arrival of the Atchison, Topeka and Santa Fe in New Mexico heralded important changes in the system of supply: transportation costs declined, wagon hauls grew shorter, and cheaper Kansas grain supplanted the local product. An immediate concern of the district quartermaster, however, was keeping pace with construction crews as they extended rails down the Rio Grande Valley. Supplies shipped from the East were invoiced to the end of the railroad tracks. Since that point constantly shifted during the two years of construction, the quartermaster's department employed civilian agents to receive and forward supplies at key points on the line.

Railroad crews worked about nine months laying track between Las Vegas and Albuquerque, the first freight trains arriving at the latter point about the first week of April 1880. Edward Webster, a clerk in the quartermaster's office in Santa Fe, was sent to Albuquerque to oversee transshipment of government supplies. He rented an office near the town's plaza, which was then more than a mile from the new depot. After about six weeks on the job, Webster asked for assistance. Desk work, loading trains, and the time consumed in going from his office to the depot—often on foot—kept him busy from early morning until late at night. Sometimes freight piled up at the depot because the contractor could not find teams to transport it. On 8 July Webster reported a total of 354,000 pounds of military supplies awaiting transportation to Forts Craig, Bliss, Apache, Bayard, Wingate, and to the post at Ojo Caliente. The contractor had located teams to move only about a third of it.[115]

By mid-August Webster had moved on to Socorro, where he rented an office and a small adjoining storeroom for ten dollars each per month. James Seymour, quartermaster's agent at Las Vegas, was ordered to Socorro to help forward government supplies that had accumulated at the end of the track. When Webster resigned his job in September, Seymour remained to oversee government shipping.[116] Subsequently Seymour moved south with the railroad. He opened an office in San Marcial, thirty miles south of Socorro, in November. But without assistance Seymour was unable to dispatch government supplies quickly. Railroad officials soon complained that railroad cars were not unloaded promptly, thus tying up rolling stock. Two cars loaded with lumber had stood untouched for three weeks; a car loaded with hay remained ten days without being unloaded. In early March 1881 Seymour moved to Grama Station, about seventy miles south of San Marcial, where he probably lived and transacted business in a hospital tent heated by a stove.[117] It was a brief stay. On 19 March he moved to Rincon, eight miles further south. Rincon already had an unsavory reputation as a railroad town. A lieutenant who preceded Seymour to Rincon in January wrote: "The R.R. Company have not an agent at Rincon, which is like all new R.R. towns full of horse thieves and bad characters who unless a strong guard was furnished to teams loading and unloading (would of necessity camp in town) would in all probability get away with Government Stock."[118] On 10 May Albert Alonzo Robinson, division superintendent and chief engineer for the railroad, telegraphed the district quartermaster that the road was open to Las Cruces. Three days later Seymour opened an office there. By mid-June the railroad had reached El Paso, Texas, where it cut directly across the parade grounds at Fort Bliss.[119]

While the Santa Fe railway completed its route down the Rio Grande Valley, it was also helping to finance the western division of the Atlantic and Pacific Railroad, starting west from Isleta, fifteen miles south of Albuquerque. The laying of track began in July 1880. Two months later William H. Decker arrived in Albuquerque to replace Webster as quartermaster's agent. A newcomer to the territory, Decker had received his appointment directly from the Secretary of War and had been promised a salary of $125 a month. Decker would have responsibility for overseeing transshipment of supplies to Fort Wingate, which by mid-December was only sixty-five miles from the end of the track. Decker struggled with multiple problems. Unable to rely upon the regular freight contractor, he had to hire independent freighters to transport supplies to Wingate at about 2 cents per mile. When the tracks reached Grants, about forty-five miles east of Wingate, Decker encountered new problems, which he described in a telegram to the district quartermaster: "Soldiers at Grants refuse to stand guard on stores there in day time and at night when placed on guard go to sleep." Perhaps he breathed a sigh of relief when posting a second telegram on 29 March 1881: "Trains run clear through to Wingate."[120]

Decker's composure would have been short-lived. Within two weeks his salary was reduced to $100 a month, and a few months later his job was in jeopardy. Decker was part-owner of an Albuquerque saloon, and the district chief quartermaster questioned his ability to carry out his official responsibilities. A defiant Decker responded: "As long as I attend to all the duties required of me as Quartermaster's Agent, I don't see any reason why I should not invest my money and make a few dollars honestly while out in this God, forsaken country." Army officials nonetheless decided to close the Albuquerque agency, and in September Decker was discharged.[121]

Still another agent of the quartermaster's department was 42-year-old J. F. Leonard, appointed in March 1881. Leonard had served the government as an enlisted man and employee for a quarter of a century. He had been wounded three times in the line of duty, once at Mesilla during the Texas invasion in 1861. When he applied for the transportation job, Leonard said he was suffering from the effects of his wounds and that an army physician had told him to work in the open air.[122] Leonard left Santa Fe on 16 March to assume duties as quartermaster's agent at Chama, New Mexico, railhead for the San Juan extension of the Denver and Rio Grande. From Chama Leonard would oversee government freight shipments to Fort Lewis, Colorado, established in July 1880 on La Plata River and adjacent to the Ute Indian Reservation. The original site of the post had been Pagosa Springs, where a camp had been established in October 1878. Supplies for the camp had been shipped by rail to Fort Garland and Alamosa, then transported overland by government and contract wagon trains. It was a difficult route, and freighters often found the mountain roads blocked by deep snow in spring and fall.[123]

In 1880 the Denver and Rio Grande connected Alamosa with Española, New Mexico, and started building an extension west of Antonito to tap the mineral-rich San Juan country in southwestern Colorado. Leonard observed firsthand these new developments. To reach his post at Chama, he traveled by stage from Santa Fe to Española, and there boarded a train for Antonito, Colorado, ninety-one miles up the track. He was snowbound four days at Antonito, but finally reached Chama by train on 19 March 1881, less than two months after regular service had been established to that point. Leonard's description of Chama is one of the best that has been preserved. Writing to Major James G. C. Lee on 7 April, Leonard observed: "There are no houses here—nothing but tents and log huts. There are probably 200 of these, ¾ of which are either whiskey saloons, dancing places, gambling and worse places still. The people here are a rough, noisy set." The best place to stay, he wrote, was the "Inter–Ocean," presumably a tent or log boardinghouse. Leonard advised Lee to travel by horseback directly from Santa Fe if he planned to visit Fort Lewis. "I would not recommend a trip over the road from Antonito to this place this summer," he wrote, "it is too dangerous. The road was built in the winter and in too great a hurry, and the consequence is, the railroad people themselves are afraid to travel over it."[124]

Leonard's stay in Chama was brief. In May he moved up the track to Amargo, New Mexico, and the following month he opened an office at Arboles, Colorado, about forty miles southeast of Durango. On 3 August he notified the district quartermaster that he had removed his office to the new town of Durango, where less than a week earlier citizens had greeted the first locomotive. He soon purchased a town lot, had a frame house built on it, and settled into community life with his wife and three small children. Fort Lewis was twelve miles west of town, and as the post expanded Leonard was kept busy forwarding supplies by contract teams. His position as transportation agent was eliminated in mid-1883 but restored by the end of the year at a lower salary. The post quartermaster described Leonard as "the most faithful and trustworthy man I have ever seen in the position."[125]

In addition to benefits mentioned earlier, railway transportation also allowed faster and more frequent shipment of army subsistence stores, reducing much of the spoilage, loss, and deterioration associated with long wagon hauls. By 1885 nearly all posts within the District of New Mexico were located within twelve miles of a railroad station. The longest wagon haul was 118 miles to Fort Stanton, which received freight from Lava, New Mexico, on the Atchison, Topeka and Santa Fe. August E. Rouiller of Paraje was the Fort Stanton freight contractor for the fiscal year ending 30 June 1885. He received $1.28 per hundred pounds for the entire distance. On the shorter hauls both government and contract transportation were used. During 1884–85 Ferdinand Schmidt of Mora County hauled supplies nine miles between Watrous Station and Fort Union for 11 cents per hundred pounds, and J. H. Crist transported supplies between Durango and Fort Lewis for 35 cents per hundred pounds.[126]

Even with railroad transportation, however, washouts, snow blockades, and mechanical mishaps sometimes delayed supplies in reaching their destination. The flood of 1884, the worst in anybody's memory, seriously disrupted railroad travel between Albuquerque and Las Cruces. The spring thaw following unprecedented winter snows in southern Colorado caused the Rio Grande to rampage down the valley, destroying hundreds of homes, injuring crops, and tearing out bridges and railroad tracks. As early as 25 May the commanding officer at Fort Craig reported that San Marcial, the railroad station five miles from the post, was almost completely under water. Four companies of the Twenty-third Infantry en route east from Bliss, Bayard, and Craig were waterbound in railroad cars at Socorro for six days. On 7 July the quartermaster at Fort Selden reported that not a single train had arrived there since 1 June because of washouts. Normal service north and south of Selden was not restored until late August.[127]

With the coming of the railroads, the army began importing cheaper forage and subsistence stores from the East. Some New Mexico posts had received eastern grain even before steel rails crossed the territory. Abraham Staab and Charles Ilfeld imported corn in 1878 to fill government contracts at Forts Stanton and Union. But corn in New Mexico had been scarce that year, causing Captain John H. Belcher, the district chief quartermaster, to advertise in posters and newspapers in Leavenworth, Kansas, inviting bids for grain and other supplies.[128]

Among the first Kansas dealers to win forage contracts in New Mexico was R. E. Thomas of Leavenworth. In June and July 1880 he received contracts to furnish corn at Fort Wingate, oats at Santa Fe and Fort Union, and corn and hay at Fort Craig. He shipped these supplies from Kansas to the end of the railroad tracks in New Mexico. At least two other contractors imported forage from Kansas that year. Thomas M. Field, who with Isaac W. Hill ran a forwarding and commission house at Alamosa, Colorado, filled his contract at Fort Lewis with Kansas corn brought in via the railroad. And Sigmund Wedeles, now a Santa Fe merchant, shipped to the quartermaster's agent in Albuquerque 100,000 pounds of corn from Eldorado, Kansas, and 100 tons of baled hay from Wichita. Wedeles also had agreed to furnish the government corral in Santa Fe with 500 tons of Kansas hay. And he may have filled his corn contracts at Santa Fe and Fort Union from the East as well.[129]

Probably even more Kansas grain was shipped to New Mexico in 1881. R. E. Thomas, the largest corn contractor in the district that year, agreed to deliver a total of 1,951,606 pounds to Forts Bayard, Craig, Cummings, Selden, and Wingate. This amounted to about 30 percent of corn requisitioned on contracts in 1881. Thomas also contracted for a total of 249,443 pounds of oats, or about 12 percent of oats received on contracts. In a private transaction the Santa Fe firm of Z. Staab and Company furnished five carloads of Kansas corn in January to the quartermaster's agent in Albuquerque. The firm also filled a Fort Wingate contract awarded in March with Kansas corn. Z. Staab and Company, in fact, was the second largest corn contractor in the district in 1881, supplying a total of 1,509,054 pounds to Forts Wingate, Stanton, Bliss, and the camp at Mescalero. Only a portion of this corn came from Kansas, however. The firm supplied Fort Stanton with American seed corn raised by local producers. It is no longer possible to determine where the other contractors obtained their grain. But it is obvious that Kansas was furnishing the district with large amounts of forage. At least four contractors in 1881 sold Kansas baled hay to the district quartermaster—Sigmund Wedeles, R. E. Thomas, Z. Staab and Company, and Jacob DeCou of Eldorado, Kansas.[130]

The trend continued in 1882, with Kansas and New Mexico dealers vying for government contracts. In May 1883 New Mexicans were jolted when all their bids for furnishing the district with corn, oats, bran, and hay were rejected. Colonel Judson D. Bingham, chief quartermaster for the Department of the Missouri, informed the district quartermaster that forage for New Mexico would be supplied from the vicinity of Fort Leavenworth. "I can purchase for about thirty thousand dollars," Bingham observed, "what will cost under bids received by you about one hundred and fifteen thousand dollars." When new forage bids were solicited in September, those from New Mexico were again rejected.[131] New Mexicans loudly protested. Abraham Staab wrote directly to Secretary of War Robert T. Lincoln, presenting figures showing the advantages that would have resulted had his own bids been accepted. Lincoln replied that Staab's figures were based "upon a misapprehension of facts, particularly as regards cost of transportation."[132]

For New Mexicans, the entire issue focused on transportation costs. Lincoln County residents had been told officially that the army appropriation would not permit the purchase of supplies at New Mexico prices. Indeed, the allotment for purchasing forage and fuel in the Department of the Missouri had been greatly reduced that year. By purchasing forage in Kansas the quartermaster's department would stay within this allotment by paying for the cost of transporting it to New Mexico out of transportation funds.[133] New Mexicans regarded this as economic nonsense. The editor of the Las Cruces *Rio Grande Republican,* in a long column entitled "A Crying Outrage," exposed the folly of government bookkeeping.

> The quartermaster at Leavenworth purchases their hay in Kansas at say $6 to $8 a ton, ships it to Las Vegas at a heavy expense, and from Las Vegas to Fort Stanton, a distance of two hundred miles, it is hauled by ox-teams. Yes, hauled by ox teams, right through a hay-growing country, at an expense of $75 a ton.

The same folly prevailed in purchasing corn.

> The corn thus imported is bought in Kansas at say 90 cents a hundred. The freight on it to Fort Stanton is $2.90; making $3.80 in all; while a better quality of corn can be bought from the ranchmen close by for $2 a hundred![134]

The protests were effective, at least in the short run. In March 1884, Quartermaster General Samuel B. Holabird issued these instructions to the chief quartermaster for the Department of the Missouri: "Preference should be given for local delivery [of forage], other things being equal; and . . . no effort should be made to reduce the legitimate and necessary expenditures for Regular Supplies by an equal or greater expenditure for Transportation of the Army."[135] During the following year New Mexico contractors furnished locally grown hay to several posts. About the only corn contracts awarded in New Mexico, however, went to residents of Lincoln County, where the cost of transporting corn from the east to Fort Stanton made the local product competitive. Elsewhere local corn was still more costly than Kansas corn with transportation costs added. And even the cost of importing baled hay declined in later months, allowing Kansas hay to replace the local commodity at most New Mexico posts.[136]

An era was passing. By 1885 the commissary department was importing almost all subsistence stores, purchasing in New Mexico only beef and small amounts of salt and potatoes. The same trend was evident in the quartermaster's department. On 2 September of that year the commanding officer in New Mexico received this message from department headquarters:

> Contracts for hay and grain for posts generally in New Mexico have been awarded, upon recommendation of the Chief Quartermaster of the Department, to non-residents of that Territory for the reason that, in each instance, the bids were lower, as shown by statements of the Chief Quartermaster.[137]

Contract freighting in west Texas involved the same problems and difficulties encountered in Arizona and New Mexico. A noticeable difference was the monopoly that one firm held on government freighting. From 1867 to 1874 Hardin B. Adams and Edwin D. Wickes hauled supplies between San Antonio and most Texas posts south of latitude 32 degrees, which included Forts Bliss, Clark, Concho, Davis, Duncan, McKavett, Quitman, and Stockton. A former employee remembered that the firm at times employed 1,500 mules and from 150 to 200 wagons. The firm also worked through subcontractors. In 1869, for instance, it contracted with three or more independent freighters to haul government supplies from coastal Indianola to San Antonio. The contract that Adams signed for fiscal year ending 30 June 1868, and which was extended through February of the following year, established rates of $1.68 per hundred pounds of freight per hundred miles. Thereafter contract prices declined. During fiscal year 1872–73 the firm received $1.06 per hundred pounds per hundred miles.[138]

With extension of railroads into Texas, the rates would decline even more. After the Civil War most military supplies had reached San Antonio via steamship to Indianola on Matagorda Bay, whence they were freighted overland. In 1873 tracks of the Gulf, Western Texas and Pacific Railway stretched from Indianola through Victoria to Cuero, Texas, about eighty-eight miles from San Antonio. At the same time the Galveston, Harrisburg and San Antonio Railway was building west from Columbus, Texas, and would reach San Antonio in February 1877, giving that town a direct rail connection with the Gulf. The wagon transportation contract awarded for fiscal year ending 30 June 1876 reflected these advances. Contractor James Callaghan of San Antonio agreed to receive supplies at San Antonio, or from agents on the above two railroads, or at Austin, which had rail connections to markets in the North and East and to Galveston on the coast, and to transport these supplies to posts in west Texas at the rate of 87 cents per hundred pounds per hundred miles. For the following two fiscal years, 1876–78, H. B. Adams again held the wagon transportation contract. He agreed to freight between the places mentioned in Callaghan's contract for 83 cents per hundred pounds per hundred miles the first year, 69 cents from July through December 1877, and 89 cents from January through June 1878.[139]

The Adams monopoly would soon end. In 1879 rival firms won contracts to carry military supplies from the rails at Fort Worth and Waco to Forts Davis and Stockton, and Callaghan won the contract to freight from Austin, San Antonio, and Cuero to the same two posts. In August 1881 both garrisons were receiving subsistence stores from Chicago and St. Louis via the Texas and Pacific Railway, which had extended its tracks from Fort Worth to Abilene, Texas. The wagon haul from Abilene to Stockton was 283 miles and to Davis, 355 miles, more than 100 miles less than the old haul from San Antonio.[140] By the end of the year a new supply route was available when the Texas and Pacific Railway joined with the Southern Pacific at Sierra Blanca, 92 miles east of El Paso. Stockton then was only 63 miles from a railroad station and Davis, about 60 miles. Supply routes would change yet again when El Paso was linked directly to San Antonio by rail in 1883. Supplies sent by this route required a 53-mile wagon haul to Stockton from Haymond Station and a 24-mile haul to Davis from Murphyville (later renamed Alpine). Daniel Murphy and son agreed to freight government supplies from the station to Fort Davis for 19 cents per hundred pounds.[141] By this date Fort Bliss was receiving all supplies directly by rail.

Before the arrival of railroads, overland freighting was a major western industry, and the federal government was one of its biggest customers. Indeed, freighting became a mixed enterprise, one in which the government combined with private investors to advance settlement and economic development. But the building of steel rails across the continent caused the demise of large freighting firms and radically changed the method of supplying the Southwest garrisons. Railroads allowed the importation of less expensive forage and subsistence stores. Supplies were shipped more frequently and arrived in better condition than before. Expensive freighting hauls were all but eliminated. Unchanged, however, was the entrepreneur's desire to profit from the army's presence. Even though local suppliers found less opportunity to sell to the army with the coming of railroads, government contracting would remain a cornerstone of the region's economy in the twentieth century. Since the army contractor has played such a prominent role in the Southwest, Chapter 8 will focus on contractors—their numbers, their ethics, and the problems they encountered while working for the government. The chapter will also clarify the amount of money the army pumped into the Southwest and its importance to the local populace.

8

Fraud, Theft, and Military Expenditures

LOOKING back on his early career as a frontier merchant, Abraham M. Franklin of Arizona gleefully remarked that "one of the chief occupations around Tucson [in the 1870s] was fleecing the Government." A similar statement was made in 1866 by Colonel James F. Rusling following an inspection tour of western military posts. Stopping at the Pima villages in Arizona, he observed traders buying Indian wheat and corn at low prices, to be sold at inflated ones to army purchasing agents. This practice, he stated, was but one of the ways in which contractors "fleeced both the Indians and the Government."[1] Over the years, other military personnel have pictured contractors as thieves, extortionists, and monopolists; men who favored their pockets at the expense of their patriotism; men who were quick to capitalize on the army's misfortunes, even prolonging Indian wars to retain lucrative contracts. As pointed out in an earlier chapter, it was General John M. Schofield, commanding the Military Division of the Pacific, who suggested that greed for contracts helped explain the infamous Camp Grant Massacre.

This image of the evil, conniving contractor has found its way into the secondary literature, even though few scholars have seriously examined the issue.[2] Presumably business ethics of the Gilded Age warrant painting government contractors in the darkest hues. A careful study of the military supply system as it developed in the Southwest in the two decades following the Civil War does not wholly support this negative image of the corrupt contractor. A more accurate image might be that of the frontier capitalist, investing money in a risky business, alternately assisted and obstructed in his operations by an economy-minded War Department. Fraud was not rampant—a dishonest contractor was the exception rather than the rule.

Still, it is not surprising that the negative image has persisted for so long. Military contractors gained an unsavory reputation during the Civil War when their corrupt business deals added to the total cost of the Union war effort. Postwar investigations left no doubt that fraud had been practiced on a wide scale.[3] With Congress threatening major cuts in army appropriation bills, the War Department had to tighten purchasing procedures and take other steps to avoid costly losses.

Army officials could not, of course, eliminate all thefts and fraud, so a case can be made for the dishonest contractor. Several examples of dishonesty, in fact, have appeared in previous chapters. One of the most common forms was "salting" supplies turned in on contracts. Captain George W. Davis, commanding officer at Camp McPherson, Arizona, reported in 1868, for example, that contractor Benjamin Block had "made a most outrageous attempt to swindle the Government by mixing with his corn some fifteen percent of sand." A decade later the post quartermaster at Fort Bliss, Texas, rejected 9,000 pounds of corn that contractor John H. Riley delivered after discovering it had been "salted to the extent of fully 20 percent with gravel." In 1877 officers at Camp Grant, Arizona, found an 11-pound stone in a sack of barley turned in on a government contract.[4] In cases such as these the offending contractor was required to make good the deficiency; he might even be prevented from bidding on future contracts. Rarely, however, was he prosecuted in a court of law, for the government found it both difficult and expensive to prove criminal intent.

And even in well-documented cases of fraud in which the government instituted legal proceedings, it had trouble securing convictions. Such a case occurred at Fort Bayard, New Mexico, in 1881 when Lieutenant Leverett H. Walker, the post quartermaster, accused grain contractors George P. Armstrong and John Brockman of attempting to defraud the government out of 30,890 pounds of corn and 2,000 pounds of oats, valued at $1,164. The alleged fraud came to light after Walker learned that his clerk, Private Charles Arey, had received money from the two contractors. Walker then checked his forage books and discovered several false entries. Confronted with the evidence, Arey confessed that Armstrong and Brockman had paid him to record deliveries that had never been made. In a preliminary investigation the contractors denied any wrongdoing. Armstrong claimed that he had paid Arey $225 for some 25,000 pounds of surplus grain at the post, which Armstrong then turned in on his contract. The case was turned over to U.S. District Attorney Sidney M. Barnes, and during the August term of the Third District Court Armstrong and Brockman were indicted for fraud. A year later one jury found Brockman not guilty; another could not agree in Armstrong's case and the jury was dismissed. The case was continued from court term to court term until March 1886, when a jury found Armstrong not guilty.[5]

More common than outright fraud against the government was the problem of failing contractors. The reasons individuals offered for their inability to complete contracts were many: drought, combinations of producers demanding inordinately high prices, Indian hostilities, failure of the Colorado River to overflow onto natural grasslands, and miscellaneous financial difficulties. For instance, William B. Hellings, hay contractor for Camp Verde, Arizona, requested release from his contract in August 1871 because Indians had burned all the grass in the vicinity of the post. In October 1877 Louis A. Stevens also asked to be relieved from his contract to supply 150,000 pounds of corn at Fort Whipple because drought had badly damaged the local crop.[6]

The government protected itself against failing contractors by requiring at least two men of recognized means to cosign as sureties, holding them equally responsible for completing the contract. When bondsmen and contractor alike refused to carry out the contract, officers purchased supplies in the open market, charging the contractor for any losses sustained in filling the contract. After Simon Filger failed on his hay contract in 1871, the quartermaster at Santa Fe purchased 248 tons in open market at an additional cost to the contractor of $1,251. When contractor James Patterson's supply of beef gave out at Camp McDowell, Arizona, in the summer of 1875, 440 pounds of fresh beef were purchased in open market at fifty cents a pound, almost forty cents above contract price. Patterson was charged with the difference.[7]

As noted earlier, oftentimes the government sustained little monetary loss when a contractor failed since bondsmen frequently honored legal obligations and payments could be withheld on goods already delivered on the contract. Such was the case in 1871 when John B. Allen, flour contractor at Tucson, failed to complete his contract. His bondsmen, Charles H. Lord and Pinckney R. Tully, subsequently purchased local flour, including some ground at Joseph Pierson's mill in Terrenate, Sonora, to "make good" the contract. After Elisha A. Dow abandoned his corn contract at Fort Stanton in 1875, his sureties, Santa Fe merchants Adolph Seligman and Lehman Spiegelberg, filled the contract themselves.[8]

Disbursing officers viewed many—but not all—defaulting contractors as unreliable. In New Mexico the chief quartermaster placed the names of some defaulters on a blacklist and barred them from submitting future bids. This practice was subsequently modified after Secretary of War Alexander Ramsey decreed in 1880 that "the bids of persons who have heretofore failed to enter into, or carry out, contracts shall not be excluded from consideration."[9]

Some failing contractors settled with the government without major protest. Within five months after his contract ended on 31 August 1870, John S. Crouch reimbursed the subsistence department $367 for his failure to keep Fort Cummings supplied with beef.[10] Failing contractors who resisted payment faced legal proceedings. Because of the difficulty in obtaining favorable court decisions, however, the government prosecuted only those cases where prospects for securing reimbursement were favorable. The army decided not to enter a suit against Mart Maloney, who failed to execute forage contracts at Fort Huachuca, Arizona, in 1882, after discovering that Maloney and his sureties were insolvent. But in the late 1870s the government successfully brought suit against two failing hay contractors at Fort Whipple, Arizona—recovering $549 from William C. Dawes and $567 from W. H. McCall.[11]

The army had less success obtaining restitution from Philip Drachman, who failed to execute contracts for delivering hay and straw at Fort Huachuca in 1885. The government estimated its loss at nearly $6,000, the difference between Drachman's bid and the actual cost of laying in supplies. Interminable delays slowed prosecution, during which time Drachman and two key government witnesses died. The case was finally dismissed in 1902 or 1903.[12]

The government lost even more money when Drachman's brother, Samuel H. Drachman, failed to complete contracts for supplying hay and straw at Forts Bowie and Huachuca in 1884. Drachman blamed his troubles on a drought that had stunted the local crop of grama grass. Having to make up deficiencies on the contracts through open market purchases, the government decided to sue Drachman and his sureties to recover an estimated loss of more than $11,300. The defendants must have secured good legal counsel, for in 1887 the government accepted a compromise settlement of $186 on one contract and in 1890 a settlement of $200 on the remaining contracts.[13]

Many failing contractors petitioned the quartermaster's department for relief from their financial obligations, but Congress alone had authority to grant their requests. John H. Marion, editor of the Prescott *Arizona Miner,* was among the failing contractors who approached Congress. After he had failed to complete a hay contract at Whipple Depot in 1884, the government initiated legal proceedings to recover the $1,042 loss it had sustained in open market purchases. The government also withheld payment for government advertising appearing in the *Arizona Miner.* In 1888 Congress approved Senate Bill 1772, relieving Marion from payment of the $1,042, but President Grover Cleveland vetoed the bill—swayed by the War Department's argument that if enacted into law it "would establish a dangerous precedent." Shortly thereafter the suit was settled by compromise—Marion agreeing to pay half the government's loss on the contract.[14]

Failing contractors contributed to the high cost of maintaining the frontier army; so, too, did the petty thievery and larceny that permeated the supply system. The most common type of theft was the pilfering from supply trains or post warehouses of small amounts of clothing, forage, canned goods, and items in the soldier's ration. Thieves also made away with horses and mules from government corrals. Civilian employees were responsible for some losses. William E. Slocum, wagonmaster at Fort Stanton, New Mexico, was discharged in 1868 for "selling and attempting to sell Government property." The same year William Mortimer, carpenter at Fort Selden, New Mexico, deserted his post, taking with him a government mule and pistol. In 1872 O. W. Dickerman and his wife Martha were accused of robbing a storehouse at Fort Davis, Texas, of about $2,000 worth of subsistence stores. The stealing occurred over several weeks, starting about the time Dickerman learned his employment as subsistence clerk would soon end.[15]

Although army officials waged constant war against petty thievery, it was impossible to stop impoverished soldiers from selling or attempting to sell their clothing, arms, and other supplies to civilians. Soldiers often exchanged these items for whiskey or sexual favors at off-reservation "hog ranches" and gambling dens. One officer described the settlement of Leasburg near Fort Selden as a town "entirely supported by selling mean whiskey to soldiers by lewd women and . . . by purchasing government arms and clothing." He estimated that the government lost fifty or sixty dollars a day because of illegal exchanges in Leasburg. Another officer claimed that enlisted men at Camp Thomas, Arizona, even stole harness, lumber, and tents to pay for liquor in nearby Maxey.[16]

Some whiskey dealers brazenly bartered their wares within the limits of a post, leaving in their wake intoxicated soldiers without weapons and clothing. The enormity of the problem is suggested by Captain James R. Kemble's request in 1867 for fifty-five revolvers to replace those lost by his company during the previous year. General James H. Carleton, then commanding the district of New Mexico, marveled that a company of eighty-eight men was left with only thirty-three pistols. It was obvious that the men had sold their weapons, for each revolver would command a much higher price on the local market than its original twelve-dollar purchase price. Carleton believed that the only way to stop this practice was to collect fifty dollars from every man who lost a pistol[17]

Soldiers accused of selling government property underwent trial by court-martial. Those judged guilty of disposing of small amounts were sentenced to forfeit pay and to confinement at hard labor for a few months. Private Isham Logan, found guilty at Fort Selden of selling a pair of infantry trousers to a citizen for one dozen eggs, was sentenced to forfeit a month's pay. Private Thomas Kreutzer was sentenced to confinement at hard labor at Fort Whipple for two months for having "lost or improperly disposed" of a Colt revolver. But Private John Devine, judged guilty of fraud and embezzlement for having sold large amounts of government clothing and equipment to a resident of Tucson, was sentenced to two years hard labor at Fort Yuma and then to be dishonorably discharged.[18]

Citizens likewise were prosecuted for buying and selling government property. Charles Van Wagoner, residing in Ancon on the Fort Craig military reservation, was arrested in 1866 for buying stolen commissary stores (bacon, coffee, and candles) from soldiers. In 1869 a jury deliberating at Mesilla, New Mexico, found Washington W. Hyde guilty of buying military arms from soldiers and ordered him to pay a $200 fine and to serve six months in the county jail. A few years later

Francisco Salmón was convicted of purchasing a pistol and ammunition from a soldier and was fined $1,000. He was then incarcerated in the Doña Ana County jail until the fine and court costs were paid.[19]

Soldiers who sold government property to civilians were often deserters. By 1871 the number of desertions in the army equaled about one third of its enlisted strength. Desertions obviously were a drain both on manpower and finances. The army paid to have men trained, equipped, fed, and transported west and then paid again for their replacements.[20] Deserters also absconded with valuable government property. In 1866 army officials in Arizona reported large numbers of soldiers crossing into Sonora, taking government animals and equipment with them. In August of the following year Colonel J. Irvin Gregg, commanding the District of Prescott, complained of mass desertions at Fort Whipple following payment of troops. Eleven men had left the night of the fifteenth, four men on the night of the seventeenth. Gregg ordered the arrest of the paymaster's entire escort—seventeen men—after learning they had offered the post guide a horse, carbine, pistol, equipment, $200, and a share of the money in the paymaster's cashbox to guide them to Sonora.[21]

Lieutenant Colonel Thomas C. Devin, commanding the Subdistrict of Prescott at Whipple in 1868, estimated that during the previous twelve months deserters had taken one hundred government horses and mules across the Colorado River. These men headed for San Bernardino, where they would find a ready market for horses and arms, as well as shelter and employment for themselves. Devin refused to send cavalry escorts in the direction of the river, fearing mass desertion and the loss of government property, which, he said, "I value more than the men."[22] Conditions were almost as bad elsewhere in the Southwest. Army Inspector Nelson H. Davis reported that during 1872 the 8th Cavalry—then garrisoning the District of New Mexico—had suffered 101 desertions. Major Andrew J. Alexander, commanding at Fort Garland, Colorado, the following year, claimed that the "harboring of deserters and receiving stolen government property" was the great "evil" of the countryside.[23]

Not all theft and illegal transactions can be blamed on enlisted men and civilians, however. During the Civil War a handful of army officers serving in the Southwest were implicated in schemes to defraud the government. Among them was Captain Prince G. D. Morton, commissary officer at Fort Sumner, New Mexico. Morton was accused of conspiring with Indian agent Lorenzo Labadie to steal government cattle. Early in 1865 a military court found Morton guilty of embezzlement, and a few weeks later he was dismissed from the service. Very few other Civil War officers suspected of wrongdoing were prosecuted, for the investigations came at the war's end when guilty parties were being mustered out of the army.[24]

In later years the quartermaster and commissary departments suffered from the lack of manpower; thus inexperienced line officers generally filled the positions of post quartermaster and commissary officers. Russell F. Weigley has pointed out that the "positions changed hands too often to allow their holders to become sufficiently acquainted with their duties." Consequently slipshod work contributed to the high cost of financing the army. Quartermaster General Montgomery C. Meigs, for example, believed that substantial savings could be achieved if trained and reliable quartermasters served at each post.[25]

Some supply officers fell victims to their own greed. A well-documented case involved Lieutenant George O. McMullin, commissary officer at Fort Marcy, New Mexico. In 1867 McMullin sold to the proprietor of the Exchange Hotel in Santa Fe ten sacks of flour and forty-four gallons of vinegar, valued at $110. He also sold $20 worth of bacon and $110 worth of corn to two other civilians. Witnesses later testified that the corn had been transported in government wagons, concealed under a layer of manure. Rumor had it that McMullin sold these and other supplies to pay his gambling debts. In November a military court found him guilty of illegally selling government property and sentenced him to repay the amount embezzled and to be dismissed from the service. President Andrew Johnson approved the findings of the court, but remitted that part of the sentence dismissing him from the army. McMullin was honorably mustered out a few years later.[26]

Another case of a dishonest supply officer was reported at Fort Craig, New Mexico, in 1871. Captain Frederick W. Coleman, the post's commanding officer, had at first believed that Lieutenant Robert E. Bradford, post quartermaster, was guilty only of careless business practices. Because of this incompetency, Coleman removed Bradford as quartermaster in September 1870. Shortly thereafter the newly appointed quartermaster reported a deficiency in commissary funds of $1,300. Bradford did not know what had become of the money; Coleman suspected that Bradford "had got his quartermaster and commissary funds mixed up and had not collected sales of stores made on [the] inspection report." Coleman allowed Bradford to make good the deficiency, which he did by securing a loan for that amount from the post trader. Subsequent developments, however, convinced Coleman of Bradford's dishonesty. After a board of survey discovered another large deficiency, this time in corn receipts, Coleman preferred charges against him. In December 1871 Bradford was dismissed from the service.[27]

The best-known case wherein a supply officer was dismissed from the army for alleged improprieties occurred at Fort Davis, Texas, and involved the nation's first black graduate of West Point. Henry O. Flipper, Class of 1877, arrived at Fort Davis in 1880 as a second lieutenant in Company A of the 10th Cavalry, a black regiment. In early December he was appointed post quartermaster and commissary of subsistence. He served in the former capacity briefly, but carried out duties as commissary officer until shortly before his arrest in August 1881 on suspicion of embezzling funds.[28]

Some time after May of that year Flipper had discovered a deficiency of $1,440 in his commissary funds. Rather than admit to this shortage, he lied about his accounts to his commanding officer, Colonel William R. Shafter, confident that the deficit would be covered in a short time. But Flipper's maneuverings were all in vain. Suspecting that something was amiss, Shafter removed Flipper as commissary officer, searched his quarters, and then placed him under arrest. Shafter then let Flipper make good the deficiency (which in fact amounted to about $2,400) through loans and contributions from townspeople who had befriended him. Despite restitution of the missing funds, Shafter preferred charges against Flipper for embezzlement and for conduct unbecoming an officer and gentleman. On 8 December 1881 a military court found Flipper not guilty of embezzlement, but guilty of the charge of conduct unbecoming an officer and gentleman and sentenced him to be dismissed from the service.[29]

Most observers agreed that Flipper had not stolen the missing funds. Colonel Benjamin H. Grierson, Flipper's regimental commanding officer, sent a testimonial letter at the time of the trial in which he praised Flipper's performance during the Victorio War, reaffirmed his faith in Flipper's honesty, and suggested that any irregularities that may have occurred resulted from Flipper's inexperience and carelessness. In reviewing the court-martial proceedings, David G. Swaim, judge advocate general of the army, concluded that Flipper had not intended to defraud the government, and he recommended that a sentence less severe than dismissal be rendered. Leniency was not forthcoming, however, for President Chester A. Arthur subsequently confirmed the verdict and sentence of the court-martial.[30] Flipper was dismissed from the army on 30 June 1882.

Corruption, theft, mismanagement, carelessness—all added to the expense of supplying the frontier army. But despite evidence of fraud and failure among contractors, military records abound with examples of honest, industrious men who not infrequently overcame severe hardships to supply the frontier posts, in some cases suffering financial loss rather than defaulting on a contract. In official correspondence, post commanders and supply officers frequently spoke with high regard for individual contractors: "His reputation for honesty and upright fair dealing throughout this section is excellent"; "Mr. [Estévan] Ochoa always fills his contract to the letter, even if he loses by it"; "the contractors are industrious and worthy men"; "Mr. Patterson having filled all his former contracts with honesty and good faith is entitled to some consideration"; "Mr. [Abraham] Staab is a most reliable business man and contractor, and . . . whatever he agrees to do he will do faithfully, to the very letter."[31] On many occasions purchasing agents simply reported that all contracts were being filled satisfactorily.

The other side of the coin was that successful contractors often had to overcome obstacles created by bungling bureaucrats and an economy-minded War Department. Some contractors waited months—and even years—for their pay. Lieutenant Colonel Henry D. Wallen, commanding the District of Arizona in 1866, reported that the indebtedness of the quartermaster's department in Arizona was $250,000, with some of the debts dating back as much as nineteen months. More than a decade later the Tucson *Arizona Citizen* claimed that the government was in arrears to officers, soldiers, and contractors south of the Gila River at least $100,000.[32]

These delays in payment hurt everyone: the contractor, who had to borrow money at a high interest rate to carry on his business; the small producer who was awaiting payment from the contractor; and the government, forced to pay higher prices because of its dismal credit record. Marcus Brunswick, a Las Vegas, New Mexico, merchant and government transportation agent, voiced a common lament: "I had no idea the Government would keep us out of our money so long for work done. It is impossible to do the work unless they pay up."[33] An Arizona newspaper correspondent made this pointed observation: "It is wonderful how prompt Government is and always has been to collect its dues, the moment they become due, and equally wonderful to us how it has the cheek to keep citizens out of money due them for weeks, yes, months and years at a time, to the ruination of business and disgust of all who are in any way dependent upon money due from Government."[34]

Disgruntled contractors did not suffer their annoyances silently, however. One who complained was William Hoberg of Cherry Valley, New Mexico, who had agreed in the fall of 1874 to build adobe corral walls at Fort Union Depot. By the time winter set in, the work was only one third done, and Hoberg was losing money on the job. Consequently the quartermaster's department requested him to give bonds for completion of the work. Hoberg refused, arguing that the government had never before required bonds in cases of this kind. With a degree of sarcasm, Hoberg wrote:

> I would have as good a right to now ask Bonds of the Government that I be paid promptly when the balance of the job is completed, as for the Government to ask me now to give bonds for the completion of the balance of the work, for the reason that in the case of [a contract Hoberg had held in 1873] I was obliged to wait about 14 months before I got my pay.[35]

Another who complained was George W. Maxwell of Las Cruces, New Mexico. In January 1873 the quartermaster at Fort McRae had rejected about 65,000 pounds of corn that Maxwell's agent attempted to deliver, claiming the grain was "unsound." Maxwell was indignant. He had been forced to haul the rejected corn to Jack Martin's ranch at Aleman, about twenty miles from the post. He demanded that a board of officers reexamine the corn, and he submitted affidavits from three Doña Ana County merchants and the county sheriff attesting to the "merchantable" quality of the corn. Maxwell requested that the grain be received at once "as money is too hard to get holt [*sic*] of to be so unjustly treated as I think I have been." Indeed, a board of officers soon submitted a favorable report on Maxwell's corn, and he was instructed to deliver it at McRae. Even though the district chief quartermaster recommended that Maxwell be reimbursed for the cost of storing the grain at Martin's and for hauling it to and from the post, records suggest that Maxwell was never paid for this added expense.[36]

Other contractors voiced complaints early in 1876 after post quartermasters at Forts Stanton, Bayard, and McRae, New Mexico, stopped receiving grain on contracts. Because of depleted allotments, General John Pope, commanding the Department of the Missouri, had ordered each quartermaster not to accept "any more grain than is necessary for the supply of the post to June 30, 1876." Thomas Bull, contractor at Fort Bayard, protested this action as a violation of his contract. He had already expended several thousand dollars on the government's behalf and had between 15,000 and 18,000 pounds of grain in transit from the Mesilla and Mimbres Valleys when he learned of Pope's order. B. J. Baca, contractor at Fort Stanton, similarly protested, stating he had purchased three fourths of the grain required on his contract and would lose $7,950 if it were terminated.[37] Eventually military officials decided that the army was obligated to take the grain as stipulated in the contracts.

Contractors like George W. Maxwell, Thomas Bull, and B. J. Baca were typical nineteenth-century entrepreneurs. They were guided by the profit motive and looked to the government to assist them in their commercial ventures. But contracting was a risky business. Bad weather, Indian hostilities, poor harvests, and stiff competition could any or all bring financial ruin. When supply officers obstructed their operations, either through delayed payment or cancellation of agreements, contractors did not hesitate to complain to higher authority. Indeed, the army chain of command, stretching from western supply offices to the War Department in Washington, D.C., was a built-in safeguard against corruption and maltreatment. It kept political influence in awarding contracts to a minimum, for a successful bid had to receive approval at several levels. And since the government published supply proposals in local newspapers, an unsuccessful bidder who felt cheated could appeal the decision. Other institutional practices that limited fraud included weekly verification of post supply funds by commanding officers, inventory of supplies and property upon a change in supply officers, and periodic investigations into the general management of a military installation by the Inspector General's department.

Despite the infrequency of outright fraud, the army could not prevent contractors from engaging in cutthroat competition. Several examples have appeared in earlier chapters, including the banding together of cattle dealers to break Andy Adam's beef contract. Such competition rarely led to violence, but an exception occurred in Lincoln County, New Mexico. Here, the competition to control government contracts helped ignite one of the most violent confrontations in the history of the American West.[38]

The Lincoln County War essentially was a struggle for economic power. In this thinly populated section of southeastern New Mexico, government contracting provided the only reliable market for local products. Historical writing about people and events leading to the violence has been colored by biased observations recorded during the war itself. A good example is the one made by Frank W. Angel, a government investigator who visited the county in 1878 during the height of the trouble:

> L. G. Murphy & Co. had the monopoly of all business in the county—they controlled government contracts and used their power to oppress and grind out all they could from the farmers and force those who were opposed to leave the county.[39]

During the early part of the decade, 1870–72, L. G. Murphy and Company held nearly all quartermaster contracts for supplying Fort Stanton with grain, hay, and charcoal, and was also the sole supplier of commissary goods to the Mescalero Indian Agency for part of this time.[40] But L. G. Murphy and Company did not monopolize the Fort Stanton supply contracts. During a twelve-year span, starting with the birth of the company in 1867 and ending with the violent deaths of John H. Tunstall and Alexander McSween in 1878, Murphy and his associates held only about 31 percent of the Fort Stanton quartermaster and commissary contracts. New Mexico's major mill owners controlled the flour contracts, and Lincoln County suppliers had to divide the beef contracts with outside cattle dealers. Fourteen different Lincoln County residents received hay contracts during these years.[41] Since L. G. Murphy and Company operated the only large store in the area, it had an advantage in filling grain contracts, for local farmers paid for supplies they needed with grain. Having less than a stranglehold on military contracts, the firm nonetheless depended on contracts with both the Indian Bureau and the military for cash (or government vouchers, which were as good as cash) to pay their bills to outside wholesale firms.

Within Lincoln County, L. G. Murphy and Company operated in a near cashless economy. The firm allowed purchases on credit and accepted payment in crops. A surviving ledger covering eighteen months in 1871–72 reveals a great deal about the firm's business practices. The prices it charged for goods, for example, do not appear exorbitant; moreover, the prices it paid to farmers for grain compare favorably with prices offered elsewhere in the territory.[42] The frequently quoted allegation that the firm ground down the local farmers is not sustained by the documents. Nor do military records show that Murphy and Emil Fritz, his partner, were involved in illegitimate transactions in filling army contracts at Fort Stanton. Lieutenant Colonel August V. Kautz testified in April 1872 that during the previous two and a half years, while he was commanding officer at Fort Stanton, "L. G. Murphy and Co. had filled all their contracts . . . to the entire satisfaction of the officers of the Post." Kautz also noted that Murphy had come to the assistance of the government in 1870 after fire destroyed the post's hay supply. Murphy had held the only surplus hay in the area, and instead of jacking up the price he sold it to the army at the regular contract price.[43]

Other contractors were not above trying to take advantage of Murphy, and Kautz's reporting of one attempt illustrates the danger of accepting without verification charges of misdeeds on Murphy's part. In July 1871 Peter Ott agreed to supply Fort Stanton with 300 tons of grama hay at $22.95 per ton. Ott left the area after delivering only about 171 tons, admitting that he could not fill the contract. The post quartermaster subsequently purchased about 129 tons of grama from Murphy at $35 per ton to complete Ott's contract. District Quartermaster Fred Myers would later charge that Murphy and Fritz had prevented Ott from filling his contract so that they could fill the remainder at an exorbitant price. Myers consequently refused to pay more than $30 per ton for hay Murphy had already delivered. Kautz defended the Lincoln County merchants. Earlier Kautz had predicted that Ott would be unable to fill his contract because "all the places where hay usually grew were taken by settlers." Kautz suspected that Ott had never intended to fill the contract but had wanted to sell out at a profit to Murphy and Fritz, who refused to take over the contract knowing that hay could not be delivered at Ott's price. Because Murphy and Fritz needed money, they were forced to accept Myers's reduction.[44]

This is by no means to suggest that Murphy and Fritz were entirely honest in all their business transactions. Evidence is overwhelming that they regularly inflated the number of Mescaleros receiving rations while the firm was supplying the agency with commissary supplies.[45] Kautz may have heard rumors of this practice and may even have had the Mescalero agency in mind when he later joined the army's efforts to have control of Indian affairs transferred from the Indian Bureau to the War Department. Kautz would argue that in supplying military troops the army kept profits low and contractors honest and would exert similar control over Indian suppliers if given the chance.[46]

Despite their success in winning contracts, selling supplies to the government did not make Murphy and Fritz wealthy men. The fault lay in their own sloppy business practices and their very small profit margin on many contracts. L. G. Murphy and Company was constantly in debt and always in need of money. And debts continued to mount after James J. Dolan and John H. Riley, Murphy's protégés, took over the firm in April 1877. The financial situation of J. J. Dolan and Company became even more precarious when two newcomers to the county, Tunstall and McSween, began building a large mercantile store in the town of Lincoln later that summer. The climate was potentially explosive, for in Lincoln County men had long settled differences with Winchester rifles and Colt revolvers. Faced with ruinous competition locally and in the larger arena of government contracting, Dolan and his associates schemed to undercut the new firm. Factions formed, and guns were hired. In the violence that soon engulfed Lincoln County, an estimated fifty or more men were killed.[47]

At the beginning of their careers as government contractors, Lawrence G. Murphy and Emil Fritz secured appointment as post traders at Fort Stanton. Post traders everywhere were closely tied to government contracting. They had firsthand knowledge of local conditions, and this gave them advantage over other bidders. And since they resided on the premises, traders made good agents for other contractors. Not all large contractors secured post traderships, but among those who did were Tully and Ochoa in Arizona and William H. Moore in New Mexico.

A good example of a less well-known entrepreneur is William V. B. Wardwell, who became post trader at Fort Craig in the fall of 1865, representing the firm of C. S. Hinckley, C. H. Blake, and W. V. B. Wardwell. In the previous year—a few months before mustering out of the California volunteers in Santa Fe—Wardwell had married Mary A. Watts, daughter of New Mexico's former delegate to Congress. Wardwell later became sole proprietor of the tradership; in the approximately seven years he served as trader, Wardwell held contracts to supply Fort Craig with corn, coal, lime, gypsum, and beef; Fort McRae with corn, lumber, adobes, oak wood, charcoal, and beef; Fort Cummings with charcoal and mesquite wood; and Fort Stanton with beef.[48]

Because of anticipated pecuniary rewards, competition to obtain post traderships was intense. Before 1867, trader nominations originated with a council of administration, composed of the second through fourth ranking officers at a post, with the Secretary of War either accepting or rejecting the nominations. For a short time thereafter the general commanding the army had total authority over appointments, but in 1870 this authority was lodged with the Secretary of War.[49]

Under each system residents solicited appointment by writing letters to the appropriate government agency. They also solicited recommendations from powerful military and political allies. After the post council system was dropped, politics played an even greater role in trader appointments. For instance, the well-established firm of William H. Moore and Company, appointed sutler at Fort Union in 1859, eventually lost its position to John C. Dent, brother-in-law of President Ulysses S. Grant. No stranger to the Southwest, Dent had first entered the territory during the Mexican War, commanding a company of cavalry in Colonel Sterling Price's regiment. In September 1869 General John M. Schofield, then commanding the Department of the Missouri, granted Dent authority to open a trading establishment at Fort Union. Moore at this time was among the largest taxpayers in New Mexico and heavily involved in government contracting. For several months the old and the new firm competed for military and civilian customers. Secretary of War William W. Belknap reaffirmed Dent's appointment on 7 October 1870 and a month later authorized "the firm of W. H. Moore & Co. to remain at Fort Union, as traders, until January 1, 1871, and *no longer.*"[50]

Scores of applicants must have had experiences similar to that of George W. Wahl, who unsuccessfully campaigned for the tradership at Fort Quitman. Born in Carlisle, Pennsylvania, in 1838, Wahl later graduated with honors from Dickinson College in Carlisle. He arrived in San Antonio in 1867 and most likely accompanied Co. F of the 9th Cavalry when it regarrisoned Fort Quitman the following year. At any rate, a document dated 5 May 1868 indicates that he had been granted a temporary contract to provide the men with medical attention. In the 1870 census Wahl and three others are listed as merchants at Fort Quitman; one, Charles M. Johnson, then served as post trader.[51] Sometime in that same year Dr. Wahl, as he was referred to both in the press and in military correspondence, applied for appointment as post trader. In October, however, the tradership was bestowed upon William E. Sweet, a former lieutenant in the 24th Infantry who had recently resigned from the service following a disabling wagon accident. In seeking the appointment Sweet had assured Secretary Belknap that he was "a firm adherent to the principles of the Great Republican party and . . . in full sympathy with the Administration of President Grant." Sweet also secured a letter of recommendation from Governor Rutherford B. Hayes of Ohio, Sweet's commanding officer during the war. In the short time he held the tradership, Sweet obtained contracts to supply Quitman with hay, wood, and charcoal.[52]

About the time Sweet became post trader, Wahl was compelled to move from the post, though he remained living in the neighborhood. In mid-1872 Wahl again applied for appointment as trader on grounds that Sweet had not been in residence at Quitman for nearly a year, that he was in fact living and working in Washington, D.C., and that he had disposed of his tradership to James Moore, from whom he received a monthly stipend of $200. A family friend in Washington approached Belknap on Wahl's behalf. But to no avail. Even though Sweet tendered his resignation on 9 January 1873, claiming that the tradership was no longer profitable, Belknap appointed Moore as the new trader.[53]

There is no evidence that either Sweet or Wahl distributed bonus money in their quest for the tradership. But the Belknap scandal that broke in the spring of 1876 revealed that several other applicants had done so. Belknap resigned to avoid impeachment, but Congress nonetheless began proceedings against him. Belknap was actually tried only in connection with selling the tradership at Fort Sill, Oklahoma. The trial ended in acquittal; many senators who believed Belknap guilty had voted against conviction, doubting that the Senate had jurisdiction over an individual no longer in office. Because of Belknap's alleged malfeasance, however, the old system allowing councils of administration to initiate trader nominations was restored in 1876.[54]

As made clear in previous chapters, the typical government contractor was a merchant. But rarely did a single firm dominate the military market, for large numbers of men competed for contracts. At least fifty submitted bids in 1868 for New Mexico's corn and bean contracts; forty-five bid on hay contracts in 1869; forty-four sought grain contracts in 1870.[55] Many contractors were among the wealthiest residents in the Southwest—men like Edward N. Fish, James M. Barney, and John B. Allen in Arizona, who in 1870 had possessions valued at $148,000, $31,000 and $42,000, respectively, and William H. Moore, Vicente Romero, and Abraham Staab in New Mexico, who claimed possessions worth $120,000, $59,735 and $50,000, respectively. Many others reported more modest holdings. John S. Thayer, wood contractor at Tucson, claimed property of $300; James Fleming, hay contractor at Mojave, $500.[56]

The overwhelming majority of military contracts went to Anglos, primarily because Hispanic ricos, at least in New Mexico, had their finances tied up in land and sheep raising. Of the fifty men mentioned above who submitted bids in 1868, twelve were Hispanos; eight of the forty-five bidding on hay contracts in 1869 were Hispanos; and nine of the grain bidders in 1870 were Hispanos.[57] During a ten-year period, 1866–75, Anglos held about 97 percent of all contracts issued for the west Texas posts, about 94 percent of contracts issued in Arizona, and roughly 88 percent of army contracts in New Mexico (including Fort Garland).[58]

This is not to suggest that all profits from government contracting went into the pockets of Anglos, for clearly contractors relied upon large number of Hispanic suppliers. It is not possible to show exactly what government contracts meant financially to local communities, but their importance can be suggested by focusing on contractors living in New Mexico's Mesilla Valley in 1870. Most of Doña Ana County's population of 5,864 lived in the valley's principal towns of Doña Ana, Las Cruces, and Mesilla. During 1870 nine county residents received military supply contracts. Mill owners Henry Lesinsky of Las Cruces and Daniel Frietze of Mesilla contracted to deliver a total of 123,000 pounds of flour to Forts Cummings, Bayard, and

McRae. Lesinsky, one of the largest merchants in southern New Mexico, also contracted to deliver a total of 10,300 pounds of beans to Forts Bayard and Selden. Merchants Louis Rosenbaum of Las Cruces, John D. Barncastle of Doña Ana, and James E. Griggs of Mesilla contracted to deliver a total of 1,440,000 pounds of corn to Forts Selden, Cummings, and Bayard. These contractors would have filled their contracts at least in part with corn, wheat, and beans raised by the county's 299 farmers. If filled entirely within the county, the contracts would have represented about 13 percent of the county's wheat crop, 26 percent of the corn crop, and about 12 percent of the bean crop—substantial amounts in a country barely producing sufficient crops to satisfy local markets.[59]

In addition to these contracts, Las Cruces farmer Pedro Serna and Mesilla merchants John Lemon and James Griggs contracted to deliver a total of 950 tons of grama hay to Forts Selden and Cummings. Conceivably each contractor may have employed forty or more local men as hay cutters. Pedro Serna also contracted to deliver 550 cords of wood at Fort Selden, which would have required additional laborers. John R. Johnson of Las Cruces, who listed his occupation as miner, agreed to supply Selden with 1,700 bushels of charcoal; perhaps six to ten local men would have helped him with the contract. Finally, George E. Blake, a San Agustín rancher, contracted to deliver 9,000 pounds of bacon at McRae. The total value of the contracts held by these nine men was $59,749, much of this amount going into the local economy.[60]

The nine Doña Ana County contractors, who were between the ages of 32 and 42, were typical of contractors elsewhere in the Southwest. The wealthiest, Louis Rosenbaum and Henry Lesinsky, were natives of Germany. The 1870 census shows that Rosenbaum had possessions valued at $60,000 and Lesinsky, $54,000. In contrast John R. Johnson's possessions were valued at $675. Pedro Serna is missing from the 1870 census; of the remaining contractors all but Rosenbaum and Blake were married and had children. Five of the six married men—Griggs, Frietze, Johnson, Lemon, and Barncastle—had married Hispanic women, a factor that helped smooth business transactions within their predominantly Hispanic communities.[61]

Gaps in financial records make it impossible to determine the total amount of money the army expended in the Southwest in any single year. The magnitude of military spending is suggested, however, in the figures that Secretary of War John M. Schofield released in 1868, showing the cost of maintaining the military in New Mexico and Arizona during the previous three years. Total expenditures for New Mexico dropped from $4,433,884 in 1865 to $2,779,294 in 1867; for Arizona the cost rose from $1,404,735 to $1,615,096.[62]

The quartermaster's department accounted for more than half the 1867 expenditures: $1,755,460 for New Mexico and $872,888 for Arizona. Most of this money—more than 80 percent for New Mexico and about 60 percent for Arizona—was spent in the territories for supplies, labor, transportation, and construction and repair of barracks and quarters. The pay department accounted for about 24 percent of the New Mexico expenditures ($677,986) and 30 percent of Arizona's ($495,566). Soldiers' pay undoubtedly had the greatest impact on the business of post traders, less on other local enterprises. Tucson merchant Lionel Jacobs reported in 1872, for instance: "Business is moderately fair, the paying off of the troops being slightly perceptible." Though the commissary department accounted for 12 percent of the 1867 expenditures for New Mexico ($345,847) and 15 percent of expenses for Arizona ($246,641), the amount spent for local supplies is not known.[63]

No figures comparable to Schofield's are available for later years, although General Edward O. C. Ord in 1869 and Schofield in 1871 testified that the cost of the military establishment in Arizona amounted to about three million dollars annually.[64] But army expenditures fluctuated from year to year depending on political and military objectives decided upon in the nation's capital. Following the Civil War a parsimonious Congress approved appropriation reductions through the early 1870s. Thereafter a vocal southern faction, angered by the use of troops in the South, joined congressional economizers in waging a vindictive economic war against the army. A persistent theme running through military correspondence is the need to economize. Typically the amount of money appropriated by Congress was not sufficient to run the various military departments so that during the year cuts had to be made in purchasing supplies and hiring employees. Ironically, when the forty-fourth Congress adjourned in March 1877 without passing an army appropriation bill, contractors remained relatively undisturbed since the army continued purchasing supplies on credit.[65]

During the early 1870s, as local suppliers increased production, a greater percentage of money spent for maintaining troops in New Mexico and Arizona would have entered the local economy. New Mexico's Chief Commissary Charles McClure reported that for the fiscal year ending 30 June 1869 his department had disbursed $167,213. Almost all of this was paid to local contractors and suppliers. Quartermaster expenditures are not available for this year, but Colonel Gordon Granger, commanding the District of New Mexico, reported that for the fiscal year ending 30 June 1871 the quartermaster's department had expended $521,326. The pay department during the same year had disbursed $702,852.[66] Even without knowing the exact amount the commissary department spent this year, which must have exceeded $100,000, the army infused directly into the New Mexico economy more than one and a quarter million dollars.

Statistics for other economic enterprises in New Mexico are not reliable. But some are available which help fix the position of military spending in the region's economy. New Mexico's farmers, for instance, produced crops in a twelve-month period (1869–70) valued at less than two million dollars. A decade later the value of crop production remained about the same. The value of livestock on farms rose in that decade from about two and a third million dollars to five million dollars.[67] Hubert H. Bancroft reports that the annual yield of gold and silver in New Mexico between 1869 and 1874 was $500,000; for the remainder of the 1870s, $400,000. He also reports that in 1876 trade over the Santa Fe Trail amounted to more than two million dollars.[68] Much of the trade was controlled by the same merchants who competed for military contracts. And as William J. Parish has pointed out: these merchants relied upon the army as their "principal source of eastern exchange, the only domestic exchange acceptable to eastern suppliers of merchandise."[69] Clearly the U.S. army must be ranked as one of the territory's leading industries—along with mining, ranching, farming, and merchandising.[70]

The national depression of 1873 seemingly had little effect on military spending in New Mexico. For the fiscal year ending 30 June 1873 the quartermaster's department in New Mexico was allotted $452,445. About 70 percent of this amount—$312,299—was allocated for fuel and forage; $39,803 for new buildings; $44,773 for transportation; and $39,712 for incidental expenses, which included hiring of civilian employees and extra-duty men. The amount the quartermaster's department received increased slightly each of the following two years but dropped to $113,661 for the 1875–76 fiscal year.[71] Even though troop strength also declined during the year, this amount was not enough to run the department. Midway into the fiscal year grain contracts were reduced by one fourth and the army stopped purchasing forage at government agencies.

During the 1870s the cost of maintaining a military force in Arizona was greater than the cost in New Mexico. More troops garrisoned the Arizona posts, but the disparity also resulted from greater transportation expenses and higher costs of labor and supplies. Incomplete records preclude a systematic comparison, but reliable figures are available for both territories in two instances. For the fiscal year ending 30 June 1874, the quartermaster's department in the District of New Mexico was allocated $485,469 and in the Department of Arizona, $907,974. For the fiscal year 1875–76 the allotments were $113,661 in New Mexico and $344,636 in Arizona.[72] No reliable figures are available for the commissary department, although in the calendar year 1873, the department issued contracts for beans and flour in Arizona valued at about $92,000 and for beans, flour, salt, and cornmeal in New Mexico valued at about $53,150.[73] Beef contracts did not specify quantities, so their value cannot be calculated.

Even though railroads made possible the importation of cheaper forage and subsistence stores, the amount of money spent locally during the first half of the 1880s remained substantial. For the fiscal year ending 30 June 1881, the quartermaster's department in Arizona was allotted $567,539, and for the 1884–85 fiscal year, $536,229. During the 1880–81 fiscal year the commissary department spent at least $41,744 in the territory, and during the 1883–84 fiscal year, $42,784. For most of these years the pay department would have disbursed at least $500,000 annually.[74]

It is clear that in Arizona, as in New Mexico, the military was a cornerstone of the economy during the years of this study. Having a much smaller non-Indian population than New Mexico, Arizona usually commanded a larger percentage of the army's appropriation. On the other hand, the value of farm products reported for the 1880 census in Arizona, $614,327, was only about one third that reported in New Mexico, and the value of Arizona livestock, $1,167,989, about one fourth. Mineral production in Arizona outdistanced that of its neighbor, however. Bancroft reports that gold and silver production in Arizona amounted to $500,000 in 1873 and 1874 and averaged $2,000,000 by the end of the decade.[75] The expansion that had occurred in all these enterprises since the Civil War was due in large part to the protection and encouragement that the military provided.

Indeed, for nearly forty years after General Kearny marched into Santa Fe, the army played a pivotal role in the economy of the Southwest. Robert Frazer has shown that between the Mexican and Civil Wars "the army was the single most significant factor" in the region's economic development.[76] During these years, as a direct result of the army's presence, corn, wheat, beans, and cattle production doubled in New Mexico. Also new mines were opened; lumber mills erected; and flour mills established.

The Civil War disrupted agriculture and other enterprises, but once the crisis had ended, the army resumed efforts to obtain supplies locally. It encouraged crop expansion directly by offering top prices for commodities and distributing improved seed for planting. Indirectly, it fostered expansion by the protection it afforded farm communities. With these strong incentives, farmers soon produced sufficient grain and forage to meet the army's requirements. Even in Arizona where agriculture had come to a standstill during the Civil War, by 1874 farmers were producing more forage than the army could absorb. For many years the government remained the only large purchaser of surplus agricultural commodities, but the market proved limited.

Still, for two decades following the Civil War, military spending stood as a major pillar in the economy of the Southwest. Largely as a result of army requirements, the flour milling industry expanded in both Arizona and New Mexico. So, too, did the cattle industry. Providing both market and protection, the army stimulated the growth of large-scale ranching, an industry that was soon tied to the national economy. Moreover, the military was the single largest employer of civilians in the Southwest. Even during economy drives, the military could not function without civilians. Large numbers were needed to build military installations and to accompany expeditions into the field. Moreover, scores of laborers found employment with military contractors, cutting hay and wood, making adobes, driving teams. The large sums of money that the army injected into the economy were widely distributed, reaching all segments of society. And in times of crisis, when the government reverted to open market purchases, the army functioned as a welfare agency, dispensing funds to poorer residents in exchange for small amounts of forage or fuel.

During an era that espoused laissez-faire capitalism, the Southwest developed a mixed economy—one in which the government assisted private enterprise to advance western settlement. Examples of the range of such assistance have appeared in earlier chapters—from purchasing local supplies and services to loaning tools to contractors and providing them with armed protection and soldier-laborers. This kind of federal aid injected limited financial vigor into a region that experienced chronic poverty. Not only did federal money help residents directly, but it also allowed contractors to invest in other enterprises: mining, banking, large-scale merchandising, community improvements. One observer of the region's growth during these years, New Mexico's Governor Miguel A. Otero, later recalled: "Government contracts in those times were the greatest business plums, and often two or three firms would join forces in securing one of them."[77] Indeed, government contracting was one of the few fields of investment available for the Southwest entrepreneur. And by affording frontier capitalists opportunity for success, the government helped assure continued occupation and growth of the region while it shunted Indian residents onto reservations. This is seen most clearly in Arizona, where during and immediately following the Civil War, when the intensity of Indian raids kept the population sparse, federal spending was nearly the only source of development funds.

Complete financial records for the quarter century under discussion have not been located, but in the late Sixties and early Seventies the army must have injected at least one and a quarter to two million dollars annually into the economies of New Mexico and Arizona. Certainly the competition to obtain military contracts shows that frontier capitalists considered them good business ventures. The point Leo E. Oliva made in his study of frontier forts in Kansas holds true for the Southwest; residents wanted garrisons to remain long after their military mission had ended—an indication of their economic importance.[78]

The coming of railroads reduced the cost of maintaining troops in the Southwest and allowed the importation of cheaper supplies. By the mid-Eighties fewer local businessmen were winning military supply contracts. Railroads also brought a wave of new emigrants so that by the end of the decade New Mexico's population stood at 153,593, a 67 percent increase over the 1870 population, and Arizona's at 59,620, a fivefold increase.[79] These new arrivals were eager to take part in the area's development, and business activity increased on farms, ranches, and in the mines. And with the end of the Indian wars, merchants who had supplied the once-bustling military posts transferred their commercial activities to other interests or markets. Their modern-day counterparts, however, have rediscovered the value of government contracts, and military and defense spending today is a major contributor to the region's economy. The wide-open spaces of the Southwest and the bloody contest for its resources were largely responsible for the presence of the Indian-fighting army in the nineteenth century. Those same wide-open spaces have proved attractive for modern military research and training, as well as for nuclear laboratories and a host of other research activities. The widespread distribution of federal money links the two eras.

Appendix 1

Fort Stanton grain contracts, 1866–80

Date	*Contractor*	*Type*	*Amount in 1,000 lbs.*	*Price per 100 lbs.*
October 1866	A. H. French	corn	250	$4.50
December 1867	Eugene Leitensdorfer	"	450	2.95
December 1868	L. Spiegelberg	"	600	2.29
December 1869	Louis Rosenbaum	"	150	3.06
	L. G. Murphy	"	250	3.11
September 1870	L. G. Murphy	"	450	3.50
		oats	85	4.00
September 1871	O. W. McCullough	corn	70	2.75
	Emil Fritz	"	900	3.50
		oats	50	3.95
November 1872	Emil Fritz	corn	360	2.00
		barley	40	3.00
		oats	30	4.25
October 1873	L. G. Murphy	corn	250	2.42
		barley	50	4.00
	L. Spiegelberg	corn	75	2.69
		"	100	2.85
September 1874	E. A. Dow	"	150	3.47
	James J. Dolan	"	200	3.47
		barley	20	4.47
October 1874	John Newcomb	corn	60	2.90
October 1875	W. Rosenthal	barley	30	5.00
	Paul Dowlin	corn	75	2.75
	B. J. Baca	"	465	2.83
		barley	30	5.00

Date	*Contractor*	*Type*	*Amount in 1,000 lbs.*	*Price per 100 lbs.*
November 1876	Abraham Staab	corn	250	1.62½
	James J. Dolan	"	300	1.69
November 1877	Abraham Staab	"	300	2.59
December 1877	Willie Dowlin	oats	30	3.50
July 1878	Charles Ilfeld	corn	50	4.25
	Frank Lesnet	oats	20	4.00
October 1878	Abraham Staab	corn	200	2.83
		"	100	3.19
		oats	100	3.67½
November 1878	David M. Easton	corn	100	3.19
	John Newcomb	"	50	2.90
		"	50	2.49
July 1879	Emil Fritz*	"	50	2.95
	Abraham Staab	"	150	2.97
	S. S. Terrell	"	187	3.00
		oats	200	3.50
	José Montaño	corn	150	2.82
July 1880	S. S. Terrell	"	300	2.65
		oats	50	3.25

Data from Reg. Cont., QMG, RG 92, NA.

* Son of Charles Fritz, the brother of L. G. Murphy's partner, Emil Fritz. The elder Emil Fritz (Murphy's partner) held contracts in 1871 and 1872.

Appendix 2

Fort Union Depot corn contracts, 1866–75

Date	*Contractor*	*Amount in 1,000 lbs.*	*Price per 100 lbs.*
November 1866	Eugene Leitensdorfer	2,500	$5.49
September 1868	B. C. Cutler	300	3.25
December 1868	John Dold	500	2.42
	Frank Chapman	300	2.25
December 1869	John Dold	25	1.65⅗
	Frank Chapman	400	1.70
	Louis Clark	50	2.44
	Henry Korte	150	1.50
October 1870	Vicente Romero	330	2.00
	Hugo Wedeles	100	1.98
	Charles E. Wesche	220	1.90 10/11
November 1870	Robert Allen	1,500	1.85
September 1871	William Kroenig	200	2.75
	John Dold	450	2.18
	Charles Ilfeld	100	2.17
November 1872	May Hays	200[a]	1.73
		100[b]	1.33
	Noa Ilfeld	100	1.74
	Sigmund Wedeles	100	1.65
	B. M. St. Vrain	300	1.69
	Abraham Staab	100	1.87½
	Henry Goeke	300	1.76
	Henry Korte	150[a]	1.74⅓
		50[b]	1.65
	M. E. Hoberg	100	1.90

Date	*Contractor*	*Amount in 1,000 lbs.*	*Price per 100 lbs.*
September 1873	John Dold	200	1.83
	Emanuel Rosenwald	100	1.94
	Henry Goeke	100	1.87
	Adolph Letcher	300	1.85
	Abraham Staab	300	1.89
September 1874	David Winternitz	100	2.07
		100	2.21
	Henry Goeke	75	2.18
	Emanuel Rosenwald	100	2.18
	Charles Ilfeld	100	1.98
	Sigmund Wedeles	150	1.85
	Henry Korte	75[a]	2.40
		75[c]	2.25
	Joseph B. Watrous	75[a]	2.00
	George W. Gregg	150[a]	2.35
November 1874	Abraham Staab	500	2.24
	Vicente Romero	400[a]	1.90
November 1875	May Hays	200	1.75
	Charles Ilfeld	200	1.71
	Willi Spiegelberg	300	1.79
		200	1.83
	George W. Gregg	250[a]	1.82
	Trinidad Romero	740	1.83

Data from Reg. Cont., QMG, RG 92, NA.
[a]American corn [b]Mexican corn [c]mixed corn

Appendix 3

Fort Union Depot oats contracts, 1868–75

Date	*Contractor*	*Amount in 1,000 lbs.*	*Price per 100 lbs.*
September 1868	John Dold	350	$3.40
December 1868	William Kroenig	90	3.40
December 1869	Henry Korte	20	1.75
October 1870	Henry Korte	100	1.70¾
September 1871	William Kroenig	50	2.50
	Charles Ilfeld	200	2.87
November 1872	May Hayes	150	1.43
September 1873	Sigmund Wedeles	150	1.44
September 1874	Sigmund Wedeles	50	2.39
	Joseph B. Watrous	75	2.00
November 1874	Vicente Romero	100	2.00
November 1875	May Hays	50	2.25
	Joseph B. Watrous	60	2.50

Data from Reg. Cont., QMG, RG 92, NA.

Appendix 4

Tucson Depot and Camp Lowell grain contracts, 1866–75

Date	*Contractor*	*Type*	*Amount in 1,000 lbs.*	*Price per 100 lbs.*
Feb. 1866	Charles T. Hayden	corn, wheat, or barley	500	$6.00 (coin)
Apr. 1866	John B. Allen	corn, wheat, or barley	250	5.00 (coin)
May 1866	John B. Allen	corn, wheat, or barley	500	6.50 (coin)
June 1867	Edw. Phelps	corn, wheat, or barley	600	2.99
Jan. 1868	Charles W. Lewis	corn	400	4.12 (coin)
Apr. 1868	Estévan Ochoa	corn, wheat, or barley	400	3.87
	Isaac Goldberg	corn, wheat, or barley	200	3.49
Apr. 1869	William Zeckendorf	corn	500	2.68
		barley	500	2.68
May 1870	Ewing and Head	corn	300	3.30
		barley	300	3.30
May 1871	Peter Kitchen	corn and barley	100	2.84
	Oscar Buckalew	corn and barley	50	3.33
	Estévan Ochoa	corn and barley	600	3.35
Oct. 1872	F. L. Austin	corn	100	3.75
		barley	100	3.75
Nov. 1872	P. R. Tully	corn	200	3.79
		barley	100	3.79
Apr. 1873	Charles T. Hayden	barley	200	2.43
		barley	100	2.68

Date	*Contractor*	*Type*	*Amount in 1,000 lbs.*	*Price per 100 lbs.*
	Fish and Hellings	corn	as required	2.98
		barley	as required	2.98
Apr. 1874	Thomas Ewing	corn	100	2.43
	Fish and Bennett	corn	as required	2.47
	Peter Kitchen	corn	100	2.47
	E. K. Buker	barley	as required	2.18
Apr. 1875	Ewing and Fish	barley	600	2.72
	Samuel Drachman	barley	100	2.25
		barley	100	2.37

Data from Reg. Cont., QMG, RG 92, NA.

Appendix 5

Camp McDowell grain contracts, 1867–76

Date	*Contractor*	*Type*	*Amount in 1,000 lbs.*	*Price per 100 lbs.*
August 1867	W. Bichard & Co.	barley, corn, or wheat	as required	$4.35 (coin)
August 1868	G. F. Hooper & Co.	barley	300	6.25 (coin)
		corn or wheat	200	4.70 (coin)
July 1869	Hooper, Whiting & Co.	barley	400	6.90
		barley	400	7.30
		barley	400	7.70
		corn	300	5.15
July 1870	W. B. Hellings	barley	75	6.20–7.40
July 1871	C. W. Beach	corn and barley	as required	4.20
December 1872	A. Barnett	barley	as required	4.94
April 1873	Charles T. Hayden	barley	500	2.43
	Fish and Hellings	barley	as required	3.18
		corn	as required	3.00
April 1874	P. M. Moore	barley	as required	1.97
April 1875	John Smith	barley	200	3.45
	Morris Goldwater	barley	100	2.74
	Michael Goldwater	barley	100	2.45
April 1876	Gideon Cornell	barley	400	3.30

Data from Reg. Cont., QMG, RG 92, NA.

Appendix 6

Fort Stockton hay contracts, 1867–76

Date	*Contractor*	*Amount*	*Price per ton*
July 1867	E. C. Dewey	475 tons	$29.30
July 1868	George Crosson	1500 tons	39.00
June 1869	W. B. Knox	as required	34.00
June 1870	A. F. Wulff	as required	26.00
May 1871	J. D. Burgess	400,000 pounds	21.50
	William Hagelseib	250 tons	20.45
	Thomas Johnson	200,000 pounds	17.40
		200,000 "	18.30
		200,000 "	19.25
July 1872	G. M. Frazer	50 tons	12.20
		100 "	13.94
		additional amounts	14.67
May 1873	Thomas Johnson	700,000 pounds	10.94
May 1874	Thomas Johnson	as required	9.84
June 1875	J. Friedlander	376 tons	9.00
July 1876	Cesario Torres	654 "	9.48

Data from Reg. Cont., QMG, RG 92, NA.

Appendix 7

Fort Davis hay contracts, 1867–76

Date	*Contractor*	*Amount*	*Price per ton*
July 1867	H. B. Adams	700 tons	$38.50
July 1868	A. F. Wulff	200 "	34.30
June 1869	A. F. Wulff	as required	28.75
June 1870	Adams and Wickes	as required	21.77
May 1871	William Hagelseib	1,300,000 pounds	19.95
1872	(purchases in open market)		
May 1873	A. W. Chaney	628,000 pounds	14.50
May 1874	A. F. Wulff	as required	15.50
June 1874	Albert W. Chaney	100 tons	14.97
June 1875	A. F. Wulff	460 "	13.50
July 1876	Joseph Sender	665 "	14.70

Data from Reg. Cont., QMG, RG 92, NA.

Appendix 8

Fort Quitman hay contracts, 1868–76

Date	*Contractor*	*Amount*	*Price per ton*
October 1868	T. A. Washington	500 tons	$23.00
June 1869	F. R. Diffenderfer	as required	18.90
June 1870	H. C. Smith	as required	17.00
September 1871	W. E. Sweet	800,000 pounds	13.90
May 1872	J. Moore	610,000 "	22.00
May 1873	Joseph Sender	85,000 "	13.75
June 1874	Geo. H. Abbott	66 tons	9.80
July 1875	Charles Wilson	65 "	11.95
July 1876	Ernest Fuhrmann	75 "	9.00

Data from Reg. Cont., QMG, RG 92, NA.

Appendix 9

Fort Bliss hay contracts, 1868–76

Date	Contractor	Amount	Price per ton
June 1868	W. P. Bacon	80 tons	$43.00
June 1869	F. R. Diffenderfer	as required	18.00
July 1870	F. R. and D. R. Diffenderfer	as required	17.00
May 1871	D. R. Diffenderfer	700,000 pounds	13.00
May 1872	Joseph Sender	265,000 "	14.94
May 1873	E. Angerstein	85,000 "	10.38
June 1874	M. Loewenstein	80 tons	8.90
1875	(no contract appears in register)		
July 1876	Ernest Fuhrmann	72 tons	6.95

Data from Reg. Cont., QMG, RG 92, NA.

Appendix 10

District of New Mexico—flour contracts, 1867

Post	*Contractor*	*Amount in lbs.*	*Price per lb. in cents*
Fort Bascom	Joseph Hersch	50,000	$6\frac{49}{100}$
Fort Bayard	W. H. Moore	40,000	8
Fort Craig	Elsberg and Amberg	100,000	$7\frac{49}{100}$
Fort Cummings	W. H. Moore	100,000	7
Fort Garland	Frederick Mueller	30,000	$8\frac{1}{2}$
	Ferdinand Meyer	90,000	$6\frac{95}{100}$
Fort Marcy	Joseph Hersch	40,000	$4\frac{7}{10}$
Camp Plummer	James Donavant	30,000	$7\frac{94}{100}$
Fort Selden	W. H. Moore	50,000	6
Fort Stanton	W. H. Moore	40,000	$7\frac{3}{4}$
Fort Sumner	Joseph Hersch	150,000	$6\frac{49}{100}$
Fort Union	Vicente Romero	500,000	7
	David Webster	250,000	$5\frac{87}{100}$
Fort Wingate	Franz Huning	80,000	$7\frac{45}{100}$

Data from Reg. Cont., CGS, RG 192, NA.

Appendix 11

Camp McDowell, flour contracts, 1869–85

Date	*Contractor*	*Amount in lbs.*	*Price per lb. in cents*
June 1869	Nicholas Bichard	300,000	$9\frac{25}{100}$
January 1870	Nicholas Bichard	101,250	$7\frac{3}{4}$
November 1870	Samuel H. Drachman	as required	$7\frac{99}{100}$
August 1871	Nicholas Bichard	59,129	$7\frac{15}{100}$
September 1871	Nicholas Bichard	as required	$7\frac{15}{100}$
May 1872	W. Bichard and Co.	as required	$5\frac{15}{100}$
March 1873	Edward N. Fish	190,000	$5\frac{24}{100}$
January 1874	Michael Goldwater	as required	$3\frac{85}{100}$
May 1875	William B. Hellings	60,000	$4\frac{44}{100}$
	William B. Hellings	as required	$4\frac{44}{100}$
February 1876	Charles H. Veil	as required	$6\frac{49}{100}$
March 1877	Smith and Stearns	as required	$5\frac{95}{100}$
November 1877	Hugo Richards	22,500	$4\frac{37}{100}$
May 1878	Charles T. Hayden	15,000	$3\frac{42}{100}$
August 1878	Charles T. Hayden	20,000	$3\frac{73}{100}$
March 1879	Bowers and Richards	25,000	$4\frac{1}{2}$
November 1879	Charles T. Hayden	70,000	$4\frac{3}{4}$
March 1880	Martin W. Kales	80,000	4
November 1880	Charles T. Hayden	10,000	$3\frac{7}{10}$
April 1881	John Y. T. Smith	18,000	$3\frac{1}{4}$
November 1881	John Y. T. Smith	20,000	$3\frac{1}{4}$
March 1882	John Y. T. Smith	20,000	$3\frac{1}{4}$
November 1882	J. R. Turman	25,000	$3\frac{23}{100}$
November 1883	John Y. T. Smith	30,000	$3\frac{1}{10}$
May 1884	Charles Goldman	56,000	$2\frac{89}{100}$

Date	*Contractor*	*Amount in lbs.*	*Price per lb. in cents*
October 1884	Charles Goldman	36,000	2 39/100
November 1884	Charles Goldman	40,000	2 39/100
April 1885	John Y. T. Smith	70,000	2 27/100

Data from Reg. Cont., CGS, RG 192, NA.

Appendix 12

Contracts for wagon transportation in Arizona—Route 1

Fiscal year	*Firm*	*From*	*To*	*Miles*[a]	*Rate per pound per 100 miles (in cents)*
1871–72	David Neahr	Yuma	Tucson	275	2
		Depot	Grant	268	2
			Crittenden	326	2
			McDowell	222	2
			Bowie	380	2
			Pinal	390	2
			Apache	491	2
1872–73	James M. Barney	Yuma	Tucson	——	2.30
		Depot	Grant	——	2.60
			McDowell	——	2.43
			Bowie	——	2⅜
1873–74	E. N. Fish	Yuma	Tucson	275	2.25
	&	Depot	Grant	388	2.25
	W. B. Hellings		McDowell	222	2.37
			Bowie	380	2.29
			Apache	491	3.25
1874–75	Estévan Ochoa	Yuma	Lowell	275	1.85
		Depot	Grant	388	1.85
			McDowell	222	1.85
			Bowie	380	1.85
			Apache	491	2.87
1875–76	Mariano G.	Yuma	Lowell	222	1.47
	Samaniego	Depot	Grant	391	1.47

Fiscal year	*Firm*	*From*	*To*	*Miles*[a]	*Rate per pound per 100 miles (in cents)*
			McDowell	281	1.69
			Bowie	380	1.47
			Apache	491	2.43
			San Carlos	471	1.47
1876–77	Estévan Ochoa	Yuma Depot	Lowell	222	1.73
			Grant	391	1.73
			McDowell	281	1.73
			Bowie	380	1.73
			Apache	491	2.45
			San Carlos	471	1.73
1877–78	James M. Barney[b]	Yuma Depot	Lowell	——	4.91[c]
			Grant	——	6.79[c]
			McDowell	——	3.93[c]
			Bowie	——	6.65[c]
			Apache	——	9.52[c]
			Thomas	——	7.52[c]
1878–79	James M. Barney	Yuma Depot	Lowell	——	1.60
			Grant	——	1.60
			McDowell	——	2
			Bowie	——	1.60
			Apache	——	2
			Thomas	——	1.60
			Huachuca	——	1.60
1879–80	Edward Hudson	Maricopa	Apache (via Verde)	432	3
			McDowell	51	1.12
		Casa Grande	Lowell	71	1.12
			Grant	171	1.12
			Bowie	169	1.12
			Apache (via Tucson)	290	3

Fiscal year	*Firm*	*From*	*To*	*Miles*[a]	*Rate per pound per 100 miles (in cents)*
			Thomas	213	1.12
			Huachuca	137	1.12
			Rucker	189	1.12

Data compiled from records in RG 92, RG 393, and Annual Reports of the Secretary of War.

[a] Mileage as specified in contracts.
[b] Barney's contracts ran from 1 July to 31 December 1877. It is not clear who held transportation contracts for January through June 1878. Barney also contracted to send goods directly from San Francisco to each post.
[c] Rate per pound per entire distance.

Appendix 13

Contracts for wagon transportation in Arizona—Route 2

Fiscal year	*Firm*	*From*	*To*	*Miles*[a]	*Rate per pound per 100 miles (in cents)*
1871–72	James M. Barney	Ehrenberg	Whipple	177	2.75
			Verde	216	2.75
			Date Creek	117	2.75
			Hualpai	204	2.85
1872–73	Jonathan M. Bryan & Abraham Frank	Ehrenberg	Whipple	——	2.35
			Verde	——	2.70
			Date Creek	——	2.50
			Hualpai	——	2.50
			McDowell	——	1.90
1873–74	Daniel Hazard	Ehrenberg	Whipple	——	2.49
			Verde	——	2.49
			Date Creek	——	2.35
			Hualpai	——	2.25
			McDowell	——	2.30
1874–75	R. B. Carley	Ehrenberg	Whipple	——	2.05
			Verde	——	2.14
1875–76	Charles W. Beach	Ehrenberg	Whipple	197	1.53
			Verde	243	1.65
			McDowell	214	1.75
			Apache	407	2.48
			Mojave	340	1.50
1876–77	Samuel C. Miller	Ehrenberg	Whipple	197	1.47
			Verde	243	1.49
			McDowell	214	1.63
			Mojave	340	1.49

Fiscal year	*Firm*	*From*	*To*	*Miles*[a]	*Rate per pound per 100 miles (in cents)*
1877–78	James M. Barney[b]	Ehrenberg	Whipple	——	3.27[c]
			Verde	——	4.25[c]
1878–79	James M. Barney	Ehrenberg	Whipple	——	2.25
			Verde	——	2.25
1879–80	Aaron Barnett	Maricopa	Whipple	125	1.60
			Verde	123	1.60

Data compiled from records in RG 92, RG 393, and Annual Reports of the Secretary of War.

[a] Mileage as specified in contracts.
[b] Barney's contracts ran from 1 July to 31 December 1877. It is not clear who held transportation contracts for January through June 1878. Barney also contracted to send goods directly from San Francisco to each post.
[c] Rate per pound per entire distance.

Appendix 14

Wagon transportation from points on railroads to any posts or places designated in the Territory of New Mexico

Year	*Firm*	*Rate per 100 pounds per 100 miles*
1871–72	Eugene B. Allen	$1.21
1872–73	Eugene B. Allen	1.00–1.30
1873–74	Henry C. Lovell	.80–1.00[a]
		.85–1.10[b]
1874–75	Eugene B. Allen	1.05
1875–76	Jacob Gross	.84¾

Year	*Firm*	*From*	*To*	*Miles*	*Rate per 100 pounds for whole distance*
1876–77	Jacob Gross	El Moro, Colo.	Union	100	$1.00
	Trinidad Romero	Fort Union	Santa Fe	100	.87
			Wingate	270	2.16
			Stanton	207	1.65⁶⁄₁₀
			Craig	260	2.08
			McRae	292	2.33⁶⁄₁₀
			Selden	345	2.76
			Bayard	438	3.50⁴⁄₁₀
July–Dec. 1877	Eugene B. Allen	El Moro	Santa Fe[c]	221	$2.05
			Stanton	328	3.05
			Craig	381	3.54
			Wingate	391	3.63

[a] Rates for services from points on the Atchison, Topeka and Santa Fe Railway.

Year	*Firm*	*From*	*To*	*Miles*	*Rate per 100 pounds for whole distance*
			Selden	466	4.33
			Bayard	559	5.19
Jan.–June 1878	Eugene B. Allen	El Moro	Stanton	328	\$3.94
	F. F. Struby	Garland	Santa Fe	143	2.00 (by mule) 1.25 (ox)
			Craig	303	4.00 (mule) 2.50 (ox)
			Wingate	313	4.50 (mule) 2.75 (ox)
			Selden	388	4.50 (mule) 3.25 (ox)
			Bayard	481	5.50 (mule) 4.00 (ox)
1878–79	Jacob Gross	El Moro	Stanton	328	\$2.85
	Eugene B. Allen	Alamosa	Santa Fe	138	1.33–1.40
			Craig	316	2.80–3.25
			Wingate	321	2.85–3.30
			Bayard	494	4.40$\frac{1}{10}$–5.13$\frac{45}{100}$
1879–80	Edward Fenlon	Las Vegas	Stanton	180	\$2.25
			Bliss (Tex.)	337	4.22
			Cummings	309	4.03
			Bayard	353	4.61
			Santa Fe	71	.94
			Albuquerque	101	1.34
			Wingate	225	2.99
		Alamosa	Ft. Lewis (Colo.)	88	2.05

Data compiled from records in RG 92, RG 393, and Annual Reports of the Secretary of War.

[b] Rates for services from points on the Kansas Pacific and the Denver and Rio Grande Railways.

[c] No freight contracts for Fort Union are listed because the government did most of its own freighting to this post.

Abbreviations Used in Notes

AG	Adjutant General
AAG	Assistant Adjutant General
AAAG	Acting Assistant Adjutant General
CCS	Chief Commissary of Subsistence
CGS	Records of the Office of the Commissary General of Subsistence (RG 192)
Cons. Corres. File	Consolidated Correspondence File (QMG)
CQM	Chief Quartermaster
Dept. Ariz.	Department of Arizona
Dept. NM	Department of New Mexico
Dist. Ariz.	District of Arizona
Dist. NM	District of New Mexico
End. S.	Endorsements Sent
LR	Letters Received
LS	Letters Sent
Mis. R.	Miscellaneous Records
NA	National Archives
NMSA	New Mexico State Records Center and Archives
OAG	Records of the Office of the Adjutant General (RG 94)
OIG	Records of the Office of the Inspector General (RG 159)
PR	Post Records
QMG	Records of the Office of the Quartermaster General (RG 92)
Reg. Cont.	Register of Contracts
Reg. Beef Cont.	Register of Beef and Fresh Meat Contracts
RG	Record Group
So. Dist. NM	Southern District of New Mexico
USACC	Records of the United States Army Continental Commands, 1821–1920 (RG 393)

Notes

Preface

1. William H. Goetzmann, *Army Exploration in the American West, 1803–1863* (New Haven, Conn.: Yale University Press, 1959); Francis Paul Prucha, *Broadax and Bayonet: The Role of the United States Army in the Development of the Northwest, 1815–1860* (Lincoln: University of Nebraska Press, 1953); Francis Paul Prucha, *The Sword of the Republic: The United States Army on the Frontier, 1783–1846* (New York: Macmillan Company, 1969); Robert M. Utley, *Frontiersmen in Blue: The United States Army and the Indian, 1848–1865* (New York: Macmillan Company, 1967).

2. Prucha, *Broadax and Bayonet.*

3. Robert W. Frazer, *Forts and Supplies, The Role of the Army in the Economy of the Southwest, 1846–1861* (Albuquerque: University of New Mexico Press, 1983).

4. Francis Paul Prucha, "Commentary," in *The American Military on the Frontier,* Proceedings of the Seventh Military History Symposium, United States Air Force Academy, 1976 (Washington: Office of Air Force History, Headquarters USAF, 1978), pp. 176–77.

Chapter 1

1. U.S., Department of Interior, *Eighth Census of the United States: 1860, Population,* vol. 1 (Washington: Government Printing Office, 1864), pp. 567, 571–72 [hereafter *Eighth Census, 1860,* vol. 1]; Robert W. Frazer, *Forts and Supplies, The Role of the Army in the Economy of the Southwest, 1846–1861* (Albuquerque: University of New Mexico Press, 1983) pp. 173, 185. The non–Indian population of Arizona County included 21 free blacks.

2. *Eighth Census, 1860,* vol. 1, pp. 568–71. Upon entering Mesilla in 1862, a Union soldier compared it favorably with a now much more famous metropolis; he described Mesilla as "a little larger than Los Angeles and built in the same style—adobes." See Darlis A. Miller, "Historian for the California Column: George H. Pettis of New Mexico and Rhode Island," *Red River Valley Historical Review,* 5 (Winter 1980): 80. Santa Fe was also

the site of Fort Marcy, headquarters for the Department of New Mexico.

3. *Eighth Census, 1860,* vol. 1, p. 568; Jay J. Wagoner, *Early Arizona, Prehistory to Civil War* (Tucson: University of Arizona Press, 1975), pp. 399–400; Frank C. Lockwood, *Pioneer Portraits, Selected Vignettes* (Tucson: University of Arizona Press, 1968), p. 36.

4. John A. Spring, *John Spring's Arizona,* ed. by A. M. Gustafson (Tucson: University of Arizona Press, 1966), p. 46.

5. *Eighth Census, 1860,* vol. 1, p. 568.

6. U.S., Congress, Senate, Annual Report of the Secretary of War, 1860, S. Ex. Doc. 1 (Serial 1079), 36th Cong., 2d sess., 1860, pp. 222–23. Fort Breckinridge would later become the site of the first Fort Grant. See Constance Wynn Altshuler, *Starting With Defiance, Nineteenth Century Arizona Military Posts* (Tucson: Arizona Historical Society, 1983), p. 28.

7. Frazer, *Forts and Supplies,* p. ix.

8. Invoices signed by Herbert M. Enos, Fort Union, 18, 30 July 1862, Mis. R., CQM, Dept. NM, United States Army Continental Commands (hereafter RG 393), NA. Vinegar was important as an antiscorbutic.

9. Frazer, *Forts and Supplies,* p. 168; Henry P. Walker, "Wagon Freighting in Arizona," *Smoke Signal,* 28 (Fall 1973): 190.

10. Martin Hardwick Hall, *Sibley's New Mexico Campaign* (Austin: University of Texas Press, 1960), p. 22; Canby to AAG, 11 June 1861, LR, OAG, RG 94, NA, M-619, roll 42.

11. Canby to AG, 22 December 1861, 4 May 1862, LR, OAG, RG 94, NA, M-619, rolls 43, 122; Semi-Monthly Report of Subsistence, Dept. NM, 15 November 1861, LR, OAG, RG 94, NA, M-619, roll 43.

12. Connelly to Seward, 11, 19 January 1862, Territorial Papers, NM, State Department, RG 59, NA, T-17, roll 2.

13. Santa Fe *New Mexican,* 7 November 1863, 2 April 1864; Circular, 25 June 1864, LS, First Cavalry, Volunteer Organizations of the Civil War, OAG, RG 94, NA.

14. Circular, 6 July 1861, *The War of the Rebellion: A Compilation of the Official Records of the Union and Confederate Armies,* 128 vols. (Washington, 1880–1901), Series I, vol. 4, p. 54 (*Official Records*); Santa Fe *New Mexican,* 2 April 1864.

15. See Miguel Antonio Otero, *My Life on the Frontier, 1864–1882* (reprint; Albuquerque: University of New Mexico Press, 1987), pp. 62–63; Proclamation by the Governor, 9 September 1861, Territorial Papers, NM, State Department, RG 59, NA, T-17, roll 2.

16. Santa Fe *Weekly Gazette,* 12 October 1861.

17. U.S., Congress, House, Letter from the Secretary of War, 24 April 1862, House Ex. Doc. 101 (Serial 1136), 37th Cong., 2d sess., 1862, pp. 81–85. Fort Lyon was established in August 1860 at Ojo del Oso

near modern Gallup. It was first named Fort Fauntleroy, in honor of Colonel Thomas T. Fauntleroy, and then renamed Fort Lyon after Fauntleroy resigned to join the Confederacy. The garrison was withdrawn in 1861; when troops returned in 1868, the post was renamed Fort Wingate. Robert W. Frazer, *Forts of the West, Military Forts and Presidios and Posts Commonly Called Forts West of the Mississippi River to 1898* (Norman: University of Oklahoma Press, 1965), p. 108.

18. McFerran to Donaldson, 19 October 1861, LS, CQM, Dept. NM, RG 393, NA.

19. The contracts are in Reg. Cont., QMG, RG 92, NA.

20. Roberts to Anderson, 26 August 1861, Roberts to Moore, 1 November 1861, Roberts to Morris, 9 December 1861, McRae to Ingraham, 10 November 1861, LS, So. Dist. NM, RG 393, NA; McFerran to Donaldson, 5 November 1861, LS, CQM, Dept. NM, RG 393, NA. A *fanega* was equal to about two and one half bushels.

21. McFerran to Donaldson, 22 April, 27 June 1861, LS, CQM, Dept. NM, RG 393, NA.

22. McFerran to Donaldson, 16 July 1861, *ibid.* (On Fort Fauntleroy, see note 17.)

23. McFerran to Donaldson, 22 August, 30 September 1861, *ibid.*

24. McFerran to Enos, 12, 15, 16 August 1861, *ibid.*

25. McFerran to Donaldson, 9 October, 5 November 1861, McFerran to Dalton, 11 November 1861, *ibid.*; Order No. 11, 22 September 1861, So. Dist. NM, RG 393, NA. Teamsters were paid $20 per month in 1861; within a year their salary was raised to $25 per month. Presumably these men would be paid more working as spies and guides. The standard pay for a principal guide was $3 per day. See Frazer, *Forts and Supplies*, p. 99.

26. See Stores Shipped by the Quartermaster's Department, 1861, Mis. R., CQM, Dept. NM, RG 393, NA; U.S., Congress, House, Letter from the Secretary of War, 24 April 1862, House Ex. Doc. 101 (Serial 1136), 37th Cong., 2d sess., 1862, p. 5.

27. McFerran to Donaldson, 3 October, 29 November, 23 December 1861, McFerran to Post Adjutant, 28 December 1861, LS, CQM, Dept. NM, RG 393, NA; Carey to Hatch, 30 August 1861, Fort Union, Cons. Corres. File, QMG, RG 92, NA.

28. Loring to Thomas, 23 March 1861, *Official Records*, Series I, vol. 1, pp. 599–600.

29. McFerran to 3rd Auditor, 29 January 1862, McFerran to Donaldson, 23 September, 22 October 1861, LS, CQM, Dept. NM, RG 393, NA.

30. McFerran to Donaldson, 21 November 1861, *ibid.*; McFerran to Meigs, 10 January 1863, LS to the Quartermaster General, CQM, Dept. NM, RG 393, NA.

31. Canby to Paymaster General, 18 November 1861, *Official Records*,

Series I, vol. 4, p. 75; Maxwell to Commanding General and St. Vrain to Commanding General, 27 November 1861, Registers of LR, Dept. NM, RG 393, NA; Clerry to McFerran, 1 December 1861, McFerran to Donaldson, 10 December 1861, 16 January 1862, LS, CQM, Dept. NM, RG 393, NA.

32. Thompson to Nicodemus, 7 February 1862, LR, Dept. NM, RG 393, NA.

33. Accounts of Pacheco, Martines, and Sandoval are in Records of the Adjutant General, NMSA.

34. McFerran to Donaldson, 2 January, 10, 22 February 1862, LS, CQM, Dept. NM, RG 393, NA; Enos to McFerran, 29 January 1862, Green to Paul, 12 February 1862, Arrott Collection (New Mexico Highlands University), "Fort Union, 1862."

35. McFerran to Donaldson, 20, 22 February 1862, McFerran to Stapp, 26 February 1862, LS, CQM, Dept. NM, RG 393, NA.

36. McFerran to Stapp, 28 February 1862, McFerran to Rudulph, 28 February 1862, *ibid.*

37. McFerran to Meigs, 21 March 1862, *ibid.*

38. McFerran to Meigs, 15 March, 28 April 1862, McFerran to Canby, 3 May 1862, *ibid.* When McFerran relieved Lieutenant Colonel James L. Donaldson as chief quartermaster in the fall of 1862, the indebtedness of the quartermaster's department in New Mexico was about $525,000. McFerran to Meigs, 26 July 1865, *Official Records,* Series III, vol. 5, pp. 444–47.

39. Connelly to Seward, 13 April 1862, *Official Records,* Series I, vol. 9, p. 663; McFerran to Canby, 5 May 1862, LS, CQM, Dept. NM, RG 393, NA; Claims Against the Government, Mis. R., CQM, Dept. NM, RG 393, NA.

40. McFerran to Hodges, 26 May 1862, McFerran to Van Vliet, 26 May 1862, McFerran to Chapin, 26 May 1862, LS, CQM, Dept. NM, RG 393, NA.

41. Purchases Made at Santa Fe, 1862, Mis. R., CQM, Dept. NM, RG 393, NA.

42. Hall to Chapin, 18 June 1862, LR, Dept. NM, RG 393, NA, M-1120, roll 16.

43. Special Order No. 27, 7 May 1862, So. Dist. NM, RG 393, NA.

44. Howe to Chapin, 1 August 1862, LS, So. Dist. NM, RG 393, NA.

45. Testimony of John C. Stedman, Charles Hinckley, Charles G. Parker, 23 July 1862, Hovey to Canby, 7 August 1862, LR, Dept. NM, RG 393, NA. Rio Abajo refers to settlements below Santa Fe.

46. Accounts of Romero, Baca y Castillo, and Torres are in Records of the Adjutant General, NMSA.

47. Howe to Chapin, 5 August 1862, LS, So. Dist. NM, RG 393, NA; Inclosure No. 2 in Canby to AG, 21 June 1862, Carleton to Drum, 22 July 1862, Steele to Coo-

per, 12 July 1862, *Official Records,* Series I, vol. 9, pp. 676–78, 554–55, 721–22.

48. Hall, *Sibley's New Mexico Campaign,* pp. 185–86; Canby to AG, 10 May, 21 June 1862, *Official Records,* Series I, vol. 9, pp. 670–71, 676–77.

49. Carleton to Drum, 21 December 1861, *Official Records,* Series I, vol. 50, pt. 1, pp. 773–80. For Carleton's earlier career in New Mexico, see Aurora Hunt, *Major General James Henry Carleton, 1814–1873, Western Frontier Dragoon* (Glendale, Ca.: Arthur H. Clark Company, 1958), pp. 113–70.

50. Frank Russell, *The Pima Indians* (reissue; Tucson: University of Arizona Press, 1975), pp. 20–21.

51. Henry F. Dobyns, "Pima and Maricopa Indian Economic Genius in 1849," paper presented at the Western History Association Conference, 1982, pp. 4, 7–11.

52. J. Ross Browne, *Adventures in the Apache Country, A Tour Through Arizona and Sonora, 1864* (re-edition; Tucson: University of Arizona Press, 1974), pp. 107–11.

53. Rigg to Cutler, 1 March 1862, *Official Records,* Series I, vol. 50, pt. 1, pp. 898–99. A *fanega* was an imprecise unit of weight. In New Mexico by 1864 the army had specified that a *fanega* of wheat was 120 pounds while a *fanega* of corn was 140 pounds. Some explanation for the variance of weights is given in an advertisement for Indian supplies appearing in the 15 November 1862 issue of the Santa Fe *Weekly Gazette:* "The corn and wheat must be of good quality, and be measured in the usual way, by heaping the corn on the measure and the wheat by level measure."

54. Wagoner, *Early Arizona,* p. 452.

55. Santa Fe *Weekly Gazette,* 26 July 1862; West to Cutler, 4 May 1862, *Official Records,* Series I, vol. 50, pt. 1, p. 1050.

56. West to Cutler, 5 May 1862, *Official Records,* Series I, vol. 50, pt. 1, p. 1052.

57. West to Cutler, 13 May 1862, *ibid.*, pp. 1070–71.

58. West to Cutler, 26 May 1862, *ibid.*, pp. 1100–101; West to Cutler, 27 May 1862, Arrott Collection, "Fort Union, 1862, California Column." Until about 1870 the army in Arizona accepted bids for army supplies both in gold coin and in U.S. currency. The value of "greenbacks" fluctuated, however. In 1864 sutlers at Fort Whipple accepted greenbacks at fifty cents on the dollar. In 1867 an officer at Camp McPherson said a greenback there was worth seventy-five cents in coin. It is not always clear in government documents whether contract prices were for coin or currency. Consequently commodity prices for these years are not always comparable. Payment in coin or in currency will be indicated whenever such information is available. After 1870 the army usually accepted

bids only in currency. See Prescott *Arizona Miner,* 5 October 1864; [name unclear] to Hobart, 10 August 1867, LR, Subdistrict of Prescott, RG 393, NA; Special Order No. 27, 20 June 1870, Fort Mojave, PR, RG 393, NA.

59. Fergusson to Kellogg, 5 September 1862, Arizona, Cons. Corres. File, QMG, RG 92, NA; White to Rigg, 21 February 1862, *Official Records,* Series I, vol. 50, pt. 1, pp. 899–900.

60. Carleton to Fergusson, 11 June 1862, Fergusson to West, 25 June 1862, *Official Records,* Series I, vol. 50, pt. 1, pp. 1133, 1159–162.

61. McFerran to Roberts, 19 June 1862, LS, CQM, Dept. NM, RG 393, NA; Canby to Carleton, 9 July 1862, Carleton to Canby, 2 August 1862, *Official Records,* Series I, vol. 9, pp. 682–83, 557–59; Stores Shipped by the Quartermaster's Department, 1862, Mis. R., CQM, Dept. NM, RG 393, NA. Each Government wagon was drawn by six yokes of oxen.

62. Santa Fe *Weekly Gazette,* 21 June 1862.

63. See Reg. Beef Cont., CGS, RG 192, NA; Santa Fe *Weekly Gazette,* 26 July 1862, 18 October 1862.

64. Entries for Moore, 22 February 1862, Winsor, 5 March 1862, and Dold, 30 April 1862, Reg. Cont., QMG, RG 92, NA; McFerran to Mills, 3 May 1862, McFerran to Donaldson, 14 June 1862, LS, CQM, Dept. NM, RG 393, NA.

65. West to Cutler, 23 September 1862, *Official Records,* Series I, vol. 50, pt. 2, pp. 132–33; West to Rohmann, 21 September 1862, LS, Dist. Ariz., RG 393, NA. The Jornada del Muerto, "Journey of the Dead Man," was a near-waterless ninety-mile stretch of the old Chihuahua–Santa Fe trade route that extended from modern Rincon in the south to San Marcial in the north.

66. Willis to Rynerson, 25 October 1862, LR, Subdistrict of Southern Arizona, RG 393, NA; West to Willis, 27 November 1862, LS, Dist. Ariz., RG 393, NA; West to Cutler, 5 February 1863, LR, Dept. NM, RG 393, NA, M-1120, roll 21.

67. McFerran to St. Vrain, 3 November 1862, LS, CQM, Dept. NM, RG 393, NA; Letter of St. Vrain, 10 November 1862, Endorsements Sent, CQM, Dept. NM, RG 393, NA.

68. McFerran to Act. Asst. Quartermaster, Fort Stanton, 5 November 1862, McFerran to Van Vliet, 27 December 1862, McFerran to Berney, 12 January 1863, LS, CQM, Dept. NM, RG 393, NA; McFerran to Meigs, 2 February 1865, LS to the Quartermaster General, CQM, Dept. NM, RG 393, NA; Entries for Barkley, Brient, Aguirre, 5 December 1862, Reg. Cont., QMG, RG 92, NA.

69. Contracts are in Reg. Cont., QMG, RG 92, NA; McMullen to West, 6 December 1862, LR, OAG, RG 94, NA, M-619, roll 123.

70. Carleton to West, 18 November 1862, *Official Records,* Series I,

vol. 15, pp. 599–601; General Order No. 24, 2 December 1862, *ibid.*, vol. 50, pt. 2, pp. 239–40.

71. Durán, Turado, and Trujillo to West, 6 December 1862, LR, Dist. Ariz., RG 393, NA.

72. West to Cutler, 7 December 1862, LS, Dist. Ariz., RG 393, NA. On the price of flour, see for example, letter of J. Hinckley, 8 September 1865, Register of LR, Inspector, Dept. NM, RG 393, NA.

73. Mullen to West, 21 December 1862, LR, Dist. Ariz., RG 393, NA.

74. West to Fergusson, 4 February 1863, and West to Rigg, 19 February 1863, LS, Dist. Ariz., RG 393, NA.

75. Carleton to St. Vrain, 9 December 1862, LR, OAG, RG 94, NA, M-619, roll 123; Cutler to Vaca, 19 December 1862, Anderson to Young, 30 May 1863, LS, Dept. NM, RG 393, NA, M-1072, roll 3; Special Order No. 1, 1 January 1863, Arrott Collection, "Fort Union, 1862"; Rigg to Carleton, 7 March 1863, LR, Dept. NM, RG 393, NA, M-1120, roll 20; Hatton to Russell, 21 March 1863, LR, Fort Craig, PR, RG 393, NA.

76. Carleton to Elsberg and Amberg, 15 December 1862, Carleton to Martin, 27 September 1862, Arrott Collection, "Fort Union, 1862"; General Order No. 4, 3 February 1863, LR, OAG, RG 94, NA, M-619, roll 283.

77. Wagoner, *Early Arizona*, p. 392; C. L. Sonnichsen, *Tucson, The Life and Times of an American City* (Norman: University of Oklahoma Press, 1982), p. 52; Carl Trumbull Hayden, *Charles Trumbull Hayden, Pioneer* (Tucson: Arizona Historical Society, 1972), pp. 1–9.

78. West to McFerran, 4 May 1863, LS, Dist. Ariz., RG 393, NA. West was promoted to brigadier general on 25 October 1862.

79. Carleton to West, 1 January, 16 October 1863, LR, Dist. Ariz., RG 393, NA.

80. Hayden to Carleton, 3 March 1863, *ibid.*; West to Cutler, 23 October 1863, LS, Dist. Ariz., RG 393, NA.

81. Carleton to West, 13 November 1863, LR, Dist. Ariz., RG 393, NA.

82. Jones to Carleton, 15 September 1863, LR, Dept. NM, RG 393, NA, M-1120, roll 19; West to Carleton, 8 October 1863, LS, Dist. Ariz., RG 393, NA.

83. Carleton to Garrison, 4 December 1863, LS, Dept. NM, RG 393, NA, M-1072, roll 3; Jones to Carleton, 15 September 1863, LR, Dept. NM, RG 393, NA, M-1120, roll 19.

84. McFerran to Archer, 28 November 1863, McFerran to Carleton, 10 February 1864, LS, CQM, Dept. NM, RG 393, NA; Testimony of F. L. Bronson, 21 December 1863 and Carleton to Davis, 18 December 1863, LR, OAG, RG 94, NA, M-619, roll 283.

85. Angerstein to McFerran, 11 December 1863, Testimony of Angerstein, 25 December 1863, Testimony of W. W. Mills, 24 Decem-

ber 1863, LR, OAG, RG 94, NA, M-619, roll 283.

86. See Testimony of W. W. Mills, 22 December 1863 and Testimony of F. L. Bronson, H. Cuniffe, M. L. Glasby, 21 December 1863, *ibid.*

87. West to Davis, 24 December 1863, *ibid.* No bid for less than 500 *fanegas* was considered.

88. Testimony of Juan N. Zubirán, 21 December 1863, Davis to Carleton, 29 January 1864, LR, OAG, RG 94, NA, M-619, roll 283.

89. DeForrest to Bowie, 6 June 1864, *ibid.*

90. Estimates of funds required for both the subsistence and quartermaster's departments for 1864 are found in Consolidated Reports of Estimates of Funds Required by Quartermasters, CQM, Dept. NM, RG 393, NA. Neither estimate included the cost of supplies that the army shipped to Fort Union from its eastern supply depots.

91. The hay contracts are in Reg. Cont., QMG, RG 92, NA, and the beef contracts are in Reg. Beef Cont., CGS, RG 192, NA. For Allen's grain contract, see Dana to Colfax, 3 March 1865, U.S., Congress, House Ex. Doc. 84 (Serial 1230), 38th Cong., 2d sess., 1865, p. 69. Arizona became a territory separate from New Mexico in February 1863.

92. Mason to Drum, 29 April 1866, LS, Dist. Ariz., RG 393, NA.

93. Ffrench to West, 15 July 1863, LR, Dist. Ariz., RG 393, NA; Brady to Whitlock, 15 March 1863, *Official Records,* Series I, vol. 50, pt. 2, p. 354.

94. Ffrench to West, 15 July 1863, LR, Dist. Ariz., RG 393, NA.

95. Fergusson to Bennett, 2 April 1863, *ibid.*; West to Coult, 8 September 1863, West to Cutler, 8 September 1863, LS, Dist. Ariz., RG 393, NA.

96. Browne, *Adventures in the Apache Country,* p. 111.

97. Davis to Coult, 1 March 1864, LS, Inspector, Dept. NM, RG 393, NA. See also Coult to Davis, 10 March 1864, White to Poston, 1 February 1864, Endorsement by Toole, 11 April 1864, enclosed in Special Investigation of Complaints made by Charles D. Poston, 1864, LR, OAG, RG 94, NA, M-619, roll 285.

98. Cutler to Woods, 12 August 1862, LR, Dept. NM, RG 393, NA, M-1120, roll 17; Davis to Beard, 28 January 1863, LR, Dist. Ariz., RG 393, NA; West to Whitlock, 29 January 1863, LS, Dist. Ariz., RG 393, NA.

99. Coult to Bennett, 14 December 1863, LR, Dept. NM, RG 393, NA, M-1120, roll 18.

100. Hinds to Rigg, 20 April 1863, LR, Fort Craig, PR, RG 393, NA.

101. Carleton to McFerran, 25 February 1863, LS, Dept. NM, RG 393, NA, M-1072, roll 3; West to Dresher, 22 April 1863, LS, Dist. Ariz., RG 393, NA; Santa Fe *Weekly Gazette,* 22 August 1863; Carleton to Brady, 27 April 1865, LR, Fort Stanton, PR, RG 393, NA.

102. Contracts are in Reg. Cont. and Reg. Beef Cont., CGS, RG 192, NA.

103. West to McFerran, 12 April 1863, LS, Dist. Ariz., RG 393, NA; Carleton to West, 3 September 1863, LS, Dept. NM, RG 393, NA, M-1072, roll 3; Cooper to West, 15 September 1862, Webb to West, 23 October 1863, LR, Dist. Ariz., RG 393, NA.

104. Carleton to Curtiss, 16 September 1864, LS, Dept. NM, RG 393, NA, M-1072, roll 3; Stevens to Carleton, 3 February 1865, LR, Dept. NM, RG 393, NA, M-1120, roll 26; Entries for Hinckley, 1 September 1864, Reg. Beef Cont., CGS, RG 192, NA.

105. Bowie to Carleton, 17 June, 8 July 1864, LS, Dist. Ariz., RG 393, NA; Chaves to Garrison, 8 June 1863, Garrison to [name unclear], 26 February 1864, LR, Dept. NM, RG 393, NA, M-1120, rolls 18, 24.

106. Contracts are in Reg. Cont., QMG, RG 92, NA; Lemon to Bennett, 29 October 1863, LR, Dist. Ariz., RG 393, NA; McDermott to Carey, 18 August 1864, LR, Dept. NM, RG 393, NA, M-1120, roll 24; Santa Fe *Weekly Gazette*, 29 August 1863.

107. Entry for Hunter, 21 December 1863, Reg. Cont., QMG, RG 92, NA.

108. McFerran to Stevens, 5 November 1863, McFerran to Butler, 10 December 1863, McFerran to Morrison, 11 December 1863, McFerran to Morton, 12 December 1863, LS, CQM, Dept. NM, RG 393, NA.

109. McFerran to Meigs, 10 December 1863, LS to the Quartermaster General, CQM, Dept. NM, RG 393, NA; McFerran to St. Vrain, 12 December 1863, LS, CQM, Dept. NM, RG 393, NA.

110. McFerran to Mitchell, 5 February, 10 March 1864, McFerran to Morton, 13 December 1863, McFerran to Dittenhoefer and Cohen, 8 January 1864, [name unclear] to Davis, 22 February 1864, LS, CQM, Dept. NM, RG 393, NA; Santa Fe *New Mexican*, 23 January 1864.

111. Carleton to Wallen, 9 March 1864, U.S., Congress, Senate, "Condition of the Indian Tribes," 1867, S. Report 156, (Serial 1279), 39th Cong., 2d sess., 1867, p. 164.

112. Proceedings of a Board of Officers Convened at Santa Fe, New Mexico, 23 June 1864, LR, OAG, RG 94, NA, M-619, roll 286; Gerald Thompson, *The Army and the Navajo, The Bosque Redondo Reservation Experiment, 1863–1868* (Tucson: University of Arizona Press, 1976), p. 42. Thompson provides a good survey of contractors who furnished supplies for the Indians at Bosque Redondo.

113. Entry for Kitchen, 19 March 1864, Reg. Cont., QMG, RG 92, NA; Enos to Davis, 27 March 1864, LS, CQM, Dept. NM, RG 393, NA; Carleton to Thomas, 3 April 1864, LS, Dept. NM, RG 393, NA, M-1072, roll 3.

114. McFerran to Enos, 18 October 1864, McFerran to Rynerson, 24, 28 October 1864, McFerran to Enos, 29 October 1864, McFerran to Enos, 3 December 1864, LS, CQM, Dept. NM, RG 393, NA.

115. Entry for Dold, 2 January 1865, Reg. Cont., CGS, RG 192, NA.

116. Sacket to Hardie, 8 July 1865, Inspection Reports, OIG, RG 159, NA; Santa Fe *New Mexican,* 9 June 1865; Entry for Dold, 28 July 1865, Reg. Cont., QMG, RG 92, NA. The anticipated grain shortage did not occur. According to the Santa Fe *Weekly Gazette* of 2 December 1865, the territorial wheat crop failed entirely, but the corn crop, "contrary to all anticipation, matured and gave a good yield. This was owing to the favorable fall weather with which we were blessed."

117. Robert W. Frazer, "The Army and New Mexico Agriculture, 1848–1861," *El Palacio,* 89 (Spring 1983): 25–29.

118. The corn that Andres Dold delivered for the Indians at Fort Sumner in 1865 cost the army more than $30 per *fanega;* that which he delivered at Fort Union cost almost $26 per *fanega.*

Chapter 2

1. Romero to Carleton, 18 March 1866, LR, Dist. NM, RG 393, NA, M-1088, roll 4.

2. Carleton to Romero, 8 April 1866, LS, Dist. NM, RG 393, NA, M-1072, roll 3.

3. Robert M. Utley, *Frontier Regulars, The United States Army and the Indian, 1866–1891* (New York: Macmillan Publishing Co., 1973), pp. 12–16; Russell F. Weigley, *History of the United States Army* (New York: Macmillan Company, 1967), p. 567.

4. Abstract from return of the Department of New Mexico (May 1865), *The War of the Rebellion: A Compilation of the Official Records of the Union and Confederate Armies,* 128 vols. (Washington, 1880–1901), Series I, vol. 48, pt. 2, p. 713 *(Official Records)*; Abstract from the return of the Department of the Pacific (April 1865), *Official Records,* Series I, vol. 50, pt. 2, p. 1217.

5. Raphael P. Thian, *Notes Illustrating the Military Geography of the United States, 1813–1880* (reprint; Austin: University of Texas Press, 1979), pp. 52, 54, 79–80, 99, 156–57. Constance Wynn Altshuler, in *Chains of Command, Arizona and the Army, 1856–1875* (Tucson: Arizona Historical Society, 1981), untangles the many changes in command structure for Arizona.

6. Position and distribution of troops can be found in the annual reports of the Secretary of War. For the figures recorded here see U.S., Congress, House, Annual Report of the Secretary of War, 1870, House Ex. Doc. 1 (Serial 1446), 41st Cong., 3d sess., 1870, pp. 68–69, 86–87; U.S., Congress, House, Annual Report of the Secretary of War, 1876, House Ex. Doc. 1 (Serial 1742), 44th Cong., 2d sess., 1876, pp. 42–

43, 56–57; U.S., Congress, House, Annual Report of the Secretary of War, 1886, House Ex. Doc. 1 (Serial 2461), 49th Cong., 2d sess., 1886, pp. 92–95. The combined troop strength for Arizona and New Mexico reached a decade low in 1877 (2,116). In 1875, however, only 682 men were assigned to the District of New Mexico, while 1,480 served in the Department of Arizona, for a combined strength of 2,162.

7. Bell to Cutler, 25 July 1865, LR, Dept. NM, RG 393, NA, M-1120, roll 26.

8. Memorial enclosed in Carter to General Grant, 16 October 1867, Arizona, Cons. Corres. File, QMG, RG 92, NA. In both Arizona and New Mexico, newspaper editors spoke eloquently on behalf of farmers, asserting that they (and not contractors) deserved all profits from government spending. Farmers themselves petitioned for an end to the contract system. Cultivators living near Fort Union wrote in 1874 to Colonel J. Irvin Gregg, commanding the District of New Mexico, protesting that they should not be compelled to sell their produce to speculators and middlemen at "ruinously low prices." See, for example, Santa Fe *Daily New Mexican,* 20 October 1874; Tucson *Arizona Citizen,* 7, 14 June 1873; Williams *et al.* to Gregg, 18 August 1874, LR, Dist. NM, RG 393, NA, M-1088, roll 23.

9. Endorsement by Devin, 7 October 1869, End. S., Subdistrict of Southern Arizona, RG 393, NA; Rucker to Allen, 17 December 1867, Arizona, Cons. Corres. File, QMG, RG 92, NA; Prescott *Arizona Miner,* 29 June 1867.

10. Hatch to [name unclear], 9 August 1877, LR, CQM, Dist. NM, RG 393, NA.

11. Foster to Stone, 30 October 1870 and Foster to Meigs, 22 February 1871, LS by Quartermaster, Subdistrict of Northern Arizona, RG 393, NA; McClure to Secretary of War, 31 January 1870, LS, CCS, Dist. NM, RG 393, NA. When the army threatened to withhold $235 from the pay of Captain Charles W. Foster, quartermaster at Fort Whipple, for expenses incurred in 1869 and 1870 by advertising in the Prescott *Arizona Miner,* Foster replied that advertising in the Prescott paper was the only means of reaching farmers north of the Gila River as none subscribed to the *Arizonan.*

12. Nash to Chief Clerk, War Department, 18 October 1872, LS, CCS, Dist. NM, RG 393, NA. The editor of the Las Cruces *Borderer* claimed that only a handful of businessmen in the Mesilla Valley subscribed to the Santa Fe *New Mexican.* But he also voiced the opinion that the government's policy of bestowing patronage on a single territorial newspaper was a conspiracy to crush Democratic journals like his own. Las Cruces *Borderer,* 21 December 1872.

13. Prescott *Arizona Miner,* 14 March 1866; Rucker to Allen, 13 November 1867, LR, Camp Crittenden, PR, RG 393, NA.

14. Santa Fe *Weekly Gazette,* 20

May 1865; Tucson *Weekly Arizonan,* 21 February 1869, 9 April 1870; Tucson *Arizona Citizen,* 1 March 1873.

15. Tucson *Arizona Citizen,* 1 March 1873; Silver City *Grant County Herald,* 18 April 1875; Eagan to Acting Assistant Commissary of Subsistence, 1 May 1878, Telegrams, Fort Bowie, PR, RG 393, NA; Belcher to Rosenbaum, 31 October 1876 and Dana to Meeson and Marriage, 27 January 1880, LS, CQM, Dist. NM, RG 393, NA. Late in 1866 posts called "forts" in Arizona became "camps" (with the exception of Fort Whipple). The camps were again designated forts in 1879. See Altshuler, *Chains of Command,* p. 62; Robert W. Frazer, *Forts of the West, Military Forts and Presidios and Posts Commonly Called Forts West of the Mississippi River to 1898* (Norman: University of Oklahoma Press, 1965), pp. 3–15.

16. Tucson *Weekly Arizonan,* 4 December 1869; Kitchen to Ludington, 12 November 1868, LR, CQM, Dist. NM, RG 393, NA; Ludington to Lesinsky, 9 February 1870 and Myers to Grimes, 17 April 1872, LS, CQM, Dist. NM, RG 393, NA. Camp Toll Gate, established in 1869, was named Camp Hualpai in 1870. See Constance Wynn Altshuler, *Starting With Defiance, Nineteenth Century Arizona Military Posts* (Tucson: Arizona Historical Society, 1983), pp. 32–33.

17. See Proposal of George P. Armstrong, 23 April 1881, Fort Bayard, Cons. Corres. File, QMG, RG 92, NA.

18. See, for example, Jacobs & Co. to Pierson, 4 August 1871 (Jacobs Manuscripts, University of Arizona); Endorsement on Hellings's letter dated 3 June 1872, LS, Dept. Ariz., RG 393, NA.

19. McClure to Randall, 9 July 1867, LR, Fort Union, PR, RG 393, NA.

20. McClure to Hunter, 14 October 1867, McClure to Jones, 11 October 1867, McClure to Rigg, 14, 28 October 1867, 16 January 1868, LS, CCS, Dist. NM, RG 393, NA. Troops were allowed to sell savings they made in their rations and in this way accumulate company funds.

21. McClure to Corbin, 15 September 1868, McClure to Crawford, 12 January 1869, LS, CCS, Dist. NM, RG 393, NA.

22. Robert W. Frazer, *Forts and Supplies, The Role of the Army in the Economy of the Southwest, 1846–1861* (Albuquerque: University of New Mexico Press, 1983), p. 72.

23. Norris to Davidson, 28 June 1862, LR, Dept. NM, RG 393, NA, M-1120, roll 17; Coues to Thompson, 24 April 1865, LS, Fort Whipple, PR, RG 393, NA. See also West to Enos, 2 April 1863, LS, Dist. Ariz., RG 393, NA.

24. Dubois to Lane, 3 July 1867, Fort Union, Cons. Corres. File, QMG, RG 92, NA; McKee to De-Forrest, 4 August 1867, LR, Inspector General, Dist. NM, RG 393, NA; McClure to Morgan, 7 October 1867, LS, CCS, Dist. NM, RG 393, NA; Price to Drum, 3 August 1867, LS, Fort Mojave, PR, RG 393, NA;

Smart to Post Adjutant, 17 March 1867, LR, Fort McDowell, PR, RG 393, NA.

25. McKee to DeForrest, 4 August 1867, LR, Inspector General, Dist. NM, RG 393, NA.

26. See War Department, Surgeon General's Office, *A Report on Barracks and Hospitals,* Circular No. 4 (Washington: Government Printing Office, 1870).

27. Cremony to Green, 1 January 1866, LR, Dist. Ariz., RG 393, NA.

28. Brown to AAA General, 10 August 1866, LR, Dist. Ariz., RG 393, NA.

29. Middleton to Reilly, 21 January 1868, LR, Camp Date Creek, PR, RG 393, NA. Camp McPherson, established in 1867, was renamed Camp Date Creek in 1868. Altshuler, *Starting With Defiance,* p. 25.

30. War Department, Surgeon General's Office, *Report on the Hygiene of the United States Army,* Circular No. 8 (Washington: Government Printing Office, 1875), pp. xxxvii–xxxviii.

31. Las Cruces *Borderer,* 14 February 1872; [name unclear] to AAG, 27 October 1875, LS, Fort Selden, PR, RG 393, NA.

32. *Report on the Hygiene of the United States Army,* p. 259; Medical Records, October 1875, August 1876, Fort Stanton, OAG, RG 94, NA.

33. *Report on Barracks and Hospitals,* pp. 226, 229–31, 235; Bentzoni to AAG, 4 May 1875, LS, Fort Quitman, PR, RG 393, NA; Shafter to Wood, 6 November 1871, LR, Fort Davis, PR, RG 393, NA; Wade to Caziarc, 9 March 1869, LS, Fort Stockton, PR, RG 393, NA. For a description of Fort Stockton's flourishing garden in 1874, see Clayton W. Williams, *Texas' Last Frontier, Fort Stockton and the Trans–Pecos, 1861–1895* (College Station: Texas A&M University Press, 1982), p. 191.

34. *Report on Barracks and Hospitals,* pp. 468, 472; *Report on the Hygiene of the United States Army,* pp. 533, 549; Endorsement by Dunn, 3 August 1870, End. S., Fort Bowie, PR, RG 393, NA.

35. *Report on the Hygiene of the United States Army,* pp. 540, 545, 553, 556; Campbell to Post Adjutant, 16 February 1878, LR, Fort Verde, PR, RG 393, NA.

36. See three letters by Carr dated 27 January 1872 to Steiner and Klauber, to Moore and Carr, and to Simpson, LS, Fort McDowell, PR, RG 393, NA.

37. Frazer, *Forts and Supplies,* pp. 61–72.

38. Special Order, 12 March 1866, Fort Goodwin, LR, Dist. Ariz., RG 393, NA; Altshuler, *Starting with Defiance,* pp. 27–28.

39. Jones to Fry, 1 June 1867, O'Beirne to Wright, 5 April 1868, LR, District and Subdistrict of Tucson, RG 393, NA.

40. Prescott *Arizona Miner,* 28 February, 11 April 1866.

41. Wallen to Drum, 6 July 1866, LS, Dist. Ariz., RG 393, NA; AAQM to Tuttle, 13 October 1866,

LR, Fort McDowell, PR, RG 393, NA.

42. Bennett to Ilges, 31 March 1867, Hancock to Grant, 22 July 1867, LR, Fort McDowell, PR, RG 393, NA. Hancock's assistant, a Mr. Thomas, received $100 a month, and twenty farm laborers (five of them Hispanos) were paid $45 each. List of Citizen Employees, August 1867, LR, Fort McDowell, PR, RG 393, NA. The salaries were calculated in coin. Bennett mustered out of the California volunteers in August 1866. He soon obtained a commission as a second lieutenant in the 6th Cavalry and was detailed to Camp McDowell as quartermaster and superintendent of the government farm. In 1867 he was transferred to the 17th Infantry and left McDowell. See Altshuler, *Chains of Command,* pp. 237–38. Hancock later settled in Phoenix, became a lawyer, and farmed extensively in the Salt River Valley. See *The Taming of the Salt* (Phoenix: Communications and Public Affairs Department of Salt River Project, 1979), pp. 29–32.

43. Jones to Fry, 8 June 1867, Inspection Reports, OIG, RG 159, NA. Prices for forage are in coin.

44. Rusling's report is enclosed in McFerran to Dana, 1 August 1867, Fort McDowell, Cons. Corres. File, QMG, RG 92, NA. Rusling does not mention Jones by name, but the calculations examined by Rusling apparently were the same given to Jones.

45. The contracts are in Reg. Cont., QMG, RG 92, NA. The rates for Hooper and Mason are specified in coin; the rates for Smith appear without specification for coin or currency. A former West Pointer, George F. Hooper retired from the company in May 1867, leaving his brothers, Joseph and William, and Augustus H. Whiting, Francis J. Hinton, and James M. Barney, to carry on the business. Major William B. Hooper had served as quartermaster in Arizona and joined the firm in 1866 after resigning from the army. See Prescott *Arizona Miner,* 14 February, 10 November 1866, 2 November 1867, 11 January 1868.

46. Hardt to Post Adjutant, 25 July 1877, LR, Fort McDowell, PR, RG 393, NA; Cunningham to Hardt, 10 August 1877, LS, Fort McDowell, PR, RG 393, NA. In 1879 the farm was rented to W. Sheridan and C. C. Eyster for $200 a year, the partners agreeing to keep the ditches and dams in good repair and to furnish enough water to irrigate the post garden. Sheridan and Eyster employed four Hispanos to work the farm, including Jesús Soto who lived there with his family. In later years the farm was leased to Benjamín Velasco, a long-time resident of Maricopa County who also farmed lands adjacent to Fort McDowell. Chaffee to President, Post Council, 30 December 1878, Corliss to AAG, 24 April 1878, Chaffee to AAG, 9 September 1882, Biddle to Chief Quartermaster, 27 January 1884, LS, Fort McDowell, PR, RG 393, NA; Sheridan and Eyster to Commanding Officer, 6 December 1878 and 10 June 1879,

LR, Fort McDowell, PR, RG 393, NA. John Smith in 1879 added the initials Y.T. to his name. Hereafter, John Smith and John Y. T. Smith refer to the same individual.

47. Gerald Thompson, *The Army and the Navajo, The Bosque Redondo Reservation Experiment, 1863–1868* (Tucson: University of Arizona Press, 1976), pp. 18, 101, 123, 141, 161.

48. Smith to Cutler, 11 November 1863, LR, Fort Stanton, PR, RG 393, NA.

49. Carleton to Carson, 26 November 1862, U.S., Congress, Senate, "Condition of the Indian Tribes," 1867, Senate Report 156 (Serial 1279), 39th Cong., 2d sess., 1867, pp. 102–3; Higdon to Fritz, 13 October 1865, LR, Dist. NM, RG 393, NA, M-1088, roll 1; Santa Fe *Weekly Gazette,* 9 June 1866. Beach moved to Arizona and later became editor of the Prescott *Arizona Miner.*

50. Rio Bonito, Socorro County, N.M., Agricultural Schedules of the Eighth Census of the United States, 1860, NMSA (hereafter Agricultural Schedules, 1860); West to Cutler, 14 September, 20 October 1862, LS, Dist. Ariz., RG 393, NA; Albuquerque *Rio Abajo Weekly Press,* 7 April 1863; Edward D. Tittman, "The Exploitation of Treason," *New Mexico Historical Review,* 4 (April 1929): 138; John P. Wilson, *Merchants, Guns, and Money, The Story of Lincoln County and Its Wars* (Santa Fe: Museum of New Mexico Press, 1987), pp. 14–16.

51. Smith to McFerran, 11 March 1864, LS, Fort Stanton, PR, RG 393, NA; Higdon to Fritz, 13 October 1865, LR, Dist. NM, RG 393, NA, M-1088, roll 1; Fritz to Ludington, 18 November 1867, LR, CQM, Dist. NM, RG 393, NA; CQM to Fritz, 6 January 1868, LS, CQM, Dist. NM, RG 393, NA. In military correspondence the Beckwith ranch was sometimes referred to as "the government ranch." The land that Fritz rented was referred to as "the government farm" or "the public farm on the Fort Stanton Reservation."

52. Albuquerque *Rio Abajo Weekly Press,* 23 June 1863.

53. See papers relating to Rafael Chávez enclosed in Butler to Hunter, 30 October 1867, LR, CQM, Dist. NM, RG 393, NA.

54. Montoya to Rigg, 6 March 1864, LR, Dept. NM, RG 393, NA, M-1120, roll 25; Abréu to DeForrest, 3 May 1866, LR, Dist. NM, RG 393, NA, M-1088, roll 2; Fort Craig, 31 March 1866, Inspection Reports, Inspector, Dept. NM, RG 393, NA; Ayres to Enos, 22 August 1867, LR, CQM, Dist. NM, RG 393, NA. The government subsequently leased the farm to Lewis F. Sanburn and William H. Ayres, veterans of the California and New Mexico volunteers, respectively.

55. Milligan to Enos, 7 November 1866, LR, CQM, Dept. NM, RG 393, NA; List of buildings on the government reservation, Post of Ft. Craig, 16 December 1869, LR, Fort Craig, PR, RG 393, NA; Milligan to Commanding Officer, Ft. Craig, [date unclear], 1870, LS, CQM,

Dist. NM, RG 393, NA. In 1866 Montoya and Milligan cultivated a total of 170 acres within the limits of the military reserve.

56. Stewart to Secretary of War, 26 February 1873, Stewart to AAAG, 15 April 1873, LS, Fort Craig, PR, RG 393, NA.

57. Sherman to Secretary of War, 20 November 1869 and Ord to Townsend, 27 September 1869, U.S., Congress, House, Annual Report of the Secretary of War, House Ex. Doc. 1 (Serial 1412), 41st Cong., 2d sess., 1869, pp. 31, 124; Schofield to Wade, 10 June 1868, U.S., Congress, Senate, Letter of the Secretary of War, Senate Ex. Doc. 74 (Serial 1317), 40th Cong., 2d sess., 1868, pp. 1–3 (medical expenses are not included in the total military expenditures).

58. Eaton to Grant, 19 October 1867, U.S., Congress, House, Annual Report of the Secretary of War, House Ex. Doc. 1 (Serial 1324), 40th Cong., 2d sess., 1867, p. 576; Eaton to McClure, 10 October 1867, LS, CCS, Dist. NM, RG 393, NA.

59. Mason to Drum, 1 September 1865, 29 April 1866, LS, Dist. Ariz., RG 393, NA.

60. Crittenden to AAG, 18 June 1867, LS, District of Tucson, RG 393, NA.

61. Settlers on the Rio Verde to Gregg, 23 September 1867, LR, Subdistrict of Prescott, RG 393, NA; Prescott *Weekly Arizona Miner,* 9, 23 October 1869; Order No. 81, 11 October 1869, Fort Verde, PR, RG 393, NA.

62. Krause to Sherburne, 26 December 1866, LS, Fort Whipple, PR, RG 393, NA.

63. Endorsement of Crook on letter from Roger Jones, 30 April 1872, LS, Dept. Ariz., RG 393, NA. Lieutenant Colonel George Crook was assigned as commander of the Department of Arizona in mid-1871 on his brevet rank of major general. See Altshuler, *Chains of Command,* p. 197.

64. Enos to DeForrest, 5 April 1867, LS, CQM, Dist. NM, RG 393, NA.

65. Ffrench to Cutler, 7 July 1865, Ffrench to Post Adjutant, 6 September 1865, LS, Fort McRae, PR, RG 393, NA; Ffrench to Cutler, 10 January 1866, Slater to Ffrench, 1 April 1866, LR, Dist. NM, RG 393, NA, M-1088, rolls 3–4.

66. García to Carleton, 27 January 1866, LR, Dist. NM, RG 393, NA, M-1088, roll 3.

67. Enis to AAG, 5 September 1867, Enis to Horn, 5 September 1867, LR, Fort Craig, PR, RG 393, NA; Hunter to Commanding Officer, 21 September 1868, LR, Fort McRae, PR, RG 393, NA.

68. Gregg to AAG, 2 October 1874, LS, Dist. NM, RG 393, NA, M-1072, roll 5. Las Animas Valley was named after the principal river of the area, "El Rio de las Animas Perdidas en Purgatorio."

69. Kautz's stand on concentration and his conflict with Safford are

summarized in Andrew Wallace, "Soldier in the Southwest: The Career of General A. V. Kautz, 1869–1886" (Ph.D. dissertation, University of Arizona, 1968), pp. 350–486. Quote is from Kautz to AAG, 23 October 1876, LS, Dept. Ariz., RG 393, NA. See also Kautz to AAG, 20 October 1875, Kautz to Adjutant General, 15 February, 9 April 1877, LS, Dept. Ariz., RG 393, NA; Kautz to AAG, 15 August 1877, U.S., Congress, House, Annual Report of the Secretary of War, House Ex. Doc. 1, pt. 2 (Serial 1794), 45th Cong., 2d sess., 1877, pp. 133–48.

70. See, for example, Mahnken to Williams *et al.*, 21 August 1874, LS, Dist. NM, RG 393, NA, M-1072, roll 5; Santa Fe *Daily New Mexican,* 20 October 1874; Tucson *Arizona Citizen,* 19 April 1873; Memorial, enclosed in Carter to General Grant, 16 October 1867, Arizona, Cons. Corres. File, QMG, RG 92, NA.

71. Frazer, *Forts of the West,* p. 95.

72. James Monroe Foster, Jr., "Fort Bascom, New Mexico," *New Mexico Historical Review,* 35 (January 1960): 32, 61; *Report on Barracks and Hospitals,* p. 255.

73. Carey to Acting Assistant Quartermaster, 22 October 1865, LS, CQM, Dept. NM, RG 393, NA. In February 1866 Fort Bascom received 100,000 pounds from Union Depot, an unspecified amount in April from Fort Union's sutler and grain dealer William H. Moore, and 100,000 pounds in August from H. B. Denman, who had purchased his corn in the States. Enos to Easton, 4 March 1866, Enos to Smith, 8 April 1866, LS, CQM, Dept. NM, RG 393, NA; Enos to Smith, 26 July 1866, LS, CQM, Dist. NM, RG 393, NA.

74. The contracts are in Reg. Cont., QMG, RG 92, NA; Entries for Frank Chapman, 31 January 1870, Register of LR by the Quartermaster, Fort Bascom, PR, RG 393, NA. Only one of the six contractors was Hispanic—Vicente Romero of La Cueva.

75. Acting Assistant Quartermaster to Ludington, 6 January 1868, Endorsements, Fort Bascom, PR, RG 393, NA; Commanding Officer to Hildebaum, 15 January 1868, Commanding Officer to AAAG, 23 January 1868, LS, Fort Bascom, PR, RG 393, NA; Santa Fe *Weekly Gazette,* 21 October 1865. See also Letcher to Ludington, 24 January 1869, LR, CQM, Dist. NM, RG 393, NA. John Watts was the son of John S. Watts, who served as New Mexico's delegate to Congress during the Civil War. In 1868 the senior Watts was appointed chief justice of the territorial Supreme Court. He owned the property on which Fort Bascom was located, receiving a nominal rent from the army. (See Chapter 6.)

76. Frazer, *Forts of the West,* p. 108.

77. Robert C. and Eleanor R. Carriker, eds., *An Army Wife on the Frontier, The Memoirs of Alice Black-*

wood Baldwin, 1867–1877 (Salt Lake City: University of Utah Library, 1975), pp. 65, 68.

78. McFerran to Thomasson, 15 January 1864, 19 January 1865, McFerran to McDermott, 20 January 1865, LS, CQM, Dept. NM, RG 393, NA; Eaton to Cutler, 1 November 1864, LR, Dept. NM, RG 393, NA, M-1120, roll 23.

79. The contracts are in Reg. Cont., QMG, RG 92, NA; Enos to Wilson, 29 May 1867, LS, CQM, Dist. NM, RG 393, NA.

80. Carriker, *An Army Wife on the Frontier,* pp. 71, 74, 78.

81. Floyd S. Fierman, "Nathan Bibo's Reminiscences of Early New Mexico," *El Palacio,* 68 (Winter 1961): 244–46, 249.

82. U.S., Department of Interior, *Eighth Census of the United States: 1860, Agriculture,* vol. 2 (Washington: Government Printing Office, 1864), p. 178–79; U.S., Department of Interior, *Ninth Census of the United States: 1870, Population,* vol. 1 and *Wealth and Industry,* vol. 3 (Washington: Government Printing Office, 1872), p. 50 (vol. 1) and pp. 208–9 (vol. 3), hereafter *Ninth Census, 1870;* Valencia County, Agricultural Schedules of the Ninth Census of the United States, 1870, NMSA. C. M. Chase provides some idea of yearly consumption among New Mexican farm families: "They calculate on 300 pounds of corn for each grown person, and 150 pounds for each child." C. M. Chase, *The Editor's Run in New Mexico and Colorado* (reprint; Fort Davis, Texas: Frontier Book Co., 1968), p. 116.

83. Bibo's contracts are in Reg. Cont., QMG, RG 92, NA; Fierman, "Nathan Bibo's Reminiscences," pp. 249–52.

84. The contracts are in Reg. Cont., QMG, RG 92, NA. Two of the thirteen contractors were Hispanos: Tranquilino Luna and Salvador Armijo.

85. John O. Baxter, "Salvador Armijo: Citizen of Albuquerque, 1823–1879," *New Mexico Historical Review,* 53 (July 1978): 224, 227–29.

86. Baldwin to Ludington, 10 May 1869, Elkins to Ludington, 15 May 1869, LS, CQM, Dist. NM, RG 393, NA.

87. Potter to Schwartz, 11 October 1870, LS, CQM, Dist. NM, RG 393, NA.

88. Staab to Belcher, 3 January 1878, Stafford to CQM, 11 January 1879, LR, CQM, Dist. NM, RG 393, NA; Stafford to Quartermaster General, 1 July 1878, Annual Reports Received from Quartermasters, RG 92, NA. Most government forage agents in the Wingate district bought their forage on the river as well.

89. Frazer, *Forts of the West,* p. 98.

90. Burkett to Cutler, 22 July 1865, LR, Dept. NM, RG 393, NA, M-1120, roll 26; *Report on Barracks and Hospitals,* p. 239.

91. The contracts are in Reg.

Cont., QMG, RG 92, NA. Fort Cummings was reoccupied during the Apache wars of the 1880s.

92. Ludington to Etling, 24 May 1869, LS, CQM, Dist. NM, RG 393, NA. Fort Selden's quartermaster also made arrangements for supplying Fort Cummings with grain in 1866. Enos to Rynerson, 26 January 1866, LS, CQM, Dept. NM, RG 393, NA.

93. Frazer, *Forts of the West,* p. 95; Frank D. Reeve, ed., "Frederick E. Phelps: A Soldier's Memoirs," *New Mexico Historical Review,* 25 (January 1950): 49–51; Mrs. Orsemus Bronson Boyd, *Cavalry Life in Tent and Field* (reprint; Lincoln: University of Nebraska Press, 1982), pp. 206, 214.

94. Darlis A. Miller, *The California Column in New Mexico* (Albuquerque: University of New Mexico Press, 1982), pp. 102–3.

95. The contracts are in Reg. Cont., QMG, RG 92, NA; Griggs letter dated 6 January 1875, LS, CQM, Dist. NM, RG 393, NA; Bull to Belcher, 13 January 1876, LR, CQM, Dist. NM, RG 393, NA. See also Custer to Ludington, 1 April 1868, LR, CQM, Dist. NM, RG 393, NA. The eleven contractors were Anglos.

96. Contracts are in Reg. Cont., QMG, RG 92, NA. French was selling corn that he had imported from Mexico to the quartermaster's department during the summer of 1866. Enos to French, 8 December 1866, LS, CQM, Dist. NM, RG 393, NA.

97. Rigg to Ludington, 29 June 1868, LR, CQM, Dist. NM, RG 393, NA; Rosenbaum to Myers, 30 June 1874, LR, CQM, Dist. NM, RG 393, NA.

98. U.S., Department of Interior, *Tenth Census of the United States: 1880, Population,* vol. 1 (Washington: Government Printing Office, 1883), p. 72, hereafter *Tenth Census, 1880.*

99. Murphy to DeForrest, 11 June 1866, LR, Dist. NM, RG 393, NA, M-1088, roll 3.

100. Kautz to Townsend, 24 January 1872, LS, Fort Stanton, PR, RG 393, NA.

101. *Ninth Census, 1870,* vol. 1 (Population), p. 50, vol. 3 (Wealth and Industry), pp. 208–9, 358.

102. Pope to Hartstuff, 31 October 1870, U.S., Congress, House, Annual Report of the Secretary of War, House Ex. Doc. 1 (Serial 1446), 41st Cong., 3d sess., 1870, pp. 15–16.

103. These figures are found in monthly reports submitted by Fort Stanton's quartermaster during 1869. The report for July has not been located. See Reports Received from Post Quartermasters, CQM, Dist. NM, RG 393, NA. The renovation program was suspended in mid-1869, and by the end of the year the post employed only four civilians. Their monthly wages totaled $316.

104. The army contracted for 450,000 pounds of corn (8,036 bushels) for use at Fort Stanton in 1870; the amount produced in Lincoln

County as listed in the 1870 census was 134,162 bushels. Since the amount produced in Lincoln County in 1871–72 is not known, I have used the 1870 census figure in calculating the percentage of corn crop required to fill Murphy's contracts. O. W. McCullough also agreed to deliver 70,000 pounds of corn at Fort Stanton in 1871–72, which, added to the amounts required on Murphy's contracts, represented about 17 percent of the county's corn crop. McCullough and the Murphy firm (the contracts were signed in the name of E. Fritz, Murphy's partner) apparently were the only Lincoln County men signing corn contracts during the fiscal year ending 30 June 1872. *Ninth Census, 1870,* vol. 3 (Wealth and Industry), pp. 208–9; Entries for E. Fritz, 1 September 1870, 15 May 1872, and for O. W. McCullough, 25 September 1871, Reg. Cont., QMG, RG 92, NA; L. G. Murphy and Co., Journal, Corn Account, 1 May 1871–December 1872 (Special Collections, University of Arizona).

105. See Lawrence L. Mehren, "A History of the Mescalero Apache Reservation, 1869–1881" (Master's thesis, University of Arizona, 1969), pp. 24–25.

106. Santa Fe *Daily New Mexican,* 16 March 1876.

107. The contracts are in Reg. Cont., QMG, RG 92, NA. See Appendix Table 1 of this volume.

108. Miller, *The California Column in New Mexico,* pp. 146–47.

109. Murphy to Myers, 22 January 1874, Murphy to Clendenin, 17 March 1874, Wittard to Chief Quartermaster, 10 March 1874, Baca to Belcher, 3 February 1876, LR, CQM, Dist. NM, RG 393, NA. The total number killed by the Horrells on their rampage through Lincoln County has never been established. See P. J. Rasch, "The Horrell War," *New Mexico Historical Review,* 31 (July 1956): 223–31.

110. Staab to Belcher, 15 March, 11 April 1878, LR, CQM, Dist. NM, RG 393, NA; Ilfeld to Belcher, 21 June, 1 July 1878, and Ilfeld to Browne, 16 July, 26 August 1878, *Letterbook No. 4* (Ilfeld Collection, Special Collections, University of New Mexico).

111. Stone and Lee to Commanding Officer, 10 October 1878, LR, Fort Stanton, PR, RG 393, NA; Catron and Thornton to Belcher, 10 August 1878 and Belcher to Acting Assistant Quartermaster, Fort Bliss, 12 August 1878, LR, CQM, Dist. NM, RG 393, NA. The lawyers also explained that a large amount of Riley's grain either had been stolen or damaged by rain after he left the county.

112. *Tenth Census, 1880,* vol. 3 (Agriculture), p. 199. The Rio Grande Valley experienced a drought in 1879, which affected harvests in that region. Information about rainfall in Lincoln County is unavailable, but even if the county experienced a drought that year, the decline in grain production was clearly a trend. The 1890 census, for

example, shows that Lincoln County produced 51,190 bushels, about 38 percent of the 1870 crop. U.S., Department of Interior, *Eleventh Census of the United States: 1890, Agriculture* (Washington: Government Printing Office, 1895), p. 377.

113. *Ninth Census, 1870*, vol. 1 (Population), p. 50, vol. 3 (Wealth and Industry), pp. 208–9, 358.

114. Fourth Annual Message of Governor Connelly, delivered December, 1865, Territorial Papers, New Mexico, Records of the Secretary of State, RG 59, NA, T-17, roll 3.

115. Report of the Governor of New Mexico Made to the Secretary of the Interior, 1879, Territorial Archives of New Mexico, NMSA, Microfilm Edition, roll 99.

116. Murphy to DeForrest, 11 June 1866, LR, Dist. NM, RG 393, NA, M-1088, roll 3.

117. James F. Meline, *Two Thousand Miles on Horseback* (reprint; Albuquerque: Horn and Wallace, Publishers, 1966), p. 163.

118. William A. Bell, *New Tracks in North America* (reprint; Albuquerque: Horn and Wallace, Publishers, 1965), pp. 123–24.

119. Mora County, Agricultural Schedules of the Ninth Census of the United States, 1870, NMSA; Brown to Getty, 15 September 1867, LR, Dist. NM, RG 393, NA, M-1088, roll 5.

120. Meyer to Granger, 27 August 1871, LR, Dist. NM, RG 393, NA, M-1088, roll 13; Letter of F. Myers, 1 August 1872, Registers of LR, Dist. NM, RG 393, NA, M-1097, roll 2; Myers to Easton, 3 February 1872, Myers to Quartermaster General, 1 August 1872, Myers to Buffum, 4 September 1872, Myers to Rosenbaum, 3 October 1872, LS, CQM, Dist. NM, RG 393, NA; Hartz to Chief Quartermaster, 2 December 1872, LR, CQM, Dist. NM, RG 393, NA.

121. Alexander to Sartle, 1 January 1873, LS, Fort Garland, PR, RG 393, NA.

122. Staab to Belcher, 15 March 1878, LR, CQM, Dist. NM, RG 393, NA.

123. The contracts are in Reg. Cont., QMG, RG 92, NA. See Appendix Tables 2 and 3 in this volume.

124. McGonnigle to Newcomb, 19 January 1875, LS, CQM, Dist. NM, RG 393, NA; Smith to AAAG, 20 June 1877, LR, Dist. NM, RG 393, NA, M-1088, roll 29; Ilfeld to Eagle, 19 March 1872, *Letter Book No. 1* (Ilfeld Collection, Special Collections, University of New Mexico).

125. William J. Parish, *The Charles Ilfeld Company, A Study of the Rise and Decline of Mercantile Capitalism in New Mexico* (Cambridge: Harvard University Press, 1961), pp. 59–60.

126. The contracts are in Reg. Cont., QMG, RG 92, NA and Reg. Cont., CGS, RG 192, NA.

127. Parish, *The Charles Ilfeld Company*, pp. 37–38. The army also was the major source of eastern drafts

needed by merchants to buy merchandise from eastern suppliers.

128. Letcher to Wilkins, 16 September 1871, Letcher to Berg, 28 October 1872, *Letterbook No. 1* (Ilfeld Collection, Special Collections, University of New Mexico); Endorsement by Enos, 25 October 1867, LR, Dist. NM, RG 393, NA, M-1088, roll 5.

129. The contracts are in Reg. Cont., QMG, RG 92, NA. See also Murphy to Enos, 27 March 1867, LR, CQM, Dist. NM, RG 393, NA; Enos to Murphy, 12 April 1867, LS, CQM, Dist. NM, RG 393, NA; L. G. Murphy and Co., Journal, Corn Account, 1 May 1871–December 1872 (Special Collections, University of Arizona). The contract signed in September 1871 was in the name of E. Fritz, Murphy's partner.

130. The contracts are in Reg. Cont., QMG, RG 92, NA; position and distribution of troops can be found in the annual reports of the Secretary of War. These figures should be taken as approximations, as all contracts may not have been registered.

131. The contracts are in Reg. Cont., QMG, RG 92, NA. Collier delivered all but about 106,000 pounds on his contract. The chief quartermaster subsequently relieved him of the balance in consideration of the low price for which the contract had been taken and because Fort Union had sufficient corn to last until new contracts were let. See Fort Union Depot Quartermaster to Chief Quartermaster, 3 July 1877, LR, CQM, Dist. NM, RG 393, NA. In commenting upon low contract prices, New Mexico's chief quartermaster wrote in 1877: "The contract prices for forage last year were very low. Contractors, especially at Fort Union, claim to be largely losers on both grain and hay." Belcher to CQM, Department of the Missouri, 8 March 1877, LS, CQM, Dist. NM, RG 393, NA.

132. Commanding officer to AAAG, 15 February 1872, LS, Fort Bliss, PR, RG 393, NA; McClure to Kobbé, 5 October 1869, McClure to Eaton, 11 October 1869, LS, CCS, Dist. NM, RG 393, NA.

133. Enos to Thomasson, 26 January 1866, LS, CQM, Dept. NM, RG 393, NA; Frazer, *Forts of the West,* p. 144; *Tenth Census, 1880,* vol. 1 (Population), p. 343.

134. The contracts are in Reg. Cont., QMG, RG 92, NA. For Bacon's background, see J. Morgan Broaddus, Jr., *The Legal Heritage of El Paso* (El Paso: Texas Western College Press, 1963), pp. 85, 110; for Knox, see Williams, *Texas' Last Frontier,* pp. 71, 97; for Samuel, Joseph, and Solomon Schutz, see W. W. Mills, *Forty Years at El Paso, 1858–1898* (El Paso: Carl Hertzog, 1962), p. 189; for Lesinsky and Angerstein, see Las Cruces *Borderer,* 13 December 1871. The sixth contractor was George Zwitzers, who delivered corn in 1874 at 2½ cents per pound. Bacon's contract price came to about 7 cents a pound. Zwitzers was discharged from the army in 1872. He later served briefly as post

trader at Fort Bliss. See C. O. to Secretary of War, 3 November 1875, LS, Fort Bliss, PR, RG 393, NA.

135. Background to the Salt War can be found in C. L. Sonnichsen, *The El Paso Salt War* (El Paso: Carl Hertzog and Texas Western Press, 1961).

136. The contracts are in Reg. Cont., QMG, RG 92, NA.

137. Blun to Dana, 3 November 1879, LR, CQM, Dist. NM, RG 393, NA; C. L. Sonnichsen, *Pass of the North, Four Centuries on the Rio Grande* (El Paso: Texas Western Press, 1968), pp. 381–82.

138. Krakauer to Davis, 30 May 1881, LR, CQM, Dist. NM, RG 393, NA.

139. Frazer, *Forts of the West,* pp. 157, 158; *Report on Barracks and Hospitals,* p. 231; Carlysle G. Raht, *The Romance of Davis Mountains and Big Bend Country* (Odessa, Texas: Rahtbooks Company, 1963), p. 121.

140. The contracts are in Reg. Cont., QMG, RG 92, NA; Bentzoni to AAG, 18 July 1875, LS, Fort Quitman, PR, RG 393, NA.

141. Floyd S. Fierman, "Jewish Pioneering in the Southwest, A Record of the Freudenthal–Lesinsky–Solomon Families," *Arizona and the West,* 2 (Spring 1960): 54–72; Las Cruces *Borderer,* 13 December 1871.

142. Frazer, *Forts of the West,* p. 148.

143. The contracts are in Reg. Cont., QMG, RG 92, NA; Rigg to Ludington, 28 October 1867, LR, CQM, Dist. NM, RG 393, NA. See also Custer to Ludington, 10 June 1868, LR, CQM, Dist. NM, RG 393, NA.

144. The contracts are in Reg. Cont., QMG, RG 92, NA; Wulff to Hatch, 6 November 1869, LR, Fort Davis, PR, RG 393, NA.

145. Shafter to Augur, 12 February 1872, LS, Fort Davis, PR, RG 393, NA.

146. Murphy to AAG, 23 October 1871, LR, Fort Davis, PR, RG 393, NA; Raht, *The Romance of Davis Mountains,* pp. 185–86.

147. Carpenter to AAG, 11 July 1879, LS, Fort Davis, PR, RG 393, NA; Williams, *Texas' Last Frontier,* pp. 195, 202.

148. Williams to Quartermaster General, 30 June 1883, Annual Reports Received from Quartermasters, RG 92, NA.

149. Frazer, *Forts of the West,* p. 162; Williams, *Texas' Last Frontier,* pp. 94, 120, 132, 137.

150. Contracts are in Reg. Cont., QMG, RG 92, NA; Hatch to Morse, 20 February 1868, LS, Fort Stockton, PR, RG 393, NA; Eldon S. Branda, ed., *The Handbook of Texas,* vol. 3 (Austin: Texas State Historical Association, 1976), p. 361.

151. Gamble to Roberts, 4 November 1868, LS, Fort Stockton, PR, RG 393, NA.

152. Carleton to AAG, 6 March 1871, Inspection Reports, OIG, RG 159, NA. (See two reports of this date, one concerning Fort Quitman and the other Fort Davis.)

153. Carleton to AAG, 6 March 1871, Inspection Reports, OIG, RG 159, NA (report on Fort Stockton).

154. Contracts are in Reg. Cont., QMG, RG 92, NA.

155. Mason to Tuttle, 7 May, 2 June 1866, LS, Dist. Ariz., RG 393, NA; Babbitt to Baker, 17 July 1866, LR, Dist. Ariz., RG 393, NA. Prices quoted did not specify coin or currency. Camp Lincoln was renamed Camp Verde in 1868. Altshuler, *Starting With Defiance*, pp. 59–60.

156. Prescott *Arizona Miner*, 26 January 1867.

157. Entries for Kendall, 2 March 1867, Grant, Postle and Brown, 12 March 1867, Reg. Cont., QMG, RG 92, NA; Prescott *Arizona Miner*, 4 May 1867.

158. Foster to Gaines, Foster to Hawley, 13 October 1870, LS by QM, Subdistrict of Northern Arizona, RG 393, NA.

159. Prescott *Weekly Arizona Miner*, 5 September 1868.

160. Prescott *Arizona Miner*, 8 February 1868.

161. Prescott *Weekly Arizona Miner*, 19 September 1868, 24 July 1869.

162. Wilkins to Chief Engineer's Office, 22 April 1878, LS, Fort Whipple, PR, RG 393, NA.

163. *Report on Barracks and Hospitals*, p. 468; Sherburne to Gregg, 4 September 1867, LR, Subdistrict of Prescott, RG 393, NA.

164. Prescott *Arizona Miner*, 8 August 1866, 13 July 1867; Krause to Sherburne, 26 December 1866, Devin to Sherburne, 1 February 1868, Wheaton to Sherburne, 4 March 1870, LS, Fort Whipple, PR, RG 393, NA.

165. U.S., Congress, Senate, Federal Census—Territory of New Mexico and Territory of Arizona, 1870, Senate Doc. 13, 89th Cong., 1st sess., 1965 (Washington: U.S. Government Printing Office, 1965), pp. 218–21.

166. Egbert to AAG, 10 November 1878, LS, Fort Verde, PR, RG 393, NA; Richard J. Hinton, *The Handbook to Arizona* (reprint; Tucson: Arizona Silhouettes, 1954) p. 300.

167. Tucson *Weekly Arizonian*, 21 February 1869. This newspaper changed its title from *Weekly Arizonian* to *Weekly Arizonan* in April 1869.

168. Tucson *Weekly Arizonan*, 27 August 1870.

169. John A. Spring, *John Spring's Arizona*, edited by A. M. Gustafson (Tucson: University of Arizona Press, 1966), pp. 145, 237–38; Tucson *Arizona Citizen*, 11 October 1873. On Pete Kitchen see Frank C. Lockwood, *Pioneer Portraits, Selected Vignettes* (Tucson: University of Arizona Press, 1968), pp. 17–27.

170. Tucson *Weekly Arizonan*, 1 October 1870; *Report on Barracks and Hospitals*, p. 462.

171. Prescott *Weekly Arizona Miner*, 6 November 1869; Letter from Tully, Ochoa, and Co., 13 January 1877, Register of LR, Dept. Ariz., RG 393, NA.

172. Thomas E. Farish, *History of Arizona*, vol. 6 (San Francisco: Filmer Brothers Electrotype Company, 1918), p. 46; Tucson *Weekly Arizonian*, 21 February 1869; Tucson *Arizona Citizen*, 19 July 1873.

173. Thomas to AAG, 2 July 1875, Arizona, Cons. Corres. File, RG 92, NA.

174. Geoffrey P. Mawn, "Promoters, Speculators, and the Selection of the Phoenix Townsite," *Arizona and the West,* 19 (Autumn 1977): 212–15. In November 1868, citizens of Phoenix asked the commanding officer at Camp McDowell, Major A. J. Alexander, to loan them 20,000 pounds of barley for seed as they were too poor to purchase it. There is no record that the request was granted, although Alexander recommended its approval. Alexander to Sherburne, 17 November 1868, LR, Fort McDowell, PR, RG 393, NA.

175. Tucson *Arizona Citizen,* 23 September 1871, 1, 29 May 1875; Prescott *Weekly Arizona Miner,* 29 June, 6, 13, July 1872; John J. Gosper, *Report of the Acting Governor of Arizona Made to the Secretary of the Interior for the Year 1881* (Washington: Government Printing Office, 1881), pp. 10–11.

176. Prescott *Arizona Miner,* 18 April 1868.

177. Alexander to Sherburne, 2 October 1868, LS, Subdistrict of the Verde, RG 393, NA; Prescott *Weekly Arizona Miner,* 26 September 1868.

178. U.S., Congress, Senate, Committee on Interior and Insular Affairs, *Indian Water Rights of the Five Central Tribes of Arizona, Hearings,* "Report for the Gila River Pima and Maricopa Tribes," by Fred Nicklason, 94th Cong., 1st sess., 1975 (Washington: U.S. Government Printing Office, 1976), pp. 608–15 (hereafter, Nicklason, "Report for the Gila River Pima and Maricopa Tribes").

179. Bennett to Green, 1 May 1866, LR, Fort McDowell, PR, RG 393, NA.

180. Nicklason, "Report for the Gila River Pima and Maricopa Tribes," pp. 619–20.

181. [Name unclear] to Post Adjutant, 24 October 1869, LR, Fort McDowell, PR, RG 393, NA; Grossman to AAAG, 16 November 1869, LR, Fort Verde, PR, RG 393, NA.

182. Nicklason, "Report for the Gila River Pima and Maricopa Tribes," p. 622.

183. Nicklason, "Report for the Gila River Pima and Maricopa Tribes," pp. 624–34. For Indian raids on grain fields, see endorsements on letter from J. H. Stout, dated 17 October 1872, LS, Dept. Ariz., RG 393, NA; Tucson *Arizona Citizen,* 1, 29 November 1873.

184. Willcox to AAG, 2, 3 December 1878 and Willcox to Frémont, 13 March 1879, LS, Dept. Ariz., RG 393, NA.

185. Nicklason, "Report for the Gila River Pima and Maricopa Tribes," pp. 640–43; Edward H. Spicer, *Cycles of Conquest, The Impact of Spain, Mexico, and the United States on the Indians of the Southwest, 1533–1960* (Tucson: University of Arizona Press, 1962), pp. 149–50.

186. *Tenth Census, 1880,* vol. 3 (Agriculture), pp. 26–27.

187. *Tenth Census, 1880,* vol. 3 (Agriculture), pp. 6, 13, 472, 497–500.

188. Green to Devin, 18 February 1870, LS, Camp Grant, PR, RG 393, NA; Endorsement by Lee [May 1870], End. S., Subdistrict of Southern Arizona, RG 393, NA; Lee to Ross, 13 July 1870, LR, Camp Crittenden, PR, RG 393, NA.

189. Prescott *Arizona Miner,* 4 April 1868.

190. Prescott *Arizona Miner,* 13 July, 21 September 1867, 28 March, 26 September, 21 November 1868.

191. Prescott *Weekly Arizona Miner,* 26 June 1869.

192. Farish, *History of Arizona,* vol. 6, p. 137; Tucson *Arizona Citizen,* 17, 31 May, 21 June 1873, 22 September 1877; Prescott *Weekly Arizona Miner,* 26 October 1872.

193. Tucson *Arizona Citizen,* 31 May 1873, 22 July 1876; Farish, *History of Arizona,* vol. 6, p. 137.

194. The contracts are in Reg. Cont., QMG, RG 92, NA. Contractor P. M. Moore has not been identified. On farmers failing to obtain contracts, see Sacket to Inspector General, 26 May 1873, Inspection Reports (Headquarters, Dept. Ariz.), OIG, RG 159, NA. See Appendix Tables 4 and 5 in this volume.

195. The contracts are in Reg. Cont., QMG, RG 92, NA. For investments of Tully and Ochoa, see Marcy G. Goldstein, "Americanization and Mexicanization: The Mexican Elite and Anglo–Americans in the Gadsden Purchase Lands, 1853–1880," (Ph.D. dissertation, Case Western Reserve University, 1977), pp. 44–48.

196. Entry for Marcus Katz, 31 December 1877, Reg. Cont., QMG, RG 92, NA; memorandum signed by L. M. Jacobs and Lord and Williams, 1 February 1878, and Calisher to Barron, 25 January 1878 (Jacobs Manuscripts, Special Collections, University of Arizona); Prescott *Weekly Arizona Miner,* 15 February 1878; Calisher to Barron, 27 July 1878 (Barron and Lionel Jacobs Business Records, Arizona Historical Society).

197. Albert to Barron, 12 November 1879 (Jacobs Manuscripts, Special Collections, University of Arizona).

198. Tucson *Arizona Citizen,* 21 March 1874. This issue carried the following statement: "Until last year, government needed rather more grain than was produced in this territory above local wants."

199. The contracts are in Reg. Cont., QMG, RG 92, NA. Prices listed in the text for 1866, 1867, and 1868 are in coin. The price of grain delivered at Forts Lowell, Bowie, and McDowell varied in 1879; some contractors received higher prices than those quoted here. There were 1,643 officers and enlisted men assigned to the Department of Arizona in 1874; 1,232 in 1877; 1,486 in 1879. See Annual Reports of the Secretary of War for these years.

200. Prescott *Arizona Miner,* 11 July 1868; Prescott *Weekly Arizona Miner,* 8 August, 5 September 1868.

201. Prescott *Weekly Arizona Miner,* 19 September, 26 December 1868, 9, 23 January, 24 July 1869.

202. Prescott *Weekly Arizona*

Miner, 6 November 1869; Entry for C. C. Bean, 1 February 1870, Reg. Cont., QMG, RG 92, NA.

203. Prescott *Weekly Arizona Miner,* 1 June 1872.

204. Tucson *Weekly Arizonan,* 25 December 1869; Tucson *Arizona Citizen,* 26 April 1873.

205. Tucson *Arizona Citizen,* 19 April 1873, 21 March, 25 April, 13 June 1874; Santa Fe *Weekly New Mexican,* 16 March 1871.

206. The contracts are in Reg. Cont., QMG, RG 92, NA. Charles Lesinsky received a contract in 1871 to deliver grain at Camp Bowie. At this time, however, the Lesinsky brothers operated a mercantile establishment in Tucson, and they probably obtained corn to fill this contract in Arizona. On the other hand, Arizona contractor C. C. Bean late in 1872 purchased 800,000 pounds of corn in New Mexico to supply Camp Verde. Prescott *Weekly Arizona Miner,* 21 December 1872.

207. The contracts are in Reg. Cont., QMG, RG 92, NA; *Tenth Census,* 1880, vol. 3 (Agriculture), p. 6.

Chapter 3

1. Erna Risch, *Quartermaster Support of the Army, A History of the Corps,* 1775–1939 (Washington: Office of the Quartermaster General, 1962), p. 379.

2. The contracts are in Reg. Cont., QMG, RG 92, NA. Two years later Captain Alexander J. McGonnigle, the depot quartermaster at Fort Union, estimated that the post and depot combined would need between 100 and 110 tons of hay per month. Endorsement on letter of Valdez to McGonnigle, 23 December 1872, LR, Dist. NM, RG 393, NA, M-1088, roll 16.

3. Davis to Inspector General, 15 December 1872, Inspection Reports, OIG, RG 159, NA (report on Headquarters District of New Mexico).

4. Enos to Alexander, 25 April 1867, LS, CQM, Dist. NM, RG 393, NA.

5. Bennett to Babbitt, 18 March 1865, LS, Fort Bowie, PR, RG 393, NA; Chamberlin to Dana, 3 April 1880, McDonald to Hunt, 28 August 1880, Shaw to Myers, 8 November 1872, LR, CQM, Dist. NM, RG 393, NA.

6. Memorandum by McGonnigle, 24 September 1871, LS, CQM, Dist. NM, RG 393, NA.

7. Murphy's contracts are in Reg. Cont., QMG, RG 92, NA; L. G. Murphy and Co., Journal, Hay Account, October 1872 (Special Collections, University of Arizona).

8. Shaw's contract is in Reg. Cont., QMG, RG 92, NA.

9. *Grant County Herald,* 4 October 1879; Bennett to AAAG, 20 November 1879, LR, Fort Wingate, PR, RG 393, NA.

10. Stapp and Hopkins to Ludington, 17 October 1868, and Day to Enos, 24 June 1867, LR, CQM, Dist. NM, RG 393, NA.

11. Garst to CQM, 19 February 1880, LR, CQM, Dist. NM, RG 393, NA. Officials believed that

frosts destroyed nutritional content of grass. By the time the first frost occurred, however, the nutrient in the grass had already been reabsorbed into the root system.

12. Proposals for Army Supplies, 14 April 1880, Fort Bayard, Cons. Corres. File, QMG, RG 92, NA.

13. Ludington to Whitman, 18 September 1869, LS, CQM, Dist. NM, RG 393, NA.

14. Proceedings of Board of Survey, 5 September 1867, Boards of Survey, Fort Stanton, PR, RG 393, NA; Vose to Sutorius, 24 June 1868, LR, CQM, Dist. NM, RG 393, NA. See note 11 above.

15. The contract is in Reg. Cont., QMG, RG 92, NA. See also Davis to Jennings, 3 August 1865, Davis to Nissen, 14 August 1865, and Davis to C. O., 24 August 1865, LS, Inspector, Dept. NM, RG 393, NA.

16. Carleton to C. O., San Elizario, 29 May 1865, and Carleton to C. O., Franklin, 29 May 1865, LS, Dept. NM, RG 393, NA; Davis to Jones, 6 June 1865, and Davis to Smith and Blanchard, 14 June 1865, LS, Inspector, Dept. NM, RG 393, NA.

17. Carleton to Enos, 17 August 1865, LS, Dept. NM, RG 393, NA.

18. Enos to Easton, 30 December 1866, LS, CQM, Dist. NM, RG 393, NA; Letter of Capt. William L. Seran, 16 April 1867, Registers of LR, Dist. NM, RG 393, NA, M-1097, roll 1.

19. Wade to CQM, 17 February 1871, LS, Fort Stockton, PR, RG 393, NA.

20. McDonald to Hunt, 28 August 1880, LR, CQM, Dist. NM, RG 393, NA; Dudley to Lee, 28 January 1881, LR, Dist. NM, RG 393, NA, M-1088, roll 42.

21. Vose to Enos, 27 July 1867, LR, CQM, Dist. NM, RG 393, NA; Santa Fe *Weekly New Mexican,* 22 June 1869.

22. *Mining Life,* 11 October 1873.

23. Stapleton to Myers, 6 June 1872, LR, CQM, Dist. NM, RG 393, NA; Hennisee to CQM, 9 November 1874, LR, Dist. NM, RG 393, NA, M-1088, roll 23.

24. Carleton to CQM, 2 August 1865, LS, Dept. NM, RG 393, NA, M-1072, roll 3; Brentlinger to McGonnigle, 2 November 1874, LR, CQM, Dist. NM, RG 393, NA; Proposals for Army Supplies, 14 April 1880, Fort Bayard, Cons. Corres. File, QMG, RG 92, NA.

25. Romero to McGonnigle, 9 October 1871, and Tanfield to Enos, 16 February 1867, LR, CQM, Dist. NM, RG 393, NA.

26. Ludington to Von Luettwitz, 1 May 1868, LR by Quartermaster, Fort Bascom, PR, RG 393, NA; Stulhammer to AAAG, 23 March 1875, LR, Dist. NM, RG 393, NA, M-1088, roll 24; McLaughlin to AAG, 5 January 1881, LS, Fort Davis, PR, RG 393, NA.

27. Belcher to Thurston, 28 May 1878, LS, CQM, Dist. NM, RG 393, NA.

28. Meinhold to Post Adjutant, 14 September 1867, LR, Fort Craig, PR, RG 393, NA.

29. Williams to AAAG, 21 Octo-

ber 1870, and Clendenin to AAAG, 7 November 1870, LS, Fort Selden, PR, RG 393, NA.

30. Farnsworth to CQM, 30 November 1872, LR, CQM, Dist. NM, RG 393, NA.

31. [Name unclear] to CQM, 3 August 1879, LR, CQM, Dist. NM, RG 393, NA.

32. Smith to McGonnigle, 10 November 1874, and Smith to CQM, 3 March 1875, LR, CQM, Dist. NM, RG 393, NA.

33. Belcher to CQM, 29 January 1876, 21 August 1876, LS, CQM, Dist. NM, RG 393, NA.

34. Santa Fe *Weekly New Mexican,* 30 October 1877.

35. Cory to CQM, 12 July 1878, LR, CQM, Dist. NM, RG 393, NA; [name unclear] to CQM, 26 April 1882, LS, Fort Selden, PR, RG 393, NA.

36. Hunt to CQM, 24 January 1880, LR, CQM, Dist. NM, RG 393, NA; Cook to CQM, 9 June 1883, LR, Dist. NM, RG 393, NA, M-1088, roll 50.

37. Getty to Pope, 9 June 1870, LS, Dist. NM, RG 393, NA, M-1072, roll 4.

38. The contracts are in Reg. Cont., QMG, RG 92, NA. See also McFerran to Nissen, 20 January 1865, LS, CQM, Dept. NM, RG 393, NA; Santa Fe *Daily New Mexican,* 22 July 1871. Prices for Hay are not strictly comparable because contract registers sometimes specify long tons (2,240 pounds), sometimes short tons (2,000 pounds), and sometimes simply tons.

39. Bristol to Enos, 16 July 1866, LR, Dist. NM, RG 393, NA, M-1088, roll 5.

40. Hilgert to Enos, 30 June 1867, LR, CQM, Dist. NM, RG 393, NA.

41. Enos to Day, 20 July 1867, LS, CQM, Dist. NM, RG 393, NA.

42. Gregg to Spiegelberg, 2 April 1874, and Mahnken to Acting Chief Quartermaster, 29 August 1874, LS, Dist. NM, RG 393, NA, M-1072, roll 5.

43. Wedeles to Potter, 2 November 1870, and McGonnigle to Potter, 28 November 1870, LR, CQM, Dist. NM, RG 393, NA.

44. Kayser to McGonnigle, 11 January 1875, Fort Union, Cons. Corres. File, QMG, RG 92, NA; Romero to Kimball, 30 December 1875, LR, CQM, Dist. NM, RG 393, NA.

45. Gregg to Spiegelberg, 2 April 1874, LS, Dist. NM, RG 393, NA, M-1072, roll 5; McEachran to McGonnigle, 5 October 1874, LR, CQM, Dist. NM, RG 393, NA.

46. Filger to McGonnigle, 5 September 1871, LR, CQM, Dist. NM, RG 393, NA; Memorandum by McGonnigle, 24 September 1871, LS, CQM, Dist. NM, RG 393, NA; Proposal of Simon Filger, 5 June 1871, Catron Papers, Miscellaneous Papers (Special Collections, University of New Mexico).

47. Gregg to AAG, 24 August 1874, LR, Dist. NM, RG 393, NA, M-1088, roll 22; Gregg to AAG, 26 August 1874, and Devin to AAG, 20 March 1875, LS, Dist. NM, RG 393, NA, M-1072, roll 5.

48. Smith to Cutler, 30 December 1863, LS, Fort Stanton, PR, RG 393, NA; Duncan to Willis, 22 August 1866, and Commanding Officer to Hunter, 22 November 1867, LS, Fort Selden, PR, RG 393, NA; Amador Collection, Box 27, Folder 3, Rio Grande Historical Collections, New Mexico State University Library.

49. Chamberlin to Dana, 3 April 1880, Bourquet to Dana, 24 April 1880, and Coghlan to Hatch, 18 April 1880, LR, CQM, Dist. NM, RG 393, NA.

50. Custer to Ludington, 10 July 1868, LR, CQM, Dist. NM, RG 393, NA.

51. McGonnigle to Acting Assistant Quartermaster, 9 November 1871, and Myers to Steelhammer, 8 January 1872, LS, CQM, Dist. NM, RG 393, NA; Medical Records, October 1871, Fort Wingate, OAG, RG 94, NA.

52. Luff to CQM, 1 September 1874, Godwin to CQM, 1 September 1874, and Hudson to CQM, 25 September 1874, LR, CQM, Dist. NM, RG 393, NA. When James J. Dolan failed to complete his contract at Fort Stanton in 1874, the post quartermaster began purchasing hay in open market at $60 per ton from residents in the Tularosa Valley. Even though these suppliers had to haul the hay more than seventy miles to the post, the district commander considered this an exorbitant price and refused to authorize the purchse of more than a small amount. Clendenin to AAAG, 31 December 1874, LS, Fort Stanton, PR, RG 393, NA; Mahnken to Post Commander, 16 January 1875, LS, Dist. NM, RG 393, NA, M-1072, roll 5.

53. Perea to Dana, 22 September 1879, and Perea to Stafford, 10 October 1879, LR, CQM, Dist. NM, RG 393, NA.

54. Dana to CQM, 10 September 1879, LS, CQM, Dist. NM, RG 393, NA.

55. Davis to Commanding Officer, 6 September 1865, LS, Inspector, Dept. NM, RG 393, NA.

56. Sophie A. Poe, *Buckboard Days* (reprint; Albuquerque: University of New Mexico Press, 1981), p. 217. Major Lawrence G. Murphy in 1866 wrote about the "good grass" found near Fort Stanton, but he added: "owing to the nature of the country, no large quantity [of hay] can be got in any one place except on the Plains fronting the Rio Pecos." Murphy to DeForrest, 11 June 1866, LR, Dist. NM, RG 393, NA, M-1088, roll 3.

57. The contracts are in Reg. Cont., QMG, RG 92, NA. José Montaño owned a store in Lincoln. Paul Dowlin was post trader at Fort Stanton and owner of a sawmill. Elisha Dow was a Lincoln merchant. Frank Lesnet owned a part interest in Dowlin's mill.

58. General Order No. 13, 14 September 1867, Fort Stanton, PR, RG 393, NA; Gerlach to Ludington, 8 June 1868, and Endorsement by Kautz, 16 April 1872 in letter of Fritz to Quartermaster General, 28 March 1872, LR, CQM, Dist. NM, RG

393, NA; and Stanwood to Hunter, 22 June 1868, LS, Fort Stanton, PR, RG 393, NA.

59. Staab and Bro. to Lee, 20 September 1881, LR, CQM, Dist. NM, RG 393, NA; José Lopez *et. al.* to Commanding Officer, 29 September 1881, LR, Fort Stanton, PR, RG 393, NA.

60. Garrett to Furey, 18 July 1884, Cavanaugh to CQM, 23 July 1884, and [name unclear] to CQM, 26 July 1884, LR, CQM, Dist. NM, RG 393, NA; Leon C. Metz, *Pat Garrett, The Story of a Western Lawman* (Norman: University of Oklahoma Press, 1974), pp. 107–12.

61. See Reg. Cont, QMG, RG 92, NA.

62. J. Ross Browne, *Adventures in the Apache Country, A Tour through Arizona and Sonora, 1864* (reprint; University of Arizona Press, 1974), p. 144.

63. Prescott *Arizona Miner,* 8 February 1868.

64. Floyd S. Fierman, "The Drachmans of Arizona," *American Jewish Archives,* 16 (November 1964): 143; Davis to Green, 13 February 1866, LR, Dist. Ariz., RG 393, NA. Contracts are in Reg. Cont., QMG, RG 92, NA. Prices quoted are in coin.

65. Jones to Fry, 19 May 1867, LR, Dist. of Tucson, RG 393, NA.

66. John A. Spring, *John Spring's Arizona,* edited by A. M. Gustafson (Tucson: University of Arizona Press, 1966), pp. 155–61; Tucson *Weekly Arizonan,* 9 April 1870.

67. Pollock to Wright, 17 December 1867, LS, Fort Bowie, PR, RG 393, NA; Clendenin to Mahnken, 25 August 1868, LR, Subdistrict of Prescott, RG 393, NA; Spring, *John Spring's Arizona,* p. 157; Fierman, "The Drachmans of Arizona," p. 143; Tucson *Weekly Arizonian,* 7 March 1869.

68. Tucson *Weekly Arizonian,* 7 March 1869.

69. Winters to Green, 13 March 1869, LS, Dist. of Tucson, RG 393, NA; Endorsement dated 1 June 1869 on letter of Tully and Ochoa, Endorsements, Fort Bowie, PR, RG 393, NA.

70. Hay to AAAG, 9 October 1873, LR, Fort McDowell, PR, RG 393, NA.

71. Thomas T. Hunter, "Early Days in Arizona," *Arizona Historical Review,* 3 (April 1930): 116; Floyd S. Fierman, "Nathan Bibo's Reminiscences of Early New Mexico," *El Palacio,* 68 (Winter 1961): 256; Endorsement on letter of Eagan, 13 August 1871, Endorsements, Fort Mojave, PR, RG 393, NA.

72. Whitman to Adjutant General, 3 March 1871, and Whitman to AAG, 17 March 1871, LS, Camp Grant, PR, RG 393, NA; Schofield to Adjutant General, 20 October 1871, U.S., Congress, House, Annual Report of the Secretary of War, House Ex. Doc. 1, part 2 (Serial 1503), 42d Cong., 2d sess., 1871, p. 67; Robert M. Utley, *Frontier Regulars, The United States Army and the Indian, 1866–1891* (New York: Macmillan Publishing Co., 1973), pp. 192–93; Constance W. Altshuler, *Chains of Command, Arizona*

and the Army, 1856–1875 (Tucson: Arizona Historical Society, 1981), pp. 190–96.

73. John G. Bourke, *On the Border With Crook* (reprint; Lincoln: University of Nebraska Press, 1971), p. 215; Tucson *Arizona Citizen*, 3 January 1874.

74. Kautz to AAG, 20 October 1875, LS, Dept. Ariz., RG 393, NA.

75. Frank C. Lockwood, ed., *Apaches and Longhorns, The Reminiscences of Will C. Barnes* (Tucson: University of Arizona Press, 1982), pp. 94–95.

76. Wheaton to AAG, 16 September 1869, and Wheaton to Sherburne, 4 March 1870, LS, Fort Whipple, PR, RG 393, NA. A writer for the Prescott *Arizona Miner* in 1872 expressed fear that the continual cutting of wild grass in the valley, "without a chance for renewal from seed," would soon end an easy way of making a living. He claimd that more than 1,600 tons of hay had been cut in Williamson Valley in 1870 and 1,000 tons in 1871. He predicted, however, that only 700 tons would be cut in 1872. "Without something is done to reclaim it," he warned, "it will soon become worthless as a meadow." Prescott *Arizona Miner*, 14 September 1872.

77. The contracts are in Reg. Cont., QMG, RG 92, NA. The contract price in 1868 was $41 per ton in coin; thereafter no designation in coin or currency qualified the contract price.

78. Nickerson to CQM, 9 February 1872, LS, Dept. Ariz., RG 393, NA; Martin to Quartermaster General, 1 August 1878, and Wotherspoon to Quartermaster General, 10 August 1880, Annual Reports Received from Quartermasters, QMG, RG 92, NA; Foster to Small, 24 March 1870, LS by Quartermaster, Subdistrict of Northern Ariz., RG 393, NA; Foster to Stone, 1 November 1870, LR, Fort Whipple, PR, RG 393, NA.

79. Thomas E. Farish, *History of Arizona*, vol. 5 (San Francisco: Filmer Brothers Electrotype Company, 1918), p. 337–38.

80. The contracts are in Reg. Cont., QMG, RG 92, NA.

81. Stacey to Andrews, 1 September 1869, Pond to AAG, 8 June 1871, Pond to CQM, 30 April 1872, LS, Fort Mojave, PR, RG 393, NA; Endorsement on letter of Eagan, 13 August 1871, Endorsements, Fort Mojave, PR, RG 393, NA.

82. The contracts are in Reg. Cont., QMG, RG 92, NA; Tucson *Weekly Arizonan*, 9 October 1869; Tucson *Arizona Citizen*, 21 April 1877; Smith to Quartermaster General, 24 August 1880, Annual Reports Received from Quartermasters, QMG, RG 92, NA.

83. The contracts are in Reg. Cont., QMG, RG 92, NA.

84. Winters to Cogswell, 21 September 1870, Inspection Reports, OIG, RG 159, NA.

85. Evans to AAAG, 29 July 1871, LS, Fort Bowie, PR, RG 393, NA.

86. Sacket to Inspector General, 18 June 1873, Inspection Reports (Camp Bowie), OIG, RG 159, NA.

87. The contracts are in Reg. Cont., QMG, RG 92, NA.

88. Smith to CQM, 6 June 1877, LS by Quartermaster, Camp Grant, PR, RG 393, NA; Proceedings of Board of Survey, 17 January 1877, LR, Camp Grant, PR, RG 393, NA.

89. See, for example, Smith to CQM, 29 May 1878, LS by Quartermaster, Camp Grant, PR, RG 393, NA; Reminiscences of A. M. Franklin, Franklin Family Papers (Arizona Historical Society, Tucson). Hoe-cut hay was known as "poor season hay."

90. The contracts are in Reg. Cont., QMG, RG 92, NA.

91. Thomas E. Farish, *History of Arizona,* vol. 6 (San Francisco: Filmer Brothers Electrotype Company, 1918), p. 138; John J. Gosper, *Report of the Acting Governor of Arizona Made to the Secretary of the Interior for the Year 1881* (Washington: Government Printing Office, 1881), p. 11.

92. Prescott *Weekly Arizona Miner,* 3 May 1878; Martin to Quartermaster General, 1 August 1878, Annual Reports Received from Quartermasters, QMG, RG 92, NA.

93. See Reg. Cont., QMG, RG 92, NA.

94. Hatch to Morse, 20 February 1868, and Wade to CQM, 17 February 1871, LS, Fort Stockton, PR, RG 393, NA; Hatch to Ekin, 5 December 1870, LS, Fort Davis, PR, RG 393, NA; Morrow to CQM, 18 December 1870, LS, Fort Quitman, PR, RG 393, NA. See Appendix Tables 6–9 in this volume.

95. Enos to Kroenig, 2, 14 April 1867, LS, CQM, Dist. NM, RG 393, NA; Kroenig to Ludington, 16 October 1868, LR, CQM, Dist. NM, RG 393, NA.

96. Romero's contract is in Reg. Cont., QMG, RG 92, NA.

97. McDermott to Inman, 9 March 1867, LS, CQM, Dist. NM, RG 393, NA.

98. Circular No. 3, 26 June 1869, enclosed in Myers to Flum, 23 March 1874, Report of N. H. Davis, 9 April 1867, Hudson to Belcher, 10 January 1877, LR, CQM, Dist. NM, RG 393, NA; Carey to Grimes, 29 August 1871, and Myers to Stevenson, 31 August 1872, LS, CQM, Dist. NM, RG 393, NA; Ludington to Hildeburn, 10 December 1867, LR by Quartermaster, Fort Bascom, PR, RG 393, NA.

99. Stedman to AAAG, 11 August 1877, LR, Dist. NM, RG 393, NA, M-1088, roll 30; Names of Agents in Fort Union District, 30 June 1877, LR, CQM, Dist. NM, RG 393, NA.

100. Tucson *Arizona Citizen,* 20 September 1873; Reynolds to Commanding Officer, 29 March 1878, LR, Fort Verde, PR, RG 393, NA.

101. Davis to Enos, 9 April 1867, LR, CQM, Dist. NM, RG 393, NA.

102. Chandler to Myers, 19 September 1873, McLure to Myers, 31 October 1873, and Rifenburg to Smith, 10 November 1873, LR, CQM, Dist. NM, RG 393, NA.

Rifenburg was appointed forage agent in November 1868.

103. Smith to Myers, 13 November 1873, LR, CQM, Dist. NM, RG 393, NA; Belcher to Rifenburg, 25 October 1875, LS, CQM, Dist. NM, RG 393, NA.

104. Engle to CQM, 4 July, 4 August 1872, and Smith to CQM, 12 February, 6 March 1875, LR, CQM, Dist. NM, RG 393, NA; Platt to CQM, 22 January 1876, LR, Dist. NM, RG 393, NA, M-1088, roll 27.

105. Ludington to Elting, 12 February 1869 and 12 June 1869, Myers to Slocum, 8 January 1872, and Myers to Chisum, 15 June 1872, LS, CQM, Dist. NM, RG 393, NA. Chisum received 2¼ cents per pound for hay.

106. Letter from Leitch, 1 August 1879, Register of LR, Dept. Ariz., RG 393, NA; Dodd to CQM, 11 January 1883, LS by Quartermaster, Camp Grant, PR, RG 393, NA.

107. Reynolds to Commanding Officer, 29 March 1878, LR, Fort Verde, PR, RG 393, NA.

108. Hunt to CQM, 14 March 1881, and Lee to Post Quartermaster, 12 March 1881, LR, CQM, Dist. NM, RG 393, NA.

109. Adams and others to Ludington, 19 March 1869, Hennisee to CQM, 21 January 1874, and Special Circular dated 30 January 1878, LR, CQM, Dist. NM, RG 393, NA; Hatch to CQM, 10 April 1880, LS, CQM, Dist. NM, RG 393, NA.

110. Feldwick to Enos, 12 August 1867, LR, CQM, Dist. NM, RG 393, NA.

111. Sachs to Spurgin, 10 February 1868, LR, CQM, Dist. NM, RG 393, NA.

112. For documentation and a more complete discussion of women's role as forage agents, see Darlis A. Miller, "Foragers, Army Women, and Prostitutes," in *New Mexico Women: Intercultural Perspectives,* ed. by Joan M. Jensen and Darlis A. Miller (Albuquerque: University of New Mexico Press, 1986), pp. 144–47.

113. Sandford to Ludington, 21 August 1868, and Trauer to Myers, 2 December 1872, LR, CQM, Dist. NM, RG 393, NA.

114. Lyon to Acting Assistant Quartermaster, 30 January 1878, LR, Dist. NM, RG 393, NA, M-1088, roll 32.

115. Endorsement of Ayers to Belcher, 2 June 1877, LR, Dist. NM, RG 393, NA, M-1088, roll 24.

116. Belcher to AAAG, 21 August 1877, LR, Dist. NM, RG 393, NA, M-1088, roll 30.

117. Darlis A. Miller, *The California Column in New Mexico* (Albuquerque: University of New Mexico Press, 1982), pp. 116–22; Enclosure in Hennisee to CQM, 10 March 1874, Casey to Belcher, 18 June 1877, Acting Assistant Quartermaster to CQM, 22 January 1879, and Gilliss to CQM, 28 August 1884, LR, CQM, Dist. NM, RG 393, NA.

118. Hobby to Lee, 29 April 1881, Walker to CQM, 17 June, 7 July 1881, and Hobby to CQM, 22 May 1882, LR, CQM, Dist. NM, RG 393, NA; Stiles to MacKenzie, 23 November 1881, and Hobby to

MacKenzie, 31 January 1882, LR, Dist. NM, RG 393, NA, M-1088, rolls 44, 45.

119. Evans to AAAG, 9 May, 12 August 1869, LS, Fort Wingate, PR, RG 393, NA. John Q. Adams was serving as forage agent at the Rio Puerco Crossing in late December 1867. On 19 December 1868, J. B. Adams received appointment there, John Q. Adams having resigned. It is likely that both men retained interest in the property, for Major Evans assumed that John Q. Adams was still agent in June 1869.

120. General Order No. 3, 10 February 1882, LR, Dist. NM, RG 393, NA, M-1088, roll 45; Letter of J. Marshall, 8 April 1882, Register of LR, Quartermaster, Fort Bliss, PR, RG 393, NA; General Order No. 11, 26 June 1883, and General Order No. 12, 20 June 1885, Dept. Ariz., RG 393, NA.

121. Inspection Report, Fort Craig, 31 March–2 April 1866, Inspector, Dept. NM, RG 393, NA; Abréu to [name unclear], 20 April 1866, Enos to Meinhold, 13 October 1866, Meinhold to Gerhart, 19 October 1866, Post Commander to AAAG, 2 November 1869, LS, Fort Craig, PR, RG 393, NA; Wardwell to Ludington, 9 September 1868, LR, Fort Craig, PR, RG 393, NA; Buffum to CQM, 28 June 1871, LR, CQM, Dist. NM, RG 393, NA.

Wardwell's contract is in Reg. Cont., QMG, RG 92, NA. Socorro residents secured coal from these mines during the 1850s. Late in 1864 the army took steps to secure the coal deposits "for the exclusive use of the garrison at Fort Craig." McFerran to Nissen, 9 December 1864, LS, CQM, Dept. NM, RG 393, NA. See Paige W. Christiansen, *The Story of Mining in New Mexico* (Socorro: New Mexico Bureau of Mines and Mineral Resources, 1974), p. 39.

122. Letterman to Hunter, 4 October 1867, LR, Dist. NM, RG 393, NA, M-1088, roll 5.

123. See Reg. Cont., QMG, RG 92, NA.

124. *Ibid.* These figures are based on contracts that appear in the registers. Some contracts may not have been registered.

125. General Order No. 97, enclosed in Townsend to Commanding General, Military Division of the Missouri, 6 May 1876, LR, Dist. NM, RG 393, NA, M-1088, roll 27.

126. See, for example, Post Commander to CQM, 26 October 1867, LS, Fort Craig, PR, RG 393, NA; Lane to AAAG, 30 March 1869, LS, Fort Selden, PR, RG 393, NA; Endorsement by Bernard, 21 September 1870, End. S., Fort Bowie, PR, RG 393, NA; Post Commander to AAAG, 10 December 1871, LS, Camp Hualpai, PR, RG 393, NA; Morrow to CQM, 18 December 1870, LS, Fort Quitman, PR, RG 393, NA.

127. Post Commander to AAAG, 5 September 1870, 6 October 1870, LS, Fort Stanton, PR, RG 393, NA.

128. See Reg. Cont., QMG, RG 92, NA. This observation is based on contracts that appear in the registers.

129. Carleton to AAG, 6 March

1871, Inspection Reports (Fort Bliss), OIG, RG 159, NA.

130. Lemon to Ludington, 23 December 1868, and Ludington to Lemon, 5 April 1869, LS, CQM, Dist. NM, RG 393, NA.

131. Carleton to AAG, 6 March 1871, Inspection Reports (Fort Bliss), OIG, RG 159, NA; Blunt to CQM, 30 July 1875, LS, Fort Stockton, PR, RG 393, NA.

132. Brown's problems with the quartermaster's department are described fully in Brown to Ludington, 30 August 1868, and Whitman to Ludington, 3 September 1868, LR, CQM, Dist. NM, RG 393, NA.

133. Brown arrived at the fee for demurrage ($1,176) by charging the government $6 per day for each of his fourteen wagons that had stood idle. The army usually charged contractors who detained a wagon $5 a day per wagon. Brown also claimed that the total bill of $5,429 was a modest sum to charge the government, for had he not received the wood contract his ox teams would have been freighting goods from the States at the ruling rates. Loss of this income he believed should properly "be included in the loss sustained by the contractor." And had it not been for the wood contract, Brown further reported, he would have bid for and received the hay contract at Fort Union, as his proposal would have been nearly $2 less than the winning bid.

134. See Reg. Cont., QMG, RG 92, NA.

135. Robert W. Frazer, *Forts of the West, Military Forts and Presidios and Posts Commonly called Forts West of the Mississippi River to 1898* (Norman: University of Oklahoma Press, 1965), pp. 36–38; Virginia McConnell Simmons, *The San Luis Valley, Land of the Six-Armed Cross* (Boulder: Pruett Publishing Company, 1979), p. 53; [Carson] to Jones, 10 June 1866, LS, Fort Garland, PR, RG 393, NA.

136. The contracts are in Reg. Cont., QMG, RG 92, NA. In 1876 Michael McCarthy, identified in the 1870 census as a farm laborer, contracted to deliver 700 cords of piñon at $1.99 per cord. This was the lowest contract price recorded for any post in the Southwest during the years of this study. The post quartermaster believed McCarthy's price was unreasonably low, for he would have to haul wood twelve to fifteen miles to the post. Waters to CQM, 20 July 1876, LR, CQM, Dist. NM, RG 393, NA. Major Andrew J. Alexander, commanding the post, reported in February 1873 that the contract price for wood was $1.91 per cord, but this contract does not appear in the registers. Alexander to AAG, 21 February 1873, LS, Fort Garland, PR, RG 393, NA. After the railroad reached Garland in 1878 the post's grain supply probably came from the East. No grain contracts for Garland appear in the registers after 1878.

137. The contracts are in Reg. Cont., QMG, RG 92, NA.

138. Baca to Lee, 29 August 1881, LR, CQM, Dist. NM, RG 393, NA; Lee to Post Commander, 13 September 1881, LS, CQM, Dist., NM, RG 393, NA; Marc Simmons, *The Little Lion of the Southwest, A Life of Manuel Antonio Chaves* (Chicago: Swallow Press, Inc., 1973), pp. 198–211.

139. The contracts are in Reg. Cont., QMG, RG 92, NA.

140. Corbett to CQM, 12 October 1867, and Ryan to Robinson, 17 June 1870, LS by Quartermaster, Fort Cummings, PR, RG 393, NA.

141. The contracts are in Reg. Cont., QMG, RG 92, NA. While filling his contract in 1868, Lemon employed Appelzoller to oversee delivery of wood at the post. Late in 1869, Appelzoller was authorized to open a trading establishment at Fort Cummings.

142. Hunt to McDonald, 23 August 1880, Bingham to CQM, 28 September 1880, and Carpenter to Lee, 30 December 1880, LR, CQM, Dist. NM, RG 393, NA.

143. Santa Fe *Daily New Mexican*, 26, 29 October 1880.

144. Wade to CQM, 17 February 1871, LS, Fort Stockton, PR, RG 393, NA; Clayton W. Williams, *Texas' Last Frontier, Fort Stockton and the Trans–Pecos, 1861–1895* (College Station: Texas A & M University Press, 1982), p. 105; see Reg. Cont., QMG, RG 92, NA.

Joseph Friedlander's wood delivered in 1875 at $5 per cord was the least expensive. Like Gallagher and Friedlander, at least three other fuel contractors at Stockton had received appointments as post traders—Thomas Johnson, W. E. Friedlander, and Michael F. Corbett.

145. The contracts are in Reg. Cont., QMG, RG 92, NA; Williams, *Fort Stockton and the Trans–Pecos,* p. 219. The average contract price for wood delivered at Fort Quitman between 1868 and 1877, the year the post was abandoned, was $6.18 per cord. The most expensive wood at $8 per cord was delivered in 1869 by Francis Rooney, a native of Ireland, who held part interest in the post trader's store. Rooney had been stationed at Quitman both before and during the Civil War. He settled near Fort Stockton in 1870. George H. Abbott delivered the least expensive wood at Quitman, receiving $3.95 per cord. Abbott served as Quitman's post trader in the mid-1870s and later became trader at Fort Davis. Williams, *Fort Stockton and the Trans–Pecos,* p. 140; Barry Scobee, *Fort Davis, Texas, 1583–1960* (Fort Davis: Barry Scobee, 1963), p. 158; Rooney to Trask, 20 August 1869, LR, Fort Quitman, PR, RG 393, NA; Bentzoni to Adjutant General, 22 March 1876, LR, Fort Quitman, PR, RG 393, NA.

146. Meyer to Grover, 16 August 1869, Merriam to AAG, 4 November 1869, LS, Fort Bliss, PR, RG 393, NA. Vernay was dismissed from the army on 18 April 1870. In 1898 he was awarded a Medal of Honor for heroic service during the Civil War. Williams was killed in a now

famous gunfight in December 1870. See Francis B. Heitman, *Historical Register and Dictionary of the United States Army, 1789–1903*, vol. 1 (reprint; Urbana: University of Illinois Press, 1965), p. 986; C. L. Sonnichsen, *Pass of the North, Four Centuries on the Rio Grande* (El Paso: Texas Western Press, 1968), p. 189. The average contract price for fuel at Fort Bliss between 1867 and 1880 was $6.46 per cord. T. A. Washington's wood in 1869 was the most expensive. The only Hispano to receive a fuel contract in west Texas during the years of this study was José María Gonzales, a justice of the peace in El Paso County, who delivered wood at Fort Bliss in 1870. He received $8.50 per cord for a year's supply, the second highest price for wood delivered at the post between 1867 and 1880.

147. Frazer, *Forts of the West*, p. 4. The contracts are in Reg. Cont., QMG, RG 92, NA; Endorsement by Bernard, 21 September 1870, End. S., Fort Bowie, PR, RG 393, NA.

148. See Reg. Cont., QMG, RG 92, NA; Evans to Russell, 14 June 1871, LS, Fort Bowie, PR, RG 393, NA; Las Cruces *Borderer*, 8 November 1871. David Dunham, who was supervising construction of Camp Lowell on the Rillito, delivered the least expensive fuel at $6.50 per cord.

149. See Reg. Cont., QMG, RG 92, NA. Missouri-born Louis B. St. James, a Yavapai County resident, delivered the least expensive wood—600 cords of soft wood in 1877 at $2.25 per cord. Will Barnes, who arrived at Fort Apache in 1880, later recalled that the government paid $12 a cord for the wood delivered at the post. Contract registers do not support this observation. See Lockwood, ed., *Apaches and Longhorns*, p. 37.

150. The contracts are in Reg. Cont., QMG, RG 92, NA. For Maish, see Richard R. Willey, "La Canoa: A Spanish Land Grant Lost and Found," *The Smoke Signal*, 38 (Fall 1979): 162; for Veil, see Altshuler, *Chains of Command*, pp. 267–68. Among the contractors delivering fuel at Fort Whipple, Veil's wood was the most expensive—$5.49 per cord in 1872. Victor Rouiller of Springerville delivered the least expensive at $3.40 per cord in 1877.

151. The contracts are in Reg. Cont., QMG, RG 92, NA; Dana to AAAG, 19 May 1880, LS, CQM, Dist. NM, RG 393, NA.

152. Goodwin to CQM, 31 May 1881, LR, CQM, Dist. NM, RG 393, NA.

153. Hunt to CQM, 31 May 1881, LR, CQM, Dist. NM, RG 393, NA.

154. See Reg. Cont., QMG, RG 92, NA; Frazer, *Forts of the West*, p. 38. Contracts issued for coal in 1880 called for a ton of 2,000 pounds. Thereafter, contracts called for a ton of 2,240 pounds.

155. See Reg. Cont., QMG, RG 92, NA; Chief Quartermaster, Dist. NM, to Quartermaster General, 10 July 1885, Annual Reports Received from Quartermasters, RG 92, NA;

Christiansen, *The Story of Mining in New Mexico,* pp. 81–84; David F. Myrick, *New Mexico's Railroads—An Historical Survey* (Golden: Colorado Railroad Museum, 1970), p. 127; C. M. Chase, *The Editor's Run in New Mexico and Colorado* (reprint; Fort Davis, Texas: Frontier Book Co., 1968), pp. 162–63.

Chapter 4

1. Robert W. Frazer, *Forts and Supplies, The Role of the Army in the Economy of the Southwest, 1846–1861* (Albuquerque: University of New Mexico Press, 1983), p. 8.

2. Quote is in Frazer, *Forts and Supplies,* p. 7.

3. Frazer, *Forts and Supplies,* pp. 44, 80, 106, 186; Robert W. Frazer, "Purveyors of Flour to the Army: Department of New Mexico, 1849–1861," *New Mexico Historical Review,* 47 (July 1972): 227; Janet Lecompte, "Ceran St. Vrain's Stone Mill at Mora," Cultural Properties Review Committee, State Planning Office, Santa Fe (copy provided by Janet Lecompte).

4. Franz Huning, *Trader on the Santa Fe Trail, Memoirs of Franz Huning* (Albuquerque: University of Albuquerque, 1973), p. 20; Marc Simmons, *Albuquerque, A Narrative History* (Albuquerque: University of New Mexico Press, 1982), pp. 153–56, 198–99; Gilberto Espinosa and Tibo J. Chavez, *El Rio Abajo* (Portales, New Mexico: Bishop Publishing Company, n.d.), p. 131.

5. Espinosa and Chavez, *El Rio Abajo,* p. 131.

6. Thomas E. Farish, *History of Arizona,* vol. 6 (San Francisco: Filmer Brothers Electrotype Company, 1918), p. 48. William and Alfred Rowlett had constructed a small, modern gristmill on the Santa Cruz River near Tucson in 1857 or 1858. Before the outbreak of the Civil War, William Grant purchased the Rowlett mill and established a second mill in the same area. Both mills were put to the torch when Union forces evacuated Arizona in 1861. Frazer, *Forts and Supplies,* pp. 138, 168–69, 178.

7. U.S., *Department of the Interior, Ninth Census of the United States: 1870, Wealth and Industry,* vol. 3 (Washington: Government Printing Office, 1872), p. 598, hereafter *Ninth Census, 1870;* Schedule 4, Products of Industry in New Mexico, Ninth Census of the United States, 1870, NMSA. Small, primitive mills apparently were not listed in the census.

8. The contracts are in Reg. Cont., CGS, RG 192, NA.

9. The contracts are in Reg. Cont., CGS, RG 192, NA. Henry Lesinsky furnished the least expensive cornmeal, agreeing in 1873 to deliver 1,000 pounds at Fort Stanton at 2 cents per pound. In 1868 between 1,000 and 4,000 pounds of cornmeal were purchased for Forts Union, Bascom, Wingate, Marcy, Stanton, and Garland, but the amounts were considered too small to issue contracts. See four let-

ters dated 18 September 1868, sent by McClure to Kroenig, Hersch, Huning, and Sandoval, McClure to Gerlach, 19 September 1868, and McClure to Meyer, 12 October 1868, LS, CCS, Dist. NM, RG 393, NA.

10. The contracts are in Reg. Cont., CGS, RG 192, NA.

11. The contracts are in Reg. Cont., CGS, RG 192, NA.

12. Whitehead to CCS, Department of the Missouri, 15 March 1877, LS, CCS, Dist. NM, RG 393, NA; Eagan to AAG, 6 September 1878 and 3 September 1879, Arizona, Cons. Corres. File, RG 92, NA.

13. McClure to Eaton, 14 March 1870, LS, CCS, Dist. NM, RG 393, NA. In Arizona in 1880 bacon was issued to troops in garrison two days in ten; pork, one day in ten. See Eagan to Macfeely, 30 August 1880, Arizona, Cons. Corres. File, RG 92, NA.

14. McClure to Blake, 9 June 1868, and McClure to Clarke, 5 August 1868, LS, CCS, Dist. NM, RG 393, NA.

15. McClure to Rigg, 12 September 1868, McClure to Blake, 24 August 1869, and McClure to Blake, 11 September 1869, LS, CCS, Dist. NM, RG 393, NA.

16. McClure to Murphy, 11 October 1869, and McClure to Slade, 27 October 1869, LS, CCS, Dist. NM, RG 393, NA.

17. McClure to Eaton, 5 January 1870, LS, CCS, Dist. NM, RG 393, NA.

18. The contracts are in Reg. Cont., CGS, RG 192, NA.

19. Nash to Commissary General of Subsistence, 14 January 1871, and Nash to CCS, Department of the Missouri, 17 April 1871, LS, CCS, Dist. NM, RG 393, NA; Nash statement included in Clarke to Eaton, 19 June 1871, New Mexico, Cons. Corres. File, RG 92, NA; Huning, *Trader on the Santa Fe Trail,* p. 101. Keerl and his wife were killed by Indians as they were returning from Santo Tomás. For an account of their deaths, see Las Cruces *Borderer,* 16 March 1871.

20. Santa Fe *Daily New Mexican,* 10 May 1871; Clarke to Morgan, 7 June 1871, enclosed in Clarke to Eaton, 19 June 1871, New Mexico, Cons. Corres. File, RG 92, NA; contracts are in Reg. Cont., CGS, RG 192, NA.

21. Santa Fe *Daily New Mexican,* 25 May 1872.

22. Santa Fe *Weekly New Mexican,* 6 January 1872.

23. Nash to Cullen, 17 July 1872, LS, CCS, Dist. NM, RG 393, NA. Cullen has not been identified. But since this letter was addressed to him at Kit Carson, Colorado, the terminus on the Kansas Pacific Railroad for freight to New Mexico, it is likely he imported bacon to fill his Fort Stanton contract. Cullen had submitted bids for each post in the district.

24. U.S., Department of the Interior, *Tenth Census of the United States: 1880, Agriculture,* vol. 3 (Washington: Government Printing Office,

1883), pp. 39, 46, hereafter *Tenth Census, 1880.*

25. *Tenth Census, 1880,* vol. 3 (Agriculture), pp. 39, 46. Figures for Yavapai County were not included in the 1870 census.

26. The contracts are in Reg. Cont., CGS, RG 192, NA. On his ranch near the Mexican border, Pete Kitchen raised several hundred hogs, selling ham and bacon on the Tucson market as well as to the army. His name does not appear in the registers as a bacon or pork contractor, however. See Frank C. Lockwood, *Pioneer Portraits, Selected Vignettes* (Tucson: University of Arizona Press, 1968), pp. 22–23; Bernice Cosulich, *Tucson* (Tucson: Arizona Silhouettes, 1953), p. 170.

27. Hardie to Marcy, 27 May 1875, Inspection Reports, OIG, RG 159, NA.

28. Hardie to Marcy, 27 May 1875, Inspection Reports, OIG, RG 159, NA.

29. Eagan to AAG, 3 September 1879, and Eagan to Macfeely, 30 August 1880, Arizona, Cons. Corres. File, RG 92, NA.

30. Davis to Inspector General, 20 September 1875, Davis Report on Fort Stockton, 26 October 1875, Davis Report on Fort Quitman, 30 November 1875, and Davis Report on Fort Bliss, 4 November 1875, Inspection Reports, OIG, RG 159, NA; Augur to Townsend, 22 May 1873, LS, Dept. Texas, RG 393, NA, M-1114, roll 2.

31. Martha Summerhayes, *Vanished Arizona* (reprint; Glorieta, New Mexico: Rio Grande Press, Inc., 1976), p. 219.

32. Davis to Inspector General, [December 1872], Inspection Reports, OIG, RG 159, NA; see entries for 1873, Reg. Cont., CGS, RG 192, NA; Eagan to Macfeely, 17 January 1876, LS, CCS, Dist. NM, RG 393, NA.

33. Whitehead to CCS, Department of the Missouri, 22 January 1877, Whitehead to Rockwell and Co., 23 March 1877, and Whitehead to Rockwell and Co., 10 December 1877, LS, CCS, Dist. NM, RG 393, NA.

34. Whitehead to Commissary General of Subsistence, 26 July 1879, New Mexico, Cons. Corres. File, RG 92, NA.

35. Letter by Thayer, 3 July 1866, Registers of LR, Dist. NM, RG 393, NA, M-1097, roll 1; Donnelly to AAG, 24 July 1871, LR, Fort Stockton, PR, RG 393, NA.

36. Post Adjutant to DeLany, 1 March 1881, LS, Fort Stanton, PR, RG 393, NA.

37. Prescott *Arizona Miner,* 9 May 1868; Prescott *Weekly Arizona Miner,* 9 October 1869.

38. Simpson to Commanding Officer, 15 January 1870, LR, Camp Lowell, PR, RG 393, NA; Foster to Simpson, 4 May 1870, LS by Quartermaster, Subdistrict of Northern Arizona, RG 393, NA.

39. Letter by Greene, and endorsements, 20 January 1877, LR, Fort Bowie, PR, RG 393, NA.

40. Eagan to AAAG, 10 August 1877, and Eagan to Macfeely, 18 Au-

gust 1877, Arizona, Cons. Corres. File, RG 92, NA.

41. Eagan to AAG, 6 September 1878, Arizona, Cons. Corres. File, RG 92, NA. Eastern butter was packed during the fall. California dairies packed butter in early summer, which meant California butter had to be stored several months in San Francisco awaiting arrival of cool weather for shipment to Arizona.

42. The contracts are in Reg. Cont., CGS, RG 192, NA.

43. McClure to Hunter, 28 February 1868, Inspection Reports, OIG, RG 159, NA; Willard to McClure, 20 August 1869, LR, CQM, Dist. NM, RG 393, NA.

44. Marian Russell, *Land of Enchantment, Memoirs of Marian Russell Along the Santa Fe Trail as Dictated to Mrs. Hall Russell* (reprint; Albuquerque: University of New Mexico Press, 1981), p. 121.

45. Carleton to Bell, 18 July 1864, LS, Dept. NM, RG 393, NA, M-1072, roll 3; Santa Fe *Weekly Gazette,* 30 July 1864; Hunter to DeForrest, 3 October 1866, LS, Fort Bascom, RG 393, NA.

46. Young to AAAG, 17 August 1873, LR, Dist. NM, RG 393, NA, M-1088, roll 20.

47. W. W. Mills, *Forty Years at El Paso, 1858–1898* (El Paso: Carl Hertzog, 1962), p. 27.

48. Marshall to AAG, 22 November 1866, LS, Fort Union, PR, RG 393, NA.

49. The contracts are in Reg. Cont., CGS, RG 192, NA. See also McClure to Haines, 20 September 1867, LS, CCS, Dist. NM, RG 393, NA.

50. The contracts are in Reg. Cont., CGS, RG 192, NA.

51. McClure to Kobbé, 5 October 1869, and Whitehead to Commissary General of Subsistence, 11 January 1878, LS, CCS, Dist. NM, RG 393, NA; Woodruff to Commissary General of Subsistence, 17 July 1882, New Mexico, Cons. Corres. File, RG 92, NA. According to Captain Frederick F. Whitehead, the district's chief commissary in 1878, salt had become "scarce and high." Since the railroad was about to enter the territory, his remarks may reflect an increased sensitivity to price and quality.

52. The contracts are in Reg. Cont., CGS, RG 192, NA.

53. The contracts are in Reg. Cont., CGS, RG, 192, NA; see Clayton W. Williams, *Texas' Last Frontier, Fort Stockton and the Trans–Pecos, 1861–1895* (College Station: Texas A & M Press, 1982), pp. 155, 176. Corbett held the Fort Stockton contracts in 1874, 1876, and 1877.

54. Frazer, *Forts and Supplies,* p. 186.

55. The contracts are in Reg. Cont., CGS, RG 192, NA; McClure to Eaton, 13 December 1866, McClure to Rynerson, 13 December 1866, McClure to Eaton, 21 January 1867, LS, CCS, Dist. NM, RG 393, NA. Hersch's bid had been lower than that of his five competitors, whose bids had ranged from 12 to 20 cents per pound.

56. See letters dated 19 February

1868 from McClure to Amador, Stafford, Watts, and Seligman, and letters dated 21 February 1868 from McClure to Willard, Murphy, Spurgin, and Pulford, LS, CCS, Dist. NM, RG 393, NA; Hersch's contracts are in Reg. Cont., CGS, RG, 192, NA. Hersch also supplied beans at Wingate, Craig, and Garland.

57. See Santa Fe *Weekly New Mexican,* 24 November 1868, 15 December 1868.

58. The contracts are in Reg. Cont., CGS, RG 192, NA; McClure to Kobbé, 11 December 1868, and McClure to Gold, 17 January 1869, LS, CCS, Dist. NM, RG 393, NA. Lemon's contracts at Forts Bayard, Cummings, and Selden together called for delivery of 19,000 pounds at between 4½ and 5¾ cents per pound. Luna agreed to deliver 7,000 pounds at Craig for 6 375/1000 cents per pound and 12,000 pounds at Wingate for 6 625/1000 cents per pound.

59. McClure to Kobbé, 5 October 1869, LS, CCS, Dist. NM, RG 393, NA. The contracts are in Reg. Cont., CGS, RG 192, NA. Staab received two bean contracts; Eldodt, four.

60. Eagan to CCS, Fort Leavenworth, 13 November 1875, LS, CCS, Dist. NM, RG 393, NA; Macfeely to Belknap, 9 October 1875, U.S., Congress, House, Annual Report of the Secretary of War, 1875, House Ex. Doc. 1 (Serial 1674), 44th Cong., 1st. sess., 1875, p. 310; Acting CCS to CCS, Department of the Missouri, 22 September 1874, LS, CCS, Dist. NM, RG 393, NA.

61. Whitehead to Commissary General of Subsistence, 11 January 1878, LS, CCS, Dist. NM, RG 393, NA; Woodruff to Commissary General of Subsistence, 24 July 1880 and 17 July 1882, New Mexico, Cons. Corres. File, RG 92, NA.

62. Hatch to Morse, 6 February 1868, LS, Fort Stockton, PR, RG 393, NA.

63. Merritt to Morse, 25 July 1868, LS, Fort Davis, PR, RG 393, NA.

64. See Reg. Cont., CGS, RG 192, NA; U.S., Congress, Senate, Letter from the Secretary of War, 31 January 1872, S. Ex. Doc. 26 (Serial 1478), 42nd Cong., 2d sess., 1872, p. 12; U.S., Congress, House, Letter from the Secretary of War, 24 January 1873, House Ex. Doc. 158 (Serial 1567), 42nd Cong., 3d sess., 1873, pp. 13, 15. Not all bean contracts for the west Texas posts were registered.

65. Whittier to Scott, 19 February 1866, Inspection Reports, OIG, RG 159, NA.

66. Alexander to Sherburne, 25 January 1869, LS, Subdistrict of the Verde, RG 393, NA; Simpson to Eaton, 19 December 1871, Arizona, Cons. Corres. File, RG 92, NA.

67. See Gerald Stanley, "Merchandising in the Southwest, The Mark I. Jacobs Company of Tucson, 1867 to 1875," *American Jewish Archives,* 23 (April 1971): 86–102; for the quotation, see Lionel to Barron, 12 September 1875 (Jacobs Manuscripts, University of Arizona).

68. The contracts are in Reg.

Cont., CGS, RG 192, NA. See John A. Spring, *John Spring's Arizona,* edited by A. M. Gustafson (Tucson: University of Arizona Press, 1966), pp. 188, 199, 208.

69. The contracts and open market purchases are in Reg. Cont., CGS, RG 192, NA. See entry for Zeckendorf, 11 November 1870. For Zeckendorf's family history, see Blaine Peterson Lamb, "Jewish Pioneers in Arizona, 1850–1920" (Ph.D. dissertation, Arizona State University, 1982), pp. 107–8.

70. Jacobs and Company to Bichard and Company, 30 December 1871, 13 January 1872, and Barron to Pa, 18 January 1872 (Jacobs Manuscripts, University of Arizona). The contracts are in Reg. Cont., CGS, RG 192, NA. Hellings received in 1871 the first bean contracts let in Arizona for Date Creek, Hualpai, Verde, and Whipple.

71. Barron to Pa, 24 April 1872 (Jacobs Manuscripts, University of Arizona).

72. Barron to Pa, 4, 7 May 1872 (Jacobs Manuscripts, University of Arizona).

73. Memo of Bean Arrangement between Otero and Jacobs, [April 1872], Barron to Pa, 23 May 1872 and 8 June 1872 (Jacobs Manuscripts, University of Arizona). Spring mentions Sabino Otero's farm in *John Spring's Arizona,* p. 188. In June of 1872 Barron and a Tucson business associate tried to corner the market in beans. They held the only large supply in town, and before the new crop was harvested they were selling beans at from 12½ to 13 cents per pound. See Barron to Pa, 22 June 1872 (Jacobs Manuscripts, University of Arizona). The registers show that the Tucson bean contract for 1872 went to W. F. Scott, but it is clear from Barron's letters that he and Otero submitted the bid and filled the contract. Some of the beans they delivered to the army before the November harvest were purchased in Sonora at 3 cents a pound.

74. The contracts are in Reg. Cont., CGS, RG 192, NA. Local contractors supplied Fort Bowie with beans for the first time in 1865 and then not again until 1872. They furnished beans at McDowell between 1870 and 1875.

75. Eagan to AAAG, 10 August 1877, Arizona, Cons. Corres. File, RG 92, NA.

76. Eagan to AAAG, 10 August 1877, Arizona, Cons. Corres. File, RG 92, NA. In October 1877 corn contractor Louis A. Stevens reported a general drought in Arizona and an almost total failure of crops. This may account in part for the lack of bids.

77. Huning's contract is in Reg. Cont., CGS, RG 192, NA. See also Eagan to AAG, 3 September 1879, Arizona, Cons. Corres. File, RG 92, NA.

78. Sacket to Inspector General, 18 June 1873, Inspection Reports, OIG, RG 159, NA.

79. Purrington to Townsend, 4 April 1869, LS, Fort Quitman, PR, RG 393, NA.

80. Fleming to AAG, 10 March

1868, LS, Fort Garland, PR, RG 393, NA. The darkness of flour was explained in part by the type of wheat raised in New Mexico. In examining flour that was a shade darker than an accepted sample, Chief Commissary Charles McClure later remarked, "I do not think that the color is objectionable as red wheat is most exclusively raised in this country." McClure to Eckles, 2 March 1870, LS, CCS, Dist. NM, RG 393, NA. The commissary department imported eastern flour for sale to officers.

81. O'Connor to Commanding Officer, 30 January 1870, and Fechet to McClure, 24 February 1870, LR, Dist. NM, RG 393, NA, M-1088, roll 11.

82. McClure to Hunter, 28 February 1868, Inspection Reports, OIG, RG 159, NA; McClure to Morgan, 20 April 1868, LS, CCS, Dist. NM, RG 393, NA.

83. Nash to AAG, 15 January 1873, Eagan to Wilson, 12 April 1876, and Eagan to Macfeely, 17 April 1876, LS, CCS, Dist. NM, RG 393, NA. Captain Samuel Cushing, who served as district chief commissary of subsistence between the terms of Nash and Eagan, reported that samples offered by the lowest bidders for flour contracts in New Mexico varied "but slightly in quality" and that bread baked from the samples was "sweet and good and apparently nutritious." Cushing to AAAG, 2 March 1874, LR, Dist. NM, RG 393, NA, M-1088, roll 21.

84. McClure to Hunter, 16 September 1868, LS, CCS, Dist. NM, RG 393, NA.

85. Nash to Kobbé, 13 January 1871, and Jones to Nash, 13 January 1871, LR, Dist. NM, RG 393, NA, M-1088, roll 14.

86. Eagan to Wilson, 12 April 1876, LS, CCS, Dist. NM, RG 393, NA.

87. Jones to Hunter, 30 March 1869, LS, CCS, Dist. NM, RG 393, NA; McKibben to AAAG, 3 July 1873, and Whitehead to AAAG, 29 July 1878, LR, Dist. NM, RG 393, NA, M-1088, rolls 20, 34.

88. McClure to Sutorius, 3 April 1868, and McClure to Eckles, 2 March 1870, LS, CCS, Dist. NM, RG 393, NA.

89. Nash to AAAG, 18 April 1871, LR, Dist. NM, RG 393, NA, M-1088, roll 14. For capacity of Romero's and St. Vrain's mills, see Schedule 4, Products of Industry in New Mexico, Ninth Census of the United States, 1870, NMSA. In July 1864, Ceran St. Vrain's mill near Taos burned to the ground. Shortly thereafter he constructed a new stone mill at Mora. Both the stone mill and the older frame mill (constructed in 1850) were still operating in Mora at the time of Ceran St. Vrain's death on 28 October 1870. Lecompte, "Ceran St. Vrain's Stone Mill at Mora."

90. Nash to AAAG, 15 May 1873, LR, Dist. NM, RG 393, NA, M-1088, roll 19.

91. See Reg. Cont., CGS, RG 192, NA. Higher contract prices were awarded in 1860, prices ranging

from 7⁸⁄₁₀ to 20½ cents per pound. Frazer, *Forts and Supplies*, p. 155.

92. McClure to Hunter, 8 October 1867, LS, CCS, Dist. NM, RG 393, NA.

93. McClure to Tanfield, 24 November 1866, and McClure to McDonald, 24 November 1866, LS, CCS, Dist. NM, RG 393, NA.

94. See Reg. Cont., CGS, RG 192, NA, and McClure to Mueller, 3 July 1867, LS, CCS, Dist. NM, RG 393, NA. See Appendix Table 10 in this volume.

95. McClure to Spurgin, 19 September 1868, LS, CCS, Dist. NM, RG 393, NA.

96. The contracts are in Reg. Cont., CGS, RG 192, NA. For capacity of Moore's mill, see Schedule 4, Products of Industry in New Mexico, Ninth Census of the United States, 1870, NMSA; for quality of Moore's flour, see McClure to Hunter, 28 February 1868, Inspection Reports, OIG, RG 159, NA.

97. The contracts are in Reg. Cont., CGS, RG, 192, NA.

98. Staab to Nash, 14 November 1871, LR, Dist. NM, RG 393, NA, M-1088, roll 14.

99. Webster to AAG, 7 October 1867, LR, Dist. NM, RG 393, NA, M-1088, roll 9; McClure to Jones, 28 October 1867, LS, CCS, Dist. NM, RG 393, NA.

100. The contracts are in Reg. Cont., CGS, RG 192, NA.

101. On Webster's freighting costs, see endorsements on Hartwell to AG, 26 May 1871, filed on 24 July 1875 with other letters relative to sale to officers of subsistence stores, LR, Dist. NM, RG 393, NA, M-1088, roll 26. On freighting costs at Cummings and Bayard, see McClure to Hunter, 16 September 1868, and McClure to Eaton, 3 October 1868, LS, CCS, Dist. NM, RG 393, NA; McClure to Hunter, 30 September 1868, LR, Dist. NM, RG 393, NA, M-1088, roll 10.

102. The contracts are in Reg. Cont., CGS, RG 192, NA; Las Cruces *Borderer*, 24 April 1872. Lesinsky received contracts at Bayard and McRae in 1870, Bayard, Cummings, and McRae in 1871 and 1872, and Bayard in 1873. Rosenbaum supplied Bayard with flour in 1874, 1876, 1877, and 1878.

103. *Deed Record* 4, Doña Ana County (Doña Ana County Court House, Las Cruces, New Mexico), pp. 74–77; Schedule 4, Products of Industry in New Mexico, Ninth Census of the United States, 1870, NMSA; Las Cruces *Borderer*, 14, 21 September 1872, 11 July 1874; *Mesilla News*, 7 March 1874. Lesinsky and his wife Mathilda sold the flouring mill to Jacob Schaublin in January 1879 for $9,200. *Deed Record E*, Doña Ana County (Doña Ana County Court House, Las Cruces, New Mexico), pp. 601–3.

104. See Schedule 4, Products of Industry in New Mexico, Ninth Census of the United States, 1870, NMSA. In 1877 Las Cruces and Mesilla each claimed one steam-powered and one water-powered

gristmill. See *Mesilla Valley Independent,* 14 July 1877.

105. Huning's contracts are in Reg. Cont., CGS, RG 192, NA.

106. Santa Fe *Daily New Mexican,* 15 January 1872; Davis to Inspector General, [December 1872], Inspection Reports, OIG, RG 159, NA.

107. See Reg. Cont., CGS, RG 192, NA. Only one contract for supplying Wingate with flour was issued between 1880 and 1885 and that went to Sigmund Wedeles in 1882.

108. Floyd S. Fierman, *Guts and Ruts, The Jewish Pioneer on the Trail in the American Southwest* (New York: Ktav Publishing House, Inc., 1985), pp. 27–31.

109. Fierman, *Guts and Ruts,* p. 38. The contracts are in Reg. Cont., CGS, RG 192, NA; Staab to Nash, 14 November 1871, LR, Dist. NM, RG 393, NA, M-1088, roll 14; Davis to Inspector General, [December 1872], Inspection Reports, OIG, RG 159, NA; Santa Fe *Daily New Mexican,* 7 September 1875.

110. Santa Fe *Daily New Mexican,* 7 September 1875, 12 March 1882.

111. The contracts are in Reg. Cont., CGS, RG 192, NA; Santa Fe *Daily New Mexican,* 23 August 1877, 17 September 1877.

112. Frazer, "Purveyors of Flour to the Army," pp. 218, 225; Frazer, *Forts and Supplies,* pp. 106–7.

113. Hatch to Anderson, 30 June 1861, LR, OAG, RG 94, NA, M-619, roll 42.

114. Hersch's contracts are in Reg. Cont., CGS, RG 192, NA. See McClure to Sutorius, 8 April 1868, and McClure to Hunter, 16 September 1868, LS, CCS, Dist. NM, RG 393, NA.

115. McClure to Jones, 13 January 1869, LS, CCS, Dist. NM, RG 393, NA.

116. U.S., Department of Commerce, Bureau of the Census, *Eighth Census of the United States,* 1860, Santa Fe County, Population Schedules, NA, M-653, roll 714; contracts for Louis and Abraham Gold are in Reg. Cont., CGS, RG 192, NA; Letter from Louis Gold, 22 July 1865, Endorsements Received and Sent By the Commissary of Subsistence, Dept. NM, RG 393, NA; McClure to Gold, 16 December 1868, 14 January 1869, and McClure to Hunter, 8 February 1869, LS, CCS, Dist. NM, RG 393, NA.

117. Jones to Gold, 17 April 1869, Jones to Baldwin, 20 April 1869, and McClure to Morgan, 26 June 1869, LS, CCS, Dist. NM, RG 393, NA; Hersch to Getty, 8 May 1869, LR, Dist. NM, RG 393, NA, M-1088, roll 11.

118. Hersch to Getty, 12 May 1869, LR, Dist. NM, RG 393, NA, M-1088, roll 11.

119. Kobbé to McClure, 16 June 1869, LS, Dist. NM, RG 393, NA, M-1072, roll 4; McClure to Acting Commissary of Subsistence, 19 June 1869, and McClure to Morgan, 26 June 1869, LS, CCS, Dist. NM, RG 393, NA. Louis Clark of Plaza Alcalde, recipient of the Fort Marcy flour contract in 1871, informed

Lieutenant Francis B. Jones, the post commissary, that he would purchase wheat along the Rio Grande and then have it ground at Louis Gold's mill in Santa Fe. This information caused Jones to recommend that Gold be barred from manufacturing any supplies intended for the government's use. "In my opinion," he wrote, "[Clark] may be able to buy excellent wheat but if Louis Gold has the grinding or anything *whatever* to do with it, there will be swindling in his transactions." Jones to Nash, 13 January 1871, LR, Dist. NM, RG 393, NA, M-1088, roll 14. Louis Gold later received contracts from the Quartermaster's Department.

120. See Gardner to Post Adjutant and endorsements, 16 February 1876, LR, Dist. NM, RG 393, NA, M-1088, roll 28.

121. Eagan to Macfeely, 17 April 1876, and Eagan to Gilman, 28 April 1876, LS, CCS, Dist. NM, RG 393, NA.

122. Hatch to Sir, 15 April 1876, and Eagan to CCS, Department of the Missouri, 2 June 1876, LS, CCS, Dist. NM, RG 393, NA.

123. The contracts for Huning and Rosenbaum are in Reg. Cont., CGS, RG 192, NA.

124. Macfeely to Cameron, 10 October 1876, U.S., Congress, House, Annual Report of the Secretary of War, 1876, House Ex. Doc. 1 (Serial 1742), 44th Cong., 2d sess., 1876, pp. 301–2; identical letters sent by Whitehead to Rosenbaum, Huning, and Mueller and Clothier, 29 December 1876, LS, CCS, Dist. NM, RG 393, NA; Whitehead to Macfeely, 20 August 1877, New Mexico, Cons. Corres. File, RG 92, NA.

125. Whitehead to Macfeely, 3 June 1878, and Whitehead to CCS, Department of the Missouri, 13 June 1878, LS, CCS, Dist. NM, RG 393, NA; Whitehead to AAAG, 8 April 1878, LR, Dist. NM, RG 393, NA, M-1088, roll 32.

126. See Whitehead to Bull, 5 July 1878, LS, CCS, Dist. NM, RG 393, NA; Santa Fe *Weekly New Mexican*, 13 July 1878.

127. Whitehead to Commissary General of Subsistence, 26 July 1879, New Mexico, Cons. Corres. File, RG 92, NA.

128. Woodruff to Commissary General of Subsistence, 17 July 1882, New Mexico, Cons. Corres. File, RG 92, NA.

129. Guthrie to Post Adjutant, 19 December 1882, LR, Dist. NM, RG 393, NA, M-1088, roll 48.

130. Hatch to Morse, 6 February 1868, LS, Fort Stockton, PR, RG 393, NA; Merritt to Morse, 25 July 1868, LS, Fort Davis, PR, RG 393, NA.

131. The contracts are in U.S., Congress, Senate, Letter from the Secretary of War, 31 January 1872, S. Ex. Doc. 26 (Serial 1478), 42nd Cong., 2d sess., 1872, pp. 11–12. The price difference for flour delivered at Stockton and Davis suggests that flour was coming from the west and not San Antonio. On Adams and Wickes, see J. Evetts Haley,

Fort Concho and the Texas Frontier (San Angelo: San Angelo Standard–Times, 1952), pp. 288–89.

132. Carleton to AAG, 6 March 1871 [Fort Stockton], and Carleton to AAG, 6 March 1871 [Fort Quitman], Inspection Reports, OIG, RG 159, NA.

133. See Inspection Reports for Forts Stockton, Davis, Quitman, and Bliss, Carleton to AAG, 6 March 1871, Inspection Reports, OIG, RG 159, NA.

134. Average prices are based on contracts appearing in Reg. Cont., CGS, RG 192, NA; see also Penrose to Commissary General of Subsistence, 19 July 1878, Texas, Cons. Corres. File, RG 92, NA; Maxon to Quartermaster General, 16 July 1883, Annual Reports Received from Quartermasters, RG 92, NA; Haley, *Fort Concho,* pp. 297, 303–7. Sender, a native of Germany, had come to the United States in 1868 and later moved to El Paso, where he joined Samuel Schutz in establishing the mercantile firm of Schutz and Sender. The two men had married sisters by the name of Siebenborn, and about 1879 Sender and his wife moved to Fort Davis where he operated a store in partnership with his wife's kinsman, Charles Siebenborn. Barry Scobee, *Fort Davis, Texas* (Fort Davis, Texas: Barry Scobee, 1963), p. 136.

135. The contracts are in Reg. Cont., CGS, RG 192, NA. El Paso contractors included Henry Cuniffe and George Zwitzers, post traders at Fort Bliss; merchants Samuel Schutz, Joseph Schutz, Albert Schutz, Joseph Sender, and Benjamin Schuster; and Adolph Krakauer, bookkeeper for Samuel and Joseph Schutz. Albert Schutz was a nephew of Samuel and Joseph Schutz.

136. C. L. Sonnichsen, *Pass of the North, Four Centuries on the Rio Grande* (El Paso: Texas Western Press, 1968), pp. 90, 122–23; Rex W. Strickland, "Six Who Came to El Paso, Pioneers of the 1840's," *Southwestern Studies,* 1 (Fall 1963): 37.

137. Frazer, *Forts and Supplies,* p. 44; J. Morgan Broaddus, Jr., *The Legal Heritage of El Paso* (El Paso: Texas Western College Press, 1963), p. 81.

138. Broaddus, *Legal Heritage of El Paso,* pp. 82, 91–92; Strickland, "Six Who Came to El Paso," p. 40.

139. Las Cruces *Borderer,* 11 May 1871.

140. Las Cruces *Borderer,* 16 March 1871. Mills's lawsuit against Hart ended in 1873; Hart died in January 1874. See Strickland, "Six Who Came to El Paso," pp. 41–42.

141. U.S., Department of the Interior, *Tenth Census of the United States: 1880, Population,* vol. 1 (Washington: Government Printing Office, 1883), p. 343; *Mesilla Valley Independent,* 23 February 1878.

142. *Mesilla Valley Independent,* 24 September 1877; C. L. Sonnichsen, *The El Paso Salt War, 1877* (El Paso: Carl Hertzog and the Texas Western

Press at the Pass of the North, 1961), pp. 31–34, 47–48.

143. The contracts are in Reg. Cont., CGS, RG 192, NA.

144. Prescott *Arizona Miner,* 1 June, 5 October 1867; Jones to Fry, 11 May 1867, Inspection Reports, OIG, RG 159, NA; contracts are in Reg. Cont., CGS, RG 192, NA. Some time before White sold his mill, William Bichard had been a copartner in the firm of A. M. White and Company. The Bichards received their first large contract for flour in August 1866 when Major John M. Taylor, chief commissary of subsistence for the Division of the Pacific, accepted Nicholas's proposal to deliver at the Pima villages 100,000 pounds at 10 cents per pound (in coin). See Taylor to Bichard, 7 August 1866, LR, Dist. Ariz., RG 393, NA.

145. Alexander to Sherburne, 13 September 1868, LS, Subdistrict of the Verde, RG 393, NA.

146. Tucson *Weekly Arizonan,* 15 January 1870; Farish, *History of Arizona,* vol. 6, p. 48; Reg. Cont., CGS, RG 192, NA.

147. Williams to Luff, 20 July 1869, LR, Fort McDowell, PR, RG 393, NA; Calhoun to Small, 4 November 1869, LS by Quartermaster, Camp Grant, PR, RG 393, NA.

148. The contracts are in Reg. Cont., CGS, RG 192, NA. Garrison agreed to deliver 75,000 pounds of flour at Tubac at 13 cents per pound (in coin). This is probably Amos F. Garrison, Jr., whose father served as chief commissary officer in New Mexico during the Civil War. The father had employed his son as clerk in Santa Fe. See Garrison to Taylor, 6 December 1861, Amos F. Garrison, Cons. Corres. File, RG 92, NA.

149. Tucson *Weekly Arizonan,* 18 April 1869. The contracts are in Reg. Cont., CGS, RG 192, NA.

150. Carter's contract is in Reg. Cont., CGS, RG 192, NA. Special Order No. 13, issued on 23 September 1869, called for a board of officers to assemble at Tucson Depot to examine flour "furnished under contract between the United States and J. P. Carter of Altar, Sonora." Special Orders, Subdistrict of Southern Arizona, RG 393, NA. See also "James P. T. Carter," Hayden File, Arizona Historical Society.

151. Fergusson to Kellogg, 5 September 1862, Arizona, Cons. Corres. File, RG 92, NA.

152. Whittier to Fry, 9 June 1868, Arizona, Cons. Corres. File, RG 92, NA. Whittier reported that Pierson had obtained his miller's education in Rochester, New York.

153. Prescott *Arizona Miner,* 18 May, 9 November 1867; Simpson to Eaton, 19 December 1871, Arizona, Cons. Corres. File, RG 92, NA.

154. Prescott *Arizona Miner,* 9 November, 21 December 1867, 3 October, 28 November 1868.

155. Prescott *Weekly Arizona Miner,* 24 July, 16, 30 October 1869.

156. Tucson *Arizona Citizen,* 16 May 1874.

157. Farish, *History of Arizona,* vol. 6, pp. 175, 217–19.

158. Farish, *History of Arizona,* vol. 6, pp. 229–32.

159. Simpson to Eaton, 19 December 1871, Arizona, Cons. Corres. File, RG 92, NA. Hellings's contracts are in Reg. Cont., CGS, RG 192, NA.

160. Simpson to Eaton, 19 December 1871, Arizona, Cons. Corres. File, RG 92, NA.

161. *Portrait and Biographical Record of Arizona* (Chicago: Chapman Publishing Co., 1901), pp. 871–73; Farish, *History of Arizona,* vol. 2, p. 289; Reg. Cont., CGS, RG 192, NA.

162. Tucson *Weekly Arizonan,* 26 February 1870; "James Lee" and "William F. Scott," Biographical Files, Arizona Historical Society; Tucson *Arizona Citizen,* 4 July 1874; *Portrait and Biographical Record of Arizona,* pp. 491–93; Reg. Cont., CGS, RG 192, NA.

163. See Reg. Cont., CGS, RG 192, NA; Tucson *Arizona Citizen,* 19 February, 17 June 1876; *Portrait and Biographical Record of Arizona,* p. 451; *The Taming of the Salt* (Phoenix: Communications and Public Affairs Department of Salt River Project, 1979), p. 18. See Appendix Table 11, in this volume.

164. Lamb, "Jewish Pioneers in Arizona," pp. 35, 69, 72; Reg. Cont., CGS, RG 192, NA.

165. See Reg. Cont., CGS, RG 192, NA; Tucson *Arizona Citizen,* 12 September 1874.

166. See Reg. Cont., CGS, RG 192, NA; Jacobs and Company to Pierson, 4 August 1871, 10 March 1872, Barron to Pa, 18, 20 April, 7, 18, 21, 25, 27 May 1872 (Jacobs Manuscripts, University of Arizona).

167. Tucson *Arizona Citizen,* 8, 15 February 1873; Reg. Cont., CGS, RG 192, NA; Farish, *History of Arizona,* vol. 6, p. 176.

168. Eagan to Macfeely, 30 August 1880, Arizona, Cons. Corres. File, RG 92, NA; Carr to AAG, 12 December 1880, LS, Camp Lowell, PR, RG 393, NA.

169. See Reg. Cont., CGS, RG 192, NA; Tucson *Weekly Arizonan,* 7 February 1869; Tucson *Arizona Citizen,* 22 February, 1 March 1873. Rosenbaum held contracts in 1871, 1873, and 1875; Lesinsky in 1872, 1874, 1877, and 1878.

170. See Reg. Cont., CGS, RG 192, NA; Floyd S. Fierman, "Jewish Pioneering in the Southwest, A Record of the Freudenthal–Lesinsky–Solomon Families," *Arizona and the West,* 2 (Spring 1960): 64–66; Abe to Uncle, 2 September 1879 (Jacobs Manuscripts, University of Arizona); I. Solomon to Jacobs, 4 February 1881, 2 March 1881, and A. Solomon to Jacobs, 22 March 1881 (Barron and Lionel Jacobs Business Records, Arizona Historical Society).

171. The contracts are in Reg. Cont., CGS, RG 192, NA.

172. Eagan to Macfeely, 30 August 1880, Arizona, Cons. Corres. File, RG 92, NA; Martin to Carr, 9 April 1880, LS, Dept. Ariz., RG 393, NA.

173. Tucson *Arizona Citizen,* 27 September 1873.

174. The contracts are in Reg. Cont., CGS, RG 192, NA.

175. Scott to CCS, 19 June 1878, Telegrams, Fort Bowie, PR, RG 393, NA.

176. Eagan to Acting Assistant Commissary of Subsistence, 18 June 1878, Telegrams, Fort Bowie, PR, RG 393, NA; Eagan to AAG, 3 September 1879, and Eagan to Macfeely, 30 August 1880, Arizona, Cons. Corres. File, RG 92, NA.

177. Eagan to Macfeely, 22 August 1881, Arizona, Cons. Corres. File, RG 92, NA.

178. See Maxwell to CCS, 12 September 1884, Arizona, Cons. Corres. File, RG 92, NA; Reg. Cont., CGS, RG 192, NA.

179. Eagan to Macfeely, 18 July 1884, Arizona, Cons. Corres. File, RG 92, NA. Forts Mojave and Yuma had long received their flour from San Francisco.

180. See Reg. Cont., CGS, RG 192, NA. Some of the imported flour probably came from Kansas as well as from California. In July 1883 Captain Eagan requested that Fort Verde be supplied with 20,000 pounds of "issue flour" from Leavenworth, Kansas. See Eagan to Acting Commissary of Subsistence, 28 July 1883, LR by the Acting Commissary of Subsistence, Fort Verde, PR, RG 393, NA.

181. McClure to Eaton, 9 December 1867, LS, CCS, Dist. NM, RG 393, NA.

Chapter 5

1. Robert W. Frazer, *Forts and Supplies, The Role of the Army in the Economy of the Southwest, 1846–1861* (Albuquerque: University of New Mexico Press, 1983), pp. 51, 186.

2. Contracts are in Reg. Beef Cont., CGS, RG 192, NA; Carleton to Curtiss, 16 September 1864, LS, Dept. NM, RG 393, NA, M-1072, roll 3; Eaton to Carleton, 9 October 1865, LR, Dist. NM, RG 393, NA, M-1088, roll 1.

3. Charles Kenner, "The Origins of the 'Goodnight' Trail Reconsidered," *Southwestern Historical Quarterly*, 77 (January 1974): 390–94. A westward trail started near Pope's Crossing, went through El Paso, thence to Arizona and California.

4. The contracts are in Reg. Beef Cont., CGS, RG 192, NA.

5. Carleton to McCleave, 11 July 1865, and Carleton to Shoemaker, 2 September 1865, LS, Dept. NM, RG 393, NA; Gerald Thompson, *The Army and the Navajo, The Bosque Redondo Reservation Experiment, 1863–1868* (Tucson: University of Arizona Press, 1976), p. 93.

6. See Reg. Beef Cont., CGS, RG 192, NA; Haines to Bell, 7 May 1866, LR, Dist. NM, RG 393, NA, M-1088, roll 3; Bell to Carleton, 10 July 1866, LR, Dist. NM, RG 393, NA, M-1088, roll 2.

7. Carleton to Bell, 9 July 1866, and Carleton to Davis, 9 October 1866, LS, Dist. NM, RG 393, NA,

M-1072, roll 3. One officer testified that William H. Moore in 1866 lost $1,400 in purchases made to fill his contract, which called for delivery of 400 head of cattle at Fort Sumner at 6¾ cents per pound. The inference was that cattle were worth more than either Moore or Connell had figured. Davis to Nichols, 31 October 1866, LS, Inspector, Dept. NM, RG 393, NA.

8. J. Evetts Haley, *Charles Goodnight, Cowman and Plainsman* (Norman: University of Oklahoma Press, 1936), pp. 127–34. Goodnight and Loving lost about 400 head of the first herd they drove to New Mexico.

9. The contracts are in Reg. Beef Cont., CGS, RG 192, NA; Davis to Nichols, 31 October 1866, LS, Inspector, Dept. NM, RG 393, NA.

10. Adams's contract is in Reg. Beef Cont., CGS, RG 192, NA; McClure to Patterson, 3 December 1866, LR, Dist. NM, RG 393, NA, M-1088, roll 4.

11. McClure to Eaton, 25 November 1866, McClure to Adams, 6, 7, December 1866, LS, CCS, Dist. NM, RG 393, NA.

12. McClure to LaRue, 16 February 1867, and McClure to McDonald, 21 February 1867, LS, CCS, Dist. NM, RG 393, NA.

13. Mills to Adams, 7 January 1867, Inspection Reports, OIG, RG 159, NA; McClure to Morgan, 10 June 1867, LS, CCS, Dist. NM, RG 393, NA.

14. Carleton to McClure, 20 February 1867, LS, Dist. NM, RG 393, NA; Carleton to Sykes, 9 March 1867, Inspection Reports, OIG, RG 159, NA; McClure to LaRue, 16 February 1867, LS, CCS, Dist. NM, RG 393, NA.

15. McClure to DeForrest, 19 February 1867, and McClure to McDonald, 21 February 1867, LS, CCS, Dist. NM, RG 393, NA.

16. Patterson and Roberts to McDonald, 4 March 1867, and McDonald to McClure, 4 March 1867, Inspection Reports, OIG, RG 159, NA.

17. Carleton to Sykes, 9 March 1867, Inspection Reports, OIG, RG 159, NA; Sykes to Carleton, 15 March 1867, LR, Dist. NM, RG 393, NA, M-1088, roll 8.

18. DeForrest to McDonald, 10 March 1867, McClure to McDonald, 10 April 1867, and McClure to Morgan, 10 June 1867, LS, CCS, Dist. NM, RG 393, NA.

19. McClure to Morgan, 10 June 1867, LS, CCS, Dist. NM, RG 393, NA; Charles L. Kenner, *A History of New Mexican–Plains Indian Relations* (Norman: University of Oklahoma Press, 1969), pp. 167–68.

20. McClure to Eaton, 23 July 1867, LS, CCS, Dist. NM, RG 393, NA.

21. McClure to Eaton, 7 July 1868, LS, CCS, Dist. NM, RG 393, NA. In September of 1867, Adams was residing in Council Grove, Kansas.

22. See two letters dated 5 August 1867, Hatch to Taylor, LS, Fort Stockton, PR, RG 393, NA.

23. Harwood P. Hinton, "John

Simpson Chisum, 1877–84," *New Mexico Historical Review,* 31 (July 1956): 188.

24. Kenner, *New Mexican–Plains Indian Relations,* p. 168.

25. Hatch to Taylor, 5 August 1867, LS, Fort Stockton, PR, RG 393, NA.

26. Kenner, *New Mexican–Plains Indian Relations,* pp. 86, 140, 150–55.

27. DeForrest to Commanding Officer, Fort Bascom, 12 March 1866, and DeForrest to Whom It May Concern, 23 June 1866, LS, Dist. NM, RG 393, NA, M-1072, roll 3.

28. McClure to Adams, 18 December 1866, LS, CCS, Dist. NM, RG 393, NA.

29. Kenner, *New Mexican–Plains Indian Relations,* pp. 160–61; Carleton to AAG, 22 January 1867, LS, Dist. NM, RG 393, NA, M-1072, roll 4.

30. McClure to Patterson, 8 June 1868, and McClure to Rosenthal, 8 June 1868, LS, CCS, Dist. NM, RG 393, NA; Santa Fe *Weekly New Mexican,* 8 December 1868; Letters by McClure dated 18 January 1869 and 2 February 1869, Registers of LR, Dist. NM, RG 393, NA, M-1097, roll 2.

31. See Reg. Beef Cont., CGS, RG 192, NA.

32. Alexander's investigation and charges are examined in Darlis A. Miller, "The Role of the Army Inspector in the Southwest: Nelson H. Davis in New Mexico and Arizona, 1863–1873," *New Mexico Historical Review,* 59 (April 1984): 151–56.

33. Miller, "The Role of the Army Inspector," pp. 154–55.

34. McClure to DeForrest, 25 September 1867, Inspection Reports, OIG, RG 159, NA; Endorsement by McClure dated 9 September 1867, LR, Dist. NM, RG 393, NA, M-1088, roll 6.

35. McClure to Hunter, 7 October 1867, Inspection Reports, OIG, RG 159, NA.

36. McClure to Hunter, 31 October 1867, Inspection Reports, OIG, RG 159, NA.

37. McClure to Hunter, 25 September 1867, Inspection Reports, OIG, RG 159, NA; McClure to DeForrest, 17 January 1867, and McClure to Eaton, 19 February 1867, LS, CCS, Dist. NM, RG 393, NA.

38. McClure to Hunter, 16 September 1867, LS, CCS, Dist. NM, RG 393, NA; Getty to Shriver, 17 October 1867, LS, Dist. NM, RG 393, NA, M-1072, roll 4.

39. Carson to Getty, 30 November 1867, Inspection Reports, OIG, RG 159, NA. But see also Lawrence R. Murphy, *Lucien Bonaparte Maxwell, Napoleon of the Southwest* (Norman: University of Oklahoma Press, 1983), p. 122.

40. Murphy, *Lucien Bonaparte Maxwell,* pp. 125, 133–35.

41. See Carson to Getty, 30 November 1867, Inspection Reports, OIG, RG 159, NA.

42. McClure to DeForrest, 28 September 1867, LS, CCS, Dist. NM, RG 393, NA.

43. McClure to DeForrest, 25 September 1867, and McClure to Maxwell, 3 December 1867, LS, CCS, Dist. NM, RG 393, NA.

44. McClure to Maxwell, 3 December 1867, LS, CCS, Dist. NM, RG 393, NA.

45. McClure to Campbell, 19 December 1867, and McClure to Eaton, 2 January 1868, LS, CCS, Dist. NM, RG 393, NA; the contracts are in Reg. Cont. and Reg. Beef Cont., CGS, RG 192, NA.

46. McClure to Eaton, 5 December 1868, and McClure to Kobbé, 21 December 1868, LS, CCS, Dist. NM, RG 393, NA; Murphy, *Lucien Bonaparte Maxwell,* p. 142.

47. Murphy, *Lucien Bonaparte Maxwell,* pp. 124–25; Carson to Getty, 30 November 1867, Inspection Reports, OIG, RG 159, NA; Getty to Davis, 12 August 1869, LS, Dist. NM, RG 393, NA, M-1072, roll 4.

48. The average price of beef appears in the Report of the Commissary General of Subsistence, included in the Annual Report of the Secretary of War.

49. McClure to Eaton, 9 December 1867, LS, CCS, Dist. NM, RG 393, NA. This figure is substantiated by Lieutenant Colonel Thomas C. Devin's statement in December 1871 that the garrison at Fort Stanton (a two-company post of about 140 men) required 50 head of cattle per month. Devin to Sartle, 7 December 1871, LR, CQM, Dist. NM, RG 393, NA.

50. McClure to Eaton, 16 July 1867, and McClure to Corbett, 19 July 1867, LS, CCS, Dist. NM, RG 393, NA.

51. McClure to Eaton, 14 August 1867, 7 August 1868, LS, CCS, Dist. NM, RG 393, NA; Post Surgeon to Post Adjutant, 30 September 1875, LR, Fort Union, PR, RG 393, NA. In the West, only the states of California and Texas raised more sheep than New Mexico in 1870.

52. McClure to Eaton, 14 August 1867, LS, CCS, Dist. NM, RG 393, NA.

53. Santa Fe *Weekly New Mexican,* 21 July 1868; McClure to Hunter, 28 February 1868, Inspection Reports, OIG, RG 159, NA.

54. McClure to Clarke, 24 June 1868, LS, CCS, Dist. NM, RG 393, NA.

55. Letter by McClure, 22 October 1868, Registers of LR, Dist. NM, RG 393, NA, M-1097, roll 2; McClure to Patterson, 29 October 1868 and 27 April 1869, LS, CCS, Dist. NM, RG 393, NA.

56. Carvello to Post Adjutant, 31 March 1879, LR, Fort Union, PR, RG 393, NA.

57. Goldbaum's contract is in Reg. Beef Cont., CGS, RG 192, NA; Porter to McClure, 28 January 1867, Duncan to Davis, 24 February 1867, and endorsements on Duncan's letter by Davis, 26 February 1867, and by Carleton, 6 March 1867, LR, Dist. NM, RG 393, NA, M-1088, roll 6.

58. Goldbaum to Davis, 26 February 1867, LR, Inspector General, Dist. NM, RG 393, NA.

59. Endorsement dated 22 April

1867 on Duncan to Davis, 24 February 1867, LR, Dist. NM, RG 393, NA, M-1088, roll 6.

60. In a letter to Inspector Davis, Goldbaum wrote: "That my knowledge of English being imperfect, I was unable, myself, to clothe facts in proper language, and employed for pay, a competent person to embody them in a statement. . . ." Goldbaum to Davis, 26 February 1867, LR, Inspector General, Dist. NM, RG 393, NA. Goldbaum and his wife Sara moved to Arizona in 1869. He was killed by Indians in 1886 while prospecting in the Whetstone Mountains. Harriet and Fred Rochlin, *Pioneer Jews, A New Life in the Far West* (Boston: Houghton Mifflin Company, 1984), p. 33.

61. Las Cruces *Borderer,* 28 February 1872; Santa Fe *Daily New Mexican,* 24 October 1876; contracts are in Reg. Beef Cont., CGS, RG 192, NA. The 1860 census for the city of Santa Fe lists Probst's age as 25 and Kirchner's as 23. The 1870 census lists Probst's age as 34 and Kirchner's as 36. U.S., Department of Commerce, Bureau of the Census, *Eighth Census of the United States,* 1860, Santa Fe County, Population Schedules, NA, M-653, roll 714 and *Ninth Census of the United States,* 1870, Santa Fe County, Population Schedules, NA, M-593, roll 896.

62. DeForrest to Probst and Kirchner, 24 September 1867, LS, Dist. NM, RG 393, NA, M-1072, roll 4; Probst and Kirchner to Gregg, 23 April 1871, LR, Dist. NM, RG 393, NA, M-1088, roll 13; Devin to Sartle, 7 December 1871, LR, CQM, Dist. NM, RG 393, NA.

63. Nash to Granger, 11, 29 May 1872, LR, CQM, Dist. NM, RG 393, NA; Nash to Davis, 2 January 1873, LS, CCS, Dist. NM, RG 393, NA. Some of the beef Probst and Kirchner delivered at Fort Union was judged unfit for issue, but this should not be labeled a swindle. See Farnsworth to Post Adjutant, 30 December 1871, LR, Fort Union, PR, RG 393, NA.

64. Contracts are in Reg. Beef Cont., CGS, RG 192, NA.

65. The conflict between Eagan and the firm of Probst and Kirchner is explained in two letters with accompanying papers: Probst and Kirchner to Devin, 24 December 1874, and Eagan to AAAG, 28 December 1874, LR, Dist. NM, RG 393, NA, M-1088, roll 23.

66. *Ibid.*

67. Shiras to Eagan, 8 January 1875, LR, Dist. NM, RG 393, NA, M-1088, roll 26; Devin to AAG, Dept. of the Missouri, 4 February 1875, LS, Dist. NM, RG 393, NA, M-1072, roll 5; Santa Fe *Weekly New Mexican,* 15 March 1879. William Breeden served as the firm's attorney.

68. McClure to Eaton, 7 August 1868, LS, CCS, Dist. NM, RG 393, NA; Kenner, "Origins of the 'Goodnight' Trail," pp. 391–93.

69. James D. Shinkle, *Fifty Years of Roswell History, 1867–1917* (Roswell, N.M.: Hall–Poorbaugh Press, Inc., 1964), pp. 10, 30–31.

70. McClure to Graham, 3 March 1870, LS, CCS, Dist. NM, RG 393, NA.

71. Contracts are in Reg. Beef Cont., CGS, RG 192, NA. Patterson won contracts to supply Forts Bascom, Sumner, Union, and Stanton with fresh beef in 1866 and Fort Bascom in 1867. He received only the Fort Wingate contract in 1869, even though he had bid for several.

72. Patterson to Kobbé, 18 December 1868, LR, Fort Stanton, PR, RG 393, NA; Letterman to DeForrest, 25 May 1867, LR, Dist. NM, RG 393, NA, M-1088, roll 5; McClure to Patterson, 7 May 1867, and McClure to Hunter, 24 February 1868, LS, CCS, Dist. NM, RG 393, NA. Patterson's former partner, Thomas L. Roberts, died 26 February 1868. See Ludington to Brown, 27 February 1868, LS, CQM, Dist. NM, RG 393, NA.

73. An officer at Fort Craig noted that since all cattle in his vicinity likewise were poor, it would be impossible to secure a better quality beef. [Name unclear] to Jones, 20 March 1869, LR, Fort Craig, PR, RG 393, NA.

74. The contracts are in Reg. Beef Cont., CGS, RG 192, NA; McClure to Eaton, 11 June 1867, and McClure to Patterson, 21 July 1869, LS, CCS, Dist. NM, RG 393, NA; Assistant Commissary of Subsistence (ACS) to CCS, 17 February 1869 and 4 May 1869, LS by Post Commissary Officer, Fort Cummings, PR, RG 393, NA.

75. Ryan to Jones, 24 May 1870, ACS to CCS, 6 June 1870, ACS to Blake, 9 August 1870, ACS to Nash, 9, 31, August 1870, LS by Post Commissary Officer, Fort Cummings, PR, RG 393, NA. Crouch's contract price was 10 cents per pound.

76. For information on Crouch, see Darlis A. Miller, *The California Column in New Mexico* (Albuquerque: University of New Mexico Press, 1982), pp. 180–91. For Patterson's contracts and subcontracts, see McClure to Eaton, 11 October 1869, LS, CCS, Dist. NM, RG 393, NA, and Reg. Beef Cont., CGS, RG 192, NA. Patterson married in Springfield, Illinois, in 1883, about the time he returned to New Mexico to take up mining in Grant County. A disgruntled former employee shot and killed Patterson on 7 August 1893. This information provided by Charles Kenner.

77. Contracts are in Reg. Beef Cont., CGS, RG 192, NA; Haley, *Charles Goodnight,* p. 147; Santa Fe *Daily New Mexican,* 27 April 1870; U.S., Department of Commerce, Bureau of the Census, *Ninth Census of the United States,* 1870, Colfax County, Population Schedules, NA, M-593, roll 893. A card advertising the services of Aaron O. Willburn, Frank's brother, and Van C. Smith as cattle dealers appeared in the Santa Fe *Daily New Mexican,* 26 November 1872. Smith had competed unsuccessfully for the beef contract at Fort Stanton that year.

78. Newsham's contracts are in

Reg. Beef Cont., CGS, RG 192, NA; *Mining Life,* 4, 11 October 1873; *Mesilla News,* 23 May 1874; *Grant County Herald,* 12 February 1881.

79. Valencia County, N.M., Agricultural Schedules of the Tenth Census of the United States, 1870, NMSA (hereafter, Agricultural Schedules, 1870).

80. Mora and San Miguel Counties, N.M., Agricultural Schedules, 1870. The contracts are in Reg. Beef Cont., CGS, RG 192, NA. Hispanic men of property more often invested in sheep than in cattle; consequently few competed for government beef contracts. Between 1865 and 1885 only five Hispanos signed contracts to furnish New Mexico's posts with beef. In addition to Rafael Chávez and Eugenio Romero, Estévan Ochoa held the Fort Selden contract in 1867, José Pino y Baca of San Marcial received the Fort McRae contract in 1876, and Rómulo Martínez of Santa Fe won contracts at Fort Wingate in 1876, 1877, and 1878 and at Fort Marcy in 1879.

81. See published abstract of beef proposals, Santa Fe *Weekly New Mexican,* 11 August 1868, 10 August 1869, *Daily New Mexican,* 3 August 1870, 1 July 1871, 10 May 1872, 2 May, 1873.

82. The following held twelve-month beef contracts at Fort Stanton: Thomas L. Roberts, 1865; James Patterson, 1866; L. G. Murphy, 1867; James Patterson, 1868; S. L. Hubbell, 1869; Willburn and Stockton, 1870; Probst and Kirchner, 1871; W. V. B. Wardwell, 1872; C. H. McVeagh, 1873; John N. Copeland, 1874; W. Spiegelberg, 1875; James J. Dolan, 1876; John H. Riley, 1878; S. S. Terrell, 1879; P. Coghlan, 1880; George W. Peppin, 1881; J. C. DeLany, 1882; J. C. DeLany, 1883; John H. Riley, 1884; J. A. LaRue, 1885. Murphy signed a three-month contract in 1870; Riley signed two six-month contracts in 1877. See Reg. Beef Cont., CGS, RG 192, NA. Emil Fritz was employed as Willburn and Stockton's agent; Fritz and Murphy were subcontractors for Probst and Kirchner and for McVeagh; and Dolan was agent for Willi Spiegelberg. For the subcontract with McVeagh, I am grateful to John P. Wilson. See also Proceedings of Board of Survey, 25 February 1871, Records Boards of Survey, Fort Stanton, PR, RG 393, NA; Devin to Sartle, 7 December 1871, LR, CQM, Dist. NM, RG 393, NA; Stewart to Dolan, 9 October 1875, LS, Fort Stanton, PR, RG 393, NA.

83. Lawrence L. Mehren, "A History of the Mescalero Apache Reservation, 1869–1881" (Master's thesis, University of Arizona, 1969), pp. 32–35, 109, 126, 144, 159.

84. The charge that L. G. Murphy and Company monopolized government contracts is examined more closely in Chapter 8.

85. Shinkle, *Fifty Years of Roswell History,* p. 27; Hinton, "John Simpson Chisum," (part 1), pp. 189, 191;

personal communication to author from Harwood Hinton, 1 October 1987.

86. Clendenin, 12 April 1875, LS, Fort Stanton, PR, RG 393, NA; Devin to Sartle, 7 December 1871, LR, CQM, Dist. NM, RG 393, NA; L. G. Murphy and Co., Journal, Livestock Account, 1 May 1871–December 1872 (Special Collections, University of Arizona).

87. Bliss to AAAG, 11 July 1872, LS, Fort Stockton, PR, RG 393, NA; Clayton W. Williams, *Texas' Last Frontier, Fort Stockton and the Trans–Pecos, 1861–1895* (College Station: Texas A & M Press, 1982), pp. 160–62.

88. Bliss to AAAG, 11 July 1872, Dodge to AAG, 13 May 1873, LS, Fort Stockton, PR, RG 393, NA.

89. McKibbin to AAAG, 21 August 1873, and Clendenin to AAAG, 5 January 1875, LS, Fort Stanton, PR, RG 393, NA.

90. Hinton, "John Simpson Chisum," (part 1), pp. 191–92.

91. Post Adjutant to Riley, 23 April 1878, LS, Fort Stanton, PR, RG 393, NA; Riley to Dudley, 7 May 1878, LR, Fort Stanton, PR, RG 393, NA.

92. Robert N. Mullin, ed., *Maurice G. Fulton's History of the Lincoln County War* (Tucson: University of Arizona Press, 1968), p. 203; Jas. J. Dolan and Co. to Dudley, 6 May 1878, LR, Fort Stanton, PR, RG 393, NA. The Dolan firm was heavily mortgaged to Catron. Mullin, p. 203. Riley's contract is in Reg. Beef Cont., CGS, RG 192, NA. Colonel Dudley reported on 10 September that Riley had not appeared at Fort Stanton for a long time—that his whereabouts were unknown. Early in 1879 Riley was in Chihuahua, Mexico, driving cattle reportedly owned by Tom Catron to the San Carlos Indian agency in Arizona. He then settled in Las Cruces where he and William L. Rynerson became partners in various cattle ventures. Dudley to AAAG, 10 September 1878, LS, Fort Stanton, PR, RG 393, NA; Walz to AAAG, 22 January 1879, LR, Dist. NM, RG 393, NA, M-1088, roll 36; Miller, *California Column,* p. 69.

93. Miller, *California Column,* pp. 76–77.

94. U.S., Department of Interior, *Ninth Census of the United States: 1870, Wealth and Industry,* vol. 3 (Washington: Government Printing Office, 1872), p. 209; U.S., Department of Interior, *Tenth Census of the United States: 1880, Agriculture,* vol. 3 (Washington: Government Printing Office, 1883), p. 198; Gerald R. Baydo, "Cattle Ranching in Territorial New Mexico" (Ph.D. dissertation, University of New Mexico, 1970), pp. 216–17.

95. The average price of beef appears in the Report of the Commissary General of Subsistence included in the Annual Report of the Secretary of War; see particularly reports for 1878 and 1881.

96. Woodruff to Commissary General of Subsistence, 17 July 1882,

19 July 1883, New Mexico, Cons. Corres. File, RG 92, NA.

97. Dudley to AAAG, 15 April 1880, Commanding Officer to Watrous and Son, 11 April 1882, Commanding Officer to County Clerk, 14 April 1882, LS, Fort Union, PR, RG 393, NA.

98. Commanding Officer to AAG, 18 August 1885, Commanding Officer to AAAG, 3 December 1885, LS, Fort Stanton, PR, RG 393, NA; Wallace to Post Adjutant, 2 December 1885, LR, Fort Stanton, PR, RG 393, NA.

99. MacArthur to AAAG, 29 June 1884, and endorsement by Woodruff, MacArthur to AAAG, 19 July 1884 and 9 August 1884, LR, Fort Selden, PR, RG 393, NA.

100. Contracts are in Reg. Beef Cont., CGS, RG 192, NA. Hooper's contracts ran from 1 August 1865 to 31 December 1865; Stevens's and Banning's from 1 January 1866 to 31 December 1866. On securing cattle in California and Sonora, see Mason to Kirkham, 17 July 1865, LS, Dist. Ariz., RG 393, NA; Davis to Green, 7 February 1866, LR, Dist. Ariz., RG 393, NA; Bert Haskett, "Early History of the Cattle Industry in Arizona," *Arizona Historical Review,* 6 (October 1935): 19.

101. Contracts are in Reg. Beef Cont., CGS, RG 192, NA. On Texas cattle supplied at Goodwin, see DeLong to Browning, 23 February 1867, LR, Camp Grant, PR, RG 393, NA. On Texas cattle supplied on Hooker's contracts, see Earle R. Forrest, "The Fabulous Sierra Bonita," *Journal of Arizona History,* 6 (Autumn 1965): 136; J. J. Wagoner, "History of the Cattle Industry in Southern Arizona, 1540–1940," *University of Arizona Bulletin,* 23 (April 1952): 33. That some cattle Hooker supplied came from Sonora is implied in Davis to AAG, 7 April 1869, LS, Camp Date Creek, PR, RG 393, NA.

102. Mason to Kirkham, 17 July 1865, LS, Dist. Ariz., RG 393, NA.

103. The contract is in Reg. Beef Cont., CGS, RG 192, NA.

104. Legislature of the Territory of Arizona, *Memorial and Affidavits Showing Outrages Perpetrated by the Apache Indians in the Territory During the Years 1869 and 1870* (reprint; Tucson: Territorial Press, 1964), pp. 26–28. Hooker testified in 1871 that Apaches had stolen 400 head of cattle from the Fresnal Rancho, apparently the site he leased in the winter of 1869–70 from Papago Indians in the Baboquivari Valley southwest of Tucson. Secondary sources suggest that Hooker reported later that friendly Papagoes had consumed the 400 cattle. See Haskett, "Early History of the Cattle Industry in Arizona," p. 20; Wagoner, "History of the Cattle Industry in Southern Arizona," p. 34.

105. Jones to Fry, 22 April 1869 [Camp Crittenden], and Jones to Fry, 11 May 1869 [Camp Grant], Inspection Reports, OIG, RG 159, NA.

106. Commanding Officer to Simpson, 23 April 1869, LS, Fort Mojave, PR, RG 393, NA.

107. Krause to Sherburne, 26 December 1866, LS, Fort Whipple, PR, RG 393, NA; [name unclear] to Grant, 29 November 1867, LR, Fort McDowell, PR, RG 393, NA.

108. Post Order No. 6, 9 September 1867, LR, District of Tucson, RG 393, NA.

109. Green to Tuttle, 3 February 1866, LS, Dist. Ariz., RG 393, NA; Baker to Hobart, 20 August 1867, LR, Subdistrict of Prescott, RG 393, NA. During the winter of 1866–67 twenty-six head of cattle died at Camp Wallen and forty at Fort Whipple. Gregg to Sherburne, [April] 1867, LS, Fort Whipple, PR, RG 393, NA; Jones to Fry, 19 May 1867, Inspection Reports [Camp Wallen], RG 159, NA.

110. Drum to Commanding Officer, 30 August 1866, LR, Dist. Ariz., RG 393, NA.

111. Calhoun to Simpson, 18 January 1870, LS by Quartermaster, Camp Grant, PR, RG 393, NA.

112. Price to Eyre, 4 August 1867, and Atchisson to Drum, 9 March 1864, LS, Fort Mojave, PR, RG 393, NA; Krause to Mahnken, 9 January 1869, LR, Fort Whipple, PR, RG 393, NA. When asked to explain why these men were not mounted, Captain David Krause, commanding at Camp Lincoln, stated that he had only infantry in his command and that cattle from the same herd earlier had been driven to the post by footmen.

113. John A. Spring, *John Spring's Arizona,* edited by A. M. Gustafson (Tucson: University of Arizona Press, 1966), pp. 58, 73–77. On hiring Hispanic vaqueros, see also Norris to Wright, 31 March 1868, LS, Camp Crittenden, PR, RG 393, NA.

114. See Bowers and Bro. to Hobart, 23 August 1867, LR, Subdistrict of Prescott, RG 393, NA; Prescott *Arizona Miner,* 18 July 1868.

115. Devin to Sherburne, 31 July 1868, LS, Fort Whipple, PR, RG 393, NA; Hobart to Post Adjutant, 29 May 1869, LR, Subdistrict of Prescott, RG 393, NA; Foster to Small, 24 March 1870, LS by Quartermaster, Subdistrict of Northern Arizona, RG 393, NA.

116. See United States Army, *Revised United States Army Regulations of 1861* (Washington: Government Printing Office, 1863), p. 250.

117. Davis to AAG, 7 April 1869, LS, Camp Date Creek, PR, RG 393, NA; Ripley to AAG, 18 May 1869, LS, Fort Bowie, PR, RG 393, NA. Hinds received payment for a total of 21,000 pounds of beef.

118. Endorsement by Bernard, 8 December 1870, End. S., Fort Bowie, PR, RG 393, NA; Endorsement by Evans, 19 April 1872, LS, Dept. Ariz., RG 393, NA.

119. Endorsements by Bernard, 19 October and 3 November 1870, End. S., Fort Bowie, PR, RG 393, NA; Gregg to Sherburne, 19 August 1867, LS, Fort Whipple, PR, RG 393, NA.

120. The contracts are in Reg. Beef Cont., CGS, RG 192, NA. At

posts where mutton contracts were in force, soldiers received mutton about one day in seven. Wilson to Eagan, 21 February 1877, and Eagan to Macfeely, 30 August 1880, Arizona, Cons. Corres. File, RG 92, NA.

121. Hooker's contracts are in Reg. Beef Cont., CGS, RG 192, NA. Epifanio Aguirre received the contract at Tucson, and José M. Redondo, the contract at Yuma Depot.

122. Commanding Officer to AAG, 1 November 1870, LS, Fort McDowell, PR, RG 393, NA.

123. Tucson *Arizona Citizen,* 8, 15 July 1871.

124. See Reg. Beef Cont., CGS, RG 192, NA; Simpson to Acting Commissary of Subsistence, 28 August 1871, LR, Fort Whipple, PR, RG 393, NA. For beef on the hoof, Kelley would receive 8 97/100 cents per pound.

125. Special Order No. 48, 20 July 1871, Fort Mojave, PR, RG 393, NA; Evans to CCS, 19 August 1871, and Smith to AAAG, 13 January 1872, LS, Fort Bowie, PR, RG 393, NA.

126. Letter by J. R. Porter and endorsements, 3 August 1872, LS, Dept. Ariz., RG 393, NA; Letter by Kraszyneski and endorsements, 25 April 1872, End. S., Fort Mojave, PR, RG 393, NA.

127. See copy of Grayson contract, 9 March 1872, in LR, Fort Verde, PR, RG 393, NA. For Hooker's return to Arizona in 1872, see Lionel to Barron, 30 June 1872 (Barron and Lionel Jacobs Business Records, Arizona Historical Society); Tucson *Arizona Citizen,* 13 July 1872.

128. Crook to AAG, 28 May 1872, LS, Dept. Ariz., RG 393, NA; Prescott *Weekly Arizona Miner,* 6 July, 24 August, 16 November 1872; Tucson *Arizona Citizen,* 28 December 1872.

129. See Sacket to Inspector General, 30 June 1873, Inspection Reports, [Camp Apache], RG 159, NA; Tucson *Arizona Citizen,* 16 August 1873.

130. Haskett, "Early History of the Cattle Industry in Arizona," p. 23; Forrest, "The Fabulous Sierra Bonita," p. 136. Secondary sources state that the Sierra Bonita was founded in 1872, but the Tucson *Arizona Citizen,* 6 June 1874, reported it was located in February 1873. See also *Arizona Citizen,* 19 December 1874. On Barney's contract with the Indian Department, see Tucson *Arizona Citizen,* 5 July 1873.

131. See Prescott *Arizona Miner,* 25 January 1873; Reg. Beef Cont., CGS, RG 192, NA.

132. Tucson *Arizona Citizen,* 31 May, 14 June, 9 August 1873; Carr to AAG, 9 December 1873, LS, Camp Lowell, PR, RG 393, NA. Curtiss later left the territory and died at his home in Cleveland, Ohio, in January 1879. Tucson *Arizona Star,* 6 February 1879.

133. Richard R. Willey, "La Canoa: A Spanish Land Grant Lost and Found," *The Smoke Signal,* 38 (Fall 1979): 162; U.S., Congress, Senate, Federal Census—Territory of New

Mexico and Territory of Arizona, 1870, Senate Doc. 13, 89th Cong., 1st sess., 1965 (Washington: U.S. Government Printing Office, 1965), pp. 170, 180; the contracts are in Reg. Beef Cont., CGS, RG 192, NA; Tucson *Arizona Citizen,* 6 June 1874.

134. Articles of Agreement between Maish and Jacobs and José M. Redondo, 14 April 1874, and between Maish and Jacobs and Head and Marks, 30 April 1874 (Jacobs Manuscripts, University of Arizona); Wood to Jacobs, 24 March 1874, and a series of letters issued by Small to Marsh [*sic*] and Jacobs between 22 October 1874 and 17 June 1875 (Barron and Lionel Jacobs Business Records, Arizona Historical Society).

135. Tucson *Arizona Citizen,* 27 December 1873, 4, 18, 25 April, 5 September 1874. The issue for 5 September 1874 reported that probably 15,000 head of cattle had been imported the previous year from New Mexico, Colorado, and Texas for military and Indian supplies.

136. Kautz to AAG, 31 August 1875, U.S., Congress, House, Annual Report of the Secretary of War, House Ex. Doc. 1 (Serial 1674), 44th Cong., 1st sess., 1875, p. 137.

137. Tucson *Arizona Citizen,* 27 January 1872, 1 November 1873, 14 April 1877.

138. Patterson's contract is in Reg. Beef Cont., CGS, RG 192, NA. The average price of beef appears in the Report of the Commissary General of Subsistence, included in the Annual Report of the Secretary of War. In July 1876 Patterson received a contract to supply fresh beef and mutton at Camp Thomas, and in March 1877 he won contracts to supply beef and mutton at Camps McDowell and Verde. Early in 1878 a Prescott newspaper announced that Patterson had purchased the Plaza Livery and Feed Stables. See "Gideon Brooke," Hayden File, Arizona Historical Society.

139. The contracts are in Reg. Beef Cont., CGS, RG 192, NA. See also Tucson *Arizona Citizen,* 19 December 1874; Carr to Gordon, 21 January 1881, LS, Fort Lowell, PR, RG 393, NA.

140. See Reg. Beef Cont., CGS, RG 192, NA; Van Horne to Mallory, 3 April 1879, LS, Fort Mojave, PR, RG 393, NA; Eagan to Macfeely, 30 August 1880, Arizona, Cons. Corres. File, RG 92, NA.

141. See Reg. Beef Cont., CGS, RG 192, NA; "Charles T. Rogers," "Orlando Allen," Hayden File, Arizona Historical Society. Rogers and Allen sold their ranch in 1882. Rogers continued ranching near Williams, Arizona, a town he helped establish. Allen later moved to Phoenix, and then in 1910 at age 73 left Arizona to live in southern California.

142. Mulford Winsor, "José María Redondo," *Journal of Arizona History,* 20 (Summer 1979): 169–92; Redondo's contracts are in Reg. Beef Cont., CGS, RG 192, NA.

143. Tucson *Arizona Citizen,* 19

May 1877; Haskett, "Early History of the Cattle Industry in Arizona," p. 24. Redondo died 18 June 1878. Only a few New Mexicans, in addition to James Patterson, held beef contracts for the Arizona posts: S. P. Carpenter at Fort Bowie in 1880, John H. Riley at Forts Grant and Thomas in 1882, and Henry Huning at Fort Apache in 1882. But other New Mexicans drove cattle to Arizona to turn in on government contracts.

144. Eagan to Macfeely, 22 August 1881 and 8 September 1882, Arizona, Cons. Corres. File, RG 92, NA.

145. See Reports of Commissary Generals of Subsistence, included in the Annual Reports of the Secretary of War; Richard J. Morrisey, "The Early Range Cattle Industry in Arizona," *Agricultural History*, 24 (July 1950): 154; "Ranch Life, The Border Country, 1880–1940, The Way It Really Was," *Cochise Quarterly*, 12 (Spring 1982): n.p.

146. The average price of beef appears in the Report of Commissary General of Subsistence, included in the Annual Report of the Secretary of War.

147. The contracts are in Reg. Beef Cont., CGS, RG 192, NA; see also letter of Shreiner, 10 July 1868, Registers of LR, 5th Military District, RG 393, NA, M-1193, roll 2. In 1871 Charles H. Mahle was employed by Charles Probst and August Kirchner as their butcher at Fort Wingate. See Santa Fe *Weekly New Mexican*, 6 January 1872.

148. See Reg. Beef Cont., CGS, RG 192, NA; Letter of Schreiner, 13 June 1868, Registers of LR, 5th Military District, RG 393, NA, M-1193, roll 2; Carpenter to Adjutant General, 11 January 1879, LS, Fort Davis, PR, RG 393, NA. The Fort Davis beef contract for 1868 has not been located.

149. Reg. Beef Cont., CGS, RG 192, NA; Barry Scobee, *Fort Davis, Texas* (Fort Davis: Barry Scobee, 1963), p. 131.

150. Williams to Quartermaster General, 30 June 1883, Annual Reports Received from Quartermasters, RG 92, NA.

151. See Reg. Beef Cont., CGS, RG 192, NA; Small to Commissary General of Subsistence, 26 August 1881, Texas, Cons. Corres. File, RG 92, NA. The army paid 8 22/100 cents per pound for beef in Texas during the fiscal year ending 30 June 1885.

152. Robert W. Frazer, *Forts of the West, Military Forts and Presidios and Posts Commonly Called Forts West of the Mississippi River to 1898* (Norman: University of Oklahoma Press, 1965), pp. 148, 162.

Chapter 6

1. Mrs. Orsemus B. Boyd, *Cavalry Life in Tent and Field* (reprint; Lincoln: University of Nebraska Press, 1982), pp. 214–16.

2. Sherman to Secretary of War,

20 November 1869, U.S., Congress, House, Annual Report of the Secretary of War, 1869, House Ex. Doc. 1, pt. 2 (Serial 1412), 41st Cong., 2d sess., 1869, p. 31.

3. Erna Risch, *Quartermaster Support of the Army, A History of the Corps, 1775–1939* (Washington, D.C.: Office of the Quartermaster General, 1962), pp. 486–87.

4. Darlis A. Miller, "Los Pinos, New Mexico: Civil War Post on the Rio Grande," *New Mexico Historical Review,* 62 (January 1987): 1–31.

5. Lease dated 28 May 1854, LR by Post Quartermaster, Fort Craig, PR, RG 393, NA; Watts to Granger, 10 July 1871, LR, Dist. NM, RG 393, NA, M-1088, roll 14.

6. Ludington to Easton, 11 September 1869, LS to Dept. of the Missouri, CQM, Dist. NM, RG 393, NA; Watts to Granger, 10 July 1871, LR, Dist. NM, RG 393, NA, M-1088, roll 14. See also Boutwell to Blaine, 21 January 1870, U.S., Congress, House Ex. Doc. 73 (Serial 1417), 41st Cong., 2d sess., 1869.

7. Watts to Granger, 10 July 1871, LR, Dist. NM, RG 393, NA, M-1088, roll 14; Lincoln to Secretary of Interior, 19 January 1885, LR, Dist. NM, RG 393, NA, M-1088, roll 56.

8. See endorsement dated 17 July 1871 in AAG, Dept. of the Missouri, to Commanding Officer, Fort Craig, 27 April 1882, LR, Dist. NM, RG 393, NA, M-1088, roll 46; Lincoln to Secretary of Interior, 19 January 1885, LR, Dist. NM, RG 393, NA, M-1088, roll 56.

9. Carleton to Watts, 10 December 1863, LS, Dept. NM, RG 393, NA, M-1072, roll 3; Lease signed by John S. Watts, 4 May 1864, LR, Dist. NM, RG 393, NA, M-1088, roll 5.

10. For the Montoya Grant, see George F. Ellis, *Bell Ranch As I Knew It* (Kansas City: Lowell Press, 1973), pp. 3–13. For the Jornada del Muerto Grant, see copy of deed by Houghton and Watts to the United States, 21 March 1865, LR, Dept. NM, RG 393, NA, M-1120, roll 26; Darlis A. Miller and Norman L. Wilson, Jr., "The Resignation of Judge Joseph G. Knapp," *New Mexico Historical Review,* 55 (October 1980): 335–44.

11. Garna Loy Christian, "Sword and Plowshare: The Symbiotic Development of Fort Bliss and El Paso, Texas, 1849–1918" (Ph.D. dissertation, Texas Tech University, 1977), p. 20; Commanding Officer to Grover, 13 August 1869, LS, Fort Bliss, PR, RG 393, NA; Martin H. Hall, *Sibley's New Mexico Campaign* (Austin: University of Texas Press, 1960), p. 20. Secondary sources claim the Confederates set fire to the post when they evacuated it. See Christian, p. 28.

12. Evans to AAG, 16 May 1863, LR, Dept. NM, RG 393, NA, M-1120, roll 18.

13. See Reports and Returns Received from Quartermasters at Fort Bliss and Franklin (January 1865), CQM, Dist. NM, RG 393, NA.

Several years later Mrs. E. Gillock would submit a claim of $15,600 for unpaid rent and damages to her property—a hotel and at least one other building—that the army had occupied in Franklin during the war. See letter of Chief Quartermaster, 6 February 1872, Register of LR, Dept. Texas, RG 393, NA.

14. Reports and Returns Received from Quartermasters at Fort Bliss and Franklin (June, July, October 1865), CQM, Dist. NM, RG 393, NA.

15. Butler to Chief Quartermaster, and endorsements, 20 July 1865, LR, Dept. NM, RG 393, NA, M-1120, roll 26.

16. Davis to DeForrest, 18 December 1866, LS, Inspector, Dept. NM, RG 393, NA; Commanding Officer to Jones, 4 August 1866, LS, Fort Bliss, PR, RG 393, NA; Cuniffe to Knapp, 27 May 1867, and Knapp to DeForrest, 1 June 1867, LR, Dist. NM, RG 393, NA, M-1088, roll 5.

17. See Robert W. Frazer, *Forts of the West, Military Forts and Presidios and Posts Commonly Called Forts West of the Mississippi River to 1898* (Norman: University of Oklahoma Press, 1965), p. 144; Mason to Morse, 8 March 1868, and Commanding Officer to Grover, 10 August 1869, LS, Fort Bliss, PR, RG 393, NA; Darlis A. Miller, *The California Column in New Mexico* (Albuquerque: University of New Mexico Press, 1982), p. 115. For Hugh Stephenson's background, see Rex W. Strickland, "Six Who Came to El Paso, Pioneers of the 1840s," *Southwestern Studies*, 1 (Fall 1963): 34–37.

18. Mason to Hartsuff, 22 November 1867 and 18 January 1868, and Merriam to Carleton, 13 January 1871, LS, Fort Bliss, PR, RG 393, NA. The lease with Zabriskie was signed 1 December 1869. See Reg. Cont., QMG, RG 92, NA.

19. West to McFerran, 7 May 1863, and Bennett to Angerstein, 28 September 1863, LS, Dist. Ariz., RG 393, NA. Cenobia M. Angerstein was married to Ernest Angerstein, the contractor who fell out of favor with General Carleton.

20. Evans to AAG, 20 May 1863, Inspection Reports Received, Inspector, Dept. NM, RG 393, NA.

21. Mayer to AAAG, 27 July 1862, LS, Fort Garland, PR, RG 393, NA; Wallen to Cutler, 30 October 1862, LR, Dept. NM, RG 393, NA, M-1120, roll 21.

22. McFerran to Meigs, 18 January 1863, LS to QMG, CQM, Dept. NM, RG 393, NA; Carleton to Thomas, 20 November 1862, LS, Dept. NM, RG 393, NA, M-1072, roll 3; see Estimate of Funds Required by Quartermaster's Department, Fort Macy, during June 1863, CQM, Dept. NM, RG 393, NA.

23. Evans to AAG, 13 July 1863, and Carson to Cutler, 30 October 1862, LR, Dept. NM, RG 393, NA, M-1120, roll 18; Haberkorn to Easton, 12 August 1866, Fort Stanton, Cons. Corres. File, RG 92, NA.

24. Evans to AAG, 13 July 1863,

LR, Dept. NM, RG 393, NA, M-1120, roll 18.

25. Enos to McFerran, 19 October 1864, Fort Stanton, Cons. Corres. File, RG 92, NA.

26. See [name unclear] to McFerran, 4 January 1870, Fort Stanton, Cons. Corres. File, RG 92, NA. Contracts are in Reg. Cont., QMG, RG 92, NA.

27. Enos to Higdon, 31 March 1864, LS, CQM, Dept. NM, RG 393, NA; Hunter to Gerlach, 11 September 1868 and entries for October and December 1868, Fort Stanton, Reports Received from Post Quartermasters, CQM, Dist. NM, RG 393, NA; Gerlach memo, 23 December 1868, Fort Stanton, Cons. Corres. File, RG 92, NA.

28. Easton to Ludington, 25 January 1869, and Gerlach to Ludington, 22 March 1869, LR, CQM, Dist. NM, RG 393, NA; Hunter to Commanding Officer, 5 February 1869, LR, Fort Stanton, PR, RG 393, NA.

29. Stanwood to Kobbé, 27 May 1869, LS, Fort Stanton, PR, RG 393, NA; Whitman to Gerlach, 30 June 1869, Fort Stanton, Reports Received from Post Quartermasters, CQM, Dist. NM, RG 393, NA.

30. Kautz to AAG, 20 October 1869, LS, Fort Stanton, PR, RG 393, NA; Andrew Wallace, "Soldier in the Southwest: The Career of General A. V. Kautz, 1869–1886," (Ph.D. dissertation, University of Arizona, 1968), pp. 247, 259.

31. Boyd, *Cavalry Life in Tent and Field,* pp. 165, 169.

32. *Ibid.*, p. 194.

33. Robert M. Utley, *Fort Union National Monument* (Washington, D.C.: National Park Service, 1962), pp. 10–11.

34. Carleton to Meigs, 3 November 1862, LS, Dept. NM, RG 393, NA, M-1072, roll 3; Santa Fe *Gazette,* 18 April 1863; Estimate of Funds Required for Use of Quartermaster's Department at Fort Union, during April 1864, CQM, Dept. NM, RG 393, NA.

35. See Estimate of Funds Required for Use of Quartermaster's Department at Fort Union, during February 1863 and April 1864, CQM, Dept. NM, RG 393, NA.

36. McFerran to Enos, 2 April 1865, LS, CQM, Dept. NM, RG 393, NA; Cutler to Shinn, 30 July 1865, LS, Dept. NM, RG 393, NA, M-1072, roll 3; Carleton to Enos, 19 September 1865, LS, Dist. NM, RG 393, NA, M-1072, roll 3; Letter of Enos, 16 September 1865, and Letter of CQM, Department of the Missouri, 6 August 1866, Registers of LR, Dist. NM, RG 393, NA, M-1097, roll 1.

37. See Description of Fort Union [1871 or 1872], LS, Fort Union, PR, RG 393, NA; contracts are in Reg. Cont., QMG, RG 92, NA. For sawmill at Fort Union, see Estimate of Funds Required for Use of Quartermaster's Department at Fort Union, during September 1864, CQM, Dept. NM, RG 393, NA.

38. Evans to AAAG, 16 June 1868, Inspection Reports, OIG, RG 159, NA.

39. Lydia Lane, *I Married a Soldier* (reprint; Albuquerque: University of New Mexico Press, 1987), p. 143.

40. Rucker to Hunter, 4 February 1868, LS, CQM, Dist. NM, RG 393, NA; Evans to AAAG, 16 June 1868, Inspection Reports, OIG, RG 159, NA.

41. Gerald Thompson, *The Army and the Navajo, The Bosque Redondo Reservation Experiment, 1863–1868* (Tucson: University of Arizona Press, 1976), pp. 15–16, 19.

42. See Reg. Cont., QMG, RG 92, NA.

43. McFerran to Barr, 26 August 1863, and McFerran to Morton, 29 November 1863, LS, CQM, Dept. NM, RG 393, NA. In constructing corrals at Fort Sumner, McFerran advised placing two or three horizontal layers of sticks in the walls to prevent them from being cut down by Indians. "This cutting down is frequently done by two persons taking a rope," McFerran wrote, "one at each end, on opposite sides of the wall, and commence sawing. They soon cut a block out of the wall, which being pushed from its place leaves an opening through which animals are gently driven, with perhaps a sentinel, or guard on the opposite side of the enclosure, in perfect ignorance of what is going on."

44. Thompson, *The Army and the Navajo,* p. 41; Bristol to AAG, 26 August 1864, LR, Dept. NM, RG 393, NA, M-1120, roll 22; Enos to Wallen, 30 March 1864, LS, CQM, Dept. NM, RG 393, NA. Shelby's contract, dated 15 April 1864, is locted in Office of the Second Controller, General Accounting Office, RG 217, NA.

45. Enos to Vroom, 4 June 1867, LS, CQM, Dist. NM, RG 393, NA; Reg. Cont., QMG, RG 92, NA; Quartermaster Reports Received from Post Quartermaster, Fort Sumner, November 1867, CQM, Dist. NM, RG 393, NA.

46. Robinson to Manderfield and Tucker, 12 May 1870, LS, CQM, Dist. NM, RG 393, NA; Lawrence R. Murphy, *Lucien Bonaparte Maxwell, Napoleon of the Southwest* (Norman: University of Oklahoma Press, 1983), p. 190.

47. Frazer, *Forts of the West,* p. 95; Bergmann to AAG, 4 June 1863, Volunteer Organizations of the Civil War, New Mexico, First Cavalry Regimental Papers, OAG, RG 94, NA.

48. Santa Fe *Weekly Gazette,* 22 August 1863; Endorsement on letter of Plymton, 30 November 1863, End. S., CQM, Dept. NM, RG 393, NA.

49. Bergmann to Carleton, 5 July 1864, LR, Dept. NM, RG 393, NA, M-1120, roll 22.

50. Bergmann to Cutler, 11 August 1865, LR, Dept. NM, RG 393, NA, M-1120, roll 26.

51. Alexander to Rawlins, 5 May 1867, and Alexander to Comstock, 14 March 1867, Inspection Reports, OIG, RG 159, NA.

52. Sandra L. Myres, ed., *Cavalry*

Wife, The Diary of Eveline M. Alexander, 1866–1867 (College Station: Texas A & M University Press, 1977), p. 114.

53. Contracts are in Reg. Cont., QMG, RG 92, NA.

54. Letter from Ludington, 31 December 1869, Register of LR by Quartermaster, Fort Bascom, PR, RG 393, NA; Endorsement by Ludington, 25 February 1870, End. of the Acting Assistant Quartermaster, Fort Bascom, PR, RG 393, NA; Contracts are in Reg. Cont., QMG, RG 92, NA.

55. Letter from Potter, 8 September 1870, Register of LR by Quartermaster, Fort Bascom, PR, RG 393, NA; Ogden's contract is in Reg. Cont., QMG, RG 92, NA.

56. Frazer, *Forts of the West*, p. 108; Chaves to Cutler, 7 November 1862, LR, Dept. NM, RG 393, NA, M-1120, roll 15.

57. Anderson to Carleton, 16 November 1862, LR, Dept. NM, RG 393, NA, M-1120, roll 15; Robert C. and Eleanor R. Carriker, eds., *An Army Wife on the Frontier, The Memoirs of Alice Blackwood Baldwin, 1867–1877* (Salt Lake City: University of Utah Library, 1975), p. 65.

58. Albuquerque *Rio Abajo Weekly Press*, 23 June 1863.

59. Eaton to AAG, 7 April 1864, Eaton to Carleton, 13 June 1864, and Eaton to Cutler, 18 August 1864, LR, Dept. NM, RG 393, NA, M-1120, roll 23.

60. McFerran to McDermott, 29 March 1865, LS, CQM, Dept. NM, RG 393, NA; Shaw to CQM, 25 May 1865, LS, Fort Wingate, PR, RG 393, NA.

61. Butler to DeForrest, 20 August 1867, LS, Fort Wingate, PR, RG 393, NA.

62. Ludington to CQM, Dept. of the Missouri, 26 April 1868, LS, CQM, Dept. of the Missouri, RG 393, NA.

63. Thompson, *The Army and the Navajo*, pp. 151, 156.

64. Rynerson to McFerran, 12 September 1864, Fort McRae, Cons. Corres. File, RG 92, NA.

65. Frank D. Reeve, ed., "Frederick E. Phelps: A Soldier's Memoirs," *New Mexico Historical Review*, 25 (January 1950): 50.

66. Mayer to AAAG, 27 July 1862, LS, Fort Garland, PR, RG 393, NA.

67. Enos to Easton, 25 July 1867, LS, CQM, Dist. NM, RG 393, NA.

68. Smith to AAAG, 23 October 1874, LR, Dist. NM, RG 393, NA, M-1088, roll 26; Smith to CQM (and endorsements), 23 June 1874, and Hoberg to Smith (and endorsements), 5 April 1875, Fort Union, Cons. Corres. File, RG 92, NA.

69. Contracts are in Reg. Cont., QMG, RG 92, NA.

70. Gerlach to Ludington, 20 June 1869, LR, CQM, Dist. NM, RG 393, NA; Fleming to McKeever, 24 June 1868, LS, Fort Garland, PR, RG 393, NA.

71. Marshall to AAAG, 15 September 1882, LR, Dist. NM, RG 393, NA, M-1088, roll 47.

72. Corbin to AAG, 9 August 1868, LS, Fort Craig, PR, RG 393, NA; Myers to Quartermaster General, 1 July 1872, LS, CQM, Dist. NM, RG 393, NA; Mahnken to Commanding Officer, 21 July 1874, LS, Dist. NM, RG 393, NA, M-1072, roll 5; Whitehead to AAAG, 8 April 1878, LR, Dist. NM, RG 393, NA, M-1088, roll 32; Smith to CQM (and endorsements), 29 July 1873, LR, CQM, Dist. NM, RG 393, NA.

73. Carriker, *An Army Wife on the Frontier,* pp. 72–73.

74. McClure to Hunter, 28 February 1868, LR, Fort Craig, PR, RG 393, NA; Lee to CQM, Department of the Missouri, 3 June 1881, LS, CQM, Dist. NM, RG 393, NA.

75. Proceedings of Board of Survey, dated 30 June 1871 and 10 April 1872, Fort Stanton, PR, RG 393, NA; Meigs to Myers, 7 June 1872, LR, CQM, Dist. NM, RG 393, NA.

76. Kobbé to Commanding Officer, 21 November 1868, LR, Fort Stanton, PR, RG 393, NA; Proceedings of Board of Survey, 19 February 1868, Fort Stanton, PR, RG 393, NA.

77. Dudley to Ludington, 31 December 1867, Smith to CQM, 28 June 1874, Proceedings of Board of Survey, 1 July 1874, LR, CQM, Dist. NM, RG 393, NA; Myers to AAAG, 29 June 1874, LS, CQM, Dist. NM, RG 393, NA.

78. Medical Records [1871], Fort Wingate, OAG, RG 94, NA; Frazer, *Forts of the West,* p. 108.

79. Ludington to Acting Assistant Quartermaster, 5 August 1868 and Ludington to Ayers, 27 September 1868, LS, CQM, Dist. NM, RG 393, NA; Ayers to Ludington, 18 December 1868, LR, CQM, Dist. NM, RG 393, NA.

80. Ayers to Ludington, 23 December 1868, LR, CQM, Dist. NM, RG 393, NA; Medical Records [1871], Fort Wingate, OAG, RG 94, NA; War Department, Surgeon General's Office, *A Report on Barracks and Hospitals,* Circular No. 4 (Washington: Government Printing Office, 1870), p. 251.

81. Ludington to Meigs, 4 June 1870, Fort Wingate, Cons. Corres. File, RG 92, NA.

82. Baldwin to Ludington, 13 April 1869, Fort Wingate, Cons. Corres. File, RG 92, NA.

83. Ludington to Meigs, 4 June 1870, Fort Wingate, Cons. Corres. File, RG 92, NA; Report of Persons Hired at Fort Wingate during June 1869, LR, CQM, Dist. NM, RG 393, NA; Dunbar's contract is in Reg. Cont., QMG, RG 92, NA.

84. Medical Records [1871], Fort Wingate, OAG, RG 94, NA; Ludington to Meigs, 4 June 1870, Fort Wingate, Cons. Corres. File, RG 92, NA; Abstract of Bids, 28 April 1870, LR, CQM, Dist. NM, RG 393, NA.

85. Easton to Robinson, 13 May 1870, LS, CQM, Dist. NM, RG 393, NA; Easton to Meigs, 25 May 1870, Fort Wingate, Cons. Corres. File, RG 92, NA.

86. Robinson to Grimes, 23 May 1870 and 30 May 1870, LS, CQM, Dist. NM, RG 393, NA; two letters

by Potter, 26 September 1870 and 12 October 1870, Registers of LR, Dist. NM, RG 393, NA, M-1097, roll 2.

87. Pope to Belknap, 31 October 1870, U.S., Congress, House, Annual Report of the Secretary of War, 1870, House Ex. Doc. 1 (Serial 1446), 41st Cong., 3d sess., 1870, pp. 11–17.

88. Medical Records [1871], Fort Wingate, OAG, RG 94, NA; Davis to Inspector General, 10 December 1872 [Fort Wingate], Inspection Reports, OIG, RG 159, NA. Army Inspector Nelson H. Davis reported that the Navajo employees worked "regularly and faithfully, and are reported as excellent laborers."

89. Lee to CQM, Department of the Missouri, 3 June 1881, and Lee to Post Quartermaster, 21 May 1881, LS, CQM, Dist. NM, RG 393, NA.

90. See Bingham to Meigs (and endorsements), 20 April 1870, Fort Bayard, Cons. Corres. File, RG 92, NA.

91. Lee to Chief Quartermaster, Department of the Missouri, 3 June 1881, LS, CQM, Dist. NM, RG 393, NA; Dale F. Giese, *Echoes of the Bugle* (n.p. Phelps Dodge Corporation, 1976), p. 16.

92. Kautz to Townsend, 24 January 1872, and Kautz to AAG [February 1872], LS, Fort Stanton, PR, RG 393, NA; the contracts are in Reg. Cont., QMG, RG 92, NA.

93. Belcher to AAAG, 14 September 1878, Dana to Post Quartermaster, 30 October 1879, Dana to AAAG, 10 March 1880, LS, CQM, Dist. NM, RG 393, NA.

94. See undated brief and notes dated 25 January 1886, pertaining to water supply, Fort Stanton, Cons. Corres. File, RG 92, NA.

95. Frazer, *Forts of the West*, pp. 100–101; *A Report on Barracks and Hospitals*, p. 257. See also Mahnken to C.O., 21 July 1874, LS, Dist. NM, RG 393, NA, M-1072, roll 5.

96. Rental fees in May 1871 amounted to $672. McGonnigle to Quartermaster General, 5 May 1871, LS to Washington, CQM, Dist. NM, RG 393, NA.

97. McGonnigle to Quartermaster General, 21 September 1871, LS, CQM, Dist. NM, RG 393, NA; Account of Expenditures on Account of New Quarters at Santa Fe, 20 April 1871, LR, CQM, Dist. NM, RG 393, NA. Fifteen of the thirty-one suppliers were Hispanos, who received a total of $2,301.

98. Carey to Meigs, 13 December 1871, and Davis to Easton, 5 November 1871, LR, Dist. NM, RG 393, NA, M-1088, roll 12; Myers to Quartermaster General, 1 July 1872, LS, CQM, Dist. NM, RG 393, NA.

99. Myers to Quartermaster General, 1 July 1872, LS, CQM, Dist. NM, RG 393, NA; *A Report on Barracks and Hospitals*, p. 257.

100. Allen to Rucker, 12 November 1872, LR, Dist. NM, RG 393, NA, M-1088, roll 16.

101. Contracts are in Reg. Cont., QMG, RG 92, NA; Eagan to AAAG, 15 June 1875, and Eagan to CQM, 25 June 1875, LS, CQM, Dist. NM, RG 393, NA.

102. Belcher to CQM, 29 May

1878, LS, CQM, Dist. NM, RG 393, NA.

103. Lee to CQM, 4 December 1880, LS, CQM, Dist. NM, RG 393, NA.

104. Endorsement by Lee, 6 January 1881, LS, CQM, Dist. NM, RG 393, NA.

105. Lee to Acting Assistant Quartermaster, 12 February 1881, LS, CQM, Dist. NM, RG 393, NA; Mackinzie to AAG, 9 February 1882, LS, Dist. NM, RG 393, NA, M-1072, roll 7. Lee had this to say about the condition of military buildings in Santa Fe: "The buildings comprising the Headquarters of the District are in fair condition. They have been well though incompletely built, and pretty well cared for. . . . The quarters of [Fort Marcy] have never been completed. They are built of adobes, and like all adobe structures are suffering from the action of the weather. The officers' quarters are in bad order, and the men's barracks in much worse." Lee to CQM, 3 June 1881, LS, CQM, Dist. NM, RG 393, NA.

106. Lee to Meigs, 20 October 1868, U.S., Congress, House, Annual Report of the Secretary of War, 1868, House Ex. Doc. 1 (Serial 1367), 40th Cong., 3d sess., 1868, pp. 866–71.

107. Hatch to Morse, 14 December 1867, LS, Fort Stockton, PR, RG 393, NA; Clayton W. Williams, *Texas' Last Frontier, Fort Stockton and the Trans–Pecos, 1861–1895* (College Station: Texas A & M University Press, 1982), p. 84.

108. Endorsement by Johnson, 9 November 1869, Endorsements, Fort Stockton, PR, RG 393, NA; *A Report on Barracks and Hospitals,* p. 225; Merritt to AAAG, 14 February 1872, LS, Fort Stockton, PR, RG 393, NA.

109. Williams, *Texas' Last Frontier,* p. 173; McLaughlin to AG, 28 August 1878, LS, Fort Stockton, PR, RG 393, NA.

110. Barry Scobee, *Fort Davis, Texas* (Fort Davis, Texas: Barry Scobee, 1963), p. 21.

111. Merritt to Commanding Officer, 1 June 1869, and Andrews to Courtney, 10 September 1877, LS, Fort Davis, PR, RG 393, NA.

112. Elvis Joe Ballew, "Supply Problems of Fort Davis, Texas, 1867–1880" (Master's thesis, Sul Ross State University, 1971), p. 90; Jerome A. Greene, *Historic Resource Study, Fort Davis National Historic Site* (Washington, D.C.: U.S. Department of the Interior, National Park Service, 1986), p. 97. A detailed account of structures and construction at Fort Davis can be found in Greene's study.

113. Bliss to AAAG, 19 November 1872, LS, Fort Stockton, PR, RG 393, NA; Andrews to AAG, 6 March 1876, Fort Davis, PR, RG 393, NA.

114. *A Report on Barracks and Hospitals,* p. 231.

115. Morrow to Assistant Inspector General, 6 July 1870, LS, Fort Quitman, PR, RG 393, NA.

116. Carleton to AAG, 6 March 1871, Inspection Reports, OIG, RG 159, NA; Morrow to AAAG, 14 February 1872, LS, Fort Quitman, PR, RG 393, NA; War Department, Surgeon General's Office, *Report on the Hygiene of the United States Army*, Circular No. 8 (Washington: Government Printing Office, 1875), p. 220; George Ruhlen, "Quitman: 'The Worst Possible Post at Which I Ever Served,'" *Password*, 11 (Fall 1966): 117. On the ownership of Fort Quitman, see George Ruhlen, "Quitman's Owners: A Sidelight on Frontier Reality," *Password*, 5 (April 1960): 54–64.

117. [Name unclear] to Post Adjutant, 29 August 1876, LS by Quartermaster, Fort Bliss, PR, RG 393, NA.

118. Bliss to Ord, 18 December 1876, LS, Fort Bliss, PR, RG 393, NA; Leon Metz, *Fort Bliss, An Illustrated History* (El Paso: Mangan Books, 1981), p. 36; see C. L. Sonnichsen, *The El Paso Salt War, 1877* (El Paso: Texas Western Press, 1961).

119. See Moore to AAAG, 14 July 1878 (with enclosure and endorsements), Pope to AAG, 5 March 1878, and Zabriskie to Drum, 21 January 1878 (with endorsements), LR, Dist. NM, RG 393, NA, M-1088, roll 34. Additional buildings had to be obtained, and by 1 January 1879 the army was paying about $408 per month in rental fees. See Gibbon to CQM, 4 December 1878, LS by quartermaster, Fort Bliss, PR, RG 393, NA.

120. [Name unclear] to Quartermaster General, 9 July 1879, LS by quartermaster, Fort Bliss, PR, RG 393, NA; Moore to AAAG, 14 July 1878, LR, Dist. NM, RG 393, NA, M-1088, roll 34.

121. [Hatch] to AAG, 5 December 1878, LS, Dist. NM, RG 393, NA, M-1072, roll 6; General Order No. 7, 12 February 1879, LR, Dist. NM, RG 393, NA, M-1088, roll 36.

122. Magoffin to Hatch, 6, 16 February 1879, LR, Dist. NM, RG 393, NA, M-1088, roll 36; Hatch to AAG, 21 March 1879, LS, Dist. NM, RG 393, NA, M-1072, roll 6.

123. Pope to Whipple, 22 September 1880, U.S., Congress, House, Annual Report of the Secretary of War, 1880, House Ex. Doc. No. 1, pt. 2, (Serial 1952), 46th Cong., 3d sess., 1880, p. 92; Dana to Cavanaugh, 30 July 1880, LS, CQM, Dist. NM, RG 393, NA.

124. Kinzie to CQM, 7 September 1880, Hunt to Lee, [n.d.] November 1880, Bean to Hunt, 4 November 1880, and Smith to CQM, 24 November 1880, LR, CQM, Dist. NM, RG 393, NA.

125. Davis to Hart, 19 February 1881, and Davis to Stout, 9 March 1881, LS by quartermaster, Fort Bliss, PR, RG 393, NA.

126. Lee to CQM, Dept. of the Missouri, 3 June 1881, LS, CQM, Dist. NM, RG 393, NA; Davis to Pierce, 15 November 1880, Davis to Quartermaster General, 21 April 1881, and Acting Assistant Quartermaster to CQM, Dept. of the Mis-

souri, 5 May 1882, LS by quartermaster, Fort Bliss, PR, RG 393, NA.

127. Platt to Lovell, 8 November 1866, LR, Dist. Ariz., RG 393, NA.

128. Crook to Schofield, 21 September 1872, U.S., Congress, House, Annual Report of the Secretary of War, 1872, House Ex. Doc. No. 1, pt. 2 (Serial 1558), 42d Cong., 3d sess., 1872, p. 74.

129. Hinds to West, 6 October 1862, LR, Dist. Ariz., RG 393, NA; Davis to Carleton, 20 February 1864, LS, Inspector, Dept. NM, RG 393, NA.

130. Hager to Choisy, 31 July 1866, LS, Fort Bowie, PR, RG 393, NA.

131. Jones to Fry, 28 April 1869, Inspection Reports, OIG, RG 159, NA; Ripley to AAG, 27 October 1868, LS, Fort Bowie, PR, RG 393, NA.

132. Bernard to Commanding Officer, Camp Lowell, 27 September 1870, LS, Fort Bowie, PR, RG 393, NA; Sacket to Inspector General, 18 June 1873, Inspection Reports, OIG, RG 159, NA.

133. [Name unclear] to AAAG, 18 February 1874, LS, Fort Bowie, PR, RG 393, NA; Kautz to AAG, 20 December 1875, LS, Dept. Ariz., RG 393, NA. Laundresses and married soldiers occupied quarters at the old post.

134. Frazer, *Forts of the West,* pp. 11–12.

135. Fitch to Drum, 20 May 1863, LS, Fort Mojave, PR, RG 393, NA; Constance Wynn Altshuler, *Starting with Defiance, Nineteenth Century Arizona Military Posts* (Tucson: Arizona Historical Society, 1983), pp. 144.

136. *A Report on Barracks and Hospitals,* pp. 467–68; Special Orders Numbers 1, 10, 11, 1 January, 6 February 1869, Fort Mojave, PR, RG 393, NA; Stacey to CQM, 16 August 1869, LS, Fort Mojave, PR, RG 393, NA.

137. Porter to Choisy, 16 July 1866, and Stacey to AAAG, 30 August 1869, LS, Fort Mojave, PR, RG 393, NA; Endorsements by Stacey, 14 May 1870 and 9 December 1870, Endorsements, Fort Mojave, PR, RG 393, NA; Altshuler, *Starting with Defiance,* p. 47; *Report on the Hygiene of the United States Army,* p. 548.

138. Endorsement by Thompson, 27 February 1873, Endorsements, Fort Mojave, PR, RG 393, NA; Letter dated 28 December 1876, Register of LR, Dept. Ariz., RG 393, NA; Endorsements on letter dated 20 January 1872, LS, Dept. Ariz., RG 393, NA; *Report on the Hygiene of the United States Army,* pp. 547–48.

139. Inspection Report of Camp Mojave, 8 April 1877, Inspection Reports, OIG, RG 159, NA; see also Reg. Cont., QMG, RG 92, NA; Van Horne to AAG, 17 August 1878, LS, Fort Mojave, PR, RG 393 NA.

140. Byrne to AAG, 24 August 1880, and [name unclear] to AAG, 8 August 1881, LS, Fort Mojave, PR, RG 393, NA; Letter dated 19 September 1881 with endorsements and Letter dated 19 February 1883, Endorsements, Fort Mojave, PR, RG 393, NA.

141. Wallace, "Soldier in the Southwest," p. 335.

142. Frazer, *Forts of the West,* p. 14; Davis to Carleton, 20 March 1864, LS, Inspector, Dept. NM, RG 393, NA. Army Inspector Nelson H. Davis reported in March 1864 that hand-sawed lumber near Whipple cost $150 per thousand feet at the sawpit.

143. *A Report on Barracks and Hospitals,* p. 457; Anderson to CQM, 16 August 1864, LR, Dept. NM, RG 393, NA, M-1120, roll 22.

144. Gregg to Sherburne, [April] 1867, LS, Fort Whipple, PR, RG 393 NA.

145. Gregg to Sherburne, 17 April 1867, LS, Fort Whipple, PR, RG 393, NA; Prescott *Arizona Miner,* 7 March 1868.

146. Prescott *Arizona Miner,* 22 August 1868, 19 December 1868, 11 December 1869; Wheaton to Sherburne, 22 August 1869, LS, Fort Whipple, PR, RG 393, NA.

147. Foster to Richardson, 31 May 1870, Memorandum of Lumber Cut at Government Saw Mill during Month of July [1870], 19 January 1871, LS by quartermaster, Subdistrict of Northern Arizona, RG 393, NA. Sometime after July 1870 the sawmill was moved to the vicinity of Camp Verde, where a new post was being built. See Foster to Richardson, 19 January 1871, LS by quartermaster, Subdistrict of Northern Arizona, RG 393, NA.

148. Foster to Tompkins, 11 January 1871, LS by quartermaster, Subdistrict of Northern Arizona, RG 393, NA.

149. Foster to Tompkins, 21 January 1871, LR, Fort Whipple, PR, RG 393, NA.

150. Foster to King, 5 February 1871, LS by quartermaster, Subdistrict of Northern Arizona, RG 393, NA.

151. Endorsement on letter dated 5 September 1872, LS, Dept. Ariz., RG 393, NA; Sacket to Inspector General, 26 May 1873, Inspection Reports, OIG, RG 159, NA.

152. Sacket to Inspector General, 26 May 1873, Inspection Reports, OIG, RG 159, NA; Goodale to Bourke, 1 May 1874, LS, Fort Whipple, PR, RG 393, NA; Reg. Cont., QMG, RG 92, NA.

153. *Report on the Hygiene of the United States Army,* p. 555.

154. Kautz to AAG, 15 August 1877, U.S., Congress, House, Annual Report of the Secretary of War, 1877, House Ex. Doc. No. 1, pt. 2 (Serial 1794), 45th Cong., 2d sess., 1877, p. 134; Meigs to Lee, 14 June 1881, LR, CQM, Dist. NM, RG 393, NA; Proceedings of a Board of Officers, 29 October 1875, LR, Fort Whipple, PR, RG 393, NA; Ray Brandes, *Frontier Military Posts of Arizona* (Globe, Arizona: Dale Stuart King, 1960), p. 79.

155. Kelton to Kautz, 17 May 1878, and Reynolds to CQM, 11 April 1878, Arizona, Cons. Corres. File, RG 92, NA; contracts are in Reg. Cont., QMG, RG 92, NA.

156. Marcy to Commanding General, 24 March 1879, Arizona, Cons. Corres. File, RG 92, NA; Wallace, "Soldier in the Southwest," pp. 555–74.

157. Kelton to Kautz, 17 May 1878, and Shriver to Adjutant General, 4 May 1878, Arizona, Cons. Corres. File, RG 92, NA; Martin to Quartermaster General, 1 August 1878, Annual Reports Received from Quartermasters, RG 92, NA.

158. Altshuler, *Starting with Defiance,* pp. 33–34; Frazer, *Forts of the West,* pp. 10–11; Mowry to Coult, 4 December 1862, LR, Dist. Ariz., RG 393, NA; series of letters to N. H. Davis, 26 February–23 April 1864, Register of LR, Inspector, Dept. NM, RG 393, NA. Most buildings the army occupied were abandoned ramshackle adobes, though J. Ross Browne, a visitor to Tucson in 1864, claimed the army had taken possession of all tenantable houses. See J. Ross Browne, *Adventures in the Apache Country, A Tour Through Arizona and Sonora, 1864* (re-edition; Tucson: University of Arizona Press, 1974), p. 134.

159. Mason to Drum, 18 June 1865, and Mason to Gibson, 19 June 1865, LS, Dist. Ariz., RG 393, NA; Report of Persons and Articles Employed and Hired at the Depot of Tucson during the month of December 1865, LR, Dist. Ariz., RG 393, NA.

160. Davis to Green, 30 January 1866, LR, Dist. Ariz., RG 393, NA.

161. Altshuler, *Starting with Defiance,* p. 34.

162. Jones to Fry, 11 May 1867, Inspection Reports, OIG, RG 159, NA; Sherburne to Crittenden, 21 August 1867, LR, Dist. Tucson, RG 393, NA.

163. Brown to Sherburne, 12 May 1869, LS, Camp Lowell, PR, RG 393, NA.

164. Tucson *Weekly Arizonan,* 9, 30 October, 6 November 1869.

165. Cogswell to Sherburne, 12 April 1870, LS, Subdistrict of Southern Arizona, RG 393, NA.

166. See Crook to AAG, 4 June 1872, LS, Dept. Ariz., RG 393, NA; Tucson *Arizona Citizen,* 22 March 1873; Thomas H. Peterson, Jr., "Fort Lowell, A. T., Army Post During the Apache Campaigns," *The Smoke Signal,* 8 (Fall 1963): 8.

167. Sacket to Inspector General, 12 June 1873, Inspection Reports, OIG, RG 159, NA; Tucson *Arizona Citizen,* 31 May, 7 June 1873; Carr to AAG, 19 June 1873, LS, Camp Lowell, PR, RG 393, NA.

168. Sacket to Inspector General, 12 June 1873, Inspection Reports, OIG, RG 159, NA. The firm of Tully and Ochoa probably owned the twenty-room building rented for storage; see Tucson *Weekly Arizonan,* 26 March 1870.

169. Ludington to Meigs, 25 September 1873, U.S., Congress, House, Annual Report of the Secretary of War, 1873, House Ex. Doc. 1, pt. 2 (Serial 1597), 43d Cong., 1st sess., 1873, p. 179.

170. Hardie to Marcy, 27 May 1875, Inspection Reports, OIG, RG 159, NA.

171. Meigs to Secretary of War, 27 October 1874, Camp Grant, Cons. Corres. File, RG 92, NA.

172. Bourke to Carr, 7 February 1874, LS, Dept. Ariz., RG 393, NA;

Reports of Inspection of Posts in the Department of Arizona made by Inspector–General Schriver in March, April, and May 1878, OIG, RG 159, NA.

173. Bill Reed, *The Last Bugle Call, A History of Fort McDowell, Arizona Territory, 1865–1890* (Parsons, West Virginia: McClain Printing Company, 1977), pp. 5–8, 14, 114; Davis to Green, 25 January 1866, LR, Dist. Ariz., RG 393, NA; *A Report on Barracks and Hospitals,* p. 460.

174. Reed, *The Last Bugle Call,* p. 116; Smart to [name unclear], 21 March 1867, LR, Fort McDowell, PR, RG 393, NA.

175. Grant to Chilson, 7 August 1867, Grant to AAG, 10 January 1868, Van Derslice to Tobey, 13 July 1868, LR, Fort McDowell, PR, RG 393, NA.

176. Sandra L. Myres, "Evy Alexander, The Colonel's Lady at McDowell," *Montana, The Magazine of Western History,* 24 (Summer 1974): 30–31; Van Derslice to Tobey, 13 July 1868, LR, Fort McDowell, PR, RG 393, NA; Reed, *The Last Bugle Call,* p. 117.

177. Endorsement of proceedings of a board of officers, convened 2 August 1872, LS, Dept. Ariz., RG 393, NA; Sacket to Inspector General, 7 June 1873, Inspection Reports, OIG, RG 159, NA. Sacket stated that if McIntyre did not receive payment soon he would be "a ruined man financially" because he had been forced to borrow money at a high rate of interest to pay his workers. Sacket also reported that laundresses' quarters then in use were "not fit abodes for women and children."

178. Sacket to Inspector General, 26 May 1873, Inspection Reports, OIG, RG 159, NA; Smith's contract is in Reg. Cont., QMG, RG 92, NA; Chaffee to AAG, 23 July 1876, LS, Fort McDowell, PR, RG 393, NA.

179. Kautz to AAG, 20 December 1875, LS, Dept. Ariz., RG 393, NA; Hay to AAG, 1 July 1873, LS, Fort McDowell, PR, RG 393, NA; Report of Inspection of Posts in the Department of Arizona made by Inspector–General Schriver in March, April, and May 1878, OIG, RG 159, NA.

180. Altshuler, *Starting with Defiance,* p. 28.

181. Ilges to Tobey, 5 September 1866, LR, Dist. Ariz., RG 393, NA; Lovell to Drum, 20 September 1866, LS, Dist. Ariz., RG 393, NA.

182. Jones to Fry, 5 June 1867, Inspection Reports, OIG, RG 159, NA; Ilges to Winters, 31 May 1868, LR, Dist. Ariz., RG 393, NA; Devin to Sherburne, 5 November 1868, LS, Dist. Tucson, RG 393, NA.

183. *A Report on Barracks and Hospitals,* p. 466; Jones to Fry, 11 May 1869, Inspection Reports, OIG, RG 159, NA.

184. John G. Bourke, *On the Border With Crook* (reprint; Lincoln: University of Nebraska Press, 1971), pp. 4–5.

185. Green to AAG, 1 January 1870, and DuBois to Sherburne, 27 May 1870, LS, Camp Grant, PR, RG 393, NA; Devin to Sherburne, 3

February 1870, LS, Subdistrict of Southern Ariz., RG 393, NA.

186. Tucson *Weekly Arizonan,* 4 June 1870.

187. Bourke, *On the Border With Crook,* p. 104.

188. Royal to AAG (and endorsement), 9 April 1873, LR, Camp Grant, PR, RG 393, NA; Hall to CQM (and endorsement), 11 April 1874, LR by quartermaster, Fort Grant, PR, RG 393, NA; Hardie to Marcy, 9 September 1875, Inspection Reports, OIG, RG 159, NA; Frazer, *Forts of the West,* pp. 4–6, 9.

189. [Name unclear] to Acting Assistant Quartermaster, 12 July 1875, LR, Camp Grant, PR, RG 393, NA; Adams to CQM, 18 August 1875, LS by quartermaster, Camp Grant, PR, RG 393, NA; Hardie to Marcy, 9 September 1875, Inspection Reports, OIG, RG 393, NA.

190. Meigs to Secretary of War, 27 October 1874, Camp Grant, Cons. Corres. File, RG 92, NA; Hardie to Marcy, 22 June 1875 and 9 September 1875, Inspection Reports, OIG, RG 159, NA. Hardie reported that the commanding officer's house was "a large rambling structure with a cupola on top—a rare architectural product."

191. Brannen's contract is in Reg. Cont., QMG, RG 92, NA; see also Smith to Quartermaster General (and enclosure), 18 August 1876, Camp Grant, Cons. Corres. File, RG 92, NA.

192. Proceedings of a Board of Officers convened by order dated 2 March 1877, Camp Grant, Cons. Corres. File, RG 92, NA.

193. Altshuler, *Starting with Defiance,* p. 30.

194. McFerran to Garrison, 28 December 1863, LS, CQM, Dept. NM, RG 393, NA.

195. McClure to Eaton, 1 December 1866, LS, CCS, Dist. NM, RG 393, NA.

196. Reports of Persons and Articles Hired during January, March, August 1866, Post Quartermaster, Fort Cummings, PR, RG 393, NA.

197. Wages obtained from several inspection reports for Arizona posts, 1866 and 1867, OIG, RG 159, NA.

198. Hunter to Bradley, 5 February 1868, LS, Dist. NM, RG 393, NA, M-1072, roll 4.

199. Bradley to Ludington, 16 May 1868, LR, CQM, Dist. NM, RG 393, NA; General Order No. 6, 29 February 1868, Dept. of the Missouri, RG 393, NA.

200. Meigs to Adjutant General, 20 October 1869, U.S., Congress, House, Annual Report of the Secretary of War, 1869, House Ex. Doc. No. 1, pt. 2 (Serial 1412), 41st Cong., 2d sess., 1869, p. 206; Ludington to AAAG, 8, 30 June 1869, LS, CQM, Dist. NM, RG 393, NA. Only a few civilians were employed in other branches of the army. In 1869 the subsistence department employed only fourteen: two in Santa Fe and twelve at the Fort Union Depot. McClure to Kobbé, 5 October 1869, LS, CCS, Dist. NM, RG 393, NA.

201. See Acting CQM to post

quartermaster (Fort Stanton), 2 May 1871, LS, CQM, Dist. NM, RG 393, NA. The eight-hour workday controversy is discussed in following pages.

202. Ludington to AAAG, 23 June 1869, LS, CQM, Dist. NM, RG 393, NA; Mitchell to Getty, 17 August 1869, LR, CQM, Dist. NM, RG 393, NA.

203. Eaton to CQM, 7 July 1871, LR, CQM, Dist. NM, RG 393, NA; Myers to CQM, Dept. of the Missouri, 18 September 1872, LS, CQM, Dist. NM, RG 393, NA; Myers to AAG, 7 October 1872, LR, Dist. NM, RG 393, NA, M-1088, roll 16.

204. Mason to Drum, 18 June 1865 and Circular Letter dated 7 October 1865, LS, Dist. Ariz., RG 393, NA.

205. See for example Commanding Officer to McKeever, 15 July 1868, LS, Fort Craig, PR, RG 393, NA; Ludington to AAAG, 12 July 1869, LS, CQM, Dist. NM, RG 393, NA; Endorsement by Bush, 6 April 1875, Endorsements, Fort Stockton, PR, RG 393, NA; Norris to Wright, 31 March 1868, LS, Camp Crittenden, PR, RG 393, NA; Post quartermaster to Potter, 7 April 1871, LR, Dist. NM, RG 393, NA, M-1088, roll 13; Ritter to Myers, 9 November 1872, LR, CQM, Dist. NM, RG 393, NA; Kautz to Chief Signal Officer, 8 May 1875, LS, Dept. Ariz., RG 393, NA; Gregg to Sherburne, 17 April 1867, LS, Fort Whipple, PR, RG 393, NA. The army lost money when it relied upon unskilled soldiers to perform tasks. Army Inspector Nelson H. Davis reported in 1872 that in the District of New Mexico many cavalry horses were being "ruined or rendered unserviceable" because troopers could not shoe horses properly. Davis offered a partial explanation: "Good shoers command in civil life so much greater pay than is allowed in the army, that few *good* and *temperate* ones will enlist." See Davis's inspection reports for Fort Union, 28 December 1872, and for Fort Wingate, 10 December 1872, OIG, RG 159, NA.

206. Smith to Tobey, 2 March 1867, LR, Dist. of Tucson, RG 393, NA.

207. See Inspection Reports for these posts, dated May and June 1867, OIG, RG 159, NA. The figure for Camp McDowell does not include twenty-three citizens employed on the government farm. Jones reported in May that only fifty-two civilians were employed in Tucson by the quartermaster's department.

208. Sherburne to Devin, 4 March 1869, LR, Dist. Ariz., RG 393, NA.

209. Prescott *Weekly Arizona Miner*, 17 April 1869.

210. Ord to Townsend, 27 September 1869, U.S., Congress, House, Annual Report of the Secretary of War, 1869, House Ex. Doc. No. 1, pt. 2 (Serial 1412), 41st Cong., 2d sess., 1869, pp. 125–26.

211. Endorsements on Consolidated Report of Persons hired in the Dept. of Arizona for November 1871, LS, Dept. Ariz., RG 393, NA.

212. Miller to Reynolds, 12 May

1869, LR, 5th Military District, RG 393, NA, M-1193, roll 22; Carleton to AAG, Dept. of Texas, 6 March 1871, Inspection Reports, [Fort Bliss], OIG, RG 159, NA.

213. Carleton to AAG, Dept. of Texas, 6 March 1871, Inspection Reports, [Fort Quitman], OIG, RG 159, NA.

214. Carleton to AAG, Dept. of Texas, 6 March 1871, Inspection Reports, [Fort Davis] and [Fort Stockton], OIG, RG 159, NA. Carleton's advice was accepted; the veterinary surgeons employed at Quitman and Davis were soon discharged.

215. See Myers to AAAG (and enclosures), 14 February 1874, LR, Dist. NM, RG 393, NA, M-1088, roll 23. Later Myers learned that only $1,455 would be allocated monthly for civilian employees. See Myers to AAAG, 23 February 1874, LR, Dist. NM, RG 393, NA, M-1088, roll 23.

216. Crook to AAG, 13 February 1874, and Circular Letter dated 23 February 1874, LS, Dept. Ariz., RG 393, NA.

217. Statement showing number of civilians employed for June 1884, U.S., Congress, House, Annual Report of the Secretary of War, 1884, House Ex. Doc. No. 1, pt. 2 (Serial 2136), 48th Cong., 2d sess., 1884, pp. 368–417. Only one superintendent of transportation was employed in New Mexico.

218. Lee to Acting Assistant Quartermaster, Santa Fe, 1 July 1881, LS, CQM, Dist. NM, RG 393, NA; [name unclear] to AAG, Dept. of the Missouri (and endorsements), 20 December 1883, LR, Dist. NM, RG 393, NA, M-1088, roll 52. Following a general reduction of employees in 1877, the depot quartermaster at Fort Union gave employment to six teamsters without pay, furnishing them a ration for each day they worked with the understanding they would be placed on the payroll when a vacancy occurred. See Depot Quartermaster to AAAG, 25 April 1877, LR, Dist. NM, RG 393, NA, M-1088, roll 30.

219. Price to Kimball, 27 March 1878, and Hatch to AAG, 30 March 1878, LR, Dist. NM, RG 393, NA, M-1088, roll 32.

220. See, for example, Inspection Report of Fort Craig, dated 31 October 1872, Inspection Reports, OIG, RG 159, NA.

221. Alexander to Sir, 23 September 1869, LS, Camp Hualpai, PR, RG 393, NA; [name unclear] to AAAG, 9 July 1882, LS, Fort Stanton, PR, RG 393, NA.

222. Enos to Butler, 17 September 1866, LS, CQM, Dist. NM, RG 393, NA; Munsford to Quartermaster General (and endorsements), 25 March 1885, LR, Dist. NM, RG 393, NA, M-1088, roll 56.

223. Jones to Fry, 22 April 1869, Inspection Reports (Camp Crittenden), OIG, RG 159, NA; Roll of men employed on extra duty for March 1869, Fort Stanton, Reports Received from Post Quartermasters, CQM, Dist. NM, RG 393, NA.

224. Meigs to Belknap, 10 October 1874, U.S., Congress, House, Annual Report of the Secretary of War, 1874, House Ex. Doc. No. 1, pt. 2 (Serial 1635), 43d Cong., 2d sess., 1874, p. 108.

225. Chandler to AAAG (and enclosure), 5 August 1874, LR, Camp Grant, PR, RG 393, NA; Willcox to AAG, 15 April 1878, LS, Dept. Ariz., RG 393, NA.

226. Belcher to AAAG (and endorsements), 29 August 1876, LR, Dist. NM, RG 393, NA, M-1088, roll 28; Van Horn to Quartermaster General, 10 August 1883, LS, Fort Stanton, PR, RG 393, NA. Because of limited funds, posts in Texas in 1875 were allowed only three extra-duty men per company, except at posts garrisoned by a single company, which would be allowed four. See Circular Letter No. 13, 26 April 1875, LS, Dept. Texas, RG 393, NA, M-1114, roll 2.

227. Ryer to [no name], 4 October 1876, LR, Camp Grant, PR, RG 393, NA.

228. Pope to Drum, 15 September 1877, U.S., Congress, House, Annual Report of the Secretary of War, 1877, House Ex. Doc. No. 1, pt. 2 (Serial 1794), 45th Cong., 2d sess., 1877, p. 67.

229. Hatch to Adjutant General, 29 December 1879, LS, Dist. NM, RG 393, NA, M-1072, roll 6; Sacket to Inspector General, 26 May 1873, Inspection Reports, OIG, RG 159, NA; Nickerson to Commanding Officer, 26 February 1872, LS, Dept. Ariz., RG 393, NA. In 1884 general service clerks were boarding in Prescott at reduced rates, supplying their landlord with food they had purchased at cost from the commissary department. See Weston to Macfeely, 3 September 1884, Arizona, Cons. Corres. File, RG 92, NA. See also General Order No. 92, 4 November 1868 and General Order No. 20, 25 January 1869, Headquarters of the Army, Washington, RG 94, NA.

230. Martin to Willcox, 6 September 1879, LS, Dept. Ariz., RG 393, NA; Loud to Wheeler, 12 March 1879, LS, Dist. NM, M-1072, roll 6. General service clerks handled all official communications forwarded to or through headquarters. The amount of work this entailed was enormous. At district headquarters in Santa Fe the endorsement book alone was assigned to a single clerk, who in a twelve-month period copied 2,845 endorsements. See Hatch to Adjutant General, 4 February 1881, LS, Dist. NM, RG 393, NA, M-1072, roll 6.

231. Martin to Willcox, 6 September 1879, LS, Dept. Ariz., RG 393, NA; Dana to CQM, 4 January 1880, LS, CQM, Dist. NM, RG 393, NA.

232. See for example Enos to Meigs, 14 November 1866, LS to Washington, CQM, Dist. NM, RG 393, NA; letter by Butler, 6 February 1866, Registers of LR, Dist. NM, RG 393, NA, M-1097, roll 1; Davis to CQM, 4 September 1866, LS, In-

spector, Dist. NM, RG 393, NA.

233. [Name unclear] to Easton, 9 June 1870, LS to Department of the Missouri, CQM, Dist. NM, RG 393, NA; Report of civilian employees in Quartermaster's Department, 21 May 1870, LR, CQM, Dist. NM, RG 393, NA. Nine clerks received $100 each per month, five received $125 each per month, and one received $150 per month.

234. General Order No. 5, 6 June 1870, Dept. Ariz., RG 393, NA.

235. McClellan to AAAG, 6 June 1869, LS, Fort Garland, PR, RG 393, NA; Endorsement by Pond dated 26 July 1871, Endorsements, Fort Mojave, PR, RG 393, NA; Gibbon to AAAG, 22 March 1878, LS by post quartermaster, Fort Bliss, PR, RG 393, NA.

236. See two letters dated 3 December 1870 from Nash to Kobbé, LS, CCS, Dist. NM, RG 393, NA; Letter of Easton, 22 March 1871, Registers of LR, Dist. NM, RG 393, NA, M-1097, roll 2.

237. See Report of civilian employees in quartermaster's department in the District of New Mexico, 21 May 1870, and Fuller to Quartermaster General, 5 April 1872, LR, CQM, Dist. NM, RG 393, NA.

238. Darlis A. Miller, "The Role of the Army Inspector in the Southwest: Nelson H. Davis in New Mexico and Arizona, 1863–1873," *New Mexico Historical Review,* 59 (April 1984): 149.

239. Dudley to Nickerson, 2 October 1871, LS, Fort McDowell, PR, RG 393, NA.

240. Bennett to Babbitt, 18 March 1865, LS, Fort Bowie, PR, RG 393, NA; John A. Spring, *John Spring's Arizona,* ed. by A. M. Gustafson (Tucson: University of Arizona Press, 1966), p. 52; Altshuler, *Starting with Defiance,* p. 19.

241. See Krause to Green, 29 May 1865, LS, Fort Whipple, PR, RG 393, NA; Jay J. Wagoner, *Early Arizona, Prehistory to Civil War* (Tucson: University of Arizona Press, 1975), pp. 252–55; Dan L. Thrapp, "Dan O'Leary, Arizona Scout, A Vignette," *Arizona and the West,* 7 (Winter 1965): 287–98; Dan L. Thrapp, *Al Sieber, Chief of Scouts* (Norman: University of Oklahoma Press, 1964), p. 42.

242. Letter of Gregg, 20 January 1872, Registers of LR, Dist. NM, RG 393, NA, M-1097, roll 2; Ludington to Baird, 25 May 1869, LS, CQM, Dist. NM, RG 393, NA; Moore to AAG, 21 August 1868, LS, Fort Cummings, PR, RG 393, NA.

243. Ludington to Von Luettwitz, 28 November 1868, LS, CQM, Dist. NM, RG 393, NA; Report of Persons and Articles Hired during August 1867, Fort Cummings, PR, RG 393, NA.

244. See, for example, Foster to AAAG, 17 November 1870, LS by quartermaster, Subdistrict of Northern Ariz., RG 393, NA.

245. See Inspection Reports for Camps Bowie, Apache, and Hualpai, dated 18 June 1873, 26 June 1873, and 23 May 1873, respectively, OIG, RG 159, NA.

246. Thrapp, *Al Sieber,* p. 94; Egbert to AAG (and endorsements), 13 November 1878, LR, Fort Verde, PR, RG 393, NA.

247. Alexander to Sherburne, 1 June 1868, LS, Subdistrict of the Verde, RG 393, NA.

248. Calhoun to AAG, 25 March 1869, LS by post quartermaster, Camp Grant, PR, RG 393, NA. As a result of improper packing, mules developed sore backs and sides. Sometimes their injuries were so severe they could no longer be used for packing. See Nelson H. Davis's inspection report of Fort Stanton, 28 November 1872, OIG, RG 159, NA.

249. Report of Persons and Articles Hired during April 1869, Fort Cummings, PR, RG 393, NA.

250. Special Order No. 1, 1 January 1869, Fort Mojave, PR, RG 393, NA; Order No. 52, 2 September 1869, Camp Hualpai, PR, RG 393, NA.

251. See endorsement by Devin, 7 March 1870, Ends. S., Subdistrict of Southern Ariz., RG 393, NA; Thrapp, *Al Sieber,* p. 76; Wagoner, *Early Arizona,* pp. 342–43.

252. Robert M. Utley, *Frontier Regulars, The United States Army and the Indian, 1866–1891* (New York: Macmillan Publishing Co., 1973), pp. 48–49; Bourke, *On the Border with Crook,* pp. 150–51.

253. Utley, *Frontier Regulars,* pp. 196–97; Crook to Adjutant General, 30 June 1873, LS, Dept. Ariz., RG 393, NA. Indian scouts received the pay and allowances of cavalry soldiers. See Blair to Merrill, 14 October 1876, LS, Dist. NM, RG 393, NA, M-1072, roll 5; Thomas W. Dunlay, *Wolves for the Blue Soldiers, Indian Scouts and Auxiliaries with the United States Army, 1860–90* (Lincoln: University of Nebraska Press, 1982), p. 44.

254. Martin L. Crimmins, "Colonel Buell's Expedition into Mexico in 1880," *New Mexico Historical Review,* 10 (April 1935): 133–42; Jack Crawford, "Pursuit of Victorio," *Socorro County Historical Society: Publications in History,* 1 (February 1965): 1–8; Dan L. Thrapp, *Victorio and the Mimbres Apaches* (Norman: University of Oklahoma Press, 1974), pp. 292–99; Hatch to AAG, 11 December 1880, LS, Dist. NM, RG 393, NA, M-1072, roll 6.

255. For the Sierra Madre Expedition, see Crook to Adjutant General, 27 September 1883, U.S., Congress, House, Annual Report of the Secretary of War, 1883, House Ex. Doc. 1, pt. 2 (Serial 2182), 48th Cong., 1st sess., 1883, pp. 159–78; Dan L. Thrapp, *General Crook and the Sierra Madre Adventure* (Norman: University of Oklahoma Press, 1972); Utley, *Frontier Regulars,* pp. 378–81. Thrapp (p. 128) says the pack train consisted of 266 mules, but Crook reported having 350.

256. Dorst to Commanding Officer, Forts Cummings and Stanton, 19 January 1882, and Bradley to Adjutant General, 10 November 1885, LS, Dist. NM, RG 393, NA, M-1072, roll 7. Colonel Luther P. Bradley, New Mexico's commanding officer, warned that packers would

not work for $42 a month since they could find work in Arizona, where the army still paid $50 a month.

257. General Order No. 54, 12 May 1866, Fort Goodwin, Regimental Papers, First Infantry, Volunteer Organizations of the Civil War, New Mexico, RG 94, NA; Anne M. Butler, "Military Myopia: Prostitution on the Frontier," *Prologue,* 13 (Winter 1981): 248. For a good overview of laundresses in the frontier army, see Patricia Y. Stallard, *Glittering Misery, Dependents of the Indian Fighting Army* (San Rafael, California: Presidio Press, 1978), pp. 53–73. See also Darlis A. Miller, "Foragers, Army Women, and Prostitutes," in Joan M. Jensen and Darlis A. Miller, eds., *New Mexico Women, Intercultural Perspectives* (Albuquerque: University of New Mexico Press, 1986), pp. 141–67.

258. Hartwell to AAG, 3 June 1871, LR, Dist. NM, RG 393, NA, M-1088, roll 14. It is not known whether Castillo's employees were male or female.

259. Endorsements on letter of Mrs. Anne Byrne, September 1872, LS, Dept. Ariz., RG 393, NA.

260. Stallard, *Glittering Misery,* pp. 63–66; Dudley to AAG, 8 February 1880, LS, Fort Union, PR, RG 393, NA; MacGowan to Post Adjutant, 12 January 1881, LR, Dept. Ariz., RG 393, NA. Laundresses continued to be supplied with room and board at some posts after 1878, though their legal right to these benefits is unclear.

261. Chaffee to AAG, 25 June 1881, LS, Fort McDowell, PR, RG 393, NA; Van Horn to Commanding Officer, 20 September 1882, LR, Fort Selden, PR, RG 393, NA; Arnold to AAG, 19 July 1881, LR, Dept. Ariz., RG 393, NA.

262. Post Surgeon to Merriam, 15 July 1865, LR, Fort Bowie, PR, RG 393, NA; Letter of post surgeon, 2 April 1869, Register of LR, Fort Wingate, PR, RG 393, NA; DeForrest to Commanding Officer, 6 May 1867, LS, Dist. NM, RG 393, NA, M-1072, roll 4; Letter of Huntington, 1 March 1869, Registers of LR, Dist. NM, RG 393, NA, M-1097, roll 2. A congressional law of 1777 defined a matron's duties as follows: "[She] shall take care that the provisions are properly prepared; that the wards, beds, and utensils be kept in neat order. . . ." No later official description of a matron's duties has been located. See Raphael P. Thian, *Legislative History of the General Staff of the Army of the United States* (Washington: Government Printing Office, 1901), p. 371.

263. Ludington to AAAG, 22 October 1868, LS, CQM, Dist. NM, RG 393, NA; Miller, "Foragers, Army Women, and Prostitutes," pp. 154–57.

264. Letter of Hobart, 3 June 1870, Registers of LR, Dist. NM, RG 393, NA, M-1097, roll 2; Lahy to Gregg, 1 November 1872, LR, Fort Union, PR, RG 393, NA; Ritter to Eagan, 24 March 1875, LR, CQM, Dist. NM, RG 393, NA;

Compton to AAG, 22 March 1878, LS, Camp Grant, PR, RG 393, NA.

265. Enos to Farnsworth, 28 January 1866, LS, CQM, Dept. NM, RG 393, NA.

266. See, for example, Roche to Enos, 20 July 1869, Defrees to McGonnigle, 16 October 1871, Cann to Belcher, 21 July 1876, McGill to Belcher, 26 January 1878, and Ingalls to Marshall, 28 June 1882, LR, CQM, Dist. NM, RG 393, NA; Brown to McKenzie, 4 April 1883, LR, Dist. NM, RG 393, NA, M-1088, roll 50.

267. Verbeskey to VanHorn, 20 May 1884, LR, Fort Stanton, PR, RG 393, NA; Dana to Moores, 15 September 1879, and Myers to Whom it may concern, 8 February 1873, LS, CQM, Dist. NM, RG 393, NA.

268. See Report of Persons and Articles employed at Fort Union Depot, June 1870, Fort Union, Cons. Corres. File, RG 92, NA; Report of Persons and Articles employed at Fort Wingate, June 1869, LR, CQM, Dist. NM, RG 393, NA.

269. Report of civilian employees retained in the Quartermaster's Department, 21 May 1870, LR, CQM, Dist. NM, RG 393, NA. There is no way of knowing how many blacks were employed in the army, but some employees are identified in the records as black. For example, John Taylor, a black, was employed by the quartermaster's department in 1872, and two black women worked at Fort Union in 1879–Emma Beeks as an officer's servant and Margaret Berry as hospital matron. Myers to Bean, 25 June 1872, LS, CQM, Dist. NM, RG 393, NA; Miller, "Foragers, Army Women, and Prostitutes," pp. 155–56. A complete list of civilian employees in the quartermaster's department in the Department of Arizona has not been located.

270. Circular dated 1 April 1865, LS, CQM, Dept. NM, RG 393, NA.

271. Myers to Depot Quartermaster, 23 December 1872, LS, CQM, Dist. NM, RG 393, NA; [name unclear] to CQM, 5 January 1878, LS, Camp Lowell, PR, RG 393, NA; Gates to CQM, 30 April 1882, LR, CQM, Dist. NM, RG 393, NA.

272. See Fritz to Quartermaster General (and enclosures), 28 March 1872, LR, CQM, Dist. NM, RG 393, NA.

273. Meigs to AG, 13 July 1868, Fort Union, Cons. Corres. File, RG 92, NA. The impact of the eight-hour law on army employees can be followed in records pertaining to the District of New Mexico. Extant records for the Department of Arizona contain little information on this topic.

274. Bradley to Meigs, 8 October 1868, and Petition of employees at Fort Union Depot, 15 August 1868, Fort Union, Cons. Corres. File, RG 92, NA.

275. See General Order No. 6, 29 February 1868, General Order No. 3, 24 March 1869, and General Order No. 4, 26 March 1869, Department of the Missouri, RG 393, NA.

276. Rodgers to Easton, 5 April 1869, LR, CQM, Dist. NM, RG 393, NA; General Order No. 3, 24 March 1869, Department of the Missouri, RG 393, NA; Ludington to Gerlach, 24 April 1869, LS, CQM, Dist. NM, RG 393, NA.

277. Ludington to Sweet, 17 July 1869, LS, CQM, Dist. NM, RG 393, NA. See also Marion Cotter Cahill, *Shorter Hours, A Study of the Movement Since the Civil War* (reprint; New York: AMS Press, 1969), pp. 67–70; Joseph G. Rayback, *A History of American Labor* (New York: The Macmillan Co., 1959), pp. 115–19.

278. Sweet to Ludington, 13 July 1869, LR, CQM, Dist. NM, RG 393, NA; Ludington to Sweet, 17 July 1869, LS, CQM, Dist. NM, RG 393, NA.

279. Cahill, *Shorter Hours*, p. 70; VanVliet to CQM, 5 December 1872, LS, CQM, Dist. NM, RG 393, NA.

280. Godwin to CQM, 4 October 1872, Brinkerhoff to CQM, 5 October 1872, Endorsement dated 10 October 1872 on Circular letter from CQM (dated 26 September 1872), Hartz to CQM, 21 October 1872, and Blair to CQM, 30 October 1872, LR, CQM, Dist. NM, RG 393, NA.

281. See McGonnigle to CQM (and endorsements), 25 February 1873, LR, CQM, Dist. NM, RG 393, NA; Myers to Post Quartermaster at Fort Garland (and enclosure), 12 June 1873, LS, CQM, Dist. NM, RG 393, NA.

282. CQM (Dept. of the Missouri) to Myers, 29 May 1874, LR, CQM, Dist. NM, RG 393, NA; Abstract of adjusted claims . . . at Fort Union, 31 July 1874, Fort Union, Cons. Corres. File, RG 92, NA; Myers to Post Quartermaster at Fort Stanton, 17 June 1874, LS, CQM, Dist. NM, RG 393, NA.

283. Belcher to Quartermaster General, 3 January 1877, and Belcher to Third Auditor, 16 February 1877, LS, CQM, Dist. NM, RG 393, NA.

284. Philip S. Foner, *History of the Labor Movement in the United States, From Colonial Times to the Founding of the American Federation of Labor* (New York: International Publishers, 1947), p. 439.

285. Rayback, *A History of American Labor*, p. 129.

286. Belcher to CQM, 18 October 1875, LS, CQM, Dist. NM, RG 393, NA; Saxton to CQM, 22 June 1877, LR, CQM, Dist. NM, RG 393, NA. On army expenditures, see Russell F. Weigley, *History of the United States Army* (New York: Macmillan Company, 1967), p. 560.

287. See United States *v.* Martin, 94 U.S. 400; for the date the Supreme Court issued its decision, see Nancy Anderman Guenther, ed., *United States Supreme Court Decisions: An Index to Excerpts, Reprints, and Discussions* (Metuchen, N.J.: Scarecrow Press, 1983), p. 95.

288. Saxton to CQM, 13 July 1877, LR, CQM, Dist. NM, RG 393, NA. Efforts to obtain an unassailable eight-hour law for govern-

ment employees are described in Cahill, *Shorter Hours*, pp. 70–82.

289. See Risch, *Quartermaster Support of the Army*, p. 491.

290. See Jack D. Foner, *The United States Soldier Between Two Wars: Army Life and Reforms, 1865–1898* (New York: Humanities Press, 1970), pp. 77, 84–87.

Chapter 7

1. Merritt to CQM, 12 June 1867, LS, Fort Davis, PR, RG 393, NA.

2. Whitehead to Commissary General of Subsistence, 26 July 1879, New Mexico, Cons. Corres. File, RG 92, NA.

3. Potter to Meigs, 15 September 1865, U.S., Congress, House, Annual Report of the Secretary of War, 1865, House Ex. Doc. No. 1 (Serial 1249), 39th Cong., 1st sess., 1865, p. 848.

4. Erna Risch, *Quartermaster Support of the Army, A History of the Corps, 1775–1939* (Washington, D.C.: Office of the Quartermaster General, 1962), p. 476.

5. Alexander, inspection report for Fort Union, 27 April 1867, OIG, RG 159, NA; Davis, inspection report for Fort Union, 28 December 1872, OIG, RG 159, NA.

6. Henry P. Walker, "Wagon Freighting in Arizona," *The Smoke Signal*, 28 (Fall 1973): 190. Walker lists three army posts in Arizona before the Civil War, choosing to ignore Fort Breckinridge, established in 1860.

7. Fergusson to West, 25 June 1862, and Fergusson to Drum, 19 August 1862, *The War of the Rebellion: A Compilation of the Official Records of the Union and Confederate Armies*, 128 vols. (Washington, 1880–1901), Series I, vol. 50, pt. 1, pp. 1159–62 and pt. 2, pp. 76–81.

8. Fergusson to Drum, 19 August 1862, *Official Records*, Series I, vol. 50, pt. 2, pp. 76–81. Distances between towns and posts in Arizona and Sonora are approximations; figures quoted in official documents are not consistent. For mileage to Lobos and Libertad, see Richard J. Hinton, *The Handbook to Arizona* (reprint; Tucson: Arizona Silhouettes, 1954), Appendix, p. xxii.

9. Fergusson to Drum, 12 November 1862, LR, OAG, RG 92, NA, M-619, roll 195.

10. Carleton to Seward, 8 March 1863, LS, Dept. NM, RG 393, NA, M-1072, roll 3; Fergusson to Bennett, 2 April 1863, LR, Dist. Ariz., RG 393, NA; Henry Pickering Walker, "Freighting from Guaymas to Tucson, 1850–1880," *Western Historical Quarterly*, 1 (July 1970): 296.

11. Davis to Carleton, 1 May 1864, *Official Records*, Series I, vol. 50, pt. 2, p. 835; Rogers to Davis, 31 May 1864, LR, Inspector, Dept. NM, RG 393, NA.

12. Coult to Colonel, 19 July 1864, LR, Inspector, Dept. NM, RG 393, NA; Proceedings of Board of Survey, 26 July 1864, LR, Dept. NM, M-1120, roll 22.

13. Walker, "Freighting from Guaymas to Tucson," p. 297. See Rodolfo F. Acuña, *Sonoran Strongman, Ignacio Pesqueira and His Times* (Tucson: University of Arizona Press, 1974), pp. 78–93, for French intervention in Sonora.

14. Smith to Meigs, 31 May 1866, and Wallen to Cole, 7 June 1866, Arizona, Cons. Corres. File, RG 92, NA.

15. Halleck to Meigs, 9 October 1866, Arizona, Cons. Corres. File, RG 92, NA. See contract dated 15 April 1867, Abstracts of Transportation Contracts, RG 92, NA.

16. Crittenden to McDowell, 13 October 1867, and Crittenden to Sherburne, 25 December 1867, LS, Dist. of Tucson, RG 393, NA; Seward to Secretary of War, 5 February 1868 (and enclosures), Mexico, Cons. Corres. File, RG 92, NA. See endorsement dated 20 August 1868 on Tully to Carleton, 31 May 1868, Arizona, Cons. Corres. File, RG 92, NA.

17. Report of Whittier, 9 June 1868, Arizona, Cons. Corres. File, RG 92, NA. Whittier was accompanied by Lt. Barnet Wagner; both men signed the report.

18. Report of Whittier, 9 June 1868, Arizona, Cons. Corres. File, RG 92, NA.

19. See contract dated 2 August 1869, Abstracts of Transportation Contracts, RG 92, NA; Tucson *Weekly Arizonan,* 25 December 1869, 22 January 1870.

20. See contract dated 15 June 1870, Abstracts of Tranportation Contracts, RG 92, NA.

21. Smith to Meigs, 31 May 1866, Halleck to Meigs, 9 October 1866, Arizona, Cons. Corres. File, RG 92, NA; Meigs to Belknap, 19 May 1871, Colorado River, Cons. Corres. File, RG 92, NA.

22. Tucson *Weekly Arizonan,* 6 November 1869; Meigs to Belknap, 19 May 1871, Colorado River, Cons. Corres. File, RG 92, NA.

23. See Tucson *Weekly Arizonan,* 16 July 1870.

24. Tucson *Weekly Arizonan,* 28 January 1871, 8 April 1871.

25. The best study of steam transportation on the Colorado River is Francis H. Leavitt's "Steam Navigation on the Colorado River," *California Historical Society Quarterly,* 22 (March, June 1943): 1–25, 151–74. See also Richard E. Lingenfelter, *Steamboats on the Colorado River* (Tucson: University of Arizona Press, 1978). Neither focuses on military supply. See Kirkham, List of Transportation Rates (no date), Wallen to Cole, 7 June 1866, and Extract from Report to Gen'l James F. Rusling, 26 January 1867, Arizona, Cons. Corres. File, RG 92, NA.

26. Lingenfelter, *Steamboats on the Colorado River,* pp. 49–57; see contracts dated 1 June 1872 and 2 June 1873, Abstracts of Transporation Contracts, RG 92, NA.

27. Sacket to Inspector General, 12 May 1873, Inspection Reports, OIG, RG 159, NA.

28. Martha Summerhayes, *Van-*

ished Arizona, Recollections of the Army Life of a New England Woman (reprint; Glorieta, N.M.: Rio Grande Press, Inc., 1976), pp. 34–46. Haskell was Lt. Harry L. Haskell, 12th Infantry, stationed at Ft. Yuma. About two years later Martha Summerhayes, returning to Arizona by stagecoach after several months absence, forwarded her clothes and personal belongings on the ill-fated *Montana.* The steamer caught fire and ran aground near Guaymas on 14 December 1876. The ship and the freight it carried, including Martha's belongings and supplies for the army, were a total loss. Tucson *Arizona Citizen,* 30 December 1876; Summerhayes, *Vanished Arizona,* p. 201; Lingenfelter, *Steamboats on the Colorado River,* p. 57.

29. Sacket to Inspector General, 12 May 1873, Inspection Reports, OIG, RG 159, NA.

30. Fitch to AAG, 24 November 1863, and Atchison to Lee, 13 August 1864, LS, Fort Mojave, PR, RG 393, NA.

31. For sketches of Jaeger, see Frank C. Lockwood, *Pioneer Portraits, Selected Vignettes* (Tucson: University of Arizona Press, 1968), pp. 9–15; Stephen N. Patzman, "Louis John Frederick Jaeger: Entrepreneur of the Colorado River," *Arizoniana,* 4 (Spring 1963): 31–37; and Janet L. Hargett, "Pioneering at Yuma Crossing, The Business Career of L. J. F. Jaeger, 1850–1887," *Arizona and the West,* 25 (Winter 1983): 329–54. Hooper to Hughes, 7 December 1878, and Kirkham to Allen, 12 June 1868, Fort Yuma, Cons. Corres. File, RG 92, NA. Fort Yuma was originally called Camp Independence. See Robert W. Frazer, *Forts of the West, Military Forts and Presidios and Posts Commonly Called Forts West of the Mississippi to 1898* (Norman: University of Oklahoma Press, 1965), pp. 34–35.

32. Kirkham to Allen, 12 June 1868, Fort Yuma, Cons. Corres. File, RG 92, NA; Jones to Fry, 1 May 1867, Inspection Reports, OIG, RG 159, NA. According to the terms of the contract that the army awarded Jaeger on 5 March 1866, he would receive $4.75 for crossing each six-mule wagon and team, seventy-five cents for crossing a horse and rider, and twenty-five cents (in coin) for each footman. The contract is in Reg. Cont., QMG, RG 92, NA.

33. See endorsements on letter from Depot Quartermaster, dated 23 May 1872, LS, Dept. Ariz., RG 393, NA.

34. Davis to Carleton, 22 January 1865, LS, Inspector, Dept. NM, RG 393, NA.

35. Abstract of Quartermaster's Contracting, 7 January 1867, House Ex. Doc. No. 28 (Serial 1289), 39th Cong., 2d sess., 1866, p. 11; Contracts made by Quartermaster's Department, May 1868, Senate Ex. Doc. 59, pt. 2 (Serial 1317), 40th Cong., 2d sess., 1868, p. 2; Prescott *Arizona Miner,* 13 October 1866.

36. Prescott *Arizona Miner,* 13 July 1867; Byrd H. Granger, *Will C.*

Barnes' Arizona Place Names (Tucson: University of Arizona Press, 1960), p. 374.

37. Prescott *Arizona Miner,* 22 February 1868; Clendenin to Sherburne, 30 September 1868, and Clendenin to Winters, 6 March 1869, LS, Fort Whipple, PR, RG 393, NA.

38. Dean Smith, *The Goldwaters of Arizona* (Flagstaff: Northland Press, 1986), pp. 22, 32; Walker, "Wagon Freighting in Arizona," p. 188; Abstracts of Transportation Contracts, RG 92, NA.

39. McFerran to Meigs, 27 December 1863, LS to the Quartermaster General, CQM, Dept. NM, RG 393, NA.

40. Mason to Drum, 7 May 1866, LS, Dist. Ariz., RG 393, NA; the contracts are in Reg. Cont., QMG, RG 92, NA.

41. Tucson *Weekly Arizonian,* 18 April 1869; Charley to Lionel, 5, 20 February 1874 (Barron and Lionel Jacobs Business Records, Arizona Historical Society.)

42. Thomas E. Sheridan, *Los Tucsonenses, The Mexican Community in Tucson,* 1854–1941 (Tucson: University of Arizona Press, 1986), p. 43; copy of Ochoa's contract enclosed with Bennett to West, 9 January 1864, LR, Dept. NM, RG 393, NA, M-1120, roll 25. Walker states that Ochoa entered into partnership with Pinckney R. Tully in 1863. Walker, "Wagon Freighting in Arizona," p. 187. Government documents show, however, that Tully and Ochoa were freighting for the military in New Mexico at the start of the Civil War. See McFerran to Ochoa, 21 August 1861, and McFerran to Enos, 21 August 1861, LS, CQM, Dept. NM, RG 393, NA.

43. The contracts are in Abstracts of Transportation Contracts, RG 92, NA; Tucson *Weekly Arizonan,* 15 May 1869, 17 July 1869; Walker, "Wagon Freighting in Arizona," p. 198.

44. Endorsement by Cogswell, 10 June 1870, End. S., Subdistrict of Southern Ariz., RG 393, NA. Tully and Ochoa freighted even more goods for the army than contract registers suggest, for they also served as subcontractors. And on at least one occasion, in 1872, a contract awarded them does not appear in the registers.

45. Abstracts of Transportation Contracts, RG 92, NA; Davis to Green, 25 January 1866, LR, Dist. Ariz., RG 393, NA; Alexander to Sherburne, 18 August 1868, and Alexander to Jones, 10 June 1869, LS, Subdistrict of the Verde, RG 393, NA.

46. Walker, "Wagon Freighting in Arizona," p. 185.

47. Abstracts of Transportation Contracts, RG 92, NA. For contractors and rates for military transportation in Arizona during the 1870s, see Appendix Tables 12 and 13 in this volume.

48. Abstract of Bids opened 15 May 1871, Papers in the Matter of Proposals for Wagon Transportation

in Arizona during the fiscal year ending 30 June 1872, Meigs to Secretary of War (and endorsements), 13 June 1871, and Andrews to Secretary of War, 21 June 1871, Arizona, Cons. Corres. File, RG 92, NA.

49. Whiting to Meigs, 22 June 1871, and Meigs to Secretary of War, 1 September 1871, Arizona, Cons. Corres. File, RG 92, NA.

50. Goldwater to Secretary of War, 30 September 1871, and Belknap to Goldwater, 24 October 1871, Arizona, Cons. Corres. File, RG 92, NA; see documents cited in footnote 48.

51. Endorsement on letter from Neahr, dated 7 October 1871, LR, Camp Lowell, PR, RG 393, NA; Endorsements on letters from Neahr, dated 24 February 1872, 13 March 1872, 10 September 1872, LS, Dept. Ariz., RG 393, NA; Jacobs and Co. to Neahr, 23 May 1872, and Barron to Pa, 23 May 1872 (Jacobs Manuscripts, University of Arizona).

52. Crook to Schofield, 21 September 1872, U.S., Congress, House, Annual Report of the Secretary of War, 1872, House Ex. Doc. No. 1, pt. 2 (Serial 1558), 42nd Cong., 3d sess., 1872, pp. 72–73; Crook to AAG, 22 September 1873, LS, Dept. Ariz., RG 393, NA.

53. See biographical sketch of James Mitchell Barney by his grand nephew James M. Barney in the George F. Hooper File, Arizona Historical Society. See also Abstracts of Transportation Contracts, RG 92, NA. Barney died in 1914 at his home in London, England. In providing transportation for the 8th Infantry, Barney received 2 cents per pound per hundred miles for baggage and 8 cents per mile for each laundress and sick soldier.

54. Kautz to AAG, 20 October 1875, LS, Dept. Ariz., RG 393, NA.

55. Robert Athearn, *The Denver and Rio Grande Western Railroad* (reprint; Lincoln: University of Nebraska Press, 1977), pp. 4, 5, 23; Eaton to [Meigs], and endorsements, 26 November 1872, LR, CQM, Dist. NM, RG 393, NA.

56. Copy of Lovell's contract is in Contracts, Fort Cummings, PR, RG 393, NA; Sacket to Inspector General, 30 June 1873, Inspection Reports, RG 159, NA.

57. Reynolds to [Meigs], and endorsements, 7 March 1877, Ingalls to AAG, 22 August 1877, and McDowell to Adjutant General, 29 August 1877, Arizona, Cons. Corres. File, RG 92, NA.

58. Hodges to Meigs, 2 September 1878, U.S., Congress, House, Annual Report of the Secretary of War, 1878, House Ex. Doc. No. 1, pt. 2 (Serial 1843), 45th Cong., 3d sess., 1878, p. 356; Eagan to Macfeely, 15 August 1878, Arizona, Cons. Corres. File, RG 92, NA. The distance between Fort Garland and Camp Apache was 456 miles.

59. Eagan to AAG, 6 September 1878, and Eagan to AAG, 3 September 1879, Arizona, Cons. Corres. File, RG 92, NA; Keith L. Bryant, Jr., *History of the Atchison, Topeka*

and Santa Fe Railway (reprint; Lincoln: University of Nebraska Press, 1982), p. 61.

60. Bert M. Fireman, *Arizona, Historic Land* (New York: Alfred A. Knopf, 1982), p. 143; Eagan to Macfeely, 30 August 1880, Arizona, Cons. Corres. File, RG 92, NA.

61. Bryant, *History of the Atchison, Topeka and Santa Fe Railway,* pp. 79–80, 84–95; Ray Allen Billington and Martin Ridge, *Westward Expansion, A History of the American Frontier,* Fifth Edition (New York: Macmillan Publishing Co., 1982), p. 586; Eagan to Macfeely, 8 September 1882, Arizona, Cons. Corres. File, RG 92, NA.

62. Wotherspoon to AAG, 29 July 1881, LR, Dept. Ariz., RG 393, NA; Reg. Cont., QMG, RG 92, NA.

63. Abstract of contracts for wagon transportation, for fiscal year ending 30 June 1883, U.S., Congress, House, Annual Report of the Secretary of War, 1883, House Ex. Doc. No. 1, pt. 2 (Serial 2182), 48th Cong., 1st sess., 1883, pp. 540–41. The abstract of contracts lists the distance between Maricopa and Whipple as 129 miles and between Maricopa and Verde as 130 miles.

64. C. L. Sonnichsen, *Tucson, The Life and Times of an American City* (Norman: University of Oklahoma Press, 1982), p. 105; Lockwood, *Pioneer Portraits,* p. 82; Sheridan, *Los Tucsonenses,* p. 48. Ochoa died in 1888 at age fifty-seven on a visit to Las Cruces, New Mexico.

65. Henry Pickering Walker, *The Wagonmasters, High Plains Freighting from the Earliest Days of the Santa Fe Trail to 1880* (Norman: University of Oklahoma Press, 1966), pp. 238–41; Abstract of Contracts by the Quartermaster's Department, 24 April 1862, U.S., Congress, House Ex. Doc. No. 101 (Serial 1136), 37th Cong., 2d sess., 1862, p. 5.

66. Abstracts of Transportation Contracts, RG 92, NA; McFerran to Carleton, 28 August 1864, LR, OAG, RG 94, NA, M-619, roll 286; McFerran to Meigs, 29 August 1864, LS to Quartermaster General, CQM, Dept. NM, RG 393, NA.

67. Santa Fe *Gazette,* 11 February 1865; Leo E. Oliva, *Soldiers on the Santa Fe Trail* (Norman: University of Oklahoma Press, 1967), p. 161; Abstracts of Transportation Contracts, RG 92, NA; Walker, *The Wagonmasters,* pp. 243–44. The name of the railroad building west from Kansas City at this time was the Union Pacific, Eastern Division. In 1869 it became the Kansas Pacific Railway.

68. Meigs to Stanton, 8 November 1865, U.S., Congress, House, Annual Report of the Secretary of War, 1865, House Ex. Doc. No. 1 (Serial 1249), 39th Cong., 1st sess., 1865, pp. 112–14; Darlis A. Miller, "Military Supply in Civil War New Mexico," *Military History of Texas and the Southwest,* 16 (Number 3, 1982): 191.

69. Enos to Carleton, 23 Novem-

ber 1865, LS, CQM, Dept. NM, RG 393, NA; McClure to Haines, 17 February 1867, LS, CCS, Dist. NM, RG 393, NA.

70. Evans to AAAG, 14 June 1868, Inspection Reports, OIG, RG 159, NA; Carleton to Eaton, 26 November 1865, LS, Dist. NM, RG 393, NA, M-1072, roll 3; Proceedings of a Board of Survey, 4 May 1878, Fort Stanton, PR, RG 393, NA.

71. Abstracts of Transportation Contracts, RG 92, NA; Rucker to Secretary of War, 30 September 1867, U.S., Congress, House, Annual Report of the Secretary of War, 1867, House Ex. Doc. 1 (Serial 1324), 40th Cong., 2d sess., 1867, pp. 533–34.

72. Abstract of Quartermaster's Contracts, 31 December 1864, U.S., Congress, House Ex. Doc. No. 84 (Serial 1230), 38th Cong. 2d sess., 1865, p. 127; Santa Fe *New Mexican,* 28 October 1864; Sheridan, *Los Tucsonenses,* p. 52.

73. Report by Card, 31 October 1865, U.S., Congress, House, Annual Report of the Secretary of War, 1865, House Ex. Doc. No. 1 (Serial 1249), 39th Cong., 1st sess., 1865, pp. 234–35; Santa Fe *New Mexican,* 20 January 1865; McFerran to Meigs, 22 April 1865, LS to Washington, CQM, Dist. NM, RG 393, NA; for Moore's contract see Reg. Cont., RG 92, NA; Yjinio F. Aguirre, "The Last of the Dons," *Journal of Arizona History,* 10 (Winter 1969): 247.

74. Enos to Farnsworth, 29 March, 8 April 1866, LS, CQM, Dept. NM, RG 393, NA.

75. Ludington to CQM, 2 February 1868, LS to Dept. of the Missouri, CQM, Dept. of the Missouri, RG 393, NA; Ludington to Rucker, 22 February 1868, LS to Washington, CQM, Dist. NM, RG 393, NA; Abstracts of Transportation Contracts, RG 92, NA; Bradley to Ludington, 30 January, 6 April 1868, LR, CQM, Dist. NM, RG 393, NA. Frank Weber, a merchant at Golondrinas, served as Berg's other bondsman.

76. Ludington to Easton, 22 January 1869, LS to Dept. of the Missouri, CQM, Dept. of the Missouri, RG 393, NA; Proceedings of Board of Survey, 20 April 1869, LS, CQM, Dist. NM, RG 393, NA.

77. Francis C. Kajencki, "Alexander Grzelachowski, Pioneer Merchant of Puerto de Luna," *Arizona and the West,* 26 (Autumn 1984): 243–47; Abstracts of Transportation Contracts, RG 92, NA.

78. Ludington to Easton, 22 January 1869, LS to Dept. of the Missouri, CQM, Dept. of the Missouri, RG 393, NA.

79. Kajencki, "Alexander Grzelachowski," p. 249; Grzelachowski to Ludington (and enclosures), 13 July 1869, LR, CQM, RG 393, NA.

80. Abstracts of Transportation Contracts, RG 92, NA; Ludington to Easton, 8 January 1870, LS to Dept. of the Missouri, CQM, Dist.

NM, RG 393, NA. Moore had held the transportation contract in his own name in 1865 and had carried out the contract of other contractors in at least four other seasons. See Ludington to Easton, 22 January 1869, LS to Dept. of the Missouri, CQM, Dept. of the Missouri, RG 393, NA.

81. Valencia to Ludington, 1 May 1868, LR, CQM, Dist. NM, NA.

82. Ludington to Post Quartermaster, Fort Craig, 8 April 1868, 11 September 1869, LS, CQM, Dist. NM, RG 393, NA; Valencia to Ludington, 1 May 1868, and Spurgin to Ludington, 2 May 1868, LR, CQM, Dist. NM, RG 393, NA.

83. Albuquerque *Semi–Weekly Review,* 21 April 1868; Ludington to Post Quartermaster, Fort Craig, 10 April 1868, Ludington to Hubbell, 10 April 1868, and Ludington to Adams, 30 November 1868, LS, CQM, Dist. NM, RG 393, NA.

84. Carleton to West, 14 August 1862, LR, Dist. Ariz., RG 393, NA; Bowie to McCleave, 5 March 1864, Bowie to Higgins, 1 July 1864, Bowie to Elkins, 15 July 1864, and Bowie to Rynerson, 25 November 1864, LS, Dist. Ariz., RG 393, NA.

85. Darlis A. Miller, "Historian for the California Column: George H. Pettis of New Mexico and Rhode Island," *Red River Valley Historical Review,* 5 (Winter 1980): 89; Rigg to Davis, 13 May 1865, Rigg to Nissen, 13 May 1865, Rigg to McFerran, 14 May 1865, and Rigg to Nissen, 22 May 1865, LS, Fort Craig, PR, RG 393, NA. An early report listed eight privates drowned, but Rigg corrected that report on 14 May, listing only three.

86. Darlis A. Miller, "Los Pinos, New Mexico: Civil War Post on the Rio Grande," *New Mexico Historical Review,* 62 (January 1987): 26; Notice dated 31 May 1866, LS, Los Pinos, RG 393, NA; Enos to Johnson, 9 June 1866, LS, CQM, Dist. NM, RG 393, NA.

87. Articles of Agreement signed by Tison, 1 January 1872 and by Creamer, 5 March 1872, LR, CQM, Dist. NM, RG 39, NA.

88. Myers to Godwin, 23 September 1872, LS, CQM, Dist. NM, RG 393, NA; Commanding Officer to AAAG, 5 October 1874, and Fountain to CQM, 9 July 1875, LS, Fort McRae, PR, RG 393, NA; Hardie to Marcy, 25 September 1875, Inspection Reports, OIG, RG 159, NA.

89. Pope to Hartsuff, 31 October 1870, U.S., Congress, House, Annual Report of the Secretary of War, 1870, House Ex. Doc. No. 1, pt. 2 (Serial 1446), 41st Cong., 3d sess., 1870, p. 17.

90. Letter by Easton, 4 October 1870, Registers of LR, Dist. NM, RG 393, NA, M-1097, roll 2.

91. A copy of Allen's contract is located in Contracts, Fort Cummings, PR, RG 393, NA; Platt to Commanding Officer, 1 March 1876, LR, Dist. NM, RG 393, NA, M-1088, roll 27.

92. Lee to AAG, 14 February 1881, LS, CQM, Dist. NM, RG 393, NA.

93. Miguel Antonio Otero, *My Life on the Frontier, 1864–1882* (reprint; Albuquerque: University of New Mexico Press, 1987), pp. 8–37. For contractors and rates for military transportation in New Mexico during the 1870s, see Appendix Table 14 in this volume.

94. Allen to Easton, 17 April 1871, LR, CQM, Dist. NM, RG 393, NA; Allen to Easton, 1 May 1872, LS, CQM, Dist. NM, RG 393, NA; Otero, *My Life on the Frontier*, pp. 69–70.

95. Brunswick and Romero to CQM, 6 June 1873, 4 August 1873, LR, CQM, RG 393, NA.

96. Fenlon to CQM, 15 November 1880, and Brunswick to Hunt, 2 August 1881 (Letterbook, Marcus Brunswick Papers, Special Collections, University of New Mexico); Copy of 1882–83 transportation contract is located in U.S. Department of War, Records of Quartermaster's Department (Special Collections, University of New Mexico); grain contracts are in Reg. Cont., RG 92, NA.

97. Maurilio E. Vigil, *Los Patrones: Profiles of Hispanic Political Leaders in New Mexico History* (Washington, D.C.: University Press of America, 1980), pp. 63–73; T. Romero and Brother, 5 June 1874, LR, CQM, Dist. NM, RG 393, NA.

98. Foote to AAG, 21 January 1875, and Townsend to Commanding General, 30 March 1875, LR, Dist. NM, RG 393, NA, M-1088, rolls 24–25; Brinkerhoff to CQM, 11 March 1875, LS, CQM, Dist. NM, RG 393, NA. For the Red River Campaign and problems with contract transportation, see Robert M. Utley, *Frontier Regulars, The United States Army and the Indian, 1866–1891* (New York: Macmillan Publishing Co., 1973), pp. 219–35. One report stated that the Romero train "was the most efficient of any in the campaign." Mahnken to CQM, 15 February 1875, LR, CQM, Dist. NM, RG 393, NA.

99. Abstracts of Transportation Contracts, RG 92, NA; Alexander to AAAG, 24 July 1875, LR, Dist. NM, RG 393, NA, M-1088, roll 24; Belcher to Staab, 4 August 1875, LS, CQM, Dist. NM, RG 393, NA.

100. Gilman to Macfeely, 14 November 1878, Fort Union, Cons. Corres. File, RG 92, NA; Vigil, *Los Patrones*, p. 64.

101. Lance Chilton, *et al.*, *New Mexico, A New Guide to the Colorful State* (Albuquerque: University of New Mexico Press, 1984), p. 203.

102. Otero, *My Life on the Frontier*, p. 234.

103. Ralph Emerson Twitchell, *The Leading Facts of New Mexican History*, vol. 2 (Cedar Rapids, Iowa: Torch Press, 1912), p. 489; Abstracts of Transportation Contracts, RG 92, NA. The Gross, Blackwell Company was succeeded in 1901 by

the firm of Gross, Kelly and Company, one of the largest and most successful of the Las Vegas mercantile firms.

104. Articles of Agreement, 10 May 1878, LR, Dist. NM, RG 393, NA, M-1088, roll 35. Earlier contracts had specified a rate of $5 per day per wagon for delays.

105. Eagan to AAAG, 24 September 1875, LS, CCS, Dist. NM, RG 393, NA. On advantages of oxen, see Utley, *Frontier Regulars*, p. 48; Walker, *The Wagonmasters*, pp. 106–8.

106. Letter by Otero, 28 September 1875, and Otero, Sellar and Company to Allen, 15 December 1877 (Gross, Kelly and Company Business Records, Special Collections, University of New Mexico).

107. Walker, *The Wagonmasters*, p. 26; Otero, *My Life on the Frontier*, pp. 146–47. Marc Simmons lists 1866 as the year the toll road opened. Marc Simmons, *Following the Santa Fe Trail, A Guide for Modern Travelers* (Santa Fe: Ancient City Press, 1984), p. 135.

108. Whitehead to AAAG, 14 May 1877, LS, CCS, Dist. NM, RG 393, NA.

109. See papers relating to transportation contract of E. B. Allen filed with Board of Survey, 4 May 1878, Fort Stanton, PR, RG 393, NA.

110. Whitehead to Macfeely, 9 August 1878, and Whitehead to Commissary General of Subsistence, 26 July 1879, New Mexico, Cons. Corres. File, RG 92, NA.

111. See Record Book 1883 (Marcus Brunswick Papers, Special Collections, University of New Mexico); William J. Parish, *The Charles Ilfeld Company, A Study of the Rise and Decline of Mercantile Capitalism in New Mexico* (Cambridge: Harvard University Press, 1961), pp. 74–75, 373. Brunswick's freighters carried a total of 377,960 pounds on this contract.

112. Sherman to Ramsey, 10 November 1880, and Sheridan to Drum, 22 October 1880, U.S., Congress, House, Annual Report of the Secretary of War, 1880, House Ex. Doc. No. 1, pt. 2 (Serial 1952), 46th Cong., 3d sess., 1880, pp. 4, 53.

113. Athearn, *The Denver and Rio Grande Western Railroad*, pp. 92–93; Whitehead to Macfeely, 9 August 1878, and Whitehead to Commissary General of Subsistence, 26 July 1879, New Mexico, Cons. Corres. File, RG 92, NA.

114. Bryant, *History of the Atchison, Topeka and Santa Fe Railway*, pp. 39, 45–46, 61; Otero, *My Life on the Frontier*, pp. 152–53.

115. Marc Simmons, *Albuquerque, A Narrative History* (Albuquerque: University of New Mexico Press, 1982), p. 219; Dana to Webster, 28 May 1880, LS, CQM, Dist. NM, RG 393, NA; Webster to Dana, 8, 16 July 1880, LR, CQM, Dist. NM, RG 393, NA.

116. Hunt to Seymour, 17 August

1880, and letter by Hunt dated 20 September 1880, LS, CQM, Dist. NM, RG 393, NA; Webster to Hunt, 19 August 1880, LR, CQM, Dist. NM, RG 393, NA.

117. Lee to Bailhache (and endorsements), 16 November 1880, Robinson to Lee, 5 January 1881, Robinson to Lee, 18 February 1881, Lee to Acting CQM, 5 March 1881, Bingham to CQM, 18 February 1881, LR, CQM, Dist. NM, RG 393, NA.

118. Seymour to CQM, 14 March 1881, and Plummer to CQM, 24 January 1881, LR, CQM, Dist. NM, RG 393, NA.

119. Robinson to CQM, 10 May 1881, and Seymour to CQM, 13 May 1881, LR, CQM, Dist. NM, RG 393, NA; Brinkerhoff to Robinson, 19, 25 June 1881, LS, Fort Bliss, PR, RG 393, NA.

120. David F. Myrick, *New Mexico's Railroads, An Historical Survey* (Golden, Colorado: Colorado Railroad Museum, 1970), pp. 29–34; Hunt to CQM, 14 September 1880, and Lee to CQM, 17 January 1881, CQM, Dist. NM, RG 393, NA; Bailhache to Lee, 19 December 1880, and Decker to Lee, 10, 29 March 1881, LR, CQM, Dist. NM, RG 393, NA.

121. Lee to Decker, 12 April 1881 and 9 September 1881, LS, CQM, Dist. NM, RG 393, NA; Decker to Lee, 18 July 1881, LR, CQM, Dist. NM, RG 393, NA. Lee later admitted that the closing of the agency was premature, and in October he appointed W. T. Giles as his agent in Albuquerque.

122. Leonard to Lee, 12 February 1881, and Bingham to CQM, 4 March 1881, LR, CQM, Dist. NM, RG 393, NA.

123. Frazer, *Forts of the West,* p. 38; [name unclear] to CQM, 9 May 1880, LR, CQM, RG 393, NA; Woodruff to Commissary General of Subsistence, 24 July 1880, New Mexico, Cons. Corres. File, RG 92, NA.

124. Leonard to Lee, 7 April 1881, LR, CQM, Dist. NM, RG 393, NA; Myrick, *New Mexico's Railroads,* p. 104.

125. Leonard to CQM, 17 May 1881, Bingham to CQM, 16 June 1881, Leonard to CQM, 3 August 1881, Leonard to Bingham, 9 July 1883, and Leonard to Furey (and endorsement), 17 July 1884, LR, CQM, Dist. NM, RG 393, NA; Myrick, *New Mexico's Railroads,* p. 104.

126. Post Quartermaster [Fort Stanton] to Quartermaster General, 1 July 1885, Annual Reports Received from Quartermasters, RG 92, NA; See abstract of contracts for wagon transportation included with the Annual Report of the Secretary of War for 1884 and 1885.

127. Simmons, *Albuquerque,* pp. 297–301; [name unclear] to AAAG, 25 May 1884, LS, Fort Craig, PR, RG 393, NA; Woodruff to Commissary General of Subsistence, 28 July 1884, New Mexico, Cons. Corres.

File, RG 92, NA; Hughes to Quartermaster General, 7 July 1884, Annual Reports Received from Quartermasters, RG 92, NA; Las Cruces *Rio Grande Republican,* 23 August 1884.

128. Staab to Belcher, 11 April 1878, LR, CQM, Dist. NM, RG 393, NA; Ilfeld to Browne and Manzanares, 26 August 1878, and Ilfeld to Belcher, 20 November 1878, Letterbook Four (Ilfeld Collection, Special Collections, University of New Mexico); Belcher to CQM, 2 May 1878, LS, CQM, Dist. NM, RG 393, NA.

129. See Reg. Cont., QMG, RG 92, NA; Field to Dana, 20 May 1880, and Wedeles to Dana, 21 June 1880, LR, CQM, RG 393, NA.

130. See Reg. Cont., QMG, RG 92, NA; Lee to Quartermaster General, 18, 27 January 1881, Lee to Post Quartermaster (Fort Selden), 25 March 1881, Lee to Smith, 27 April 1881, and Lee to Staab, 23 July 1881, LS, CQM, Dist. NM, RG 393, NA; DeCou to Lee, 26 October 1881, LR, CQM, Dist. NM, RG 393, NA.

131. Bingham to CQM, 29 May 1883 and 4 October 1883, LR, CQM, Dist. NM, RG 393, NA.

132. Lincoln to Staab, 20 September 1883, LR, CQM, Dist. NM, RG 393, NA.

133. Platt to Commanding Officer, 14 November 1883, LR, Dist. NM, RG 393, NA, M-1088, roll 52; Bingham to CQM, 26 June 1883, and Willett to CQM (and endorsements), 20 December 1883, LR, CQM, Dist. NM, RG 393, NA.

134. Las Cruces *Rio Grande Republican,* 10 November 1883.

135. Holabird to CQM, 1 March 1884, LR, CQM, Dist. NM, RG 393, NA.

136. See Reg. Cont., QMG, RG 92, NA. In October 1884, R. E. Thomas received 57 cents per hundred pounds for corn delivered on board railroad cars at Burlington, Kansas; and in July another contractor received 87½ cents per hundred pounds for baled hay delivered on board railroad cars at Leavenworth, Kansas.

137. Martin to Commanding Officer, 2 September 1885, LR, Dist. NM, RG 393, NA, M-1088, roll 58.

138. Abstracts of Transportation Contracts, RG 92, NA; Clayton W. Williams, *Texas' Last Frontier, Fort Stockton and the Trans–Pecos, 1861–1895* (College Station: Texas A & M University Press, 1982), p. 130; Letter of Hooker, 19 February 1869, Registers of LR, 5th Military District, RG 393, NA, M-1193, roll 2. The wagon haul from San Antonio to Fort Stockton was 394 miles, to Fort Davis, 466 miles, to Fort Quitman, 611 miles, and to Fort Bliss, 691 miles.

139. Abstracts of Transportation Contracts, RG 92, NA; Ira G. Clark, *Then Came the Railroads, The Century from Steam to Diesel in the Southwest* (Norman: University of Oklahoma Press, 1958), pp. 69, 71; S. G. Reed, *A History of the Texas Railroads* (Houston: St. Clair Publishing Co., 1941), pp. 191–95, 261.

140. Abstracts of Transportation Contracts, RG 92, NA; Small to Commissary General of Subsistence, 26 August 1881, Texas, Cons. Corres. File, RG 92, NA.

141. Clark, *Then Came the Railroads,* pp. 141, 152; Sweitzer to AG, 5 October 1883 (two letters of this date), Inspection Reports, RG 159, NA; See abstract of contracts for wagon transportation included with the Annual Report of the Secretary of War for 1882 and 1884. Supplies for Fort Davis could also be freighted from the railroad station at Marfa—a distance of twenty miles.

Chapter 8

1. Undated, unidentified newspaper clipping under "A. M. Franklin," clipbook, Arizona Historical Society, Tucson; James Fowler Rusling, *Across America; or, The Great West and the Pacific Coast* (New York: Sheldon and Co., 1874), p. 370.

2. Dan Thrapp refers to some government contractors as "vultures of the frontier who made a living out of army contracts and were unscrupulous about it, and who saw in any successful endeavor to pacify and settle the Indians an economic loss to themselves in a lessened need for garrisons and troops. Their determination to keep the war flames burning is evident right up to the very end of the Indian wars. . . ." Dan L. Thrapp, *Conquest of Apacheria* (Norman: University of Oklahoma Press, 1967), p. 86.

3. See Erna Risch, *Quartermaster Support of the Army, A History of the Corps, 1775–1939* (Washington: Office of the Quartermaster General, 1962), pp. 340–48, 469.

4. Davis to [name unclear], 27 September 1868, LS, Camp Date Creek, PR, RG 393, NA; Gibbon to CQM, 13 May 1878, LR, CQM, Dist. NM, RG 393, NA; [name unclear] to Post Adjutant, 19 March 1877, LR, Camp Grant, PR, RG 393, NA.

5. See Walker to Lee, 22 April 1881 (and accompanying papers), LR, Dist. NM, RG 393, NA, M-1088, roll 42; Third Judicial District, New Mexico, Record Book B, Federal Records Center, Denver Colorado.

6. Endorsement on letter from W. B. Hellings, 29 August 1871, LS, Dept. Ariz., RG 393, NA; Stevens entry, 9 October 1877, Register of LR, Dept. Ariz., RG 393, NA.

7. Proposal of Simon Filger, 5 June 1871, Catron Papers, Miscellaneous Papers (Special Collections, University of New Mexico); Corliss to AAG, 1 August 1875, LS, Fort McDowell, PR, RG 393, NA.

8. Jacobs and Co. to Pierson, 4 August 1871 (Jacobs Manuscripts, University of Arizona); McGonnigle to Seligman and Spiegelberg, 16 January 1875, and McGonnigle to Hennisee, 9 March 1875, LS, CQM, Dist. NM, RG 393, NA.

9. Nash to CQM, 11 May 1872, and Valdez to Kimball, 12 September 1875, LR, CQM, Dist. NM, RG

393, NA; Weeks to Post Quartermaster, 23 March 1880, LR by Quartermaster, Fort Grant, PR, RG 393, NA.

10. Nash to Commissary General of Subsistence, 14 January 1871, LS, CCS, Dist. NM, RG 393, NA. The army had scaled down its loss on Crouch's contract from $527 to $367 (see Chapter 5).

11. See entries for Maloney, Dawes, and McCall in Register of Contractors and Bidders Failing to Fulfill Contracts, RG 92, NA.

12. See entry for Philip Drachman in Register of Contractors and Bidders Failing to Fulfill Contracts, RG 92, NA.

13. See entry for Samuel H. Drachman in Register of Contractors and Bidders Failing to Fulfill Contracts, RG 92, NA; Floyd S. Fierman, "The Drachmans of Arizona," *American Jewish Archives,* 16 (November 1964): 154–59.

14. See entry for John H. Marion in Register of Contractors and Bidders Failing to Fulfill Contracts, RG 92, NA; *Congressional Record,* 12 June 1888, pp. 5150–51.

15. Gerlach to Ludington, 17 July 1868, LR, CQM, Dist. NM, RG 393, NA; Tilford to Commanding Officer, 21 April 1868, Fort Selden, PR, RG 393, NA; Andrews to AAG, 20 October 1872, LS, Fort Davis, PR, RG 393, NA.

16. Tilford to Hunter, 26 February 1868, LS, Fort Selden, PR, RG 393, NA; Perry to AAG, 14 January 1881, LR, Dept. Ariz., RG 393, NA.

17. [Name unclear] to Adjutant General, 30 March 1868, LS, Fort Union, PR, RG 393, NA; Carleton to McKeever, 26 January 1867, LS, Dist. NM, RG 393, NA, M-1072, roll 4.

18. Entry for 11 January 1868, Court Martials, Fort Selden, PR, RG 393, NA; Cases Tried, [Sept. 1880], LR, Camp Grant, PR, RG 393, NA; Miles to Post Adjutant, 20 February 1873 (and accompanying papers), LR, Fort McDowell, PR, RG 393, NA.

19. Letter of James Woodall, 26 March 1866, Registers of LR, Dist. NM, RG 393, NA, M-1097, roll 1; Darlis A. Miller, *The California Column in New Mexico* (Albuquerque: University of New Mexico Press, 1982), p. 213; Third Judicial District, New Mexico, Record Book A, Federal Records Center, Denver, Colorado; *Mesilla News,* 19 June 1875.

20. Jack D. Foner, *The United States Soldier Between Two Wars: Army Life and Reforms, 1865–1898* (New York: Humanities Press, 1970), pp. 6–7.

21. Brown to AAAG, 10 August 1866, LR, Dist. Ariz., RG 393, NA; Gregg to Sherburne, 31 August 1867, LS, Fort Whipple, PR, RG 393, NA.

22. Devin to Sherburne, 10 March 1868, LS, Fort Whipple, PR, RG 393, NA.

23. Inspection Report of Fort Union, 28 December 1872, OIG, RG 159, NA; Alexander to Franco, 2 May 1873, LS, Fort Garland, PR, RG 393, NA.

24. Gerald Thompson, *The Army*

and the Navajo, The Bosque Redondo Reservation Experiment, 1863–1868 (Tucson: University of Arizona Press, 1976), pp. 75–76; Risch, *Quartermaster Support of the Army*, p. 471.

25. Russell F. Weigley, *Quartermaster General of the Union Army, A Biography of* M. C. *Meigs* (New York: Columbia University Press, 1959), p. 349.

26. McClure to Monahan, 16 November 1867, LS, CCS, Dist. NM, RG 393, NA; General Court Martial Orders No. 7, 8 February 1868, LR, Fort Union, PR, RG 393, NA.

27. Coleman to AAAG, 16 July 1871, LS, Fort Craig, PR, RG 393, NA; Robert E. Bradford Court Martial File, PP 2108, RG 153, NA; Francis B. Heitman, *Historical Register and Dictionary of the United States Army, 1789–1903*, 2 vols. (reprint; Urbana: University of Illinois Press, 1965), vol. 1, p. 238.

28. Bruce J. Dinges, "The Court-Martial of Lieutenant Henry O. Flipper, An Example of Black–White Relationships in the Army, 1881," *The American West*, 9 (January 1972): 12–15; Barry C. Johnson, *Flipper's Dismissal, The Ruin of Lt. Henry O. Flipper, U.S.A., First Coloured Graduate of West Point* (London: Privately Printed, 1980), pp. 6–8.

29. Dinges, "The Court-Martial of Lieutenant Henry O. Flipper," pp. 15–17, 59–60. See Johnson, *Flipper's Dismissal*, p. 39 for the amount of missing funds.

30. Johnson, *Flipper's Dismissal*, pp. 7, 73, 84–86. Modern-day observers believe that Flipper was the victim of nineteenth-century racial prejudice and that white officers found guilty of similar charges would have remained in the service. Following a review of the case in 1976, the army removed the stain on Flipper's record by issuing a certificate of honorable discharge in his name—thirty-six years after his death. Edward M. Coffman, *The Old Army, A Portrait of the American Army in Peacetime, 1784–1898* (New York: Oxford University Press, 1986), p. 229; William H. Leckie and Shirley A. Leckie, *Unlikely Warriors, General Benjamin Grierson and His Family* (Norman: University of Oklahoma Press, 1984), p. 280.

31. Dudley to Tompkins, 27 April 1871, LS, Fort McDowell, PR, RG 393, NA; Cogswell to Dept. Ariz., 10 June 1870, End. S., and Devin to Sherburne, 3 October 1869, LS, Subdistrict of Southern Arizona, RG 393, NA; McClure to Graham, 3 March 1870, LS, CCS, Dist. NM, RG 393, NA; Lee to Bingham, 24 August 1881, LS, CQM, Dist. NM, RG 393, NA.

32. Wallen to Drum, 18 June 1866, LS, Dist. Ariz., RG 393, NA; Tucson *Arizona Citizen*, 17 March 1877.

33. Gregg to Sherburne, 5 September 1867, LS, Fort Whipple, PR, RG 393, NA; Brunswick to Wilson and Fenlon, 28 November 1880, Marcus Brunswick Letterbook (Special Collections, University of New Mexico).

34. Prescott *Weekly Arizona Miner*, 21 December 1872.

35. Smith to CQM, 23 June 1874 (and accompanying papers), Hoberg to Smith, 5 April 1875, and Eagan to Smith, 6 May 1875, Fort Union, Cons. Corres. File, RG 92, NA.

36. Farnsworth to CQM, 20 January 1873, Maxwell to Myers, 28 January 1873, Farnsworth to Maxwell, 21 February 1873, and Shorkley to AAAG, 22 February 1873, LR, CQM, Dist. NM, RG 393, NA; Myers to Maxwell, 6 March 1873, LS, CQM, Dist. NM, RG 393, NA; AAG to Commanding Officer, 20 June 1873, LR, Dist. NM, RG 393, NA, M-1088, roll 18.

37. Belcher to CQM, 20 December 1875, Smith to Bull, 8 January 1876, Bull to Belcher, 13 January 1876, Spiegelberg to Hugo, 1 February 1876, and Baca to Belcher, 3 February 1876, LS, CQM, Dist. NM, RG 393, NA.

38. See Robert M. Utley, *High Noon in Lincoln, Violence on the Western Frontier* (Albuquerque: University of New Mexico Press, 1987), pp. 171–72.

39. Quoted in Harwood P. Hinton, "John Simpson Chisum, 1877–84," *New Mexico Historical Review,* 31 (July 1956): 193.

40. See Chapter 5 and Reg. Cont., QMG, RG 92, NA.

41. See Reg. Cont., QMG, RG 92, NA; Reg. Beef Cont. and Reg. Cont., CGS, RG 192, NA.

42. L. G. Murphy and Co., Journal (Special Collections, University of Arizona).

43. Fritz to Quartermaster General (and enclosures), 28 March 1872, LR, CQM, Dist. NM, RG 393, NA.

44. *Ibid.*

45. John P. Wilson, *Merchants, Guns, and Money, The Story of Lincoln County and Its Wars* (Santa Fe: Museum of New Mexico Press, 1987), pp. 34–41. I am grateful to John P. Wilson for the many discussions that helped me arrive at this assessment of L. G. Murphy and Company.

46. Kautz to AAG, 15 August 1877, U.S., Congress, House, Annual Report of the Secretary of War, 1877, House Ex. Doc. 1, pt. 2 (Serial 1794), 45th Cong., 2d sess., 1877, pp. 143, 146.

47. For an analysis of Murphy-Dolan finances, see Wilson, *Merchants, Guns, and Money,* pp. 27–41, 49–76, and p. 125 for men killed in the Lincoln County War. For violence in Lincoln County, see Utley, *High Noon in Lincoln,* pp. 20–23.

48. Miller, *California Column in New Mexico,* p. 145; Santa Fe *Weekly Gazette,* 26 March 1864; Reg. Cont., QMG, RG 92, NA; Reg. Beef Cont., CGS, RG 192, NA.

49. W. N. Davis, "Post Trading in the West," *Explorations in Entrepreneurial History,* 6 (October 1953): 31.

50. See Dale F. Giese, "Soldiers at Play, A History of Social Life at Fort Union, New Mexico" (Ph.D. dissertation, University of New Mexico, 1969), pp. 71, 76; Santa Fe *Weekly Gazette,* 7 July 1866; General Order No. 177, 23 September 1869, Dent

to Gregg, 9 July 1870, Townsend to Commanding Officer, 8 November 1870, Dent to Wade, 30 August 1876, LR, Fort Union, PR, RG 393, NA. General Grant had requested a tradership for Dent as early as 1867. See William S. McFeely, *Grant, A Biography* (New York: W. W. Norton and Company, 1981), p. 435. In 1871 Secretary of War William W. Belknap awarded the tradership at Camp McDowell, Arizona, to his brother-in-law, James A. Tomlinson. Tucson *Arizona Citizen,* 8 April 1876.

51. J. Morgan Broaddus, Jr., *The Legal Heritage of El Paso* (El Paso: Texas Western College Press, 1963), p. 118; Letter by Henry Carroll, 5 May 1868, Register of LR, Fifth Military District, RG 393, NA, M-1193, roll 1; U.S., Department of Commerce, Bureau of the Census, *Ninth Census of the United States,* 1870, El Paso County, Population Schedules, NA, M-593, roll 1583.

52. Wood to Morrow, 27 June 1870, LS, Dept. Texas, RG 393, NA, M-1114, roll 1; Personal File of William E. Sweet (2565 ACP 1872), RG 94, NA.

53. Personal File of William E. Sweet (2565 ACP 1872), RG 94, NA; AAG to Commanding Officer, 8 February 1873, LR, Fort Quitman, PR, RG 393, NA. Wahl moved to San Elizario and then to Ysleta, Texas, where he was elected deputy county surveyor for El Paso County. See Broaddus, *Legal Heritage of El Paso,* p. 118.

54. See McFeely, *Grant,* pp. 435–36; Davis, "*Post Trading in the West,*" pp. 31, 34–36.

55. Santa Fe *Weekly New Mexican,* 15 December 1868, 27 July 1869; Santa Fe *Daily New Mexican,* 13 January 1870.

56. See Population Schedules for 1870 census.

57. See Note. 55.

58. See Reg. Cont., QMG, RG 92, NA; Reg. Cont. and Reg. Beef Cont., CGS, RG 192, NA. Because inconsistencies exist in the manner of registering contracts during this ten-year period, these percentages should be looked upon as approximations.

59. See Reg. Cont., QMG, RG 92, NA; Reg. Cont., CGS, RG 192, NA; U.S., Department of the Interior, *Ninth Census of the United States: 1870, Wealth and Industry,* vol. 3 (Washington: Government Printing Office, 1872), pp. 208–9, 358. Historian Marc Simmons notes that traditionally New Mexican farmers "almost never tried to produce and store a surplus." Personal communication to author, 31 July 1981.

60. See Reg. Cont., QMG, RG 92, NA; Reg. Cont., CGS, RG 192, NA.

61. U.S., Department of Commerce, Bureau of the Census, *Ninth Census of the United States,* 1870, Doña Ana County, Population Schedules, NA, M-593, roll 893.

62. Schofield to Wade, 10 June 1868, U.S., Congress, Senate, Letter of the Secretary of War, Senate Ex. Doc. 74 (Serial 1317), 40th

Cong., 2d sess., 1868, pp. 1–3 (medical expenses are not included in the total military expenditures).

63. *Ibid.*; Lionel to Pa, 6 January 1872 (Jacobs Manuscripts, University of Arizona). On soldiers' pay, see also Robert W. Frazer, *Forts and Supplies, The Role of the Army in the Economy of the Southwest, 1846–1861* (Albuquerque: University of New Mexico Press, 1983), pp. 190–91. As reported in the annual report of the Secretary of War in 1867, troop strength in New Mexico was 1,608 and in Arizona, 1,704.

64. Ord to Townsend, 27 September 1869, U.S., Congress, House, Annual Report of the Secretary of War, 1869, House Ex. Doc. 1, pt. 2 (Serial 1412), 41st Cong., 2d Sess., 1869, p. 124; Schofield to AG, 20 October 1871, U.S., Congress, House, Annual Report of the Secretary of War, 1871, House Ex. Doc. 1, pt. 2 (Serial 1503), 42nd Cong., 2d sess., 1871, p. 68.

65. Robert M. Utley, *Frontier Regulars, The United States Army and the Indian, 1866–1891* (New York: Macmillan Publishing Co., 1973), pp. 59–68.

66. McClure to Kobbé, 5 October 1869, LS, CCS, Dist. NM, RG 393, NA; Granger to AAG, 21 September 1871, LS, Dist. NM, RG 393, NA, M-1072, roll 4.

67. *Ninth Census, 1870,* vol. 3 (Wealth and Industry), pp. 208–209; U.S., Department of Interior, *Tenth Census of the United States: 1880, Agriculture,* vol. 3 (Washington: Government Printing Office, 1883), p. 4.

68. Hubert Howe Bancroft, *History of Arizona and New Mexico, 1530–1888* (San Francisco: History Company, Publishers, 1889), pp. 750, 771.

69. William J. Parish, *The Charles Ilfeld Company, A Study of the Rise and Decline of Mercantile Capitalism in New Mexico* (Cambridge: Harvard University Press, 1961), p. 37.

70. The Indian Bureau expended substantial sums for maintaining agencies and fulfilling promises to reservation Indians in both New Mexico and Arizona. The regular appropriation for Arizona in the fiscal year ending 30 June 1875, for example, was $452,900, and for New Mexico, $176,000. What percentage of these funds was spent locally is not known, however. Appropriations are reported in *U.S. Statutes at Large.* A careful study of Indian contracting in the Southwest is needed. Brief attention to the topic is given in Thomas G. Alexander's *A Clash of Interests, Interior Department and Mountain West, 1863–96* (Provo: Brigham Young University Press, 1977), pp. 98–99, 188.

71. Analysis of the Estimates of Funds Required by Officers of the Quartermaster's Department (see entries for Dist. NM, 1873–76), RG 92, NA.

72. For Dist. NM allocations and 1875–76 allocations to Dept. Ariz., see *ibid.* For 1873–74 Arizona allocation, see Bingham to Quarter-

master General, 3 September 1874, U.S., Congress, House, Annual Report of the Secretary of War, 1874, House Ex. Doc. 1, pt. 2 (Serial 1635), 43 Cong., 2d sess., 1874, pp. 144–46.

73. See Reg. Cont., CGS, RG 192, NA.

74. Analysis of the Estimates of Funds Required by Officers of the Quartermaster's Department (see entries for Dept. Ariz., 1881, 1885), RG 92, NA; Macfeely to Secretary of War, 10 October 1881, U.S., Congress, House, Annual Report of the Secretary of War, 1881, House Ex. Doc. 1, pt. 2 (Serial 2010), 47th Cong., 1st sess., 1881, p. 470; Eagan to Macfeely, 18 July 1884, Arizona, Cons. Corres. File, RG 92, NA. The estimate for the pay department is probably conservative.

75. *Tenth Census, 1880,* vol. 3 (Agriculture), p. 4; Bancroft, *History of Arizona and New Mexico,* p. 583.

76. Frazer, *Forts and Supplies,* p. ix.

77. Miguel Antonio Otero, My *Life on the Frontier, 1864–1882* (reprint; Albuquerque: University of New Mexico Press, 1987), p. 164.

78. Leo E. Oliva, "Frontier Forts and the Settlement of Western Kansas," in *Kansas and the West,* eds. Forrest R. Blackburn *et. al.* (Topeka: Kansas State Historical Society, 1976), pp. 69–70.

79. U.S., Department of the Interior, *Eleventh Census of the United States: 1890, Population,* part 1 (Washington: Government Printing Office, 1895), p. 2.

Bibliography

Manuscript Materials

Arizona Historical Society, Tucson, Arizona.
Barron and Lionel Jacobs Business Records.
Biographical Files.
Franklin Family Papers.
George F. Hooper File.
Hayden File.
Doña Ana County Courthouse, Las Cruces, New Mexico.
Deed Books.
Federal Records Center, Denver, Colorado.
Third Judicial District, New Mexico, Record Books.
National Archives of the United States, Washington, D.C.
RG 29. Records of the Bureau of the Census.
Population Schedules of the Eighth Census, 1860, M-653, New Mexico, rolls 712–716; Population Schedules of the Ninth Census, 1870, M-593, New Mexico, rolls 893–897, Texas, roll 1583; Population Schedules of the Tenth Census, 1880, T-9, New Mexico, rolls 802–804.
RG 59. Records of the Secretary of State.
Territorial Papers, New Mexico, T-17, rolls 1–4.
RG 92. Records of the Office of the Quartermaster General.
Abstracts of Transportation Contracts.
Analysis of the Estimates of Funds Required by Officers of the Quartermaster's Department.
Annual Reports Received from Quartermasters.
Consolidated Correspondence File.
Register of Contractors and Bidders Failing to Fulfill Contracts.
Register of Contracts.
RG 94. Records of the Office of the Adjutant General.
General Orders.
Letters Received, M-619.

Personal Files.
Post Medical Records.
Volunteer Organizations of the Civil War, Regimental Papers.
RG 153. Records of the Office of the Judge Advocate General (Army).
Court Martial Files.
RG 159. Records of the Office of Inspector General.
Inspection Reports.
RG 192. Records of the Office of the Commissary General of Subsistence.
Register of Contracts.
Register of Beef Contracts.
RG 217. Records of the United States General Accounting Office.
Contracts.
RG 393. Records of the United States Army Continental Commands, 1821–1920.
Department of Arizona.
Department of New Mexico.
Department of Texas.
Department of the Missouri.
District of Arizona (and subdistricts).
District of New Mexico.
Fifth Military District.
Post Records, Arizona, Colorado, New Mexico, Texas.
New Mexico Highlands University, Las Vegas, New Mexico.
Arrott Collection.
New Mexico State Records Center and Archives, Santa Fe, New Mexico.
Agricultural Schedules of the Eighth, Ninth, and Tenth Census of the United States, 1860, 1870, 1880, New Mexico.
Products of Industry Schedules of the Ninth Census of the United States, 1870, New Mexico.
Records of the Adjutant General.
Territorial Archives of New Mexico, Microfilm Edition.
New Mexico State University (Special Collections), Las Cruces, New Mexico.
Amador Collection.
State Planning Office, Santa Fe, New Mexico.
Janet Lecompte, "Ceran St. Vrain's Stone Mill at Mora," Cultural Properties Review Committee.
University of Arizona (Special Collections), Tucson, Arizona.
Jacobs Manuscripts.
L. G. Murphy and Company, Journal.

University of New Mexico (Special Collections), Albuquerque, New Mexico.
Catron Papers.
Gross, Kelly and Company Business Records.
Ilfeld Collection.
Marcus Brunswick Papers.
U.S. Department of War, Records of the Quartermaster's Department.

Government Publications

Gosper, John J. *Report of the Acting Governor of Arizona Made to the Secretary of the Interior for the Year 1881*. Washington: Government Printing Office, 1881.

Legislature of the Territory of Arizona. *Memorial and Affidavits Showing Outrages Perpetrated by the Apache Indians in the Years 1869 and 1870*. Reprint. Tucson: Territorial Press, 1964.

United States Army. *Revised United States Army Regulations of 1861*. Washington: Government Printing Office, 1863.

United States Congress. *Congressional Record.*

United States Congress. House of Representatives.
1862. 37th Cong., 2d sess., House Ex. Doc. 101, Serial 1136, Abstract of Subsistence and Quartermaster's Contracts.
1865. 38th Cong., 2d sess., House Ex. Doc. 84, Serial 1230, Abstract of Quartermaster's Contracts.
1866. 39th Cong., 2d sess., House Ex. Doc. 28, Serial 1289, Abstract of Quartermaster's Contracts.
1869. 41st Cong., 2d sess., House Ex. Doc. 73, Serial 1417, Regarding Pedro Armendaris claim.
1873. 42nd Cong., 3d sess., House Ex. Doc. 158, Serial 1567, Abstract of Subsistence and Quartermaster's Contracts.

United States Congress. Senate.
1867. 39th Cong., 2d sess., S. Report 156, Serial 1279, Condition of the Indian Tribes.
1868. 40th Cong., 2d sess., S. Ex. Doc. 59, pt. 2, Serial 1317, Abstract of Quartermaster's Contracts.
1868. 40th Cong., 2d sess., S. Ex. Doc. 74, Serial 1317, Military Expenses in New Mexico and Arizona.
1872. 42nd Cong., 2d sess., S. Ex. Doc. 26, Serial 1478, Abstract of Subsistence Contracts.
1965. 89th Cong., 1st sess., S. Doc. 13, Federal Census—Territory of New Mexico and Territory of Arizona, 1870.

United States Congress. Senate. Committee on Interior and Insular Affairs.
Indian Water Rights of the Five Central Tribes of Arizona, Hearings, "Report for the Gila River Pima and Maricopa Tribes," by Fred Nicklason, 94th Cong., 1st sess., 1975. Washington: U.S. Government Printing Office, 1976.

United States Department of Interior.
Eighth Census of the United States: 1860, Population, vol. 1, *Agriculture,* vol. 2. Washington: Government Printing Office, 1864.
Ninth Census of the United States: 1870, Population, vol. 1, *Wealth and Industry,* vol. 3. Washington: Government Printing Office, 1872.
Tenth Census of the United States: 1880, Population, vol. 1, *Agriculture,* vol. 3. Washington: Government Printing Office, 1883.
Eleventh Census of the United States: 1890, Population, part 1, *Agriculture.* Washington: Government Printing Office, 1895.

United States Secretary of War.
Annual Reports, 1860–1886.

United States v. Martin, 94. U.S. 400.

United States War Department. Surgeon General's Office.
A Report on Barracks and Hospitals. Circular No. 4. Washington: Government Printing Office, 1870.
Report on the Hygiene of the United States Army. Circular No. 8. Washington: Government Printing Office, 1875.

The War of the Rebellion: A Compilation of the Official Records of the Union and Confederate Armies. 128 vols. Washington: Government Printing Office, 1880–1901.

Newspapers

Arizona Citizen, Tucson.
Arizona Miner (and *Weekly Arizona Miner*), Prescott.
Arizona Star, Tucson.
Borderer, Las Cruces.
Grant County Herald, Silver City.
Mesilla News.
Mesilla Valley Independent, Mesilla.
Mining Life, Silver City.
Rio Abajo Weekly Press, Albuquerque.
Rio Grande Republican, Las Cruces.
Santa Fe *New Mexican* (also *Daily New Mexican* and *Weekly New Mexican*).
Santa Fe *Weekly Gazette.*

Semi–Weekly Review, Albuquerque.
Weekly Arizonan (and *Weekly Arizonian*), Tucson.

Personal Communications

Marc Simmons, 31 July 1981.
Harwood Hinton, 1 October 1987.
Charles Kenner, 12 January, 1988.

Dissertations and Theses

Ballew, Elvis Joe. "Supply Problems of Fort Davis, Texas, 1867–1880," Master's thesis, Sul Ross State University, 1971.

Baydo, Gerald R. "Cattle Ranching in Territorial New Mexico," Ph.D. dissertation, University of New Mexico, 1970.

Christian, Garna Loy. "Sword and Plowshare: The Symbiotic Development of Fort Bliss and El Paso, Texas, 1849–1918," Ph.D. dissertation, Texas Tech University, 1977.

Giese, Dale F. "Soldiers at Play, A History of Social Life at Fort Union, New Mexico," Ph.D. dissertation, University of New Mexico, 1969.

Goldstein, Marcy G. "Americanization and Mexicanization: The Mexican Elite and Anglo–Americans in the Gadsden Purchase Lands, 1853–1880," Ph.D. dissertation, Case Western Reserve University, 1977.

Halla, Frank Louis, Jr. "El Paso, Texas, and Juárez, Mexico: A Study of a Bi–Ethnic Community, 1846–1881," Ph.D. dissertation, University of Texas at Austin, 1978.

Lamb, Blaine Peterson. "Jewish Pioneers in Arizona, 1850–1920," Ph.D. dissertation, Arizona State University, 1982.

Mayer, Arthur James. "San Antonio, Frontier Entrepot," Ph.D. dissertation, University of Texas at Austin, 1976.

Mehren, Lawrence L. "A History of the Mescalero Apache Reservation, 1869–1881," Master's thesis, University of Arizona, 1969.

Wallace, Andrew. "Soldier in the Southwest: The Career of General A. V. Kautz, 1869–1886," Ph.D. dissertation, University of Arizona, 1968.

Books

Acuña, Rodolfo F. *Sonoran Strongman, Ignacio Pesqueira and His Times*. Tucson: University of Arizona Press, 1974.

Alexander, Thomas G. *A Clash of Interests, Interior Department and Mountain*

West, 1863–96. Provo: Brigham Young University Press, 1977.

Altshuler, Constance Wynn. *Chains of Command, Arizona and the Army, 1856–1875.* Tucson: Arizona Historical Society, 1981.

———. *Starting With Defiance, Nineteenth Century Arizona Military Posts.* Tucson: Arizona Historical Society, 1983.

Athearn, Robert. *The Denver and Rio Grande Western Railroad.* Reprint. Lincoln: University of Nebraska Press, 1977.

Bancroft, Hubert Howe. *History of Arizona and New Mexico, 1530–1888.* San Francisco: History Company, Publishers, 1889.

Bell, William A. *New Tracks in North America.* Reprint. Albuquerque: Horn and Wallace, Publishers, 1965.

Billington, Ray Allen and Martin Ridge. *Westward Expansion, A History of the American Frontier.* Fifth Edition. New York: Macmillan Publishing Co., 1982.

Bourke, John G. *On the Border With Crook.* Reprint. Lincoln: University of Nebraska Press, 1971.

Boyd, Mrs. Orsemus Bronson. *Cavalry Life in Tent and Field.* Reprint. Lincoln: University of Nebraska Press, 1982.

Branda, Eldon S., ed. *The Handbook of Texas.* vol. 3. Austin: Texas State Historical Association, 1976.

Brandes, Ray. *Frontier Military Posts of Arizona.* Globe, Arizona: Dale Stuart King, 1960.

Broaddus, J. Morgan, Jr. *The Legal Heritage of El Paso.* El Paso: Texas Western College Press, 1963.

Browne, J. Ross. *Adventures in the Apache Country, A Tour Through Arizona and Sonora, 1864.* Re-edition. Tucson: University of Arizona Press, 1974.

Bryant, Keith L., Jr. *History of the Atchison, Topeka and Santa Fe Railway.* Reprint. Lincoln: University of Nebraska Press, 1982.

Cahill, Marion Cotter. *Shorter Hours, A Study of the Movement Since the Civil War.* Reprint. New York: AMS Press, 1969.

Carriker, Robert C. and Eleanor R., eds. *An Army Wife on the Frontier, The Memoirs of Alice Blackwood Baldwin, 1867–1877.* Salt Lake City: University of Utah Library, 1975.

Chase, C. M. *The Editor's Run in New Mexico and Colorado.* Reprint. Fort Davis, Texas: Frontier Book Co., 1968.

Chilton, Lance, *et al. New Mexico, A New Guide to the Colorful State.* Albuquerque: University of New Mexico Press, 1984.

Christiansen, Paige W. *The Story of Mining in New Mexico.* Socorro: New Mexico Bureau of Mines and Mineral Resources, 1974.

Clark, Ira G. *Then Came the Railroads, The Century from Steam to Diesel in the Southwest.* Norman: University of Oklahoma Press, 1958.

Coffman, Edward M. *The Old Army, A Portrait of the American Army in Peacetime, 1784–1898.* New York: Oxford University Press, 1986.

Cosulich, Bernice. *Tucson.* Tucson: Arizona Silhouettes, 1953.

Dunlay, Thomas W. *Wolves for the Blue Soldiers, Indian Scouts and Auxiliaries with the United States Army, 1860–90.* Lincoln: University of Nebraska Press, 1982.

Ellis, George F. *Bell Ranch As I Knew It.* Kansas City: Lowell Press, 1973.

Espinosa, Gilberto and Tibo J. Chavez. *El Rio Abajo.* Portales, New Mexico: Bishop Publishing Company, n.d.

Farish, Thomas E. *History of Arizona.* 8 vols. San Francisco: Filmer Brothers Electrotype Company, 1915–1918.

Fierman, Floyd S. *Guts and Ruts, The Jewish Pioneer on the Trail in the American Southwest.* New York: Ktav Publishing House, Inc., 1985.

Fireman, Bert M. *Arizona, Historic Land.* New York: Alfred A. Knopf, 1982.

Foner, Jack D. *The United States Soldier Between Two Wars: Army Life and Reforms, 1865–1898.* New York: Humanities Press, 1970.

Foner, Philip S. *History of the Labor Movement in the United States, From Colonial Times to the Founding of the American Federation of Labor.* New York: International Publishers, 1947.

Frazer, Robert W. *Forts and Supplies, The Role of the Army in the Economy of the Southwest, 1846–1861.* Albuquerque: University of New Mexico Press, 1983.

———. *Forts of the West, Military Forts and Presidios and Posts Commonly Called Forts West of the Mississippi River to 1898.* Norman: University of Oklahoma Press, 1965.

Giese, Dale F. *Echoes of the Bugle.* n.p. Phelps Dodge Corporation, 1976.

Goetzmann, William H. *Army Exploration in the American West, 1803–1863.* New Haven, Conn.: Yale University Press, 1959.

Granger, Byrd H. *Will C. Barnes' Arizona Place Names.* Tucson: University of Arizona Press, 1960.

Greene, Jerome A. *Historic Resource Study, Fort Davis National Historic Site.* Washington: U.S. Department of the Interior, National Park Service, 1986.

Guenther, Nancy Anderman, ed. *United States Supreme Court Decisions: An Index to Excerpts, Reprints, and Discussions.* Metuchen, N.J.: Scarecrow Press, 1983.

Haley, J. Evetts. *Charles Goodnight, Cowman and Plainsman.* Norman: University of Oklahoma Press, 1936.

———. *Fort Concho and the Texas Frontier.* San Angelo: San Angelo Standard–Times, 1952.

Hall, Martin Hardwick. *Sibley's New Mexico Campaign.* Austin: University of Texas Press, 1960.

Hayden, Carl Trumbull. *Charles Trumbull Hayden, Pioneer.* Tucson: Arizona Historical Society, 1972.

Heitman, Francis B. *Historical Register and Dictionary of the United States Army, 1789–1903.* 2 vols. Reprint. Urbana: University of Illinois Press, 1965.

Hinton, Richard J. *The Handbook to Arizona.* Reprint. Tucson: Arizona Silhouettes, 1954.

Huning, Franz. *Trader on the Santa Fe Trail, Memoirs of Franz Huning.* Albuquerque: University of Albuquerque, 1973.

Hunt, Aurora. *Major General James Henry Carleton, 1814–1873, Western Frontier Dragoon.* Glendale: Arthur H. Clark Company, 1958.

Johnson, Barry C. *Flipper's Dismissal, The Ruin of Lt. Henry O. Flipper, U.S.A., First Coloured Graduate of West Point.* London: Privately Printed, 1980.

Kenner, Charles L. *A History of New Mexican–Plains Indian Relations.* Norman: University of Oklahoma Press, 1969.

Lane, Lydia. *I Married a Soldier.* Reprint. Albuquerque: University of New Mexico Press, 1987.

Leckie, William H. and Shirley A. Leckie. *Unlikely Warriors, General Benjamin Grierson and His Family.* Norman: University of Oklahoma Press, 1984.

Lingenfelter, Richard E. *Steamboats on the Colorado River.* Tucson: University of Arizona Press, 1978.

Lockwood, Frank C., ed. *Apaches and Longhorns, The Reminiscences of Will C. Barnes.* Tucson: University of Arizona Press, 1982.

———. *Pioneer Portraits, Selected Vignettes.* Tucson: University of Arizona Press, 1968.

McFeely, William S. *Grant, A Biography.* New York: W. W. Norton and Company, 1981.

Meline, James F. *Two Thousand Miles on Horseback.* Reprint. Albuquerque: Horn and Wallace, Publishers, 1966.

Metz, Leon. *Fort Bliss, An Illustrated History.* El Paso: Mangan Books, 1981.

———. *Pat Garrett, The Story of a Western Lawman.* Norman: University of Oklahoma Press, 1974.

Miller, Darlis A. *The California Column in New Mexico.* Albuquerque: University of New Mexico Press, 1982.

Mills, W. W. *Forty Years at El Paso, 1858–1898.* El Paso: Carl Hertzog, 1962.

Mullin, Robert N., ed. *Maurice G. Fulton's History of the Lincoln County War.* Tucson: University of Arizona Press, 1968.

Murphy, Lawrence R. *Lucien Bonaparte Maxwell, Napoleon of the Southwest.* Norman: University of Oklahoma Press, 1983.

Myres, Sandra L., ed. *Cavalry Wife, The Diary of Eveline M. Alexander, 1866–1867.* College Station: Texas A & M University Press, 1977.

Myrick, David F. *New Mexico's Railroads—An Historical Survey.* Golden: Colorado Railroad Museum, 1970.

Oliva, Leo E. *Soldiers on the Santa Fe Trail.* Norman: University of Oklahoma Press, 1967.

Otero, Miguel Antonio. *My Life on the Frontier, 1864–1882.* Reprint. Albuquerque: University of New Mexico Press, 1987.

Parish, William J. *The Charles Ilfeld Company, A Study of the Rise and Decline of Mercantile Capitalism in New Mexico.* Cambridge: Harvard University Press, 1961.

Poe, Sophie A. *Buckboard Days.* Reprint. Albuquerque: University of New Mexico Press, 1981.

Portrait and Biographical Record of Arizona. Chicago: Chapman Publishing Co., 1901.

Prucha, Francis Paul. *Broadax and Bayonet: The Role of the United States Army in the Development of the Northwest, 1815–1860.* Lincoln: University of Nebraska Press, 1953.

———. *The Sword and the Republic: The United States Army on the Frontier, 1783–1846.* New York: The Macmillan Company, 1969.

Raht, Carlysle G. *The Romance of Davis Mountains and Big Bend Country.* Odessa, Texas: Rahtbooks Company, 1963.

Rayback, Joseph G. *A History of American Labor.* New York: The Macmillan Co., 1959.

Reed, Bill. *The Last Bugle Call, A History of Fort McDowell, Arizona Territory, 1865–1890.* Parsons, West Virginia: McClain Printing Company, 1977.

Reed, S. G. *A History of the Texas Railroads.* Houston: St. Clair Publishing Co., 1941.

Risch, Erna. *Quartermaster Support of the Army, A History of the Corps, 1775–1939.* Washington: Office of the Quartermaster General, 1962.

Rochlin, Harriet and Fred. *Pioneer Jews, A New Life in the Far West.* Boston: Houghton Mifflin Company, 1984.

Rusling, James Fowler. *Across America; or, The Great West and the Pacific Coast.* New York: Sheldon and Co., 1874.

Russell, Frank. *The Pima Indians.* Re-edition. Tucson: University of Arizona Press, 1975.

Russell, Marian. *Land of Enchantment, Memoirs of Marian Russell Along the Santa Fe Trail as Dictated to Mrs. Hal Russell.* Reprint. Albuquerque: University of New Mexico Press, 1981.

Scobee, Barry. *Fort Davis, Texas, 1583–1960.* Fort Davis: Barry Scobee, 1963.

Sheridan, Thomas E. *Los Tucsonenses, The Mexican Community in Tucson, 1854–1941.* Tucson: University of Arizona Press, 1986.

Shinkle, James D. *Fifty Years of Roswell History, 1867–1917.* Roswell, N.M.: Hall–Poorbaugh Press, Inc., 1964.

Simmons, Marc. *Albuquerque, A Narrative History.* Albuquerque: University of New Mexico Press, 1982.

———. *Following the Santa Fe Trail, A Guide for Modern Travelers.* Santa Fe: Ancient City Press, 1984.

———. *The Little Lion of the Southwest, A Life of Manuel Antonio Chaves.* Chicago: Swallow Press, Inc., 1973.

Simmons, Virginia McConnell. *The San Luis Valley, Land of the Six–Armed Cross.* Boulder: Pruett Publishing Company, 1979.

Smith, Dean. *The Goldwaters of Arizona.* Flagstaff: Northland Press, 1986.

Sonnichsen, C. L. *The El Paso Salt War.* El Paso: Carl Hertzog and Texas Western Press, 1961.

———. *Pass of the North, Four Centuries on the Rio Grande.* El Paso: Texas Western Press, 1968.

———. *Tucson, The Life and Times of an American City.* Norman: University of Oklahoma Press, 1982.

Spicer, Edward H. *Cycles of Conquest, The Impact of Spain, Mexico, and the United States on the Indians of the Southwest, 1533–1960.* Tucson: University of Arizona Press, 1962.

Spring, John A. *John Spring's Arizona.* Edited by A. M. Gustafson. Tucson: University of Arizona Press, 1966.

Stallard, Patricia Y. *Glittering Misery, Dependents of the Indian Fighting Army.* San Rafael, California: Presidio Press, 1978.

Summerhayes, Martha. *Vanished Arizona.* Reprint. Glorieta, New Mexico: Rio Grande Press, Inc., 1976.

The Taming of the Salt. Phoenix: Communications and Public Affairs Department of Salt River Project, 1979.

Thian, Raphael P. *Legislative History of the General Staff of the Army of the United States.* Washington: Government Printing Office, 1901.

———. *Notes Illustrating the Military Geography of the United States, 1813–1880.* Reprint. Austin: University of Texas Press, 1979.

Thompson, Gerald. *The Army and the Navajo, The Bosque Redondo Reservation Experiment, 1863–1868.* Tucson: University of Arizona Press, 1976.

Thrapp, Dan L. *Al Sieber, Chief of Scouts.* Norman: University of Oklahoma Press, 1964.

———. *Conquest of Apacheria.* Norman: University of Oklahoma Press, 1967.
———. *General Crook and the Sierra Madre Adventure.* Norman: University of Oklahoma Press, 1972.
———. *Victorio and the Mimbres Apaches.* Norman: University of Oklahoma Press, 1974.
Twitchell, Ralph Emerson. *The Leading Facts of New Mexican History.* 5 vols. Cedar Rapids, Iowa: Torch Press, 1911–1917.
———. *Old Santa Fe, The Story of New Mexico's Ancient Capital.* Reprint. Chicago: The Rio Grande Press, 1963.
Utley, Robert M. *Fort Union National Monument.* Washington, D.C.: National Park Service, 1962.
———. *Frontier Regulars, The United States Army and the Indian, 1866–1891.* New York: Macmillan Publishing Co., 1973.
———. *Frontiersman in Blue: The United States Army and the Indian, 1848–1865.* New York: Macmillan Company, 1967.
———. *High Noon in Lincoln, Violence on the Western Frontier.* Albuquerque: University of New Mexico Press, 1987.
Vigil, Maurilio E. *Los Patrones: Profiles of Hispanic Political Leaders in New Mexico History.* Washington, D.C.: University Press of America, 1980.
Wagoner, Jay J. *Early Arizona, Prehistory to Civil War.* Tucson: University of Arizona Press, 1975.
Walker, Henry Pickering. *The Wagonmasters, High Plains Freighting from the Earliest Days of the Santa Fe Trail to 1880.* Norman: University of Oklahoma Press, 1966.
Weigley, Russell F. *History of The United States Army.* New York: Macmillan Company, 1967.
———. *Quartermaster General of the Union Army, A Biography of M. C. Meigs.* New York: Columbia University Press, 1959.
Williams, Clayton W. *Texas' Last Frontier, Fort Stockton and the Trans–Pecos, 1861–1895.* College Station: Texas A & M University Press, 1982.
Wilson, John P. *Merchants, Guns, and Money, The Story of Lincoln County and Its Wars.* Santa Fe: Museum of New Mexico Press, 1987.

Articles and Papers

Aguirre, Yjinio F. "The Last of the Dons," *Journal of Arizona History,* 10 (Winter 1969): 239–55.
Baxter, John O. "Salvador Armijo: Citizen of Albuquerque, 1823–1879," *New Mexico Historical Review,* 53 (July 1978): 219–37.

Butler, Anne M. "Military Myopia: Prostitution on the Frontier," *Prologue,* 13 (Winter 1981): 233–50.

Crawford, Jack. "Pursuit of Victorio," *Socorro County Historical Society: Publications in History,* 1 (February 1965): 1–8.

Crimmins, Martin L. "Colonel Buell's Expedition into Mexico in 1880," *New Mexico Historical Review,* 10 (April 1935): 133–42.

Davis, W. N. "Post Trading in the West," *Explorations in Entrepreneurial History,* 6 (October 1953): 30–40.

Dinges, Bruce J. "The Court–Martial of Lieutenant Henry O. Flipper, An Example of Black–White Relationships in the Army, 1881," *The American West,* 9 (January 1972): 12–17, 59–61.

Dobyns, Henry F. "Pima and Maricopa Indian Economic Genius in 1849," paper presented at the Western History Association Conference, 1982.

Fierman, Floyd S. "The Drachmans of Arizona," *American Jewish Archives,* 16 (November 1964): 135–60.

———. "Jewish Pioneering in the Southwest, A Record of the Freudenthal–Lesinsky–Solomon Families," *Arizona and the West,* 2 (Spring 1960): 54–72.

———. "Nathan Bibo's Reminiscences of Early New Mexico," *El Palacio,* 68 (Winter 1961): 231–57; 69 (Spring 1962): 40–60.

Forrest, Earle R. "The Fabulous Sierra Bonita," *Journal of Arizona History,* 6 (Autumn 1965): 132–46.

Foster, James Monroe, Jr. "Fort Bascom, New Mexico," *New Mexico Historical Review,* 35 (January 1960): 30–62.

Frazer, Robert W. "The Army and New Mexico Agriculture, 1848–1861," *El Palacio,* 89 (Spring 1983): 25–29.

———. "Purveyors of Flour to the Army: Department of New Mexico, 1849–1861," *New Mexico Historical Review,* 47 (July 1972): 213–38.

Hargett, Janet L. "Pioneering at Yuma Crossing, The Business Career of L. J. F. Jaeger, 1850–1887," *Arizona and the West,* 25 (Winter 1983): 329–54.

Haskett, Bert. "Early History of the Cattle Industry in Arizona," *Arizona Historical Review,* 6 (October 1935): 3–42.

Hinton, Harwood P. "John Simpson Chisum, 1877–84," *New Mexico Historical Review,* 31 (July–October 1956): 177–205, 310–37; 32 (January 1957): 53–65.

Hunter, Thomas T. "Early Days in Arizona," *Arizona Historical Review,* 3 (April 1930): 105–20.

Kajencki, Francis C. "Alexander Grzelachowski, Pioneer Merchant of Puerto de Luna," *Arizona and the West,* 26 (Autumn 1984): 243–60.

Kenner, Charles. "The Origins of the 'Goodnight' Trail Reconsidered," *Southwestern Historical Quarterly,* 77 (January 1974): 390–94.

Leavitt, Francis H. "Steam Navigation on the Colorado River," *California Historical Society Quarterly,* 22 (March, June 1943): 1–25, 151–74.

Mawn, Geoffrey P. "Promoters, Speculators, and the Selection of the Phoenix Townsite," *Arizona and the West,* 19 (Autumn 1977): 207–24.

Miller, Darlis A. "Foragers, Army Women, and Prostitutes," in *New Mexico Women: Intercultural Perspectives,* ed. by Joan M. Jensen and Darlis A. Miller. Albuquerque: University of New Mexico Press, 1986.

———. "Historian for the California Column: George H. Pettis of New Mexico and Rhode Island," *Red River Valley Historical Review,* 5 (Winter 1980): 74–92.

———. "Los Pinos, New Mexico: Civil War Post on the Rio Grande," *New Mexico Historical Review,* 62 (January 1987): 1–31.

———. "Military Supply in Civil War New Mexico," *Military History of Texas and the Southwest,* 16 (Number 3, 1982): 177–97.

———. "The Role of the Army Inspector in the Southwest: Nelson H. Davis in New Mexico and Arizona, 1863–1873," *New Mexico Historical Review,* 59 (April 1984): 137–64.

Miller, Darlis A. and Norman L. Wilson, Jr. "The Resignation of Judge Joseph G. Knapp," *New Mexico Historical Review,* 55 (October 1980): 335–44.

Morrisey, Richard J. "The Early Range Cattle Industry in Arizona," *Agricultural History,* 24 (July 1950): 151–56.

Myres, Sandra L. "Evy Alexander, The Colonel's Lady at McDowell," *Montana, The Magazine of Western History,* 24 (Summer 1974): 26–38.

Oliva, Leo E. "Frontier Forts and the Settlement of Western Kansas," in *Kansas and the West,* ed. by Forrest R. Blackburn *et al.* Topeka: Kansas State Historical Society, 1976.

Patzman, Stephen N. "Louis John Frederick Jaeger: Entrepreneur of the Colorado River," *Arizoniana,* 4 (Spring 1963): 31–37.

Peterson, Thomas H., Jr. "Fort Lowell, A.T., Army Post During the Apache Campaigns," *The Smoke Signal,* 8 (Fall 1963): 1–20.

Prucha, Francis Paul. "Commentary," in *The American Military on the Frontier,* Proceedings of the Seventh Military History Symposium, United States Air Force Academy, 1976. Washington: Office of Air Force History, Headquarters USAF, 1978.

"Ranch Life, The Border Country, 1880–1940, The Way It Really Was," *Cochise Quarterly,* 12 (Spring 1982): n.p.

Rasch, P. J. "The Horrell War," *New Mexico Historical Review,* 31 (July 1956): 223–31.

Reeve, Frank D., ed. "Frederick E. Phelps: A Soldier's Memoirs," *New Mexico Historical Review,* 25 (January–October 1950): 37–56, 109–35, 187–221, 305–27.

Ruhlen, George. "Quitman's Owners: A Sidelight on Frontier Reality," *Password,* 5 (April 1960): 54–64.

———. "Quitman: 'The Worst Possible Post at Which I Ever Served,'" *Password,* 11 (Fall 1966): 106–26.

Stanley, Gerald. "Merchandising in the Southwest, The Mark I. Jacobs Company of Tucson, 1867–1875," *American Jewish Archives,* 23 (April 1971): 86–102.

Strickland, Rex W. "Six Who Came to El Paso, Pioneers of the 1840's," *Southwestern Studies,* 1 (Fall 1963): 3–48.

Thrapp, Dan L. "Dan O'Leary, Arizona Scout, A Vignette," *Arizona and the West,* 7 (Winter 1965): 287–98.

Tittman, Edward D. "The Exploitation of Treason," *New Mexico Historical Review,* 4 (April 1929): 128–45.

Wagoner, J. J. "History of the Cattle Industry in Southern Arizona, 1540–1940," *University of Arizona Bulletin,* 23 (April 1952): 1–132.

Walker, Henry Pickering. "Freighting from Guaymas to Tucson, 1850–1880," *Western Historical Quarterly,* 1 (July 1970): 291–304.

———. "Wagon Freighting in Arizona," *The Smoke Signal,* 28 (Fall 1973): 182–204.

Willey, Richard R. "La Canoa: A Spanish Land Grant Lost and Found," *The Smoke Signal,* 38 (Fall 1979): 154–71.

Winsor, Mulford. "José María Redondo," *Journal of Arizona History,* 20 (Summer 1979): 169–92.

Index